Sexual Selection

MONOGRAPHS IN
BEHAVIOR AND ECOLOGY

Edited by John R. Krebs and
Tim Clutton-Brock

Five New World Primates: A Study in Comparative Ecology, *by John Terborgh*

Reproductive Decisions: An Economic Analysis of Gelada Baboon Social Strategies, *by R.I.M. Dunbar*

Honeybee Ecology: A Study of Adaptation in Social Life, *by Thomas D. Seeley*

Foraging Theory, *by David W. Stephens and John R. Krebs*

Helping and Communal Breeding in Birds: Ecology and Evolution, *by Jerram L. Brown*

Dynamic Modeling in Behavioral Ecology, *by Marc Mangel and Colin W. Clark*

The Biology of the Naked Mole-Rat, *edited by Paul W. Sherman, Jennifer U. M. Jarvis, and Richard D. Alexander*

The Evolution of Parental Care, *by T. H. Clutton-Brock*

The Ostrich Communal Nesting System, *by Brian C. R. Bertram*

Parasitoids: Behavioral and Evolutionary Ecology, *by H.C.J. Godfray*

Sexual Selection, *by Malte Andersson*

Sexual Selection

MALTE ANDERSSON

Princeton University Press
Princeton, New Jersey

Published by Princeton University Press,
41 William Street,
Princeton, New Jersey 08540
In the United Kingdom: Princeton University Press,
Chichester, West Sussex

Library of Congress Cataloging-in-Publication Data

Andersson, M. B.
Sexual selection / Malte Andersson.
p. cm. — (Monographs in behavior and ecology)
Includes bibliographical references (p.) and indexes.
ISBN 0-691-03344-7 (CL) — ISBN 0-691-00057-3 (PA)
1. Sexual selection in animals. I. Title. II. Series.
QL761.A53 1994
591.56—dc20 93-33276

This book has been composed in Adobe Times Roman

Princeton University Press books are
printed on acid-free paper and meet the guidelines
for permanence and durability of the Committee
on Production Guidelines for Book Longevity
of the Council on Library Resources

Printed in the United States of America

10 9 8 7 6 5 4 3 2 1

10 9 8 7 6 5 4 3 2 1

To Åsa, Karin, Johanna, Gunilla

Contents

Preface

"Of the branches of biological science to which Charles Darwin's life-work has given us the key, few, if any, are as attractive as the subject of sexual selection." This view of Fisher (1915), in one of the most original works on sexual selection after Darwin (1871), apparently is shared by many biologists today. The study of sexual selection is in an eruptive phase, with new exciting results being published almost every week. Sexual selection is now recognized as an important evolutionary agent, with far-reaching consequences for morphology, behavior, mating system, and life history evolution, perhaps even for the species composition of biological communities, as sexual selection may influence speciation as well as extinction.

Although his theory of evolution by natural selection gave rise to more public debate, Darwin's (1871) ideas on sexual selection in *The Descent of Man, and Selection in Relation to Sex* were even more controversial among scientists, especially his ideas on female choice. Strong skepticism remained with regard to sexual selection long after natural selection became accepted as a major evolutionary factor. During the last two decades, however, the theory of sexual selection has gained support from genetic and game theory models, and from a rich variety of empirical studies. Most of *The Descent of Man* has stood up against time amazingly well: it is crammed with enlightening insights, anticipates several presently active research areas in evolutionary biology, such as sex ratio theory, reproductive tactics, and mate choice, and gives an admirable presentation of sexual selection based on the data then available. Most of Darwin's concepts and ideas on sexual selection have proven more useful and lasting than the alternative views presented by, for example, Alfred Wallace and Julian Huxley. There will often be occasion in this book to return to Darwin's (1871) work, and not only for historical reasons.

Sexual selection is the mechanism that Darwin arrived at when he tried to understand a conspicuous class of traits that defied explanation by ordinary natural selection for improved survival. The traits are "sexual differences . . . such as the greater size, strength, and pugnacity of the male, his weapons of offence or means of defence against rivals, his gaudy colouring and various ornaments, his power of song and other such characters." Darwin (1871) suggested that these traits are favored by competition over mates, and he termed the process "sexual selection." The classic main goal of the theory of sexual selection is to explain the existence of such traits, the detailed mechanisms by which they are favored, and their occurrence and variation among organisms.

In spite of admirable early work by a few pioneers, sexual selection did not become a lively field of research until the 1970s, when it was stimulated by important books by Ghiselin (1974), Williams (1975), and Maynard Smith (1978b). The many puzzling secondary sex traits that prompted Darwin to formulate his theory were then provoking explanation from a new generation of evolutionarily inclined biologists. The empirical and theoretical study of sexual selection has since grown into one of the major parts of behavioral and evolutionary ecology, engaging many evolutionary biologists of various brands (see, e.g., Bradbury and Andersson 1987; Krebs and Davies 1987; Cockburn 1991). This book attempts to review the theory and the main developments that have taken place, and to point to some of the many remaining unsolved problems.

Its main subject is the classical problems addressed by Darwin and Fisher: the selection and evolution of secondary sex traits, including mating preferences. Intensive research during two decades has added many new theoretical and empirical insights. Limited capacity and time prevent me from covering some areas that are part of sexual selection theory or at least have close connections with it. One such omission is the evolution of mating systems, which is discussed here mainly as one of the background factors that influence sexual selection. The influence goes both ways, however, and explanations of mating systems necessarily include sexual selection as one important component. Another omission is sperm competition, which is only briefly dealt with here. Reviews of this rapidly growing field are found in, for example, R. L. Smith (1984), Eberhard (1985), and Birkhead and Møller (1992). Another area that receives little space is sex allocation theory; Charnov's (1982) monograph treats many problems that also involve sexual selection.

A vast number of studies bearing on sexual selection in different organisms have now been published, and the number increases rapidly. I have not tried to make a complete review, which would probably more than double the number of references. The book by Thornhill and Alcock (1983) on sexual selection and mating systems in insects alone filled over 500 pages already a decade ago, and a large number of studies have appeared since then. Many important relevant studies are not mentioned here; I can only regret the omissions and ask for the indulgence of authors and readers.

The presentation here assumes some acquaintance with sexual selection as introduced by, for example, Alcock (1984b), Krebs and Davies (1987), or Cockburn (1991). The contents are as follows. Chapter 1 outlines the theory of sexual selection, its concepts, development, and main scope. The theory is full of debated, insufficiently tested ideas. The development of these often far from self-evident ideas, which partly live their own lives and are not always directly related to existing empirical knowledge, is almost as fascinating as the biological reality the ideas attempt to reflect. I therefore pay some attention to the evolution of the theory.

Chapters 2 and 3 review two main classes of genetic models of the evolution of mating preferences for secondary sex traits: Fisherian self-reinforcing mate choice, and indicator mechanisms. Empirical methods in the study of sexual selection are briefly described in chapter 4. Chapter 5 describes and discusses in some detail several studies of various animals, and the empirical approaches and results upon which conclusions about sexual selection in these studies are based. Chapter 6 briefly reviews empirical work that demonstrates sexual selection of a trait by some identified mechanism.

Chapter 7 treats the factors that determine the strength and effects of sexual selection in the two sexes. Primary sex roles, patterns of parental care, and the mating system are major influences that combine to make sexual selection usually stronger in males. Special attention is paid to lek mating systems, where males provide females with no material resources besides genes, and to species with reversed sex roles and stronger sexual selection in females than males.

Material benefits from mate choice, the subject of chapter 8, can be crucial for the evolution of mating preferences. So can species recognition advantages, reviewed in chapter 9. Sexual selection may also influence species divergence and speciation. Factors that constrain the evolution of secondary sex traits, such as predation and other mortality risks, are dealt with in chapter 10, as are physiological and other costs of sexually selected traits. Their relation to life history is discussed.

So far, focus has been mainly on the mechanisms of sexual selection. Next follows a group of chapters that instead review the selection of various secondary sex traits. As the aim is to understand these traits, the scope is broadened to include other selection mechanisms and factors besides sexual selection that may explain these traits. They include sexual dimorphism in body size, probably the most often studied sex-dimorphic trait (chapter 11), horns and other weapons (12), coloration and other visual signals (13), song and other acoustic signals (14), and chemical signals (15). Chapter 16 reviews alternative mating tactics: the different ways in which individuals of a species attempt to achieve success in competition over mates.

Sexual selection theory is now also being applied and tested in botany, one of the presently most vigorous and fascinating areas of research in sexual selection, reviewed in chapter 17. Finally, chapter 18 summarizes conclusions, points to open questions, and discusses problems of shared interest with other fields.

Göteborg
19 March 1993

Acknowledgments

I am very grateful for help and constructive criticism from many colleagues and friends, which greatly improved the manuscript, although I did not always take their advice. The entire text was read by Andrew Cockburn, John Endler, Frank Götmark, Andrew Pomiankowski, and Bill Searcy. Comments on one or more chapters were given by Tim Clutton-Brock, Nick Davies, Åke Norberg, Nick Waser, and David Winkler. The manuscript was also read and discussed by the participants in a series of seminars on sexual selection at the Department of Zoology, University of Göteborg: Staffan Andersson, Björn Arvidsson, Conny Askenmo, Torgny Bohlin, Jonas Lemel, Mats Olsson, Raimo Neergaard, Jan-Erik Svensson, and Mikael Tholleson. I was also helped by discussions with, among others, Jack Bradbury and Stephen Emlen. Many colleagues kindly sent manuscripts or papers in press. Staffan Andersson, Ulf Jondelius, Mikael Tholleson, and Matti Åhlund generously let me tap their macological insights. Magnus Neuendorf and Jimmy Persson advised me on plant illustrations. Birgit Lundell helped prepare the reference list; Johanna, Karin, and Åsa Andersson-Wilde helped with indexes. Without the kind help and services of the staff at the biomedical library, University of Göteborg, this study would not have been possible. My work on sexual selection has been supported by the Swedish Natural Sciences Research Council and by the University of Göteborg.

Figure Credits

For permission to reproduce published figures on which they hold the copyright, I thank the authors and artists (whose names appear below the relevant figures), and Academic Press (Figs. 7.5.1, 7.5.6, 8.2.3, 9.2.4, 10.4.1, 11.8.3, 11.8.4, 11.9.3, 12.4.2, 13.3.1, 13.6.2, 13.6.3, 13.6.5, 14.2.1, 14.3.2, 16.1.1, 16.2.1, 16.2.4); The American Association for the Advancement of Science (Figs. 5.2.2, 10.1.1, 17.6.3); The American Ornithologist's Union (Figs. 11.8.1, 16.2.5); The American Philosophical Society (Fig. 11.8.2); The American Scientist (Figs. 16.3.1, 17.1.1); The American Society of Ichtyologists and Herpetologists (Fig. 11.6.1); Annual Reviews, Inc. (Fig. 13.2.2); The Biological Bulletin (Fig. 10.3.1); Blackwell Scientific Publications (Figs. 16.2.1, 17.4.2); E.J. Brill (Figs. 12.2.3, 14.1.1); The Carlsberg Foundation (Fig. 11.2.1); Elsevier Science Publishers (Figs. 9.2.3, 9.3.1); Cambridge University Press (Fig 2.3.1, 5.5.1); Walter de

Gruyter & Co. (Fig. 15.1.3); The Ecological Society of America (Fig. 17.4.2); Harvard University Press (Fig. 12.4.1); A.R. Gantner Verlag (Fig. 17.7.1); The Genetical Society of Great Britain (Fig. 3.5.1); International Society for Behavioral Ecology (Fig. 7.5.2), The Linnean Society of London (Figs. 1.2.1, 11.8.4); Macmillan Magazines Ltd. (Figs. 4.3.2, 5.3.1, 5.4.2, 7.4.1, 7.5.3, 10.2.1, 11.9.4, 12.2.1, 12.2.5, 13.3.2, 13.6.1, 14.2.1, 14.3.1, 16.2.4, 17.7.1, 17.8.2); The New York Botanical Garden (Figs. 17.6.1, 17.7.1); The New York Zoological Society (Fig. 15.1.1); Oxford University Press (Figs 2.2.1, 2.4.1, 3.2.1, 8.2.3, 15.3.1, 16.2.5); Paul Parey Scientific Publishers (Fig. 13.4.1); Plenum Publishing Corp. (Fig. 15.2.1); Princeton University Press (Figs. 7.6.2, 11.1.1, 16.3.2, 17.9.1); The Royal Society, London (Fig. 17.6.2); Scientific American Inc. (Figs. 8.2.1, 12.4.3, 16.2.2); Sinauer Associates, Inc. (17.9.2); The Society for the Study of Evolution (Figs. 3.3.1, 4.5.2, 4.5.3, 5.2.1, 8.4.1, 9.1.1, 9.2.1, 9.2.2, 11.6.2, 12.2.4, 16.2.3); Springer-Verlag (Figs. 4.2.1, 7.5.2, 7.6.1, 12.4.4, 13.6.4, 15.1.2, 17.4.1); University of California Press (Fig. 5.4.1); University of Chicago Press (Figs. 3.5.2, 5.1.1, 5.5.1, 5.6.1, 8.2.1, 8.4.2, 11.1.2, 11.9.1); University of Washington Press (Fig. 17.6.2); Van Nostrand Reinhold Co. (Fig. 13.2.1).

Sexual Selection

1 The Theory of Sexual Selection

1.1 Sexual Selection

Among the many problematic animals that Linnaeus (1758) classified were two distinctive ducks. One species had mottled brown plumage and a blue wing patch; he named it *Anas platyrhynchos*. The other was pale gray except for a chestnut breast, a metallic green head and neck, and a blue wing patch; Linnaeus called it *A. boschas*. Later on, the two ducks became recognized as one species, *A. boschas* being nothing but the male mallard *A. platyrhynchos*. This is not the only time that a prominent taxonomist has mistaken the sexes for separate species (Darwin 1871; Mayr 1963). Why are males and females so different in appearance? What are the selective pressures behind sex differences in size, shape, coloration, and behavior? And, most puzzling, how do extravagant male traits evolve, such as bright colors, huge feather plumes or fins, and other conspicuous male attributes that can hardly improve survival? It was primarily to solve these problems that Darwin (1859, 1871) developed his theory of sexual selection.

This chapter introduces the main concepts in sexual selection theory, and describes the basic reasons why there will be sexual selection. The development of the theory has been rich in controversies and alternative explanations, briefly reviewed in sections 1.3–1.5, and discussed in much of the rest of the book (outlined in the Preface). A major issue concerns the mechanisms that favor conspicuous secondary sex ornaments and signals. The debate persists today; Maynard Smith (1991a) remarked that "no topic in evolutionary biology has presented greater difficulties for theorists."

According to Darwin (1871), sexual selection arises from differences in reproductive success caused by competition over mates. It therefore requires sexual reproduction: the combination of genetic material from two parents in the progeny. It does not necessarily require different sexes. Demonstrations of sexual selection in bacteria may not be soon to come, but competition over mates in principle can occur also in unisexual organisms that exchange genetic material. The most obvious results of sexual selection in animals and plants depend, however, on there being two different sexes with their defining gamete properties, with females making large, nutritious eggs and males small, mobile sperm. This anisogamy, with gametes of two different sizes, underlies the evolution of sex differences in behavior and morphology.

How and why sexual reproduction arose are questions outside of sexual selection theory, whose interest is the forms and consequences of competition over mates. The origin and maintenance of sexual reproduction is still a major debated problem in evolutionary biology.[1]

Anisogamy

Explaining anisogamy may be a more tractable problem than that of sex. It seems likely that sexual reproduction, with two different mating types, preceded anisogamy. Most models for the evolution of anisogamy assume two basic selection pressures: for increasing zygote size (which improves zygote survival), and for increasing gamete number (reviewed by Hoekstra 1987). As the resources available for reproduction are limited, these two pressures oppose each other. A compromise solution is the evolution of two different sexes, one of which produces few, large gametes, the other many small gametes.

Parker et al. (1972) suggested that once sexuality has arisen, anisogamy is likely to evolve from isogamy through disruptive selection. If there is variation in gamete size, and if zygote survival increases with size, selection should favor gametes that fuse with large partners. For certain relationships between zygote size and survival, the evolutionarily stable strategy (ESS) is anisogamy, with one large and one small type of gamete in the population (see, e.g., Maynard Smith 1982).

Knowlton (1974) made a start at testing these ideas, finding a correlation between degree of coloniality and anisogamy in Volvocidae. Anisogamy may be favored when zygote survival increases markedly with its size, and optimal zygote size is large. This seems likely in multicellular forms such as *Volvox* colonies, where the zygote must give rise to many cells. In a more extensive analysis, gamete dimorphism was correlated with the level of cellular organization among algae and protozoa, and anisogamy was correlated with large zygote size among seventeen orders of algae (Bell 1982; Madsen and Waller 1983). These comparative studies lend qualitative support to the ideas of Parker et al. (1972), but there are many exceptions, and more aspects need to be tested, especially the relationship between zygote size and survival (Hoekstra 1987).

Another set of ideas points to advantages of small sperm size in pre-

[1] See, for example, Ghiselin 1974, Williams 1975, Maynard Smith 1978, Bell 1982, B. Charlesworth 1989, Hamilton et al. 1990, Hurst and Hamilton 1992. The volumes edited by Stearns 1987 and Michod and Levin 1988 treat in detail the ideas and evidence raised to explain sexual reproduction.

venting transmission of cytoplasmic organelles and parasites to the zygote. There is a possibility of harmful conflict among cytoplasmic organelles such as mitochondria and chloroplasts (Cosmides and Toby 1981). Whereas the nuclear genes from both parents are required for proper functioning of the diploid zygote, this is not necessary for genes in the organelles. Such a gene that codes for destruction of the organelles from the other gamete, for example through digestion by DNA restriction enzymes, might be favored by selection. This could lead to intracellular conflict with deleterious consequences for the zygote and its nuclear genome. The conflict may be prevented by stripping sperm from the cytoplasma and its attendant organelles (e.g., Cosmides and Toby 1981; Hurst 1992). In addition, small sperm size probably reduces the risk of transmitting cytoplasmic parasites. Compared to isogamous sex, this should reduce the diversity of parasites in the zygote (Hurst 1990). The possible roles of intragenomic conflict in the evolution of sexual reproduction and anisogamy are discussed by Cosmides and Toby (1981), Hoekstra (1987), Hurst (1992), and Hurst et al. (1992). There is evidence that such aspects may have been crucial in the evolution of two sexes (Hurst and Hamilton 1992).

Regardless of how sex and anisogamy evolved, once they exist the stage is set for sexual conflict, within and between the sexes. Females produce large macrogametes, rich in energy: eggs. Males make small, highly motile microgametes: sperm. Other things being equal, a male can make many more gametes than a female, and males will compete to mate with as many or as fecund females as possible. On the other hand, the greater investment in fewer gametes by females should make them more careful in their choice of mating partner (also see section 7.2). The nature and consequences of competition over mates is the subject matter of the theory of sexual selection.

Competition over Mates

One of the greatest problems facing Darwin's (1859) theory of evolution by natural selection concerned conspicuous male traits, such as song and other display, bright colors, and horns and other weapons. These and other extravagant male characters would seem to reduce survival, and so should be opposed by ordinary natural selection (figure 1.1.1). How, then, can such traits be explained?

Darwin's solution to the problem was his perhaps most controversial idea. He made a distinction between sexual and other natural selection: "Sexual Selection . . . depends, not on a struggle for existence, but on a struggle between the males for possession of the females; the result is not death to the unsuccessful competitor, but few or no offspring" (Darwin 1859, p. 88). Male ornaments, according to Darwin, evolve through sexual

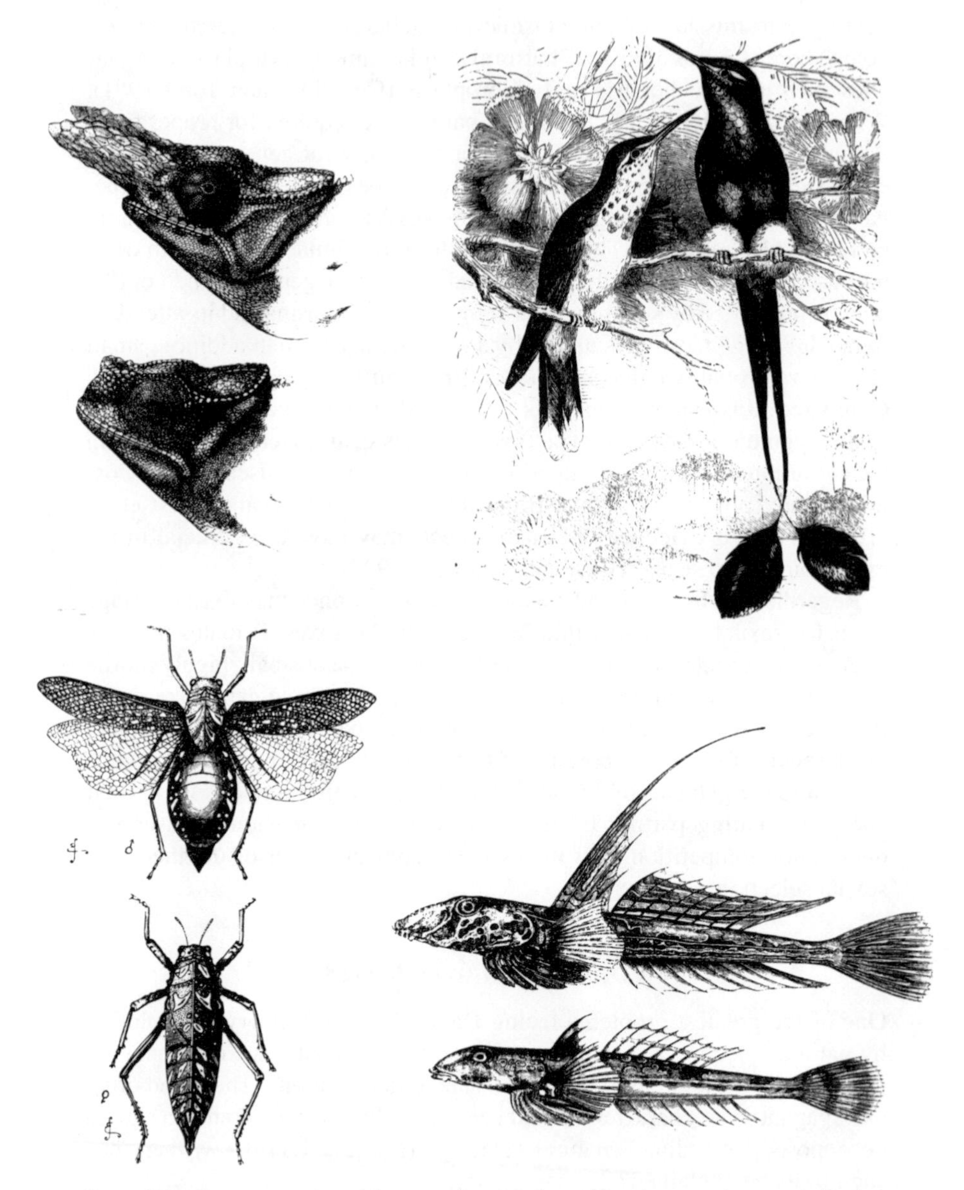

Figure 1.1.1 Some examples of the conspicuous dimorphism in secondary sex traits that provoked Darwin to develop his theory of sexual selection. In *Chamaeleon bifurcus*, only the male is horned. In the hummingbird *Spathura underwoodi*, the male is more brightly colored than the female, and has long ornamental tail feathers. The male gemmeous dragonet *Callionymus lyra*, with elongated dorsal fin, is also more brightly colored than the female. In the winged male of the *Pneumora* grasshopper, the air-filled abdomen is distended into a resonance chamber that probably amplifies his stridulation song. (From Darwin 1871)

selection by female choice of mate, and male weapons evolve through sexual selection by contests over females.

From the previous and other passages, competition over mates emerges as the key aspect of Darwin's concept of sexual selection. (Competition is here taken in a broad sense, and includes not only contests, but also other mechanisms such as mate choice by the other sex; see below.) For instance, in *The Descent of Man, and Selection in Relation to Sex* (p. 257), Darwin talks about male organs "perfected through sexual selection, that is, by the advantage acquired by certain males over their rivals." More highly developed sensory or locomotory organs in males "serve only to give one male an advantage over another, for the less well-endowed males, if time were allowed them, would succeed in pairing with the females; and they would in all other respects, judging from the structure of the female, be equally well adapted for their ordinary habits of life. In such cases sexual selection must have come into action, for the males have acquired their present structure, not from being better fitted to survive in the struggle for existence, but from having gained an advantage over other males, and from having transmitted this advantage to their male offspring alone. It was the importance of this distinction which led me to designate this form of selection as sexual selection."

SEXUAL SELECTION OF A TRAIT can therefore be viewed as a shorthand phrase for *differences in reproductive success, caused by competition over mates, and related to the expression of the trait*. Such differences can arise by many forms of direct or indirect mating competition. It may concern mate quality as well as numbers of mates, and can be brought about by mate choice, scrambles, contests, competition based on endurance, or any other form of rivalry over mates. Note that mate choice by one sex can suffice for there to be competition over mates in the other sex, even if there are no aggressive interactions (see below, this section).

Darwin did not define sexual selection in relation to reproductive success in general. Many traits that raise fertility and reproductive success have nothing to do with mating competition. For instance, the brood patch in birds improves incubation, and parental alarm calls help offspring avoid predators. Both traits increase the chances of successful reproduction, but probably neither improves success in competition over mates; they are therefore naturally but not sexually selected. Many aspects of courtship may also have functions other than improving success in competition over mates. For example, it may reduce escape responses or aggression in the mate, synchronize endocrine reproductive functions, or coordinate the behavior of mates in space and time for copulation (e.g., Bastock 1967).

Darwin (1871) noted that in borderline cases, such as sensory or locomotory organs that help a male find his way to a mate, it may not be possible in practice to distinguish between sexual and other natural selection. But such subtle exegetic problems do not concern the most puzzling traits mentioned above: advertising traits that are unlikely to improve survival. Such characters motivated the distinction between sexual and other natural selection.

A common cause of natural selection is the external environment, physical or biological; the agents of sexual selection are sexual rivals and mates (Ghiselin 1974). Contrasted with artificial selection, sexual selection in the wild is a subset of natural selection (J. L. Brown 1975). Yet Darwin's distinction between sexual and other natural selection is often useful; the evolutionary effects of competition over mates often differ in remarkable ways from those of other natural selection (see Endler 1986a for discussion and additional aspects of the relations between sexual and natural selection).

Conspicuous secondary sex traits may often be favored only or mainly by sexual selection and counteracted by other natural selection. Many other traits will be favored by both sexual and other natural selection, for example general metabolic efficiency, pathogen resistance, and any characteristic that improves both survival, mating, and reproductive success (figure 1.1.2). Although sexual selection is here viewed as a subset of natural se-

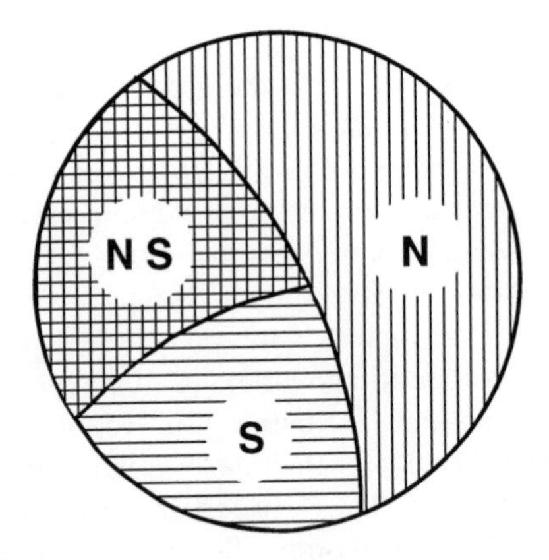

Figure 1.1.2 Relations between sexual and natural selection. The circle includes all traits favored by selection in the wild: natural selection. A subset of those traits is favored by competition over mates: sexual selection (horizontal hatching; areas NS and S). Some traits are favored by other natural selection (vertical hatching) as well as by sexual selection (area NS), or by other natural selection only (area N). Finally, some traits are favored only by sexual selection, but not by other natural selection (area S). This latter set of traits, favored only by competition over mates, contains secondary sex traits that reduce survival; they prompted Darwin to coin the term "sexual selection." Note, however, that many traits are probably favored by both sexual and other kinds of natural selection. (The relative sizes of the areas are arbitrary. The figure concerns only traits favored by selection in the wild, not those maintained by artificial selection or genetic drift.)

lection, for brevity I will sometimes write of traits as being favored by sexual selection and disfavored by natural selection—that is, by natural selection other than that involving competition over mates. In spite of many suggestions to the contrary by leading biologists, from Wallace (1889) to Huxley (1938b) to Mayr (1982), the term sexual selection is here restricted to competition over mates. Darwin's concept still appears simpler and more useful than the alternatives.

Competition is here used in a similar sense as in ecology: competition occurs whenever the use of a resource (in this case, mates) by one individual makes the resource harder to come by for others. This is so whether or not the rivals meet in actual contests; the only requirement is that a user makes the resource less available to others. Mate choice by one sex therefore usually implies (indirect) competition over mates in the other sex, even if rivals never meet each other. When accepting and mating with a male that fertilizes her eggs, a female becomes unavailable to other males, at least temporarily. There is a similar distinction in evolutionary ecology between *interference competition*, often with direct aggressive struggles over some resource, with the competitors trying to exclude each other, and *exploitation competition*, where rivals consume the same resource, but do not necessarily meet each other (e.g., Pianka 1978; Futuyma 1986). For instance, they may be active at different parts of the day. Contest competition in sexual selection therefore corresponds to interference competition in ecology, whereas mate choice leads to exploitation competition over mates in the other sex. Among the mechanisms of mating competition, I therefore include not only contests, scrambles, endurance rivalry, and sperm competition, but also mate choice (table 1.1.1).

Competition over mates has been demonstrated in many studies. For example, if successful males are removed, their territories are often taken over by other males, and the mating success of previously less successful males rises (e.g., Watson 1967; Alatalo et al. 1991). Experimental increase of an ornament in some males often raises their success in attracting mates over that of other males, implying competition over mates (table 6.A).

In polygynous species where some males mate with several females, others with none, success in competition over mates is crucial for the fitness of males, and their weapons and ornaments are often highly developed. Sexual selection can, however, work also in monogamous species if the sex ratio is skewed, or if mates differ in quality (Darwin 1871; Fisher 1958; section 7.3 below). For example, among female birds, those in best condition are ready to breed first in the season, and so will have most time for raising young, hence producing most offspring (reviewed by Price et al. 1988). The first males to mate are therefore likely to get the most productive females. In addition, variation in mate quality might also affect the

Table 1.1.1
Mechanisms of Competition over Mates, and Traits Likely to Be Selected in the Competing Sex

Mechanism	*Characters Favored in the Competing Sex*
I. Scrambles	Early search and swift location of mates; well-developed sensory and locomotory organs
II. Endurance Rivalry	Ability to remain reproductively active during a large part of the season
III. Contests	1. Traits that improve success in fights, such as large size, strength, weaponry, agility, or threat signals
	2. Alternative mating tactics of inferior competitors, avoiding contests with superior rivals
IV. Mate Choice	1. Behavioral and morphological traits that attract and stimulate mates
	2. Offering of nutrition, territories, nest sites, or other resources needed by the mate for breeding
	3. Alternative mating tactics, such as forced copulation
V. Sperm Competition	1. Mate guarding, sequestering, frequent copulation, production of mating plugs, or other means of preventing rivals from copulating with the mate
	2. Ability of displacing rival sperm; production of abundant sperm to outcompete those of rivals

partner's survival. The better a parent provides food and guards against predators, the better its family should fare, including its mate. Such effects should augment the advantage of having a high-quality mate in species with biparental care. Traits that improve the ability to compete over mates should therefore be favored in monogamous as well as polygynous species (O'Donald 1987; Price et al. 1988; Kirkpatrick et al. 1990). Owing to more similar parental roles in monogamous than polygynous species, female competition over mates should also often be stronger in monogamous forms.

In the minority of species with mainly paternal care, the sex differences in parental roles can override the effects of anisogamy and lead to a reversal of other aspects of sex roles and sexual dimorphism (chapter 7). Limits to the number of sperm (ejaculates) that a male can deliver may also favor male mate choice (Dewsbury 1982b). Depending on the mating system, parental roles, and reproductive ecology, there can be competition and mate choice in both sexes, but competition should usually be most pronounced in males, and mate choice in females (see chapter 7).

Although Darwin (1871) did not include plants in his review, they can also be sexually selected; this is presently one of the most active and exciting applications of sexual selection theory (reviewed in chapter 17).

Forms of Mating Competition

Competition over mates can take several forms and favor a wide range of attributes (table 1.1.1; also see Ghiselin 1974; Otte 1979). Scrambles to first find a mate are often important; such scrambles may partly explain why males in many arthropods have larger eyes, chemorecepting antennae, or locomotory organs than females (reviewed by Thornhill and Alcock 1983). In many animals and plants, scrambles to fertilize females may select for "protandry," for instance earlier male than female hatching, maturation, or appearance on the breeding ground (e.g., Ghiselin 1974; Wiklund and Fagerström 1977; Wang et al. 1990; Baughman 1991). Competition over fertilization in plants selects for conspicuous corollas and other pollinator attractants (chapter 17). Searching for females apparently favors better male than female spatial memory in some mammals, and is associated with larger male hippocampus, a part of the brain that plays an important role in spatial learning (Gaulin and FitzGerald 1986, 1989; Jacobs et al. 1990).

Sexual selection will often favor traits that improve the endurance of a male, enabling him to remain longer at a breeding site and mate with females that otherwise would mate with other males. Male mating success is then correlated with the length of time spent at such sites. Endurance should increase with foraging efficiency and several other factors that are also naturally selected (Koenig and Albano 1986). In some cases, however, selection for endurance in competition over mates may lead to consequences not favored by ordinary natural selection. For example, male insectivorous marsupials of the genus *Antechinus* remain at the mating site without feeding until they die from stomach ulcers, disease, or parasites. All males die at the end of the breeding season, whereas some females survive and reproduce a second year (A. Cockburn, in prep.).

Fights over mates select for strength, often achieved by large size, and for weapons such as antlers, horns, and spurs. This idea was accepted by many of Charles Darwin's colleagues; such a function of male weapons had been suggested already by his grandfather Erasmus Darwin (1794) and other zoologists (see Aiken 1982). Males fight over females in a variety of animals, including some intuitively unlikely cases such as a gastropod, the fighting conch *Strombus pugilis* (Bradshaw-Hawkins and Sander 1981).

In addition to large sensory or locomotory organs and weapons, males in many animals have conspicuous ornaments or behavioral signals. Being well aware of the amazing results from artificial selection of traits such as plumage in domestic pigeons, Darwin (1871) suggested that female choice of mate in the wild can have similar effects: "It appears that in a state of nature female birds, by having long selected the more attractive males, have added to their beauty." Female choice became the most controversial part of the theory of sexual selection, and the detailed mechanisms by which mate choice works still remains so (e.g., Ghiselin 1974; Heisler et al. 1987; Kirkpatrick 1987a; discussion of criteria for identifying female choice are found in, e.g., Searcy and Andersson 1986; Heisler et al. 1987; Ryan and Keddy-Hector 1992). It has been the subject of much modeling (chapters 2–3), whereas the "law of battle" between males is more easily observed and is probably therefore less controversial, and less attractive to theoreticians (but see section 11.1).

The main forms of competition over mates are here termed contests, mate choice, scrambles, and endurance rivalry. Since Huxley (1938b,c), contests and scrambles have often been called "intrasexual" selection; competition by mate choice in contrast is often called "intersexual" selection. Huxley also used "epigamic selection" and several other technical terms. They seem to offer little advantage over the simple everyday words of Darwin (1871), who seems to have liked playing with dogs but not engaging in ludic activity with canine companions (O'Donald 1980a; Bonner and May 1981; J. L. Brown 1983). Here, the term *contest* is used where rivals display to or fight each other in competition over mates (or resources needed to attract mates); *scramble* where rapid location of the mate is crucial for success; *endurance rivalry* where persistence, for example the length of stay at a breeding site, affects mating success; and *mate choice* where the mate at stake determines, or at least influences, which rival will win. This broad definition of mate choice includes behavior that would not be called choice in a human context; it refers to external events, not mental processes (see Halliday 1983; Maynard Smith 1987). Darwin (1882) discussed this aspect in his last defense of sexual selection, written a few months before he died and read at the Zoological Society of London only hours before his death (Bajema 1984): "It would, however, be more correct to speak of the females as being excited or attracted in an especial degree by the appearance, voice, &c. of certain males, rather than deliberately selecting them." Today, mate preference and choice are often defined broadly, for example as behavior patterns that make the female (or male) more likely to mate with some potential partners than with others (Halliday 1983; Pomiankowski 1988; Kirkpatrick and Ryan 1991).

Female choice has sometimes been put in contrast with male competition, but there is competition among males also when females choose their mate, even if rivals never meet. Competition for mates is the defining aspect of all forms of sexual selection, including that based on mate choice, because then "the struggle is likewise between the individuals of the same sex, in order to excite or charm those of the opposite sex, generally the females, which no longer remain passive, but select the most agreeable partners" (Darwin 1871, II, p. 398; also see Ghiselin 1974; J. L. Brown 1983).

In practice, all the mechanisms in table 1.1.1. may occur together; determining their relative importance is a challenging empirical task. For example, after males fight over dominance status or territories, females might choose among the winners. Contests and choice may favor the same or different attributes (table 1.1.1.).

Sexual competition does not always come to an end at mating. It can continue in several forms, one of which is sperm competition between males that mate with the same female (e.g., Parker 1970b; R. L. Smith 1984; Eberhard 1985, 1991; Birkhead and Møller 1992). Common male tactics of reducing the risk that a rival fertilizes the female is to guard or sequester and often mate with her during the receptive period (reviewed, e.g., by Birkhead and Møller 1992). Other tactics are blocking the genital opening of the female with a mating plug, applying "anti-aphrodisiac" substances that reduce her attractiveness (for example, by making her smell like a male), and reducing her pheromonal output (e.g., Happ 1969; Gilbert 1976; Frankie et al. 1980; McLain 1980; Sillén-Tullberg 1981; Thornhill and Alcock 1983). For instance, before proceeding further with the affair, a male spider *Linyphia litigiosa* that finds a virgin female will reduce the evaporation of attractive pheromones from her web by packing it into a tight mass (Watson 1986). Eberhard (1985) reviewed postcopulation competition over fertilization in animals.

Another form of sexual competition after copulation is induced abortion or infanticide that makes a female receptive to a new male.[2] Here, however, I will focus on the stages of mating competition up to copulation, that is, the realm of classical Darwinian sexual selection.

As they are likely to play major roles in the evolution of conspicuous secondary sex traits, mate choice and contests have received by far the most attention. The importance of endurance rivalry has only recently become clear. Scramble competition is important, for example, in many anurans and invertebrates where females are receptive for only brief periods (e.g., Wells 1977b; Thornhill and Alcock 1983). It has been clarified in detail by Schwagmeyer (e.g., 1988) in the thirteen-lined ground squirrel *Spermophilus tridecemlineatus*, where the most mobile males with largest ranges during the mating season are most successful (section 7.2).

1.2 Sex Traits and Sexual Dimorphism

Darwin (1871) distinguished three kinds of traits that differ between males and females: primary, secondary, and what might be termed "ecological" sex traits. Primary sex traits are "directly connected with the act of repro-

[2] See, e.g., Trivers 1972, Mallory and Brooks 1978, Hrdy 1979, Schwagmeyer 1979, Labov

duction": gonads and copulatory organs. They are necessary for breeding and hence are favored by natural selection. But males and females also differ in secondary sex traits with no direct mechanical role in insemination; Darwin suggested that most such traits raise the success of the possessor in competition over mates. Examples are "the greater size, strength, and pugnacity of the male, his weapons of offence or means of defence against rivals, his gaudy colouring and various ornaments, his power of song, and other such characters." According to Darwin, these secondary sex traits evolve by sexual selection, often in opposition to natural selection. He noted that it is often not possible to separate clearly between primary and secondary sex traits. For example, prehensile organs such as the modified legs and antennae of males in oceanic crustaceans help ensure successful copulation, but they also make it harder for a rival to dislodge the male and take over the female. Such organs may be favored by sexual as well as other natural selection.

Also, copulatory organs and reproductive glands can be shaped in part by competition over mates or fertilizations. For example, the penis of the damselfly *Calopteryx maculata* carries stiff hairs and other structures that help remove sperm from males that have previously mated with the female (figure 1.2.1; Waage 1979; see Birkhead and Hunter 1990 for review). Testes size among primates provide other evidence of sperm competition (Harcourt et al. 1981; figure 4.3.2 below). Even within species, variation in testes size can reflect sexual selection. In the bluehead wrasse *Thalassoma bifasciatum*, large territorial males have smaller testes than small males that spawn in groups, where strong sperm competition puts a premium on massive sperm production (Warner and Robertson 1978).

In his review of animal genitalia, Eberhard (1985) suggested sexual selection by female choice has helped shape male copulatory organs among insects and other taxa. Animals with internal fertilization often show rapid and divergent evolution of male genitalia and other structures used in copulation. Patterns of variation suggest that earlier explanations may be inadequate. For example, the "lock and key" hypothesis states that complex, species-specific genitalia evolve because they reduce mismatings with other species. It fails, however, to explain for instance why male genitalia often vary much more than female genitalia among closely related species. Given the probably stronger selection of females to avoid mismatings, the opposite pattern is expected from the lock and key hypothesis. Eberhard (1985) instead proposed that male genitalia in part are "internal courtship

---

1981, Sherman 1981, Freed 1987, Hoffman et al. 1987, Emlen et al. 1989, Wolff and Cicirello 1989, Veiga 1990. Reviews of postcopulatory competition are found in, e.g., Thornhill and Alcock 1983, Hausfater and Hrdy 1984, R. L. Smith 1984, Clutton-Brock 1991, Birkhead and Møller 1992.

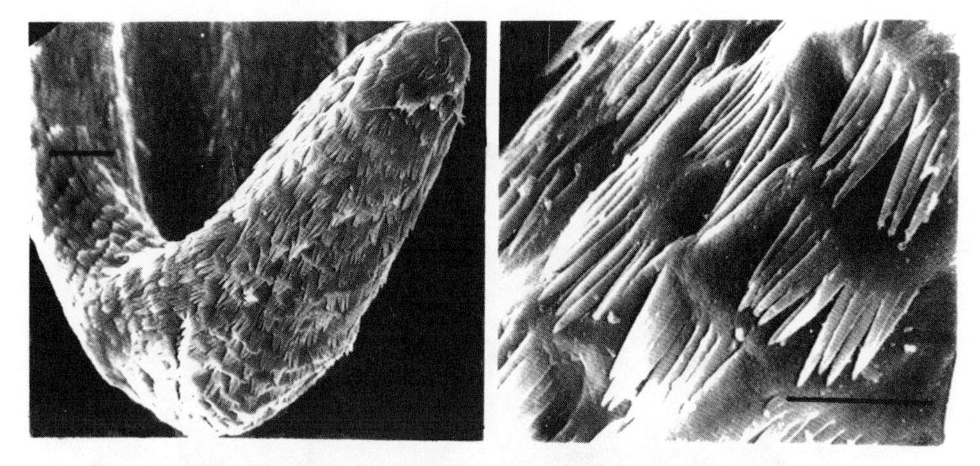

Figure 1.2.1 *Left*: Tip of the distal penis segment of the damselfly *Argia moesta*. *Right*: The segment carries spines used for scooping out sperm of previous males mated to the female. (From Waage 1986b)

devices" which increase the chances that the female will let the male's sperm fertilize her eggs. These new ideas render the distinction between primary and secondary sex traits less clear; perhaps it will eventually be abandoned.

A special, apparently less common kind of sex difference is "connected with different habits of life, and not at all, or only indirectly, related to the reproductive functions" (Darwin 1871, II, p. 254; see Shine 1989 for review). Such *ecological sex traits* are usually adapted to foraging differences between the sexes (which ultimately are probably related to anisogamy and sexual selection). Two examples are the mouth parts of mosquitos in which females suck blood and males live on flowers, and the bill of the extinct New Zealand huia (*Neomorpha acutirostris*; figure 1.2.2) and of certain woodpeckers in which the sexes use different foraging methods. In *Centurus striatus* on Hispaniola, and in several other island populations of woodpeckers with little or no competition from similar species, the sexes have evolved extreme dimorphism in bill morphology. Coupled with foraging differences, this should reduce food competition and expand the food niche of a pair (Selander 1966; Ligon 1968). Evidence that intraspecific foraging competition can create disruptive selection on the feeding apparatus comes from the African finch *Pyrenestes ostrinus*, in which two bill morphs (unrelated to sex) use partly different food resources (Smith 1990a,b, 1993). There is evidence that the morph difference may depend on a single gene locus (Smith 1993).

Slatkin (1984) concluded from quantitative genetic models that ecological sex differences may evolve under several conditions. For example,

Figure 1.2.2 A pair of the extinct New Zealand huia *Neomorpha acutirostris*. The contrasting bill shapes probably allowed males (lower right bird) and females to take different kinds of prey, broadening the total food niche of the pair, which foraged together. (From Doflein 1914)

ecological selection pressures can differ between the sexes, or they may compete over some resource (Darwin 1871; Lande 1980; Shine 1989). Ecological sex differences may first arise owing to sexual selection, later on being enlarged by natural selection for separation in diet or other aspects (Clutton-Brock and Harvey 1977; Clutton-Brock, Harvey, and Rudder 1977; Power 1980; Caro 1985). For example, in the isopod *Idothea baltica*, sex differences in coloration may in part be adaptations to different habitats and risks of predation in males and females (Jormalainen and Tuomi 1989).

For ecological sex differences to evolve, it may not be necessary that the initial difference in the trait arises through sexual selection. Even if the sexes are similar to start with, imperfect genetic correlations between them can permit natural selection to create sexual dimorphism (Slatkin 1984). For a particular type of sex dimorphism to have the same direction in many species that independently evolved the dimorphism, however, some initial sex-related bias in the trait seems necessary (Power 1980; also see Trivers 1972).

Probably most of the genes that influence secondary sex differences in higher animals are not sex linked, but sex limited in their expression. They are not located on sex chromosomes but on ordinary autosomes (for exceptions, see, e.g., Farr 1983). Both sexes therefore carry the genes. The difference in expression by males and females is usually brought about by sex-specific hormones produced by the gonads, whose sex differences do depend on genes in the sex chromosomes. In mammals, a newly identified Y-linked gene triggers the development of male rather than female gonads (Koopman et al. 1991; also see McLaren 1988). Testosterone is necessary for the development of the secondary sex traits of the adult male, and estrogen for those of the adult female. In birds, where females are the heterogametic sex, female gonadal hormones lead to the early development of female rather than male characters. It seems that the gonads of the heterogametic sex in vertebrates produce the signal substance that triggers sex differentiation of genitalia and other early sex traits (see, e.g., Daly and Wilson 1983; Renfree and Short 1988).

1.3 Criticism of Darwin's Theory of Sexual Selection

The idea that conspicuous male display, colors, feather plumes, and other secondary sex ornaments evolve through female choice met much early skepticism (reviewed, e.g., by Kellogg 1907; Ghiselin 1969b, 1974; Otte 1979; Ruse 1979; Kottler 1980, 1985; Aiken 1982; West-Eberhard 1983; Bajema 1984; Pomiankowski 1988; Cronin 1992). Darwin (1871, I, p. 259) seemed to assume a similar sense of beauty in other higher vertebrates as in man. But this assumption, right or wrong, is not necessary for female choice: discrimination among males in relation to size, shape, color, or other aspects should suffice. Yet the assumption was often held up against the entire idea of sexual selection by female choice (e.g., Morgan 1903; Dewar and Finn 1909; Huxley 1938b,c).

After first accepting female choice, Alfred Wallace (e.g. 1889) came to doubt its relevance, and a debate arose between him and Darwin over problems that are not yet fully resolved (also see section 13.6). Wallace objected against the entire concept of sexual selection: agreeing that male weapons might evolve through contests over mates, he argued that this is ordinary natural selection that increases the vigor, fighting power, and survival of males. Modern studies have shown, however, that sexual selection can lead to much higher mortality in males than in females (chapter 10).

Several influential biologists were even more negative to Darwin's ideas on sexual selection. One example is T. H. Morgan (1903), Nobel laureate and explorer of the Mendelian machinery with *Drosophila* as test organism. He concluded that "the theory meets with fatal objections at every turn," and he regarded it as definitely refuted. Morgan (1932) considered sexual dimorphism fully explained by a proximate mechanism—hormonal differences between males and females—and failed to see that it also requires an evolutionary explanation. A contributing reason for the strong early resistance to the idea of female choice was probably the rise of the mutation school of thought at the turn of the century, which largely dismissed the role of selection in evolution (Provine 1971).

Julian Huxley (e.g., 1923, 1938b,c) accepted several but criticized others of Darwin's ideas on sexual selection. Even Lack (1968) argued against the importance of female choice, referring to Huxley's views. Yet, as O'Donald (1980) points out, Huxley (1938b,c) confused sexual selection with natural selection, and natural selection with group selection; he also obscured the issue with a thicket of complicated terminology. Surprisingly, Huxley's confusing views for decades came to be regarded as superior to Darwin's (1871) clear insights.

Except for some important empirical or theoretical studies, for example by Huxley (1914), Fisher (1915, 1930), and Noble (e.g., 1934, 1936), almost a century passed before Darwin's ideas on sexual selection gained ground. Yet the stage seemed set for rapid growth of the field already in the 1930s, when it was seeded by exciting theoretical and empirical contributions. Fisher's (1930) important but difficult *Genetical Theory of Natural Selection* was, however, for long apparently not read or understood by many, and Noble's (1934, 1936; chapter 13 below) pioneering experimental approach in the field for some reason was not taken up in large scale by other workers. One reason why the study of sexual selection did not catch on from the 1930s was perhaps the disregard for, or neglect of, Darwin's and Fisher's ideas on the subject by the leaders of the "evolutionary synthesis," and their tendency to explain conspicuous secondary sex traits by species isolation (e.g., Dobzhansky 1937; Huxley 1942; Mayr 1942; see West-Eberhard 1983). In spite of early work by Sturtevant (1915), it also took many years for research on *Drosophila* mate choice to gain momentum (e.g., Bateman 1948; Merrell 1949; Reed and Reed 1950; Petit 1954; Maynard Smith 1956).

The publication in 1958 of the revised edition of Fisher's (1930) book roughly marks the birth of a new generation of work on sexual selection, two early examples being Maynard Smith (1958) and O'Donald (1962). If Darwin was the architect, Ronald Fisher clearly was his main assistant in the construction of sexual selection theory. Although his name is associated with self-reinforcing mate choice and the celebrated "runaway process" (section 1.5), Fisher also put forth two other major hypotheses for conspicuous secondary sex ornaments: contest competition and indicator processes (sections 7.4 and 1.5). Since Fisher (1958), interest in sexual selection has grown continuously, resulting in today's eruption of studies. This is part of a general move toward testing natural selection in the wild, reviewed by Endler (1986a). The development is also closely related to the shift of emphasis in ecology toward evolutionary problems, individual selection, and reproductive tactics, which began in the 1960s (e.g., Williams 1966; see Cronin 1992). But still, sexual selection is very incompletely understood, especially in species where females choose mates and males apparently contribute nothing but genes.

A main thesis of *The Descent of Man* was that sexual selection is important in *Homo sapiens*, and that mate choice has played a major role in the evolution of some conspicuous differences among human populations, differences that are not obviously related to survival. After a massive, but not necessarily correct, critique by the sociologist E. Westermarck (1891) in his *History of Human Marriage*, this idea of Darwin's has been largely neglected (Thornhill 1986b). On the other hand, speculations about sexual selection of various human traits are common but often difficult to test rigorously. Substantial indications are found in studies that put our own species in relation to the other primates in a quantitative, comparative framework, for example as regards sexual size dimorphism and testes size (sections 4.3 and 11.9). Human mate choice and the possible importance of sexual and natural selection in the morphological differentiation of human populations was reviewed by Diamond (1991b).

1.4 Selection of Secondary Sex Signals

A main goal of sexual selection theory is to explain the often conspicuous secondary sex traits. Studies during the last two decades have shown that sexual selection is common in nature (chapter 6), but the selective mechanisms behind conspicuous sex ornaments and behavioral signals are still debated. A host of alternative explanations has been suggested, summa-

Table 1.4.1

Explanations for Conspicuous Male Ornaments and Female Preferences

A. *Male ornaments* have evolved because of:
1. Pleiotropic gene effects
2. Selection of ecological sex differences
3. Males being unprofitable prey for predators
4. Male contests
5. Female choice and mating preferences

B. *Female preferences* for male ornaments (section 1.5) have evolved because of:
1. Fisherian self-reinforcing selection
2. Indicator mechanisms
3. Selection for species recognition
4. Direct phenotypic benefits to choosy females
5. Selection of the sensory system in other contexts (sensory bias)
6. Advantages in the timing of reproduction (mating sychronization)

rized in table 1.4.1. They form a heterogeneous set of ideas, some aiming to explain secondary sex signals, others also explaining mating preferences.

These hypotheses aim to give ultimate, evolutionary explanations of male traits and female preferences (B.5 and B.6 in table 1.4.1 may not, however, suffice to do so). Few of the explanations are mutually incompatible, so they may not be open to "strong inference" testing (Platt 1964), where all the alternatives except for one can be refuted. Several mechanisms may work in concert, presenting the more difficult problem of assessing their relative roles.

Ghiselin (1974) discussed and refuted a number of additional, less plausible hypotheses, for instance some that require group selection (but see Seger and Trivers 1986).

Pleiotropic Gene Effects

In an otherwise excellent developmental study that helped clarify sex-limited and sex-linked inheritance, Morgan (1919) suggested that secondary sex traits are "only by-products of genes whose important function lies in some other direction." One way of testing this idea is to examine the consequences of secondary sex traits for mating success. If a trait enhances mating and reproductive success when other factors are controlled for, the trait is not a neutral by-product of pleiotropy. Many such cases have now been demonstrated (table 6.A). Even if pleiotropy may play a role in the origin of secondary sex traits, Morgan's hypothesis is refuted as a sufficient general explanation for their spread and maintenance, which requires some form of selection (also see Endler 1986a; Endler and MacLellan 1988).

Selection of Ecological Sex Differences

Ecological differences between the sexes seem to explain some sex dimorphic traits used for feeding, for instance the bill in some birds already discussed above (section 1.2). However, most ornamental or other sex traits used mainly for signaling do not seem explicable by ecological sex differences. Possibly, such differences will sometimes influence the traits selected and the direction taken by other mechanisms, such as Fisherian processes.

Males Are Unprofitable Prey

Baker and Parker (1979) proposed that some secondary sex traits evolve because the bearer is unsuitable as prey for predators. From their analysis of coloration in Western Palearctic birds, they concluded that bird coloration has evolved mainly in response to predation, not sexual selection. Evidence on this point is discussed in section 13.6. Even if this mechanism may sometimes contribute, it does not seem to be the main reason for conspicuous sex-dimorphic coloration, either in birds or in other animals.

Male Contest Competition

An important alternative to Darwin's idea of female choice was raised by several biologists in the 1930s. They suggested that male contests over mates select not only for horns, tusks, spurs, and other physical weapons, but also for conspicuous signals. Fisher (1930) proposed that ornament and display may be psychological weapons: "A sprightly bearing with fine feathers and triumphant song are quite as well adapted for war propaganda as for courtship." A similar idea, that bright plumage in male birds functions as a "gaudy uniform of battle" in contests over females, was put forth by R. Hingston (1933, p. 114). Noble (e.g., 1934) drew a similar conclusion from his work on the use of colors in display by lizards.

None of these authors explained how such badges of status might evolve. Huxley (1938a), Peek (1972), Smith (1972), and Borgia (1979) proposed that conspicuous male coloration and display in territorial species might be favored in part because it advertises occupancy and presence of the owner on the territory. Rohwer (e.g., 1975, 1982) suggested several other mechanisms, including the possibility that badges arise by natural selection, for instance in contests over winter food. Status-signaling badges that indicate strength could be selected for if they make good fighters recognizable and memorable, and spread in the population by mimicry. Under certain circumstances, this might give rise to sexual color dimorphism even without direct sexual selection of conspicuous colors (see Rohwer 1982; Butcher and Rohwer 1989; Rohwer and Røskaft 1989). Similar ideas were presented by West-Eberhard (1979, 1983, 1984), who emphasized that other forms of social competition than sexual selection can favor bright colors and other conspicuous signals (also see Ghiselin 1974; Weldon and Burghardt 1984). Members of a social group or population can limit each other's access to many kinds of resources, not only mates.

Social competition might sometimes lead to rapid open-ended evolution of signals. In contrast with traits adapted to ecological functions, such as the bill of a bird for which there may be a "best" shape depending on diet, a social competitive function in itself need not put a limit to the development of the trait (Darwin 1871). "Each successive improvement sets a new standard which the next can profitably surpass" (West-Eberhard 1979; also see Zahavi 1981). Sexual and other social selection can therefore lead to extreme traits, and some conspicuous signals may be socially selected through competition over resources other than mates. Social selection of signals used in contests might lead to rapid divergence in coloration between populations and species (Rohwer 1982; West-Eberhard 1983; W. S. Moore 1987). Mathematical models by Parker (1979, 1983a) and Maynard Smith and Brown (1986) suggest that contest competition can easily lead to evolutionary instability in favorable traits (section 11.1).

Zahavi (1977a), Borgia (1979), and Baker and Parker (1979) suggested a form of "indicator mechanism" (also see sections 1.5 and 13.6 below): "An evolutionary stable strategy (ESS) can exist in which only the opponent with the greater strength or fighting ability is prepared to fight, whereas the one with less ability withdraws. . . . A more brightly coloured male may have avoided predators under handicap and hence may also be stronger and more likely to win against a more cryptic rival" (Baker and Parker 1979). This situation could lead to a form of agonistic indicator process, if a male in good condition benefits from having larger ornaments than other males (Andersson 1982b; Parker 1982). The idea of condition-dependent badges is free from some of the problems of indicator mechanisms for female choice (see section 1.5); in particular, fitness need not be heritable.

Female Choice and Mating Preferences

There is now much evidence that females often choose their mate, and that such choice favors conspicuous male traits (e.g., table 6.A). The exact ways in which female choice selects for such traits are still debated, and so are the ways in which female preferences evolve, which remain a main controversial issue in the theory of sexual selection (e.g., Searcy 1982; Kirkpatrick and Ryan 1991; Maynard Smith 1991a; Williams 1992).

Some male traits may evolve simply because they make it easier for females to find the male. Natural selection will favor such traits if they reduce the time during which males and females are vulnerable to predation (Darwin 1871; Wallace 1889; Mayr 1972). A natural selection advantage is by no means inevitable, however: signals used for mate attraction can make the sender easier to locate for parasites or predators (chapter 10). In such cases, sexual selection has probably favored the signal in opposition to natural selection.

Mating usually contains an element of discrimination, at least as regards species identity. Otherwise, mismatings between members of different species should be much more common than they are. Among the acceptable signals, sexual selection should favor those which most effectively stimulate the recipients, that is, intense, persistent, or otherwise conspicuous signals (reviewed by Ryan 1990b). In Darwin's (1871, I, p. 418) words about acoustic insects, "individuals which were able to make the loudest or most continuous noise would gain partners before those which were less noisy."

Even though female choice based on visits to and comparison of several males does occur and is probably common (e.g., Robertson 1986a; Gibson et al. 1991), such "comparing" choice is not necessary to favor conspicuous signaling traits. A male with a strong or frequent signal might attract more mates simply because he is noticed more quickly or is noted farther away than other males (e.g., Otte 1974; Wells 1977b; Lloyd 1979; Parker 1982, 1983b). A game theory model by Parker (1982) suggests that such direct attraction can lead to conspicuous male traits even if there is no genetic variation in female preferences. In contrast with Fisherian self-reinforcing selection and indicator models (section 1.5 below), the female preference in this case may have evolved for reasons not related to fitness advantages for males with the most far-reaching signals. Other factors, for example an initial sensory bias, or selection for species recognition, might help explain such mate choice (Ryan 1990b). Direct attraction may be favored if, compared to other rules, it reduces female expenditure of energy or time, or exposure to predators (Wilson and Hedrick 1982; Searcy and Andersson 1986; Arak 1988c).

This kind of direct mechanism has sometimes been called "passive" attraction, in contrast with "active" choice based on examination of several potential mates and rejection of some of them. The terms active and passive choice may, however, confound proximate and ultimate causation, and passive attraction is no evolutionary explanation for the mating preference and its characteristics (see Pomiankowski 1988; Sullivan 1989; Ryan 1990b; Wiley 1991).

Direct attraction to the strongest signal might sometimes evolve into a more subtle, discriminating preference (Parker 1982). If males differ in their range of attraction, females that search for and compare several males and mate with the most effective signaler should also tend to bear sons with higher than average mating success; this could lead on to a Fisherian process (section 1.5).

There are many possible tactics of female choice among males, based for instance on sequential comparisons or threshold decisions. What tactic is most favorable should depend on, among other things, the variation among males, and costs of choice in females (e.g., Janetos 1980; Parker 1983b; Wittenberger 1983; Real 1990, 1991).

1.5 Evolution of Female Preferences for Male Traits

The Fisher Process

Although much of the early criticism of Darwin's theory was mistaken, the theory had a large gap: it did not explain the origin of female choice. Darwin simply pointed to evidence that it exists (but in the second edition of *The Descent*, 1874, II, p. 495, he added a short section dealing with the evolution of preferences). This void was ridiculed by Morgan (1903), in words that suggested an explanation which he did not, however, take seriously: "Shall we assume that still another process of selection is going on, . . . that those females whose taste has soared a little higher than that of the average (a variation of this sort having appeared) select males to correspond, and thus the two continue heaping up the ornaments on one side and the appreciation of these ornaments on the other? No doubt an interesting fiction could be built up along these lines, but would anyone believe it, and, if he did, could he prove it?"

Morgan's (1903) sarcastic suggestion apparently came under the eyes of Ronald Fisher, who develped it into a coherent theoretical explanation for the evolution of female choice (see below). It is not fully certain that this is the source of Fisher's (1915, 1930) idea, as he did not refer to Morgan (1903). He knew Morgan's book, however, and cited it (Fisher and Stock 1915) in a paper published shortly before the article on mate choice (Fisher 1915). This, and the resemblance between their hypotheses, suggests that Morgan (1903) invented the idea of self-reinforcing evolution of female preferences and male traits ("the two continue heaping up the ornaments . . . and the appreciation of these ornaments"). Ironically, it therefore seems that a skeptical comment by Morgan (1903) became the seed of one of the most celebrated hypotheses for the evolution of female preferences and conspicuous male traits.

Fisher (1930) suggested that "a sexual preference of a particular kind may confer a selective advantage, and therefore become established in the species." He envisaged a two-step process. Suppose there arises genetic variation in a male trait such as tail length, and that males with, say, a longer than average tail have a slight survival advantage, for example owing to improved agility. Assume also that females choose mates, and that there is genetic variation among females in their tendency to mate with males of different tail lengths. Females preferring males with long tails tend to bear sons with high survival. Hence alleles that code for longer tails in males will spread, and so will alleles that make females prefer long-tailed males, as the two types of alleles become associated in their off-spring.

When this process continues, a new effect grows in importance: males with long tails are favored not only by better survival, but also by higher mating success as the preference for long tails spreads among females. The higher mating success of long-tailed males helps carry the associated alleles for the long tail and the female preference to yet higher frequency, and a feedback "runaway process" develops at accelerating pace. Although Fisher did not explain fully how this would happen, he suggested that females will prefer tails of ever-increasing length, until it finally becomes so long that higher mortality balances the mating advantage of long-tailed males. This brings the process to rest. (For further discussion, see O'Donald 1977, and Arnold 1983.)

Fisher (1930) provided the capstone for Darwin's theory of sexual selection and made it a coherent if untested explanation for the evolution of secondary sex ornaments by mate choice. He did not, however, clarify in detail how male traits evolve to extremes that apparently reduce survival. O'Donald (1977) suggested that male ornaments can become supernormal stimuli for females, which then prefer larger than natural ornaments (also see Halliday 1978; Lande 1981). This might explain how male traits evolve to extremes such as the tail of the peacock (section 2.2).

Eberhard (1985, 1991, 1993) proposed that not only ornamental traits, but also male copulation behavior, genitalia, and other contact organs used in copulation, have been affected by Fisherian runaway processes of female choice (section 1.2 above). Genital structures seem less likely than conspicuous ornaments or displays to incur costs such as increased risk of predation, perhaps leaving more freedom for runaway evolution of genitalia.

Fisher presented his idea on the evolution of female choice in words, even though it assumes changes in gene frequencies, which he treated mathematically in many other contexts. O'Donald (e.g., 1962, 1967) pioneered the mathematical modeling of sexual selection (chapter 2 below), verifying the logical consistency of the Darwin-Fisher theory (summarized in O'Donald 1980a). He showed that a preference allele can spread together with an allele for a preferred trait that improves survival. When the preference is sufficiently common, the allele for the trait can increase even if it reduces survivorship, as Fisher suggested. Sexual selection in few-locus systems was further clarified, for example, by Kirkpatrick (1982), who found that a strong and common female preference may even lead to fixation of a male trait that reduces survival (box 2.2.1 below).

Using quantitative genetics, Lande (1981) showed that male trait and female preference under certain conditions can evolve in runaway fashion, the outcome partly depending on chance events at the start (box 2.3.1 below). Such processes in small founder populations subject to genetic drift might help explain why related species often differ most in male secondary sex traits (Lande 1981; West-Eberhard 1983).

The previous genetic models of female choice assumed that mating preferences carry no cost. Mate choice often requires time and energy, or may increase the risk of predation, which can greatly affect the outcome of sexual selection (e.g., Parker 1983a; Kirkpatrick 1987b; Pomiankowski 1988; Iwasa et al. 1991). Most links in the chain of genetic events suggested by the Darwin-Fisher theory remain to be tested (chapter 2).

If females differ in fecundity or parental ability, and if mating reduces a male's chances of fertilizing other females, male preferences might evolve for the most fecund females, or for females that complement the male as regards size or other aspects. Male choice of mate is also likely to occur in role-reversed species, such as pipefish and certain waders, where females compete strongly over males (e.g., Williams 1975; Ridley 1983). There is increasing empirical evidence of female as well as male mate choice (chapters 6 and 8 and section 7.6).

Indicator Mechanisms

Besides Fisherian self-reinforcing mating preferences, a much debated idea is that conspicuous, costly male traits become targets of female choice because such traits indicate high heritable viability. After mating with a highly ornamented male, a female might then bear offspring that survive well.

In some respects, already Wallace (1889, p. 295) foreshadowed this idea, proposing that female choice concerns male traits that indicate high general fitness: "This extremely rigid action of natural selection must render any attempt to select mere ornament utterly nugatory, unless the most ornamented always coincide with the 'fittest' in every other respect." Wallace did not, however, explain how females would come to have such preferences.

Although he is not usually associated with indicator models but with self-reinforcing "runaway selection," Fisher (1915) also gave the first outline of an indicator mechanism. "In every animal there are a few noticeable points or features which readily attract and arrest the attention; and of these it may be expected . . . that some will be more conspicuous among the healthy, active and biologically fit. . . . Consider, then, what happens when a clearly-marked pattern of bright feathers affords . . . a fairly good index of natural superiority. A tendency to select those suitors in which the feature is best developed is then a profitable instinct for the female bird, and the taste for this 'point' becomes firmly established among the female instincts. . . . Let us suppose that the feature in question is in itself valueless, and only derives its importance from being associated with the general vigour and fitness of which it affords a rough index."

Fisher's (1915) indicator idea received even less attention than his other work on sexual selection, and was largely forgotten. The same hypothesis was, however, redeveloped fifty years later by George Williams in his influential book, *Adaptation and Natural Selection*, a rich mine of ideas for evolutionary biologists: "It is to the female's advantage to be able to pick the most fit male available for fathering her brood. Unusually fit fathers tend to have unusually fit offspring. One of the functions of courtship would be the advertisement, by a male, of how fit he is. A male whose general health and nutrition enables him to indulge in full development of secondary sexual characters, especially courtship behavior, is likely to be reasonably fit genetically. . . . In submitting only to a male with such signs of fitness a female would probably be aiding the survival of her own genes" (Williams 1966, p. 184). Later, Williams (1975, 1992) came to doubt whether the heritability of fitness is high enough for such a process to work, but the passage cited probably helped inspire many versions of what has been termed "handicap," "indicator," or "good genes" models.

Williams' (1966) idea was taken up, for example, by Trivers (1972), Emlen (1973), and especially Zahavi (1975, 1977b), who presented his "handicap principle" as an exclusive alternative to Fisherian sexual selection by self-reinforcing mating advantages. Early genetic models indicated, however, that Zahavi's (1975) handicap principle is unlikely to work on its own. For this and other reasons, it was severely criticized, but later models that combine heritable viability differences with a mating advantage suggest that indicator mechanisms might contribute to the evolution of male ornaments. Many authors have put forth reasons why some form of indicator mechanism might be important (section 3.1). The idea of Hamilton and Zuk (1982) based on host-parasite cycles has received most attention. It suggests that genetic cycles of changing resistance in hosts, and virulence in parasites, maintain heritability of resistance in hosts. The degree of resistance is reflected in secondary sex traits such as bright colors, which might then be used as a cue in mate choice. Even if such coevolutionary cycles should turn out to be uncommon, parasites and pathogens may still make indicator mechanisms work; health is one of the aspects likely to be reflected by indicator traits (Fisher 1915; Williams 1966; Hamilton and Zuk 1982).

Models of genetic indicator mechanisms have been found to work most easily if ornament development depends on the phenotypic condition and overall genotype of the male, as implied by Fisher (1915) and Williams (1966) (chapter 3 below). Several empirical studies have provided support for indicator models, but the evidence can be interpreted in different ways. Indicator traits may be correlated with some direct (nongenetic) material benefit to female or offspring, such as food, protection, or parental care. It has not yet been convincingly shown that an indicator process based on genetic benefits for offspring is involved in the selection of any secondary sex trait, a similar situation as for Fisherian self-reinforcing selection.

Species Recognition

To increase in frequency, it is not necessary that an ornament initially reflects higher survival: improvement of species recognition is a plausible alternative (e.g., Fisher 1930; Mayr 1963, 1972; Trivers 1972; Halliday 1978; Maynard Smith 1978). Already Wallace (1889) discussed traits that enable "the sexes to recognize their kind and thus avoid the evils of infertile crosses." He suggested that this might explain "the wonderful diversity of colour and of marking that prevails, especially in birds and insects." Fisher (1930) pointed out that the "grossest blunder in sexual preference which we can conceive of an animal making would be to mate with a species different from its own."

The leaders of the evolutionary synthesis around the middle of this century often pointed to a species isolation function of secondary sex signals and mating preferences, that is, a form of natural selection. This emphasis on species isolation, and the neglect of choice among mates within the species, was apparently a consequence of the focus of the synthesis on speciation and related problems (e.g., Dobzhansky 1937; Huxley 1942; Mayr 1942).

Even if species recognition is involved, it appears unlikely to explain fully the most conspicuous secondary sex traits such as the peacock's tail, which are much more extreme than necessary for species recognition; some additional process seems likely to be involved. In several taxa, however, there is evidence that selection for species recognition has played a role in the evolution of secondary sex traits or mating preferences (chapter 9). Mating discrimination against members of another species can be viewed as a means of avoiding matings that lead to genetically inferior offspring.

Sex dimorphic traits may also function in sex recognition (e.g., Noble and Vogt 1935; Noble 1936; sections 13.5–13.6 below). This seems unlikely to be the main explanation for many conspicuous sex traits, however, which are developed far beyond what should be needed for sex recognition.

Direct Phenotypic Benefits

Fisherian and indicator models have been suggested to explain extreme ornamentation especially in cases where males provide females with nothing but genes, for example in lekking species (section 7.5). But males have conspicuous sex traits also in many species where females get direct material benefits from their mates. Heywood (1989) and Hoelzer (1989) concluded from genetic models that nonheritable variation in parental ability may lead to the evolution of male traits that advertise high parental quality (see also Grafen 1990a). The hypothesis does not require heritable variation in fitness among males. As a likely example, Hoelzer (1989) pointed to courtship feeding in birds (section 8.2 below). Male song rate is another trait that, owing to a correlation with food abundance in the territory, may give a clue to the male's future performance as a parent, and to the food situation for the female and her offspring (chapter 14). There are many other possibilities for direct selection of female preferences (e.g., Williams 1975; Kirkpatrick and Ryan 1991; chapter 8 below), for example asymmetrical fitness distribution of female choice in relation to a male trait that reflects his fertility. Williams (1992) suggested that such a mechanism might suffice for the evolution of extreme male traits and female preferences also in species where females obtain nothing but sperm from males.

Sensory Bias and the Origins of Preferences

Fisher (1930) suggested that new mating preferences can arise by mutations that change female responsiveness to aspects of male behavior or morphology. But the table can also be turned around: a new male trait might be favored because it happens to fit an already existing bias in the female sensory system.[3] If, say, the foraging ecology of a species has led to high sensitivity to certain colors, this bias might favor the evolution of male ornaments with such colors. The peacock's tail with its many "eyespots" might catch and hold the attention of females by exploiting a widespread responsiveness to eyes in animal cognition and communication (Ridley 1981). In some insects, males attract females by pheromones that

[3] This idea has been suggested in one form or other by many authors, e.g., West-Eberhard 1979, 1984, Ridley 1981, Davison 1983b, Burley 1985, Ryan 1985, 1990b, Borgia 1987, Kirkpatrick 1987a,b, Endler and MacLellan 1988, Endler 1989, Enquist and Arak 1993; for review, see Endler 1992a,b, Ryan and Keddy-Hector 1992, Reeve and Sherman 1993, and chapter 10.7 below.

are also present in their fruit food (Baker and Cardé 1979; Löfstedt et al. 1989). Although there may be alternative explanations, in such cases it is possible that "by using responses strongly selected in other contexts, the signal in effect creates a sensory trap to manipulate behaviour in the signaller's own favour" (West-Eberhard 1984). This possibility was referred to as "sensory exploitation" by Ryan (1990) in the context of sexual selection (see section 10.7 below for examples).

Sensory bias may vary with aspects of the environment such as food, predation, and light conditions, influencing the direction of evolution of sexual signals, and sometimes leading to divergence in signals among populations. Such processes were termed "sensory drive" by Endler and MacLellan (1988) and Endler (1989, 1992a). He emphasized that it, together with runaway sexual selection, may lead to rapid divergence of mate recognition systems between species. The "preference" or bias may be common or even fixed when the male trait arises, which would permit rapid spread of the trait even if there is no self-reinforcing evolution of the female preference. The evolution of the male trait might occur without appreciable change in the female preference, for example if it lacks sufficient genetic variation (West-Eberhard 1984; Rowland 1989a; Ryan 1990b). Later on, trait and preference might coevolve further. Based on analysis of artificial neural networks, Enquist and Arak (1993) suggested that signal recognition mechanisms will have inevitable biases that impose selection of signal form, and that the two can coevolve to a state where extreme male traits are favored.

Mating Synchronization and Stimulation

Another idea from the era of the evolutionary synthesis is mating synchronization. Marshall (1936) suggested that "birds which have brighter colours, more elaborate ornamentation, and a greater power of display must be supposed to possess a superior capacity for effecting by pituitary stimulation a close degree of physiological adjustment between the two sexes so as to bring about ovulation and the related processes at the most appropriate time." Moreover, this is the "value of sexual display and of the adornment which in many species is taken advantage of to render the display more effective." According to Marshall, sex ornaments and courtship function after pair formation and are not involved in competition over mates, but are favored by natural selection (also see Morgan 1919; Huxley 1938b). Even if it has a long-term stimulatory function, however, male courtship behavior is also likely to be favored by sexual selection. It will be so, for example, if females leave males with poorly developed courtship signals for males that provide more effective stimulation. Examples of long-term priming effects of male signals are given in chapters 13–15 below.

Marshall's (1936) proximate hypothesis does not explain why females should be stimulated by conspicuous male traits. To provide an ultimate explanation, the idea must be supplemented with a mechanism for the evolution of female responsiveness. One possibility is that the development of male traits is correlated with reproductive condition, for example via testosterone levels. A well-developed male trait might indicate full gonadal development and high sperm numbers, with maximum probability that the female will have all her eggs fertilized (e.g., M. B. Williams 1978; G. C. Williams 1992). Other possible advantages to females were discussed above.

1.6 Summary

Sexual selection is the differences in reproduction that arise from variation among individuals in traits that affect success in competition over mates and fertilizations. Scrambles, contests, endurance rivalry, and mate choice are the main forms of premating sexual selection; sperm competition in many species also influences fertilization success. Anisogamy and greater female than male parental effort is the likely reason for stronger male than female competition over mates. In Darwin's theory of sexual selection, mating competition favors male secondary sex traits, such as weapons and conspicuous signal traits. His ideas on female choice have been much criticized, but they receive support from formal models and many recent empirical studies. Some male signals and ornaments may be favored by male contests rather than female choice, however, or by both mechanisms. The evolution of female preferences remains controversial. Some of the alternative explanations, which are all compatible and may apply in combinations, are Fisherian self-reinforcing selection, genetic indicator (handicap) mechanisms, avoidance of hybrid matings, direct material advantages for discriminating females, and male exploitation of female sensory bias.

2 Genetic Models of Fisherian Self-Reinforcing Sexual Selection

2.1 Introduction

The fundamental mechanism in Fisher's (1930) process is that female preference alleles for a new male trait become coupled genetically with alleles for the trait. This occurs in the offspring from matings between females with the preference, and males with the trait (section 1.5 above). The mating advantage of preferred males therefore carries to higher frequency (or expression) not only the male trait, but also the female preference, which leads to an even higher mating advantage for the male trait, and so on. This positive feedback loop can lead to rapid self-reinforcing increase in trait and preference in a "runaway process." Stability can also be the result, however, depending on the biological assumptions.

Fisher presented his idea verbally, but the evolution of trait and preference in this and other kinds of sexual selection can be so complex that formal models are needed to explore the possibilities and outcomes. As a background for later empirical chapters, models of sexual selection by mate choice are discussed here and in the next chapter. Much of the theoretical progress from O'Donald (1962) onwards has been based on genetic models.

Mathematical models have contributed to the theory of sexual selection in several important respects. Compared to verbal arguments, they require more precise formulation of the assumptions and clarify what must be measured for the behavior of the system to be understood and predicted. Mathematical models therefore help focus empirical work on crucial aspects. As simple causes often have complex outcomes, such models may be necessary to make sure what follows from the assumptions; the results are sometimes nonobvious and lead to new hypotheses. Mathematical formulation also helps show similarities between problems that first seem unrelated, leading to broader insight and exchange of ideas between fields. Not least, comparison of assumptions and predictions with empirical results shows if some crucial aspect is not understood. It also indicates how the mistake can be corrected and insight gained through modification of the model, followed by renewed testing.

Levins (1966) suggested that models in population biology should have high generality, precision, and realism. To this might be added simplicity, permitting an intuitive grasp of the behavior of the model. Not all these goals are compatible. Most genetic models of sexual selection aim for generality and simplicity; they are logical tools that help show consequences of the assumptions and derive new theoretical insights. They are not realistic and testable in the sense of yielding, as they stand, precise predictions in specific real cases (Andersson 1987). Probably none of the models discussed here can be tested quantitatively without modification and elaboration. Some qualitative assumptions or predictions can be tested, however, and the models help clarify sexual selection by showing what is and what is not possible under various assumptions, and by sometimes predicting outcomes that were not foreseen. In most respects, quantitative models of sexual selection remain to be critically tested.

This chapter and the next deal mainly with genetic models. Game theory models have gained increasing use in evolutionary biology, but they may be less suitable for exploring the dynamics of sexual selection by genetically variable mating preferences (Maynard Smith 1982). One reason is that the fitness of offspring will often depend on the genetic covariance that develops through gametic phase disequilibrium (Falconer 1989) between genes for the preferred trait and the preference. Understanding the dynamics of such a system seems to require genetic modeling. Game theory models have been used to clarify many other aspects of sexual selection (e.g., Parker 1979, 1982, 1983a,b; Hammerstein and Parker 1987; Sutherland and De Jong 1991), and Grafen (1990a,b) showed that game theory can also be useful for exploring the evolutionary equilibria of male traits and female preferences in sexual selection (section 3.4).

Genetic models of sexual selection are often complex. Moreover, owing to rapid theoretical progress, they tend to quickly become obsolete. The following review therefore sketches only their main features. Details are found in the papers cited; general approaches in genetic modeling are described, for example, by Roughgarden (1979) and Maynard Smith (1989).

2.2 Two-Locus Models

O'Donald's pioneering models of sexual selection, based on two-locus diploid genetics and polygyny, are summarized in his book from 1980; some of the results were mentioned in section 1.5 above. Here, the two-locus modeling approach is introduced by the more recent, transparent model of Kirkpatrick (1982).

A common approach, followed in many few-locus models of sexual selection, is to assume starting frequencies of the different alleles and genotypes in males and females in generation 1. Next, the mating frequencies of the different genotypes in both sexes are calculated, based on the model's assumptions about mating preferences and other aspects. (For review of the many assumptions of some genetic models of sexual selection, see Andersson 1987). This leads on to the genotype frequencies among zygotes in the following generation (see, e.g., O'Donald 1967, 1980a). Any differences in natural selection (survival) among the genotypes are then assumed to occur, leading to new genotype frequencies among the adults in the second generation. From these genotype frequencies, the allele frequencies can be calculated and compared with those of the previous generation. The process is repeated for the desired number of generations, showing what happens with the frequencies of the different alleles in the long term. (An alternative approach is to start with and compare gene frequency changes at the zygote stage.)

Kirkpatrick (1982) studied the evolution of male trait and female preference with a haploid two-locus model (also verifying that a corresponding diploid model gave the same qualitative results; see also Gomulkiewicz and Hastings 1990). His analysis greatly clarified the dynamics and equilibrium conditions of simple two-locus models of male traits and female preferences. It is probably the most pedagogical, albeit not most realistic, model of such a system (box 2.2.1).

The equilibria of trait and preference alleles in Kirkpatrick's model are not stable points in the usual mathematical sense. Instead, they fall on a line of equilibria (figure 2.2.1). The system, if perturbed away from the line, will return to it, but there is no selection against changes of gene frequencies along the line, so the system is liable to drift along it.

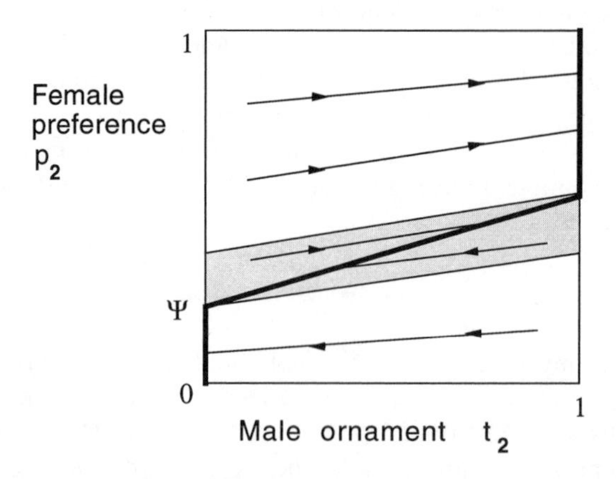

Figure 2.2.1 The line of equilibria (solid line) in Kirkpatrick's (1982) two-locus model of the Fisher process. In the speckled area, deviations from the equilibrium line return to another internal equilibrium. Deviations outside the speckled area lead to boundary equilibria where the ornament allele is lost ($t_2 = 0$) or fixed ($t_2 = 1$). The threshold frequency of female preference needed for the Fisher process to favor the male trait is $p_2 = \Psi$. (After Pomiankowski 1988.)

BOX 2.2.1 TWO-LOCUS HAPLOID MODEL OF FISHERIAN SEXUAL SELECTION

Variation in the male trait in Kirkpatrick's (1982) model is determined by the T (trait) locus, where allele T_1 produces males without the trait. T_2 males carry the trait, which reduces their survival to $1 - s$ compared to 1 in T_1 males. Female mating preferences are coded at the P (preference) locus. P_1 females mate at random. P_2 females show a preference of strength a_2 for T_2 males, which obtain a proportion $a_2 t_2'/(t_1' + a_2 t_2')$ of their matings; t_i' is the frequency of adult T_i males. (In choices between two males, a P_2 female would mate with a T_2 male a_2 times more often than with a T_1 male.) There is no direct selection of the preference alleles; they do not affect the fecundity or survival of the female, an assumption of importance for the conclusions. Kirkpatrick (1982) did not formulate his model in relation to any explicit choice situation; a more concrete situation was modeled by Seger (1985, described later in this box).

A crucial feature of this and many other models of the evolution of genetic mating preferences and secondary sex traits is the development of gametic phase disequilibrium between alleles at the two loci. Even if they first occur at random in relation to each other, the preference allele P_2 over the generations becomes associated with the trait allele T_2. This is because P_2 females tend to mate with T_2 males. As a consequence, even though there is no direct selection of the preference allele P_2, it spreads owing to coupling in their offspring with the T_2 allele, which is favored by sexual selection.

Defining the frequency of the T_2 allele after zygote formation as t_2 (= combined frequency of the $T_2\,P_1$ and $T_2\,P_2$ zygotes), and for the P_2 allele as p_2 (= $T_1\,P_2$ and $T_2\,P_2$ zygotes), Kirkpatrick derived the evolutionary equilibrium of the system, where gene frequencies do not change (figure 2.2.1). The main conclusion is that for any values of the mating preference a_2, selection coefficient s, and frequency p_2 of the preference allele, there is a corresponding equilibrium frequency t_2 of the male trait. If the mating preference is strong enough relative to the selection coefficient, so that $s < 1 - 1/a_2$, the male trait can have non-zero frequency; the equilibrium points fall along a line (figure 2.2.1). If gene frequencies are displaced away from the line, they tend to return to some point on it, but usually not to the original point. If the displacement is large, the gene frequencies may end up at boundary equilibria, where the trait allele T_2 is either fixed ($t_2 = 1$) or lost ($t_2 = 0$).

As long as a moderately strong preference for a deleterious trait is still rare, the equilibrium frequency of the trait is zero (figure 2.2.1). Unless the preference becomes more common for some reason, such as drift or pleiotropy, the male trait then will not spread. The system is neutrally stable along the line: none of the processes included in the model will change the gene frequencies after a point on the line is reached. Other things, such as random genetic drift, selection of pleiotropic effects of the alleles, or mutation pressure, can lead to further changes.

BOX 2.2.1 CONT.

The line of equilibria in Kirkpatrick's (1982) model represents stability only in the sense that the population evolves back to the line if displaced away from it. Seger (1985) showed that this need not be the case under somewhat different rules of mate choice. He modeled female choice according to a "best of *N*-rule," where each female compares the males in a group of size *N*, say, at a lek. A nonchoosing (P_1) female accepts one of the males at random, whereas a choosy (P_2) female is more likely to mate with one of the ornamented (T_2) males, if any is present. Male fitness can then be positively frequency-dependent near the line of equilibria, displacement away from the line triggering runaway selection farther away from it, as in Lande's (1981) polygenic model (section 2.3 below). This is particularly likely if *N* is small, i.e., if a female compares only few males before accepting one. The reason can be intuitively understood as follows. When males without the trait (T_1) are common, most leks contain no ornamented (T_2) male. Most P_2 females therefore must accept one of the T_1 males, which hence mate almost in proportion to their frequency in the population. When T_1 males are rare and T_2 males are common, however, most leks contain one or more such males, so most P_2 females can follow their preference, and the success of T_1 males is reduced. The success of both types of male for this reason will be positively frequency-dependent. Under certain circumstances, this can lead to instability and rapid evolution also in a two-locus model. Note, however, that even if smaller *N* increases the chances for instability, this need not increase the likelihood of exaggeration of the male trait, for reasons discussed by Pomiankowski (1988).

The existence of a line of equilibria can be explained as follows. At any equilibrium point, the mating advantage of preferred over other males is balanced by their survival disadvantage, so the two types of male have identical fitness. The frequency and strength of the female preference and the frequency of the male trait affect the mating success of the males. If the preference allele spreads, the preferred male trait is selected to higher frequency, until it reaches the point where the mating advantage has declined enough to be balanced by reduced survival. The relative mating advantage of preferred males is negatively frequency-dependent, decreasing as their proportion in the population goes up (for an opposite case, see Seger 1985, and box 2.2.1). When the trait comes close to fixation, nearly all males carry the trait, and the relative mating advantage tends to zero.

Contrary to what has sometimes been predicted, a new female preference for a male trait that improves survival does not necessarily eliminate an earlier male trait that reduces survival or the original preference for it (Kirkpatrick 1982). The more deleterious the earlier trait, however, the wider the range of preference frequencies under which the trait will vanish. In this sense, therefore, the chances that the male trait disappears increase with its survival disadvantage. As suggested by Fisher (1915), competitive replacement of sexually selected traits may therefore take place, though it is no necessary outcome.

Like Fisher, O'Donald (e.g., 1980a, his chapter 8.7) assumed that the preferred male trait to begin with improves survival. In the next stage, the preference has undergone a peak shift, the trait now being a supernormal stimulus for females that prefer a trait larger than the original modification. Ethologists have described many examples of supernormal stimuli, such as birds preferring to incubate eggs of much larger than normal size (Tinbergen 1951). The reason for such preferences may be asymmetric selection of responsiveness (Staddon 1975). If females that mate with strongly ornamented males bear offspring of higher than average fitness, selection can favor a shift in peak responsiveness toward larger than natural ornaments (O'Donald 1977; also see Halliday 1978). (An alternative possibility is sensory bias, if larger objects generally are more stimulating to visual receptors; J. Endler, pers. comm.; see section 10.7 below.) Fisher (1958) apparently had something similar in mind when suggesting that there is selection for higher "intensity" of female preferences. Females with a shifted peak preference are more likely than others to mate with males carrying an extreme trait, and the cost of the peak shift initially may be low. When females preferring supernormal ornaments become more common, modifier alleles that amplify the male trait are favored. Finally, the trait becomes so large that it reduces survival enough to offset the male mating advantage. Selection of the female preference then is no longer asymmetrical, or at least less so than before, and the balance between the mating advantage and the survival disadvantage stabilizes the size of the male trait (O'Donald 1977, 1980).

2.3 Polygenic Models

An unrealistic feature of the few-locus approach is that probably most secondary sex signals and mating preferences depend on alleles at many different loci. Quantitative genetics then is more appropriate. In a series of papers, Lande (1980–1982a) developed such models of Fisherian sexual selection (also see, e.g., Arnold 1983, 1987; Heisler 1984b, 1985; Barton and Turelli 1991; Pomiankowski et al. 1991).

Lande's (1981) study was the first to demonstrate Fisherian runaway selection in a genetic model. It can be intuitively explained as follows. If there is initial equilibrium but the mean female preference for some reason, for example genetic drift, is moved away from the equilibrium line, which

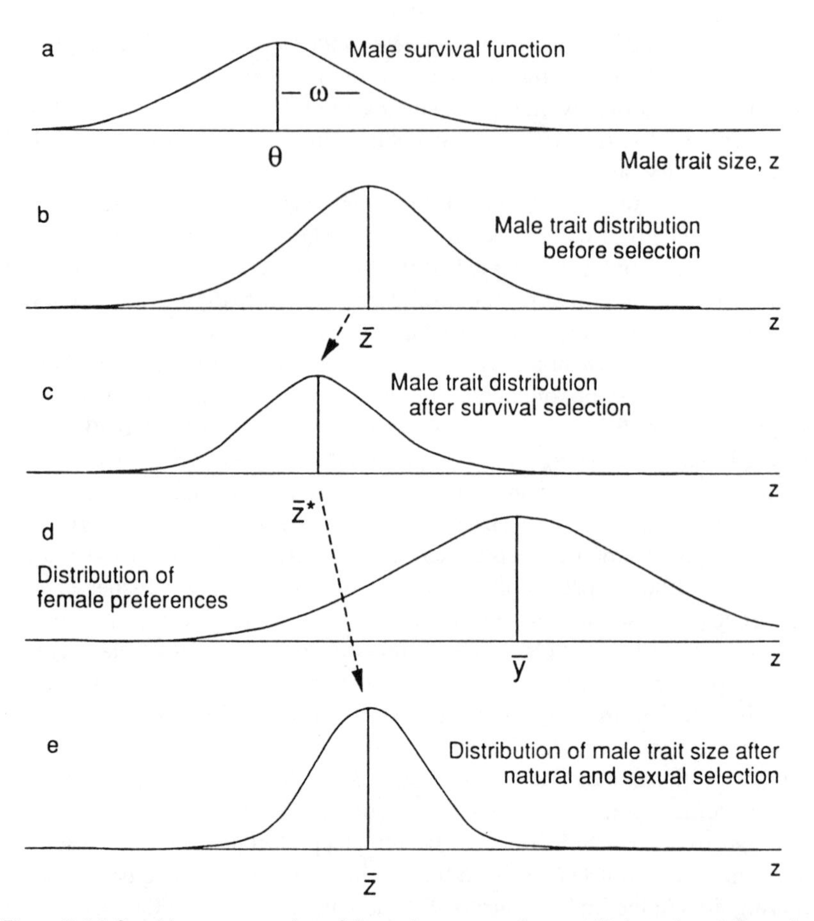

Figure 2.3.1 Graphic representation of the maintenance of an equilibrium size of the male trait by a balance between natural and sexual selection in Landes (1981) polygenic model of the Fisher process. For explanation, see text. (Modified from Maynard Smith 1982)

occurs also in this model (figures 2.3.1, 2.3.2), the male trait as a consequence evolves in the direction of the new preference mean. Because of the genetic correlation between trait and preference, the mean preference will then also move farther away from the equilibrium line. If the genetic correlation is sufficiently strong, the female preference can move so far away that the male trait is unable to catch up, and the system will "run away" until arrested by, for example, insufficient genetic variation (see box 2.3.1 for details).

Heisler (1984b, 1985) clarified the initial conditions under which female preferences might evolve by a Fisherian process, showing, among other things, that an initial viability advantage for the male trait greatly facilitates such evolution, and that mating preferences that first are favored because

BOX 2.3.1 FISHERIAN SEXUAL SELECTION IN A POLYGENIC MODEL

Lande's (1981) model is based on two sex-limited quantitative characters: a male trait and a female mating preference, both subject to genetic and environmental influence. The characters depend on many autosomal genes and have normal distributions with means $\bar{z}$ and $\bar{y}$, and additive genetic variances G and H, respectively. There is additive genetic covariance B between trait and preference owing to pleiotropy and nonrandom association of alleles at different loci (see, e.g., Bulmer 1980; Falconer 1989; over the generations, alleles that code for a large male trait become associated in offspring with alleles that make females prefer such males, and alleles for a small trait become associated with preference alleles for a small trait). The mating system is polygynous, and males provide no resources but genes; the expected number of newborn offspring is the same for all females. Generations do not overlap.

The female mating preference does not affect fecundity or survival, but evolves only as a correlated response to selection of males, as in the few-locus models above. The male trait is subject to weak stabilizing selection: the risk of mortality increases with the deviation of the trait from its optimum size, θ. The strength of stabilizing selection is described by the characteristic width, ω, of the Gaussian selection function around the optimum; smaller ω implies stronger selection (figure 2.3.1a).

Lande examined three different kinds of female preferences. With *absolute* preference, each female shows maximum preference for some size y of the male trait, the maximum differing among females. The preference falls off around y according to a Gaussian function with characteristic width ν (and the same general shape as in figure 2.3.1a); the smaller ν, the more critical is the female in her choice of mate. Lande also examined *relative* preferences, with maximum strength for a male trait that is a certain amount larger than the male average. As a third alternative, he assumed that the preference is *open-ended* and exponentially increases with the size of the male trait. The main qualitative conclusions were similar for the three forms of preference.

At equilibrium, the following events take place in the population during each generation of selection. Consider the distribution of genotypes for male ornament size z among the zygotes (figure 2.3.1b). Because the male trait is subject not only to natural selection towards θ, but also to sexual selection for larger size, the mean genotypic size $\bar{z}$ will be greater than the optimum with respect to natural selection, θ. The genotypes are subject to natural selection before mating, which reduces the proportion of males with larger than average trait size; this shifts the distribution to that in figure 2.3.1c, closer to θ. Females make their choice of mates from the distribution of adult males. As the average peak preference $\bar{y}$ among females (figure 2.3.1d) exceeds $\bar{z}^*$ (mean trait size among males after natural selection), the differences in mating success among males shifts the proportional frequency

BOX 2.3.1 CONT.

distribution of mating males towards larger male trait sizes. If the process has reached equilibrium, the male trait will then be back at the same mean size $\bar{z}$ as in the population of zygotes, but with smaller variance owing to the selective removal of extremes that has taken place (figure 2.3.1e). The variance is then restored to its original value through recurrent mutations (Lande 1981).

If the process has not reached equilibrium, the mean size of the male trait, $\bar{z}$, as well as the average female preference, $\bar{y}$, changes between generations. This happens because of the genetic covariance that develops between male trait and female preference. As a consequence, if sexual selection leads to larger mean size of the male trait, this will also carry the mean female preference $\bar{y}$ to higher values.

The process reaches an equilibrium when the male trait gets so large that it reduces survival enough to just offset the mating advantage. The mean male trait, $\bar{z}$, will then have a size somewhere between the optimum with respect to survival, θ, and the mean, $\bar{y}$, of the female preference peak. At equilibrium, sexual selection balances natural selection, and the average male trait size $\bar{z}$ remains the same between generations. As there is no direct selection of the female preference, its population mean, $\bar{y}$, also remains constant. In general, $\bar{z}$ at equilibrium will not coincide with the survival optimum, θ.

In figure 2.3.1, the mean of the female preference, $\bar{y}$, is larger than the mean of the male trait, $\bar{z}$. Sexual selection therefore favors larger male traits. If the mean peak of female preferences, $\bar{y}$, were instead smaller than $\bar{z}$, sexual selection would favor a reduced male trait.

Also in this model, the points of possible equilibria for the trait and the preference form a line (figure 2.3.2); the point reached depends on the starting conditions. Lande (1981) and Kirkpatrick (1982) therefore drew similar

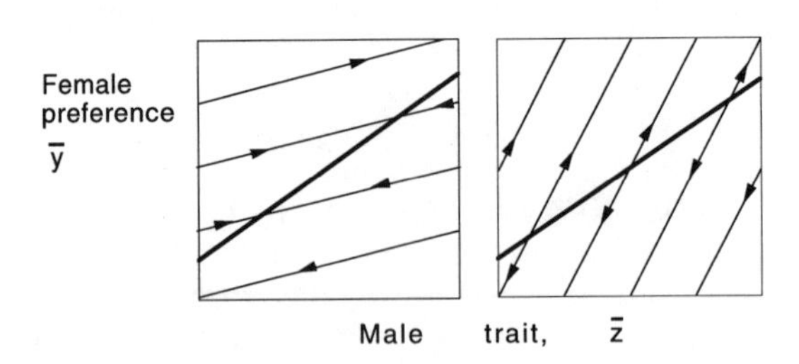

Figure 2.3.2 The joint evolution of mean male trait size and mean female mating preference in Lande's (1981) model is described by evolutionary trajectories (arrowed lines). The slope of the trajectories depends on the genetic regression, *B*/*G* (see text). When this slope is smaller than that of the equilibrium line (solid; given by $(\nu^2/\omega^2) + 1$), the equilibria are stable (a). In the opposite case they are unstable (b), leading to Fisherian runaway evolution of trait and preference away from the equilibrium line. (After Lande 1981)

BOX 2.3.1 CONT.

conclusions from two rather different models (but see Barton and Turelli 1991). Lande (1981) in addition showed that the line can represent stable or unstable equilibria, with the system, if unstable, evolving away from the equilibrium after disturbance. Whether the equilibrium line represents stability or not depends on its slope, $(\nu^2/\omega^2) + 1$, relative to the slope B/G of the trajectories along which the average male trait and female preference evolve. If these trajectories have smaller slope than the line of equilibria, it represents stability (figure 2.3.2).

Hence the system is unstable if females are sufficiently critical (small ν) in their choice of mate, and if natural selection against deviations from the survival optimum θ for the male trait is weak enough (large ω). Accelerating evolution away from the line of equilibria can then occur toward larger (or smaller) traits, as in Fisher's (1930) runaway process. It will cease when natural selection against the male trait increases more rapidly with deviations from θ than does a Gaussian function. Alternatively, females preferring extreme (and hence rare) males might have lower fecundity, which could also arrest the process.

they increase offspring viability may evolve further and lead to reduced survival.

Seger (1985) found that lines of unstable equilibria may occur also in two-locus models under apparently realistic rules of mate choice (see box 2.2.1). Few-locus as well as polygenic models can therefore give lines of stable or unstable equilibria, depending on the conditions. Stability is promoted by strong natural selection of the male trait, weak mating preferences, and weak genetic covariance between male trait and female preference. The two kinds of model are compared in detail by Pomiankowski (1988).

The models of Kirkpatrick (box 2.2.1) and Lande (box 2.3.1) greatly clarified theoretical aspects of sexual selection by mate choice, but some of the conclusions depend on unrealistic assumptions, for example no costs of mate choice. If there is such a cost, no line of equilibria arises, but only a single equilibrium point (Lande 1981; Kirkpatrick 1985; Pomiankowski 1987b; Bulmer 1989a; Barton and Turelli 1991). Mutations can also influence sexual selection in several important ways (Bulmer 1989a; Maynard Smith 1991a; Barton and Turelli 1991; Pomiankowski et al. 1991). Effects of choice costs and mutations are discussed in sections 2.4 and 3.3 below.

A controversial aspect of Lande's (1981) model concerns the maintenance of genetic variation and its consequences. Lande showed that, if the characters depend on numerous loci with mutation and many alleles at each locus, under certain circumstances a stable ratio B/G can arise between the genetic covariance and variance (box 2.3.1). The assumptions behind this result are difficult to test; for discussion see Arnold (1983), O'Donald (1983a), Engen and Saether (1985), Lande (1987), Pomiankowski (1988), and Barton and Turelli (1991). There is evidence that genetic covariances may sometimes change markedly over a few generations in the laboratory (Cohan and Hoffman 1989), and it is not clear whether a stable ratio of B/G will often obtain. Complexity has so far prevented mathematical modeling of evolution of the covariance matrix, and the maintenance of genetic variation by a balance between mutation and selection is a debated, difficult issue (e.g., Lande 1976, 1981; Turelli 1984, 1988; Barton and Turelli 1989, 1991; Bulmer 1989b; Bürger 1989; Charlesworth 1990).

All models discussed so far assume polygyny. Darwin (1871) and Fisher (1958) suggested that sexual selection may favor conspicuous male traits also in monogamous species (sections 1.2, 7.3 below). If well-fed females that breed early in the season have high fecundity and mate with the most attractive males, such males get more offspring than other males. Other factors can also give rise to sexual selection in monogamous species, for example biased adult sex ratios (see, e.g., Darwin's finches, section 4.5). O'Donald (1980a,b) supported this idea using single-locus models of the male trait. Using a quantitative genetic model with variation in preference as well as trait, Kirkpatrick et al. (1990) showed that conspicuous male traits that reduce survival can evolve in a monogamous system. In accordance with empirical results, the model indicates that such traits in monogamous animals tend to be less extreme than in polygynous species. The suggested reason is that variation in male reproductive success under monogamy is limited by the variation in fecundity among females, whereas in polygynous species, some males obtain many mates, others none, leading to larger variation in male reproductive success. Direct selection of the female preference, favoring females that mate with the males providing the best material resources, might also limit sexual selection in monogamous species (Kirkpatrick et al. 1990). Another possibility is that secondary sex differences are less extreme in monogamous species because males and females have more similar sex roles, and therefore are subject to more similar natural as well as sexual selection than in polygynous species (chapter 7).

2.4 Direct Selection of Preferences, and Deleterious Mutations in the Preferred Trait

The previous few-locus as well as polygenic models of the Fisherian process assumed that there is no direct selection of female preferences. Direct selection may, however, be common (see Kirkpatrick 1987b). For example, female choice often involves evaluation of several males (Janetos 1980), which likely takes time and energy (e.g., Parker 1979, 1983; Real 1990). When preferred males are rare, choosy females usually have to examine many males before finding one of the preferred type. Preferences therefore may have a cost that is inversely related to the frequency of the preferred type of male. The dynamics and stability conditions of the system then change drastically (Lande 1981; Kirkpatrick 1982, 1985, 1987a,b; Pomiankowski 1987b, 1988; Bulmer 1989a; Heisler and Curtsinger 1990; Kirkpatrick and Ryan 1991; Maynard Smith 1991; Barton and Turelli 1991; Pomiankowski et al. 1991).

For example, Bulmer (1989a) showed that direct selection, no matter how weak, of the female preference, or mutation, however rare, in trait or preference, replaces the line of equilibria by a single equilibrium point (or a small set of disconnected equilibria). He therefore considered the idea of a neutrally stable line of equilibria misleading. As rare mutation and weak direct selection are likely to be present in any real system, stable equilibrium lines may not occur in reality. Bulmer (1989a) concluded that Fisherian sexual selection is likely to be dominated by direct selection of the female preference, such as delays in mating for very discriminating females, and mismatings with other species by nondiscriminating females. If one allele at the preference locus has a direct advantage over the other in the two-locus model of Bulmer (1989a), it goes to fixation, also dragging along the gene for the male trait.

These results suggest that equilibrium lines of the kind found in earlier models (Lande 1981; Kirkpatrick 1982) may not arise in real systems. This does not mean that Fisherian sexual selection of male traits and female preferences may not occur (see below), but it does suggest that evolutionary exaggeration of male traits by random drift along a neutrally stable line of equilibria is unlikely. The evolution of the male trait may be governed in part by direct selection of the female preference.

Two types of direct selection of mating preferences are (1) costs in terms of reduced fecundity or survival as a consequence of searching for and examining several potential mates; and (2) direct benefits in terms of improved fecundity or survival as a consequence of choosing a particular mate.

Costs of Mate Choice

Costs may take many forms, for example predation risk or loss of time or energy, leading to lower survival or fecundity (reviewed by Reynolds and Gross 1990). For example, searching for a mate reduces the time available for feeding in female pied flycatchers *Ficedula hypoleuca* (Slagsvold et al. 1988): their breeding success declines rapidly with later laying date (e.g.,

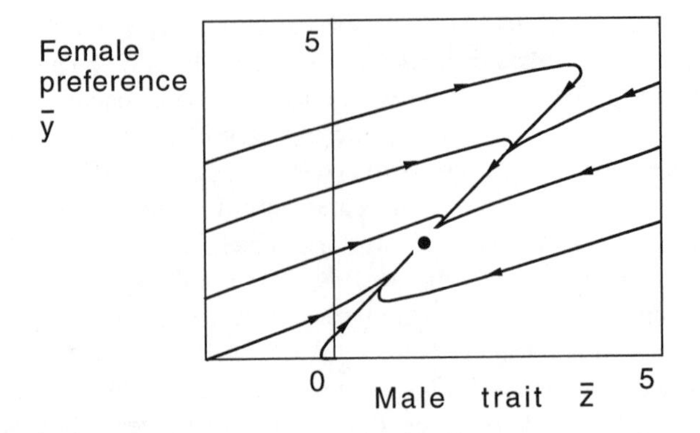

Figure 2.4.1 Evolution of the male trait and the female preference in a polygenic model when the male trait is subject to biased mutations that tend to reduce it. The female preference may then evolve even if it is costly. At the equilibrium point, the cost of the female preference is offset by the advantage of having more attractive male offspring. (After Pomiankowski et al. 1991)

Alatalo and Lundberg 1984; also see Alatalo et al. 1988a). In the seeweed fly *Coelopa frigida*, choosy females may have reduced fertility when the adult population density is low (Engelhardt et al. 1989). Various consequences of costs of mate choice have been explored theoretically by Parker (1983b), Kirkpatrick (1985, 1987b), Andersson (1986a), Pomiankowski (1987b), Bulmer (1989a), and Pomiankowski et al. (1991).

In the previous models of Fisherian sexual selection (sections 2.2–2.3), the process can maintain a preference for an extreme trait only if the preference has no cost (e.g., Parker 1983b; Pomiankowski 1987b). In another polygenic model, Pomiankowski et al. (1991; also see Bulmer 1989a; Barton and Turelli 1991) pointed to a mechanism by which costly preferences can evolve. They added the assumption of biased (deleterious) mutations affecting the male trait, tending to reduce rather than enhance it. Such a bias in the effects of mutations seems likely, as there will be many more mutational ways of disrupting a complex trait than improving it. Under this assumption, costly preferences can evolve and be maintained by the Fisherian process alone (figure 2.4.1). At the equilibrium point, the cost of mate choice is balanced by the advantage of mating with an attractive male and bearing attractive sons. A Fisherian process might therefore establish costly female preferences for ornamented males, contrary to what was previously thought (Pomiankowski et al. 1991).

Direct Benefits of Mate Choice

Kirkpatrick (1985, 1987b) explored cases where female fecundity depends on her choice of mate, for example species in which the male feeds the female or helps raise the offspring, or defends a territory in which the female forages. Using a polygenic model, Kirkpatrick (1985) found that an evolutionary equilibrium is reached only when females on average mate with males of the phenotype that maximizes female fecundity. This result seemed to refute the "sexy son hypothesis" of Weatherhead and Robertson (1979), which was inspired by work on polygynous red-winged blackbirds *Agelaius phoeniceus* (section 5.4). The idea is that females may prefer to mate with an attractive, already mated male, even if he provides less parental care than monogamous males. Females that mate with attractive males are compensated for reduced fecundity by bearing "sexy sons" with higher than average mating success. Kirkpatrick's (1985) results cast doubt on the potency of such a mechanism. Using a two-locus diploid model, Curtsinger and Heisler (1988, 1989) and Heisler and Curtsinger (1990) found, however, that a sexy-son mechanism under certain circumstances can work, and so did Pomiankowski et al. (1991), using a polygenic model with deleterious mutations affecting the male trait (also see Barton and Turelli 1991). Whether the mechanism works in practice remains to be shown.

There are many possible kinds of direct benefits from mate choice to the female and her offspring (e.g., Reynolds and Gross 1990; Kirkpatrick and Ryan 1991), reviewed in chapter 8 below. As costs of mate choice also seem likely to be common, results from the models discussed in this section may be particularly relevant for the understanding of sexual selection by mate choice.

2.5 The Evolution of Sexual Dimorphism by Fisherian Sexual Selection

The previous models assumed that male secondary sex traits are not expressed in females. In reality, male traits are often partly developed in adult females, as the underlying genes are not fully sex-limited in their expression. Lande (1980) and Lande and Arnold (1985) modeled the evolution of sex dimorphism and mating preferences in such situations, assuming that the evolving trait is subject to stabilizing natural selection in both sexes. In the simplest case, there is sexual selection only in males. Lande (1980) did not specify the mechanism but assumed that the male trait is subject to constant sexual selection toward a certain size, larger than that favored by natural selection. The population evolves toward an equilibrium point where natural and sexual selection exactly balance each other. Evolution toward the equilibrium first proceeds by a rapid change in character means for both sexes, followed by a slow growth of dimorphism as the genetic correlation of the trait between males and females is gradually broken up, a process that may perhaps take millions of generations. Analyzing human data, Rogers and Mukherjee (1992) concluded that strong genetic correlations between male and female size makes sexual size dimorphism respond much more slowly to selection than does the mean size of each sex. A possible example, where genetic correlations between the sexes may explain partial expression in females of a trait favored by selection in males, is the red wing epaulet of red-winged blackbirds *Agelaius phoeniceus* (Muma and Weatherhead 1989; also see Whittingham et al. 1992).

Lande and Arnold (1985) found that, if there is no pleiotropy between the preference and the preferred trait, partial expression of the male trait in females has little effect on the dynamics of sexual selection and the evolution of sex dimorphism; there may still be runaway evolution. Sexual selection of one trait may also lead to changes in genetically or functionally related traits that are not directly subject to sexual selection (Lande 1980; Arnold 1987). The correlation, if it is purely genetic and maladaptive, will diminish as the pleiotropic alleles are gradually replaced by others. If the correlation is adaptive, however, it may persist (Arnold 1987). Consider, for example, the wing of a bird. With respect to flight performance, there may exist an optimal ratio of wing to tail size. If the tail, because of sexual selection, gets larger, natural selection might favor a new wing size that harmonizes with the larger tail (see Andersson and Andersson 1994).

2.6 Speciation by Sexual Selection?

Small differences in starting conditions between populations can make them end up at very different equilibria in Lande's (1981) model, especially if the process at some stage becomes unstable, as in figure 2.3.2. Founder effects, genetic drift, and geographic differences in natural selection might greatly influence the outcome, causing large contrasts in secondary sex traits between populations or closely related species. Lande (1981; also see Rohwer 1982; West-Eberhard 1983; Thornhill and Alcock 1983) suggested that Fisherian processes might explain why closely related species of higher animals often differ most in male secondary sex traits, in apparently nonadaptive or random fashion in relation to natural selection (Darwin 1871).

Striking examples are the plumage of birds of paradise, ducks, and pheasants, color patterns among certain lizards, and calls used for mate attraction in some anurans, birds, and insects, which often differ markedly between closely related species. Eberhard (1985) presented evidence that this may also be the case with animal genitalia for similar reasons. On the other hand, Alatalo et al. (1988b) found that a conspicuous ornament, the long tail of several passerine birds, showed more variation within but not between geographical areas than did other body size characters.

Male secondary sex traits in some cases are more stable than corresponding traits in females. For example, plumage color in the North American purple martin *Progne subis* varies geographically in females and subadult males, but adult males are uniformly steel blue over the entire range (Johnston 1966). Other examples are coloration in many butterflies and lizards, females varying more than males (Stamps and Gon 1983). There is evidence that conspicuous male coloration in these cases may be favored by male contest competition rather than female choice (chapter 13). As Fisherian runaway selection by genetic coupling cannot arise by male contest competition, there is no reason to expect greater geographical variation in males in such cases. Empirical evidence even suggests that male contest competition may be a stabilizing factor that reduces the variability of the signals (section 13.2).

Lande (1982a) explored the potential for sexual selection to promote speciation along a geographic cline (also see Fisher 1958). He added spatial structure and dispersal to his previous models of female choice, and assumed that geographic variation in mating preferences evolves through coupling with the evolving geographic variants of the male trait. The results showed that runaway sexual selection can amplify geographic variation and increase differences in male traits between populations, perhaps leading to premating isolation and even parapatric speciation (see Endler 1977). There was not, however, any cline reversal in the ecological boundary zone, so Lande's results do not suggest that reproductive character displacement (section 9.1) will arise in this situation. There may also be direct selection of female preferences, for instance if hybrids have reduced fitness and less discriminating females run a risk of mating with a male of another (sub)species. Lande (1982a) suggested that Fisherian runaway selection is more powerful for initiating reproductive isolation than is selection against hybridization.

A partly different type of "ecological speciation" through female choice of mate, based on ecologically important characters, was modeled by Lande and Kirkpatrick (1988). Relations between speciation and sexual selection are discussed in detail in chapter 9. For tests of these ideas, more studies are needed of the variability of secondary sex traits within and among populations, also identifying the detailed mechanisms of selection favoring the trait. Does female choice promote rapid divergence of preferred secondary sex traits among populations, whereas male contests counteract it? These possibilities remain to be explored.

2.7 Empirical Tests

Genetic models have shown that the Darwin-Fisher theory is logically plausible, in spite of its dismissal by eminent biologists such as Wallace, Morgan, Huxley, and Lack (section 1.5). The models paved the way for a more open attitude toward sexual selection by mate choice, making it an attractive field for empirical research. As in several other areas of evolutionary and behavioral ecology, such as foraging theory and social evolution, empirical work has been stimulated by mathematical models. There is now abundant evidence that mate choice often does occur in the wild (chapter 6).

This does not mean that Fisherian sexual selection has been adequately tested. Even fundamental aspects, for instance that females choosing the most ornamented males bear sons with larger than average ornament size and mating success, remain to be tested. Costs of mate choice is another aspect that needs much more empirical study, as it can have important consequences for sexual selection. A main challenge is to design and carry out critical tests in a variety of species to provide a general picture of how sexual selection works in the wild.

Fisherian and other models of sexual selection are so complex, and contain so many assumptions, that testing of an entire model may not be feasible. Compared to, for example, more readily testable foraging models (e.g., Pyke 1984; Stephens and Krebs 1986), critical quantitative tests of sexual selection models remain scarce or nonexistent. A special difficulty is that Fisherian runaway selection is likely to occur in brief episodes, so most species at any given time may be in a stage of relative stability. Nonetheless, some important aspects should be amenable to tests with suitably chosen organisms. In addition, less direct evidence needs to be explored, such as assumed necessary conditions or long-lasting effects of Fisherian selection (Andersson 1987; Heisler et al. 1987).

Some assumptions or predictions of Fisherian models that have been tested are the following:

1. Individuals with large secondary sex traits have higher mating success but lower survival than those with smaller than average traits.
2. The trait and the preference show heritable genetic variation.
3. Owing to their joint evolution, there is genetic covariance between the trait and the preference.

Evidence on these points is discussed below.

Natural Selection Opposing Sexually Selected Traits

It is now clear that secondary sex traits are often penalized by natural selection, as assumed by Darwin and Fisher (chapter 10 below). But this is expected also from other hypotheses, such as indicator models (chapter 3), and is not exclusive support for a Fisherian process.

Heritability of Trait and Preference

Genetic analyses have shown that secondary sex traits are often determined by several or many gene loci (e.g., Danforth 1950; Templeton 1981; Coyne 1983). For example, results of crosses between the pheasants *Chrysolophus pictus* and *C. amherstiae* suggested that the morphology of the crest feathers depends on several loci, and their color on at least three loci (Danforth 1950). With reservation for the problems of obtaining unbiased estimates (e.g., Boag and van Noordwijk 1987; van Noordwijk 1987), sexually selected traits seem to have substantial heritabilities, as large as those of other morphological or behavioral traits.[1]

This indicates that selection in the wild, which may often vary and perhaps change direction, has not depleted the genetic variance of sexually selected traits. Estimated heritability values > 0.5 have been reported in several cases (e.g., Cade 1981; Carson and Lande 1984; Butlin and Hewitt 1986; Hedrick 1988; Houde 1992). For example, Hedrick (1986, 1988) found that calling-bout length in male field crickets *Gryllus integer* is strongly heritable, even though female choice favors long calling bouts. (It is not clear how heritability is maintained in spite of this directional selection.) Another example is the size of the nutritious spermatophore transferred by male orthopterans during copulation. Although it is important for male fertilization success, the relative investment in spermatophore shows heritable variation in *Gryllodes supplicans* (Sakaluk 1986a; Sakaluk 1988).

Mate choice has now been demonstrated in a variety of animals (table 6.A). Genetic variation in mating preferences as well as in preferred male traits, required by Fisher's (1930) hypothesis, has also been found in some species (Bakker 1990), for example the cockroach *Nauphoeta cinerea*, where female preferences for male olfactory cues show genetic variation (Moore 1989, 1990b). In *Drosophila melanogaster*, some mutant females show altered mating preferences, and selection for premating isolation between mutant strains leads to further differences in mating preferences (Crossley 1974; Heisler 1984a). Preferences are therefore genetically variable and can be altered by selection.

There is also evidence from the polymorphic ladybird *Adalia bipunctata* for genetically variable mating preferences. Selection over 14 generations in the laboratory increased the frequency of females preferring black males from 18% to 63%; selection for reduced frequency almost eliminated the

[1] See, e.g., for *insects*: Cade 1981a, 1984, Carson and Lande 1984, Butlin and Hewitt 1986, McLain 1987, Simmons 1987a, Hedrick 1988, Sakaluk 1988, Moore 1990b; *fish*: Farr 1983, Endler 1983, Houde 1992; *birds*: Boag 1983, Price 1984a, Grant 1986; also see Boag and van Noordwijk 1987.

preference (Majerus et al. 1982, 1986; O'Donald and Majerus 1985). The presence, nature, and variation of the preference seems to vary among populations, however, and its genetic basis is not clear (Kearns et al. 1992; O'Donald and Majerus 1992). Genetic variation in female mating preferences has also been inferred in the seaweed fly *Coelopa frigida* (Engelhard et al. 1989). For review of several experiments on genetic mating preferences, and the difficulties encountered in such experiments, see Bakker (1990) and Ritchie (1992).

In other cases, genetic variation in mating preferences may be absent. For example, some attempts at producing reproductive isolation by selection in laboratory populations have failed, perhaps owing to lack of genetic variation (reviewed by Kirkpatrick 1987a).

Covariance between Trait and Preference

Genetic covariance between male trait and female preference is predicted by Fisherian models (e.g., Lande 1981). If there is genetic variation in trait as well as preference, yet no positive covariance between them, a Fisherian process seems unlikely to explain the trait and the preference. Several kinds of results suggest that male traits and female preferences have sometimes evolved together. For example, neighboring populations of cricket frogs *Acris crepitans* differ in female preferences for male call properties. Local males are preferred, and the result is not explained by population differences in body size (Capranica et al. 1973; Ryan and Wilczynski 1988).

A correlation between sender (trait) and receiver (preference) might be achieved by pleiotropic gene effects in sender as well as receiver, or by reciprocal selection of genetically separate sender and receiver mechanisms. In at least one case—pheromonal communication in the European cornborer *Ostrinia nubilalis*—sender and receiver properties depend on separate loci (Löfstedt 1990). In other cases, for example in acoustic communication in crickets and frogs, sender and receiver mechanisms may be under common genetic control, but the evidence is less clear (reviewed by Colgan 1983; Doherty and Hoy 1985; Butlin and Ritchie 1989; Boake 1991).

Other suggestive evidence comes from the stickleback *Gasterosteus aculeatus*. Males during the breeding season develop red belly color, which is preferred by females (e.g., Semler 1971; McPhail 1969; Milinski and Bakker 1990). At the same time of the year, the sensitivity to red of the female visual system increases markedly, but not that of the male visual system, suggesting that the female preference has evolved in relation to male coloration (Cronly-Dillon and Sharma 1968). Moreover, in a breed-

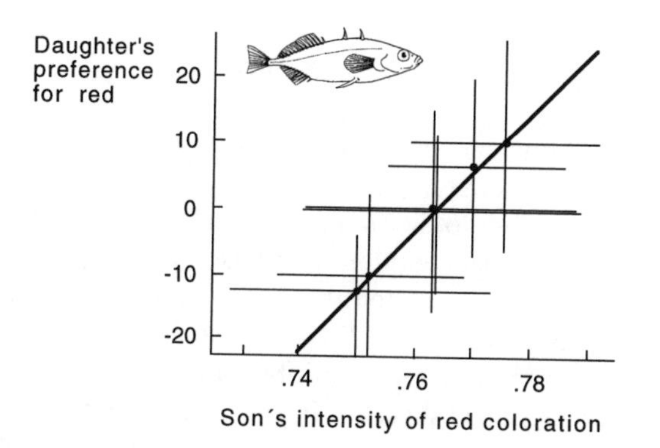

Figure 2.7.1 Correlation among stickleback *Gasterosteus aculeatus* fathers between the mean intensity of red coloration in sons, and the mean preference in daughters for red males. (The regression line is based on six fathers; horizontal and vertical bars are standard deviations for the offspring from each father. After Bakker 1993.)

ing experiment where intensely red males and dull males were crossed with females from the same population, Bakker (1993) showed that the male trait and the female preference are genetically correlated (figure 2.7.1). There was variation in both traits in the offspring generation, and the intensity of red coloration in sons was positively correlated with the degree of preference for red in their sisters (Bakker 1993; also see section 13.3).

Strong evidence for coevolution of trait and preference also comes from the guppy *Poecilia reticulata*. Female mating preferences in relation to male coloration differ among Trinidad populations (e.g., Breden and Stoner 1987; Houde 1988a). Comparing seven populations, Houde and Endler (1990) found that female preference strength for orange color spots in males was correlated with the amount of orange coloration in the males from the female's population. The differences among populations persisted after several generations in the laboratory, suggesting that the male trait and the female preference have diverged genetically in parallel among populations (also see section 5.2).

Unfortunately, neither genetic variation in trait and preference, nor a genetic correlation between them, is conclusive support for a Fisherian mechanism. Such variation is expected from several other hypotheses based on female choice (table 1.4.1, hypotheses B.1–4). To distinguish among the different mechanisms and estimate their relative importance empirically is a major problem. Solving it will probably require as much careful thinking as the development of any theoretical model, combined with detailed knowledge of suitable test species, and realistic judgment of what can be achieved with them. This is one of the greatest present challenges in the field of sexual selection.

2.8 Summary

Genetic models show that the Darwin-Fisher process, based on female choice of mate, with self-reinforcing selection of the mating preference through genetic coupling to the preferred trait, is a feasible mechanism of sexual selection. It has been explored with two-locus as well as polygenic models. In the simplest cases, both types of model depending on the biological assumptions can result in lines of stable or unstable equilibria in plots of trait versus preference. At equilibrium, the mating advantage of the trait is balanced by reduced survival. If the line represents stable equilibria, the system returns to the line after a disturbance. In the unstable case, "runaway" selection may rapidly lead to marked differences between populations in secondary sex traits. The relevance of these equilibrium properties in practice is doubtful. When, as often seems likely, there are costs to female choice, the lines of equilibria shrink to points, and direct selection of the preference dominates the outcome. These results suggest that evolutionary exaggeration of male traits by random drift along a line of neutrally stable equilibria is unlikely. Costly female preferences for male ornaments can, however, evolve by a Fisherian process if the male trait is subject to deleterious mutations. Genetic models also suggest that Fisherian sexual selection can be important in the evolution of sexual dimorphism, and in speciation.

Empirical studies have verified some assumptions of Fisherian models. Secondary sex traits usually, and preferences sometimes, have considerable heritability and can be altered by selection. There is also some evidence for coevolution of trait and preference, as predicted by these and other models. No critical test has been performed that supports Fisherian sexual selection and excludes the alternatives, or estimates their relative importance. The role of Fisherian sexual selection by mate choice in population divergence and speciation also remains to be clarified.

3 Genetic Models of Indicator Mechanisms

3.1 Introduction

If sexually selected traits are expressed in proportion to the condition of the possessor, a well-developed trait should indicate good condition and viability. To the extent that condition and viability are heritable, a female choosing such a male then makes it likely that her offspring will inherit high viability from their father, with "the best chance of overcoming the enormous variety of dangers and difficulties with which every species is surrounded" (Fisher 1915).

Essentially the same indicator idea was raised again by Williams (1966), who probably inspired many later versions (section 1.7). Zahavi (1975, 1977) dismissed Fisherian self-reinforcing selection entirely, and argued that the "handicap principle" is the main mechanism of sexual selection by female choice. He emphasized that, as Fisher (1915) and Williams (1966) implied, indicator traits must reduce viability, that is, handicap the survival of the male. Otherwise, all males could develop the trait, and it would not indicate high general viability (but see Borgia 1993). Zahavi presented no formal model, and so did not show that the advantage of inheriting genes for high viability can override the cost of also inheriting the handicapping ornament. Moreover, Fisher's "fundamental theorem of natural selection" predicts that fitness at equilibrium has zero heritability (section 3.4 below). If so, females that mate with highly ornamented and viable males will not have offspring of higher than average viability, and indicator mechanisms cannot work. The handicap principle was strongly criticized for these reasons. The prediction of zero fitness heritability does not seem realistic, however (section 3.4). Later genetic models have also shown that indicator mechanisms may work, particularly if the indicator trait depends on the phenotypic and overall genetic constitution of the male, as suggested by Williams (1966).[1]

[1] For criticism, see, e.g., Davis and O'Donald 1976, Maynard Smith 1976, 1978a, Halliday 1978, O'Donald 1980, and Kirkpatrick 1986a,b. For genetic models suggesting that indicator mechanisms may work, see, e.g., Andersson 1982b, 1986a,b, Pomiankowski 1987a,b, 1988, Heywood 1989, Hoelzer 1989, Grafen 1990a, Michod and Hasson 1990, Iwasa et al. 1991, Maynard Smith 1991a, Price et al. 1993.

Although the verbal ideas of Williams (1966) and Zahavi (1975, 1977a) did not make it clear exactly which genetic assumptions were involved, they drew attention to a possible alternative to Fisher's (1930) "runaway process" based on self-reinforcing mating advantage, and helped revive interest in sexual selection. Many other authors have since suggested that some form of indicator mechanism may play a role in sexual selection.[2]

This chapter briefly reviews genetic models and empirical tests of indicator mechanisms. These models are often called "good genes" models or, after Zahavi (1975, 1977b), "handicap" models, referring to the survival cost of the male trait. Such a cost is, however, not unique to this kind of model; already Darwin (1871) and Fisher (1930) emphasized that conspicuous secondary sex traits will reduce survival. Also, the term "handicap" draws attention to a side aspect: a female preference for males with a large sex trait is favored not because the trait is a handicap, but primarily because it is an index (Fisher 1915) of high general viability or condition, leading to survival advantages for the offspring. The word "indicator," which puts this primary aspect in focus, is used here. I prefer this term to "good genes" because Fisherian mechanisms are also based on favorable genes (with respect to mating success), and because some indicator models are based not on genetic but on direct material advantages (see section 3.4 below).

3.2 Three-Locus Models

Most genetic models of indicator mechanisms have been based on three haploid loci for mathematical convenience. The general modeling approach is usually similar to that described in section 2.2. Three variants of the mechanism can be distinguished, differing mainly in the expression and mortality cost of the male trait (Maynard Smith 1985, 1987, 1991a; also see Pomiankowski 1988; Tomlinson 1988) (box 3.2.1). Like two-locus Fisherian models, these three-locus models make some unrealistic assumptions. For example, most secondary sex traits and mating preferences are probably polygenic. Few-locus models are, however, more tractable than polygenic models in some respects. The development of genetic coupling between alleles at different loci is easily included in a few-locus but not in a polygenic model.

[2] E.g., Trivers 1972, 1976, Emlen 1973, Bell 1978, Eshel 1978, Halliday 1978, Borgia 1979, 1987, West-Eberhard 1979, Endler 1980, Thornhill 1980d, Hamilton 1982, Hamilton and Zuk 1982, Motro 1982, Parker 1982, 1983b, Dominey 1983, Eshel and Hamilton 1984, Kodric-Brown and Brown 1984, Nur and Hasson 1984, Manning 1985, Seger and Trivers 1986, B. Charlesworth 1988, Butcher and Rohwer 1989, Hasson 1989–1991, Folstad and Karter 1992, Reynolds 1993.

Some further properties of the models are outlined below. There is no elegant analytical result comparable to Kirkpatrick's (1982) discovery of the equilibrium line in two-locus haploid Fisherian models (section 2.2); three-locus models yield no analytical solution. As with Fisherian models, rapid progress in modeling also tends to render earlier models obsolete. Only some major features are therefore sketched here; see Pomiankowski (1988) for a thorough treatment of three-locus indicator models. Other reviews of interest are found in, for example, Maynard Smith (1985, 1991a), Heisler et al. (1987), Kirkpatrick (1987a), and Tomlinson (1988).

Pure Epistatic Indicators

In Maynard Smith's (1976) model of Zahavi's (1975) mechanism, allele *A* makes males but not females develop the ornament that reduces survivorship; *a* males as well as females are cryptic. Allele *C* makes females prefer ornamented (*AB* and *Ab*) males, whereas *c* females mate at random. The model assumes polygyny, and viability pattern I of box 3.2.1. Can the male trait and the female preference increase in frequency under these conditions? There is a problem of interpretation: Is an observed increase caused by the viability-based indicator mechanism, or by the Fisherian mating advantage for ornamented males? One way of tackling this problem is to start with the preference allele *C* at low frequency compared to the trait allele *A*; the Fisherian mating advantage of *A* males is then slight. Doing so, Maynard Smith (1976) found no set of parameter values for which allele *A* or *C* increased. If the epistatic viability interactions are strong enough (u is sufficiently large in box 3.2.1, case I), the preference allele *C* may temporarily increase (also see Eshel 1978), but *A* always decreases. Maynard Smith (1985) therefore concluded that the indicator mechanism in this form does not work.

It is theoretically possible for the epistatic indicator mechanism in conjunction with Fisherian selection to increase the rate of spread of the ornament as well as the preference, and to expand the conditions under which the ornament will be fixed (Bell 1978; Andersson 1982b, 1986b; Pomiankowski 1987a, 1988). This seems, however, to require unrealistically large viability differences. Purely epistatic indicators therefore appear unlikely to be important (also see Iwasa et al. 1991; Maynard Smith 1991a).

Condition-dependent Indicators

In many species there is conspicuous variation in the development of male secondary sex traits. Williams (1966, 1978; also Fisher 1915) suggested that ornaments are condition-dependent and develop in proportion to the

BOX 3.2.1 OUTLINE OF THREE-LOCUS INDICATOR MODELS

In these models,
locus A codes for the male trait,
locus B codes for general viability,
locus C codes for the female preference.
Two alleles are segregating at each locus:
allele *A* codes for the male trait, allele *a* for absence of the trait,
allele *B* codes for higher viability than does allele *b*;
allele *C* females prefer males with the trait coded for by *A*, and allele *c* females mate at random.
There are three main versions of three-locus models (also see the main text):

I. *Pure epistatic indicators.* All males with allele *A* develop the ornament (Maynard Smith 1976, 1978a; Bell 1978; Eshel 1978; Andersson 1982b, 1986b; Pomiankowski 1987a, 1988; Tomlinson 1988; Heywood 1989).

II. *Condition-dependent indicators.* Males with the ornament allele *A* develop the ornament only if also possessing the allele for high viability, *B* (Maynard Smith 1985; Andersson 1986a; Pomiankowski 1987b, 1988; Tomlinson 1988; Heywood 1989).

III. *Revealing indicators.* All males with allele *A* develop the ornament to full size, but it is in visibly poorer state in *b* than in *B* males (Maynard Smith 1985; Kirkpatrick 1986b; Pomiankowski 1987a, 1988; Tomlinson 1988; Hasson 1989).

In other words, the difference between revealing and condition-dependent indicators is that with condition-dependence, low-quality males develop a small ornament or none at all, and so have no cost for it, whereas a revealing indicator trait develops fully in all males carrying the allele for it. Low-quality males can be identified as such (and avoided) if they carry the revealing indicator, for example because it is poorly groomed or otherwise reflects poor condition (Maynard Smith 1985; Pomiankowski 1988).

The viabilities of the different male genotypes in these three main versions of three-locus indicator models are as in the table (modified from Maynard Smith 1985).

Model	Allele, locus B	Allele, locus A	
		A	*a*
I. Pure epistatic indicator	*B*	(1+s) (1–t)	1+s
	b	(1–t) (1–u)	1
II. Condition-dependent indicator	*B*	(1+s) (1–t)	1+s
	b	1	1
III. Revealing indicator	*B*	(1+s) (1–t)	1+s
	b	1–t	1

BOX 3.2.1 CONT.

The boxes indicate males preferred by choosy (C) females; s, t, and u are positive selection coefficients (u and $t < 1$) that determine genotypic relative viabilities until mating. s is the viability advantage conferred by the B over the b allele; t is the viability cost for developing the ornament by B males in all three models, and also by b males in the revealing indicator model. In model II, b males do not develop the ornament and so have no cost for it. In model I, there are epistatic viability interactions such that b males have a higher cost (the extra term $1 - u$) for the ornament than do B males.

The nature of the mating preference varies among models, also within each category (I–III). The models have in common that a C female upon encountering a male is more likely to accept him as a mate if he is ornamented than if not (see Seger 1985; Maynard Smith 1985; Pomiankowski 1988, for reviews of assumptions concerning the preference; one example is described in detail in box 2.2.1). Females with allele c mate at random. The preference locus is assumed not to influence viabilities; all B females have viability $1 + s$, and b females 1.

In the terms of these models, an indicator mechanism can favor the evolution of costly male traits (allele A) and female preferences (allele C) for such traits as follows. Consider an allele B, arising, for example, through mutation or immigration, which raises viability. The indicator models suggest various ways (see I–III in the main text) in which the ornament allele A becomes associated with viability allele B during its increase to fixation in the population. For example, the reduction in survivorship caused by development of the ornament may be lower in males that carry allele B than in other males (I), or the ornament may be expressed only in males of high condition and viability (II). Surviving ornamented males are then more likely than other males to carry allele B. C females, with a preference for the ornament (A), therefore are more likely to mate with males carrying the viability allele B than are other females.

For these reasons, alleles A, B, and C tend to become associated in sons as well as daughters from matings between A males and C females. The result is positive gametic phase disequilibria (Falconer 1989) among these alleles. Therefore, as allele B, owing to its survival advantage, rises in frequency, the alleles for the ornament (A) and preference (C) can hitchhike with it to higher frequency as well. Repeated occurrence of advantageous viability alleles might then make the ornament and the preference common in the population. Conversely, deleterious alleles that arise owing to mutation, immigration, or environmental changes should tend during their decline to become linked with alleles a and c, reducing their frequency in the population and increasing that of the alleles for the ornament and the preference, A and C.

The development of gametic phase disequilibrium is important in these models; whether the process works or not often depends on the strength of the disequilibrium (e.g. Andersson 1986a; Pomiankowski 1987a; Tomlinson 1988; but see Grafen 1990a).

phenotypic quality of the male (see section 1.5 above). This idea has since been taken up by many authors.[3]

Critics of early genetic models of indicator mechanisms emphasized that the models did not clearly distinguish between Fisherian and indicator selection (e.g., Kirkpatrick 1986b). It was therefore hard to know whether the indicator mechanism, and not only the Fisherian mechanism, produced the evolutionary effects (changes in gene frequencies). One way of seeing whether an indicator mechanism can work alone is to isolate it from the Fisherian process. This can be done by using a monogamous system where all surviving males pair and get the same number of offspring (Bell 1978; Maynard Smith 1978a). Any increase of the male trait or the female preference can then be ascribed to the indicator mechanism. For this reason, Andersson (1986a) studied selection of a condition-dependent male trait in a model that assumes monogamy. If an indicator process can favor costly male traits and female preferences in such a system, it is likely to have effects also in a polygynous species, together with Fisherian self-reinforcing selection, with which genetic-indicator processes seem likely to cooccur.

All females are assumed to be cryptic, and so are males with allele *a*, whereas males with allele *A* develop the ornament if and only if they also possess the allele for high viability, *B*. Females carrying allele *C* prefer ornamented (*AB*) males, whereas *c* females choose mates at random among all males remaining after *C* females have made their choices. (For viabilities, see Andersson 1986a; they differ from box 3.2.1, case II, but can be cast in that form.) The model was used to explore whether alleles *A* (flexible ornament) and *C* (preference) can increase from low frequency. Allele *B* was also introduced at low frequency, corresponding to a mutation or immigration, or a change in the environment that makes a previously rare allele beneficial. Simulations showed that preference and ornament can grow in frequency over many generations also with genotypic differences in viability of only a few percent (Andersson 1986a). Continued presence of the ornament required sufficient input of beneficial alleles through mutation, immigration, or temporal fluctuation in selection.

Tomlinson (1988) repeated the simulations with a corresponding diploid model. Diploidy made the process slower and the increase in trait and preference smaller, and he did not find realistic values of the selection coefficients that gave rise to fixation of the male trait. He therefore concluded that the indicator process alone, without any Fisherian mating advantage, is unlikely to make a costly male trait and a female preference for it evolve

[3] E.g., Zahavi 1977b, Halliday 1978, Borgia 1979, West-Eberhard 1979, Andersson 1982b, Hamilton and Zuk 1982, Parker 1982, Dominey 1983, Kodric-Brown and Brown 1984, Nur and Hasson 1984, Zeh and Zeh 1988, Reynolds 1993. For genetic models, see Maynard Smith 1985, Andersson 1986a, Pomiankowski 1987a,b, 1988, Tomlinson 1988, Heywood 1989, Hoelzer 1989, Grafen 1990a, Michod and Hasson 1990, Iwasa et al. 1991, Price et al. 1993.

from low frequencies (also see Heisler et al. 1987). Like Andersson (1982b, 1986a,b) and Pomiankowski (1987a,b, 1988), Tomlinson found that an indicator mechanism in conjunction with the Fisherian process could have important effects, facilitating the evolution of trait and preference.

Revealing Indicators

A third type of three-locus "revealing" indicator model was stimulated by the idea of Hamilton and Zuk (1982; section 1.7 above). Male variation in resistance to parasites is assumed to be reflected in the secondary sex ornaments. Females can judge the degree of past or present parasitism by the condition of the male ornament. By choosing a male with an ornament in good shape, a female is likely to have offspring that inherit parasite resistance; this might favor the evolution of female preference and male ornament (Hamilton 1982; Hamilton and Zuk 1982; Eshel and Hamilton 1984).

Revealing indicators were studied with three-locus haploid models by Maynard Smith (1985), Kirkpatrick (1986a), and Pomiankowski (1987a, 1988). The viabilities in the model of Maynard Smith are shown in box 3.2.1, case III. The models all assume polygyny, so the Fisherian mating advantage for ornamented males is also at work. Simulations showed that if the preference allele C is above a certain threshold frequency, the trait allele A goes to fixation, and otherwise to extinction. The model does not explain how the C allele reaches the threshold, so Fisher's (1930) or some other mechanism is needed to explain its initial evolution. This is the case also with condition-dependent indicators.

Another model of revealing indicators was developed by Hasson (1989, 1990, 1991), showing that a viability-based process can favor the evolution of "amplifying" male traits that enable females to discriminate more accurately among males in relation to their general viability. Hasson suggested that such selection may explain, for example, aspects of plumage patterns in birds, such as feather bars that facilitate evaluation of the condition of the plumage.

Comparison of Three-Locus Models

Pomiankowski (1987a,b, 1988) compared the three kinds of indicator models, using host-parasite cycles as well as recurrent deleterious mutations as sources of heritable viability differences. Deleterious mutations seem a particularly likely source of fitness heritability (Partridge 1983; Charlesworth 1987; section 3.5 below). Pomiankowski concluded that an indicator mechanism in itself cannot make a rare female preference spread for a male trait that reduces survival. As in a pure Fisherian model of a deleterious male trait, there is a threshold frequency of the female preference that must

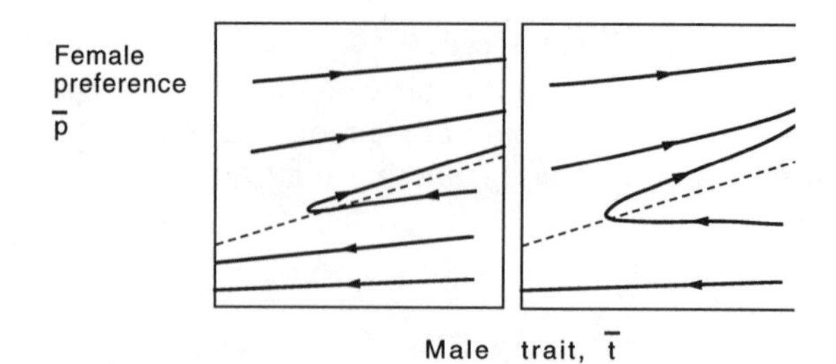

Figure 3.2.1 Results from simulations with two three-locus indicator models with recurrent deleterious mutations: pure epistatic indicators (*left*) and condition-dependent indicators (*right*). See text for explanation. (After Pomiankowski 1988)

be reached for the process to work. This threshold did not change substantially if any of the indicator mechanisms was added, nor did the rate of evolution of trait and preference above the threshold.

Pomiankowski (1987a,b, 1988) discovered, however, that addition of an indicator mechanism eliminates the line of equilibria in Fisherian models (chapter 2). A population that was previously on the line or in its zone of attraction may evolve to fixation of the male ornament. With pure epistatic indicators, the female preference may end up at intermediate frequency; with condition-dependent or revealing handicaps the preference goes to fixation as well (figure 3.2.1). In this important respect, an indicator process when added to a Fisherian self-reinforcing process can enhance the evolution of costly secondary sex traits in males—and female preferences for them. Tomlinson (1988) and Heywood (1989) reached similar results. Indicator mechanisms might therefore play important roles in sexual selection of male ornaments that reduce survival, and of female preferences for such traits.

The analyses show that condition-dependent and revealing indicators are much more potent than pure epistatic indicators in the selection of traits and preferences (Andersson 1986a,b; Pomiankowski 1987a,b, 1988; Tomlinson 1988; Heywood 1989). The pure epistatic models also seem biologically less realistic. Using diploid models, Tomlinson (1988) verified that results from haploid models with polygynous mating were not artifacts arising from haploidy. He also showed that dominant alleles for indicators and preferences evolve under a wider range of conditions than recessives.

3.3 Polygenic Models

A quantitative genetic model was formulated by Iwasa et al. (1991) to examine the evolution of costly male traits and costly female preferences by Fisherian and indicator mechanisms (also see Heisler 1984b, 1985; Kirkpatrick 1985, 1986a; Price et al. 1993). The model in some respects uses similar assumptions as Lande (1981; box 2.3.1 above). In addition, selection is weak, with individual fitness functions that change little over the range of phenotypic variation. Owing to the problems of calculating changes in genetic variances and covariances, they are assumed constant, an assumption with consequences that are uncertain at present (e.g., Heywood 1989, and section 2.3 above).

The model follows the evolutionary dynamics of the male trait *t*, the female preference *p*, and the general viability trait *v*, which all are polygenic (figure 3.3.1). For simplicity, nongenetic variation was excluded. Male fitness is the product of mating success and survivorship, the latter depending on his general viability *v* and the size of his trait *t*, which reduces viability. Female fitness is a function of her general viability *v* and preference *p*. Random mating has no viability cost, but discriminating mate choice reduces viability in (exponential) proportion to the strength of the preference.

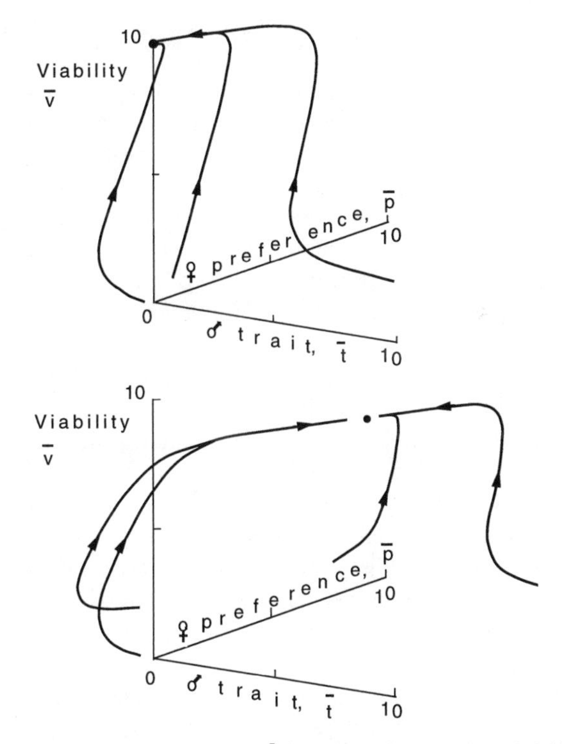

Figure 3.3.1 Evolution of mean male trait $\bar{t}$, female preference $\bar{p}$, and viability trait $\bar{v}$ in a polygenic indicator model. *Top*: When there are no deleterious mutations, no costly preference or trait evolves (the equilibrium point is zero for trait as well as preference). *Bottom*: When there are deleterious mutations that reduce viability, the trait and the preference evolve to a non-zero equilibrium point. (After Iwasa et al. 1991)

A crucial assumption is that deleterious mutations constantly reduce the viability v below its maximum value (figure 3.3.1). This is necessary for the process to work, that is, for costly female preferences to evolve. There is also a second necessary condition. Let s be the realized size of the male ornament; s is a function of the male genotype both as regards male trait genes, t, and viability, v; that is, $s = g(v, t)$. Iwasa et al. (1991) showed that for costly female preferences to evolve, it is necessary that $\partial s/\partial v > 0$. That is, ornament size must increase with general viability, as assumed by models of condition-dependent and revealing indicator traits. This second point helps explain why condition-dependent indicators also in other models evolve more easily than ornaments that do not depend on general viability or condition (Iwasa et al. 1991). Several kinds of models therefore suggest that condition dependence of the preferred trait makes an indicator process more likely to work. For comparison of these and related models, see Price et al. (1993).

3.4 Indicators and Direct Phenotypic Benefits

Heywood (1989), Hoelzer (1989), and Price et al. (1993) generalized indicator models in another respect, using genetic models based on variation in nonheritable paternal resources correlated with the expression of the male trait (see section 1.7 above). The offspring of choosy females in these models benefit from direct material resources provided by the male, not from genes for high viability. Yet the qualitative results were similar to those of models based on heritable viability.

Grafen (1990a) showed that an indicator mechanism can favor costly male ornaments and female preferences without any self-reinforcing Fisherian process. To make the model analytically tractable, he assumed that male ornament development as well as female preference are coded for by the same alleles at a single haploid locus; the main results were also verified with a two-locus haploid model. In contrast with most other models, males here differ in phenotypic but not genetic quality, and female fecundity depends on the phenotypic quality of the male. The model therefore assumes direct selection of the female preference (section 2.4 above). Ornament size also depends on the phenotypic quality of the male, and on his genetically determined rule of advertisement in relation to quality. Grafen's approach was then to find, by Herculean mathematical modeling efforts, an uninvasible allele from the set of possible advertising and preference rules, such that when the allele is common, no allele specifying another pair of strategies can invade the population.

The model, which uses continuous functions for male trait and female preference, predicts two possible equilibria, one of which is no ornamentation or preference. At the other equilibrium, the degree of ornamentation is a continuous, increasing function of male phenotypic quality, and females show costly preferences of a strength that increases with the size of the male trait, favoring high-quality males. There is no "cheating," since a male that advertises too much will reduce his viability more than he raises his mating success (Grafen 1990a,b; also see Zahavi 1975, 1977b; Andersson 1982b; Parker 1982, 1983b; Kodric-Brown and Brown 1984; Nur and Hasson 1984). Using a game-theory approach, Parker (1982) reached similar conclusions as regards the male trait; the female preference was assumed to be fixed.

Grafen (1990a) found that an indicator mechanism can suffice for the evolution of extravagant male ornaments and corresponding female preferences, without Fisherian self-reinforcement of the mating preferences. The assumption that male trait and female preference are determined by the same allele by definition excludes self-reinforcement, as it depends on linkage disequilibrium between loci. On the other hand, the model does not represent a genetic indicator or handicap process of the kind discussed above, which is based on heritable viability advantages for offspring from ornamented males (e.g., Zahavi 1975; Andersson 1986a; Pomiankowski 1988; Kirkpatrick 1992). There is no heritable viability advantage in Grafen's (1990a) model, but a direct phenotypic benefit in form of higher fecundity for a female choosing a high-quality male. The model also includes a direct mating advantage, favoring males in proportion to the development of their ornament. Males can mate with several females, and highly ornamented males are most likely to do so. Therefore, although the model works without the self-reinforcing aspect of the Fisherian process, like Fisherian models it does contain a direct mating advantage (highly ornamented males obtaining more mates than others), which seems crucial for its function.

As Grafen (1990a) pointed out, the assumption that male trait and female preference are determined by the same allele is unrealistic; trait and preference may often be determined by different loci (e.g., Löfstedt 1990). A Fisherian self-reinforcing process might then develop. This need not necessarily affect the equilibrium outcome, however. Grafen found that a corresponding two-locus model, with one locus for the trait and one for the preference, has the same uninvasible equilibria as the single-locus model, even if linkage disequilibrium is set to zero in each generation and hence cannot develop.

Grafen (1990a) therefore concluded that the self-reinforcing aspect of Fisherian processes does not have much effect on the equilibrium outcome in this kind of model. This conclusion was criticized by Kirkpatrick (1992), who questioned whether the equilibrium in Grafen's (1990a) model is always stable. Results from Fisherian models suggest that apparently realistic assumptions about mutation pressure and costs of choice will influence the equilibrium (e.g., Pomiankowski et al. 1991; section 2.4 above). Perhaps they will do so also if combined with an indicator process.

It also seems likely that the rate of evolution of trait and preference can increase markedly with positive linkage disequilibrium. This might enhance the relative effect of sexual selection compared to other processes that influence secondary sex traits and mating preferences. Like most other current models of sexual selection, Grafen's (1990a,b) analysis focuses more on evolutionary equilibria than on rates and dynamics. Evolutionary rates may, however, be important for the distribution of male traits and female preferences in real populations, if conditions are rarely so stable that equilibria will be closely approached. A Fisherian mechanism may be important in sexual selection, for example if it increases the rate at which sexual selection drives a trait in a certain direction. Such dynamic properties have received scant attention in the modeling of sexual selection; they may be important, and should be possible to study by simulation models, if not analytically.

3.5 The Heritability of Fitness

Theory

A common objection against genetic indicator models is that they require heritable fitness. Otherwise, a female cannot improve the genetic fitness of her offspring by choosing a particularly fit male (Maynard Smith 1978a). Theoretical calculations by Fisher (1930) suggested that fitness heritability tends to vanish, and empirical evidence from domestic organisms and laboratory populations suggests that it is indeed low. Yet there is reason to believe that fitness in natural populations has some heritability. The question is whether it is large enough for indicator mechanisms to have effects compared to other mechanisms of sexual selection.

The heritability h^2 is the ratio between the additive genetic variance of a trait and its total phenotypic variance, V_p, which can be partitioned into a number of components (e.g., Falconer 1989; Maynard Smith 1989):

$$V_p = V_a + V_d + V_i + V_e,$$

where V_a is the additive genetic variance of the trait, the main cause of resemblance between relatives; V_d is the dominance variance, arising from dominance between alleles; V_i is the variance due to interactions among loci; and V_e is the variance caused by environmental factors. In addition, there may be genotype-phenotype correlation and interaction, which can bias heritability estimates (see Bulmer 1980; Falconer 1989).

In the previous symbols, the ("narrow sense") heritability $h^2 = V_a/V_p$.

One among several common methods of estimating h^2 is to calculate the regression coefficient b of offspring on one parent; $2b$ is an estimator of h^2. This requires that there are no substantial influences of a common environment, or of parental phenotypic effects on offspring Charlesworth (1987) discusses measures of fitness heritability in the context of indicator models.

The heritability h^2 is important not only because it quantifies the resemblance between parent and offspring, but also because it determines the rate of response to selection (see section 4.5 below; Bulmer 1980; Falconer 1989; Maynard Smith 1989). A trait that is important for fitness and has high heritability will evolve rapidly, which tends to fix the most favorable alleles and reduce genetic variation in the trait. Fisher's (1930) "fundamental theorem of natural selection" states that the rate of increase of fitness in a population is equal to the additive genetic variance in fitness. A consequence is that the additive genetic variance V_a tends to decline; when the population reaches maximum fitness, its additive genetic variance in fitness vanishes. The heritability of a trait should therefore be inversely related to its importance for fitness, with little heritable variation for important traits.

Although stable selection tends to deplete additive genetic variance, counteracting processes might, however, keep fitness heritable to some extent. Three likely processes are temporal and spatial variation in selection, and mutation pressure (e.g., Charlesworth 1987; Kondrashov 1988; Rice 1988).

TEMPORALLY VARYING SELECTION

Haldane and Jayakar (1963), Gillespie (1973), and Ewing (1979) concluded from genetic models that temporal changes in selection pressures can sometimes preserve genetic variation in traits that influence fitness. For example, coadaptational cycles of hosts and parasites might maintain additive genetic variance in host resistance to parasites (Hamilton and Zuk 1982; Eshel and Hamilton 1984; Hamilton et al. 1990).

SPATIALLY VARYING SELECTION

If populations are genetically adapted to different local conditions, immigration might maintain some additive genetic variance in fitness (e.g., Jain and Bradshaw 1966; Antonovics 1968; Felsenstein 1976; Endler 1977; Slatkin 1978; Ewing 1979; Bulmer 1980; Partridge 1983). An example of a secondary sex trait is male coloration in the guppy *Poecilia reticulata*, which is influenced by geographic differences in selection, combined with gene flow (Endler 1978, 1983).

MUTATION PRESSURE

Harmful as well as beneficial mutations might perhaps create enough heritable variance in fitness to confer a selective advantage on mate choice

(Maynard Smith 1978b; Partridge 1983; Manning 1984; Charlesworth 1987; Kondrashov 1988; Rice 1988). Based on theory and data on the variance maintained by a balance between deleterious mutations and selection at many loci (e.g., Mukai et al. 1974; Crow and Simmons 1983), Charlesworth (1987) estimated the additive genetic variance in egg-to-adult viability of *Drosophila* to be on the order of 0.003 in the laboratory, summing over all loci with mildly deleterious mutations. In addition, differences in mating success may add important variance to that of viability. On the other hand, environmental variation might be much greater in the wild than in the lab, reducing the chances to find mates with heritably high fitness (Charlesworth 1987).

Using a partly different approach, Rice (1988) estimated the effect of deleterious mutations to be as large as a roughly 20% reduction of relative fitness. Kondrashov (1988) came to similar conclusions with another method of calculation. If these estimates are realistic, there will be substantial variation in total fitness, perhaps large enough to lend importance to indicator mechanisms.

Another estimate of mutational contributions to fitness heritability can be derived from the theory of mutation: selection balance for quantitative traits under stabilizing selection (Lande 1976; Bulmer 1989b). Owing to genetic correlations and interactions among such traits, it is difficult to estimate their total contribution to additive fitness variance. Charlesworth (1987) suggested it is likely to be at least of the order of 10^{-3}. The complex underlying theory is debated (e.g., Lande 1975; Turelli 1984; Barton and Turelli 1989, 1991; Bulmer 1989b; Charlesworth 1990).

Empirical Results

The heritability of total fitness seems not to have been measured accurately for any organism, neither in the lab nor in the wild, where the necessary sample sizes would usually require an unrealistic amount of field work. The heritability of traits related to fitness have, however, been studied in some laboratory and domestic species. Two conclusions are of particular interest here. First, although several components of fitness show considerable heritability, there may be negative genetic correlations between such components (see, e.g., Rose and Charlesworth 1981; Rose 1984; Reznik 1985). Total fitness therefore need not necessarily show much heritability, even if some fitness components do so.

Second, as expected from theory, the highest heritability is found for traits with probably little effect on fitness, such as number of abdominal bristles in *Drosophila*, and tail length in mice. Life history traits more closely related to fitness, such as fecundity, have relatively lower heritability (figure 3.5.1; Falconer 1989; Mousseau and Roff 1987; Roff and Mous-

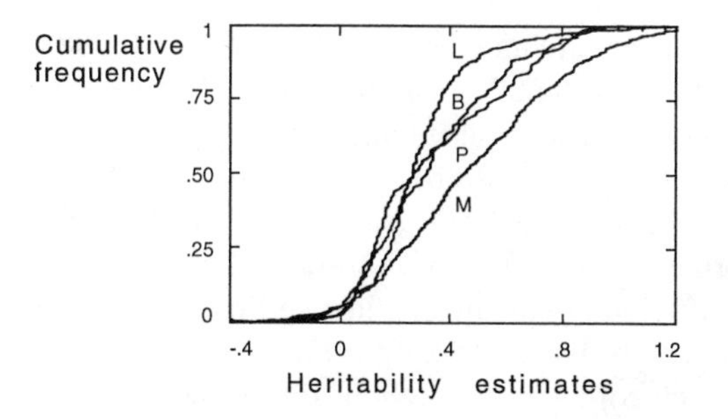

Figure 3.5.1 Cumulative frequency distributions of heritabilities of four trait categories (L, life history; B, behavior; P, physiology; M, morphology). Life-history traits have lowest heritability. Based on 1120 heritability estimates from 75 species. (From Mousseau and Roff 1987)

seau 1987). A field study of collared flycatchers *Ficedula albicollis* at the island of Gotland by Gustafsson (1986) is particularly relevant. Owing to the unusually strong tendency for birds in this population to return and breed near their birth site, he could estimate the heritability of a number of traits, including lifetime reproductive success, which should be closely related to fitness (figure 3.5.2). As expected, traits weakly correlated with fitness have high heritabilities, such as aspects of body size, whereas lifetime breeding success has low heritability. The standard errors of the mean

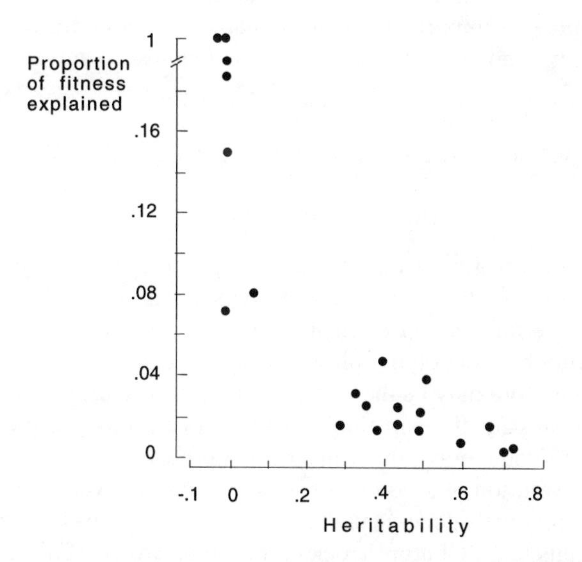

Figure 3.5.2 Heritabilities of various traits in relation to their importance for fitness (lifetime reproductive success) in male and female collared flycatchers *Ficedula albicollis*. Traits that are important for fitness have low heritability. (After Gustafsson 1986)

were wide, however, and included zero heritability: 0.008 ± 0.13 for males, and –0.014 ± 0.16 for females. Lifetime reproductive success and its main components also had low heritability in a population of song sparrows *Melospiza melodia* (Smith 1988).

This illustrates a general problem: for indicator mechanisms to play a role, it is perhaps enough if fitness shows a few percent heritability. Estimates with sufficiently narrow confidence limits to exclude zero heritability will then require back-breaking sample sizes of perhaps thousands of individuals, each followed throughout life in the wild in enough detail to accurately measure lifetime reproductive success; this must also be done for their offspring. In most cases this is out of the question for practical reasons. Possible realistic exceptions can perhaps be found, for example, among insects or other short-lived animals. But more direct tests of indicator mechanisms also need to be done, for example by exploring the fitness consequences of female choice of particular mates (see below). Perhaps a tractable approach is to concentrate on extremes in the population, that is, on the males with the largest and with the smallest ornaments. Are there significant differences between the offspring from the two kinds of male in viability, mating, and lifetime reproductive success, when differences in the phenotypic condition of their mothers are also controlled? If there are no such differences in sufficiently large samples, genetic-indicator as well as Fisherian mechanisms can probably be refuted.

Whether fitness has high enough heritability for genetic indicator mechanisms to play an important role in sexual selection is, for these reasons, still not clear; innovative empirical work is required to solve the problem.

3.6 Condition Dependence of Secondary Sex Traits

Phenotypic Condition

Theory and observations suggest that many sex traits develop in proportion to the condition of the male. Their growth can be viewed as part of the reproductive effort, which often increases in absolute terms with nutritional condition.[4] Phenotypic plasticity and flexible growth of secondary sex traits therefore may be the normal outcome of widespread physiological mechanisms for the partitioning of resources among growth, maintenance, and reproduction—mechanisms that probably often existed already before the evolution of secondary sex ornaments and weapons.

Antlers are a well-documented example of condition-dependent secondary sex traits (e.g., Clutton-Brock et al. 1982; Brown 1983; Goss 1983).

[4] E.g., Williams 1966, West-Eberhard 1979, Andersson 1982b, 1986a, Dominey 1983, Kodric-Brown and Brown 1984, Nur and Hasson 1984, Reznik 1985, Bell and Koufopanou 1986, Zeh and Zeh 1988, Partridge 1989, Folstad and Karter 1992.

Starvation reduces relative antler size in red deer *Cervus elaphus*. Compared to fossil skeletons from Scotland, and red deer elsewhere in Europe, Scottish deer have smaller body and relative antler size. This is likely an effect of poor nutrition, for Scottish deer released in New Zealand after one generation grew antlers as large as those in fossil Scottish and extant mainland European populations. As deer became common in New Zealand, body and relative antler size declined again (Huxley 1932). In a nine-year study of red deer at Rhum, the size of antlers in both juvenile and adult bucks declined drastically in parallel with a growth of the population (Clutton-Brock et al. 1982). Captive Scottish deer fed ad libitum grew as big as continental animals (Suttie 1980), and deer with supplemental feed during the winter grew larger antlers than a control group. Both categories showed similar body mass just after antler production. Even for a given body mass, nutrition therefore affects antler size (Suttie and Kay 1983).

Studies of other deer also point to a strong effect of nutrition on relative antler growth. Antler size seems to depend in part on special loci, in part on the phenotypic condition of the male, as assumed by models of condition-dependent indicator traits (reviewed by Andersson 1986a). Whether antler size directly affects female choice of mate in deer is, however, not clear (Clutton-Brock 1982).

In the African elephant *Loxodonta africana*, bulls seem to come into the "musth" rutting stage only if they are in good condition. Bulls not in musth are avoided as mates by females (Moss 1983; Poole 1987, 1989a). Also in many other animals, secondary sex traits develop in proportion to the condition of the male; they seem to have low growth priority compared to other organs. For example, the nuptial crest of male newts regresses rapidly under starvation, and male relative tail height increases with body condition (Green 1991; Baker 1992). In male guppies *Poecilia reticulata*, orange carotenoid color spots raise the attractiveness of males to females. The brightness of such spots increases with the carotenoid content of the diet, and may reflect male food-finding ability (e.g., Endler 1980, 1983; Kodric-Brown 1985; section 5.2 below). The red coloration of male sticklebacks *Gasterosteus aculeatus* is also condition-dependent (e.g., Milinski and Bakker 1990).

Bright carotenoid colors in birds are related to diet (e.g., Ralph 1969). This has been demonstrated in detail for the red ventral coloration of male house finches *Carpodacus mexicanus* in a study that also showed the importance and consequences of male coloration for female choice of mate (Hill 1990–1993). The length of the tail feathers in male swallows *Hirundo rustica* and Jackson's widowbird *Euplectes jacksoni*, which are important for male mating success, also depends on male condition (Møller 1988a, 1989a; Andersson 1989, 1992; see chapter 13 for examples involving coloration).

Among invertebrates, larval nutrition influences many aspects of the adult phenotype, including the size of secondary sex traits (Zeh and Zeh 1988). One example is horn length in the beetle *Podischnus agenor* (Eberhard and Gutiérres 1991; also see section 16.2). Male display performance is also likely to be condition-dependent; for additional examples, see Price et al. (1993).

Hormonal regulation of secondary sex traits has been demonstrated in many vertebrates (reviewed by Andersson 1986a; Ligon et al. 1990; Folstad and Karter 1992). Hormone production depends on nutrition; animals in poor condition tend to have low levels of sex hormones, and probably partly for this reason develop relatively small secondary sex traits. For example, comb size in cock red grouse *Lagopus l. scoticus* and red jungle fowl *Gallus gallus* depends on testosterone levels (Moss et al. 1979; Ligon et al. 1990). So does the nuptial plumage in phalaropes (Johns 1964). Testosterone production is sensitive to nutritional status in birds and falls rapidly during food shortage (Wingfield 1980). Testosterone level seems to act as a link relating the expression of secondary sex traits to male condition.

Age Dependence

Many male sex traits reflect age (Trivers 1972, reviewed by Manning 1985). This may in part be a consequence of improved condition, if foraging and other skills increase with age, as demonstrated in several species (e.g., Marchetti and Price 1989; Wunderle 1991). For example, the ornamental feather plumes of male argus pheasant *Argusianus argus*, peacock *Pavo cristatus*, and lyrebird *Menura superba* may increase in size for five to ten years (Smith 1965, 1982; Davison 1981; Manning 1989). The adult plumage of male long-tailed manakins, a tiny 20 g bird, is attained at four years age, and males usually do not become dominant, with chances of mating, until at least ten years old (D. McDonald 1989b)! Age-dependent expression of secondary sex traits may be an important aspect of indicator mechanisms in some species. Females that choose males with large secondary sex traits will then get mates that have survived well. To the extent that survival is heritable, their offspring should also tend to live long (e.g., Trivers 1972; Halliday 1978, 1983; Manning 1985).

In some species, the expression of secondary sex traits may be reduced late in life. For example, in pied flycatchers *Ficedula hypoleuca*, the oldest males become browner and more female-like (Winkel et al. 1970; Potti and Montalvo 1991). Males with poorly developed sex traits will, however, much more often be young rather than senile.

Female preferences for well-developed secondary sex traits therefore will often discriminate against young males. For example, in the peacock *Pavo cristatus*, the preferred males with large tails should be among the oldest in the population (Manning 1989; Petrie et al.1991). It is not known, however, whether females mating with old males give birth to offspring with higher than average survival. In species with parental care, an advantage of mating with an old, experienced partner might be better parental care or other direct benefits (e.g., Rowley 1983; Alatalo et al. 1986a).

Fluctuating Asymmetry

A recent discovery supports the idea that large ornaments may indicate a harmonious genome. There is a relationship between ornament size and "fluctuating bilateral asymmetry," which may reflect underlying phenotypic and genotypic constitution (Møller 1990b, 1992a; Møller and Höglund 1991). Such asymmetry refers to random deviations between left and right in traits that tend to be symmetric. As both sides are produced by the same genome, random deviations from symmetry reflect how accurately the genome can maintain developmental homeostasis in the face of environmental variation and stress (e.g., Parsons 1992). The degree of asymmetry is, for example, lower in heterozygotes than in inbred, homozygous individuals. It is increased by hybridization, and by new mutants with major effects; both events may disrupt previously coadapted gene complexes (reviewed by Leary and Allendorf 1989; Watson and Thornhill 1994).

Fluctuating asymmetry in the long tail of the male swallow *Hirundo rustica*, a trait favored by female choice, decreases with increasing tail length, contrary to what would be expected from the greater stress that development of a larger trait should impose (Møller 1990b, 1992a, Møller and Höglund 1991). A long tail should therefore reflect good developmental homeostasis, which might in part be heritable. Natural as well as experimentally induced asymmetry reduces flight performance in swallows. Long-tailed males tend to have lower asymmetry than males with shorter tail, and have offspring less affected by parasitic mites (Møller 1990a,b, 1991a–d). In an experiment manipulating tail length and asymmetry independently, Møller (1992a) found that female choice favors both symmetry and long tail (also see section 7.3, and Balmford et al. 1992b).

Perhaps even more surprisingly, in the scorpionfly *Panorpa japonica*, females favor the pheromone of a male with symmetric body over that of an asymmetric male. Males that win contests tend to be more symmetric than losers (Thornhill 1992a–c; also see Thornhill and Sauer 1992, Watson and Thornhill 1994). More studies of a variety of organisms are needed to clarify how general such patterns may be in sexual selection and other contexts.

Møller and Höglund (1991) compared sixteen pairs of bird species as regards ornament asymmetry (long tail in males). Each pair consisted of one species with large male sex ornaments and another related species without. The ornaments showed larger absolute as well as relative asymmetry than did wing length or the homolog of the ornament in females and in the non-ornamented species. The degree of asymmetry for tail ornaments in general decreased with increasing tail length. The result contrasts with expectations from Fisherian models of "arbitrary" sex ornaments, but it accords with what is expected from genetic indicator models (Møller and Höglund 1991; Møller 1992a). This approach offers an interesting new perspective; the results will perhaps help distinguish between the roles of Fisherian and indicator mechanisms. There may, however, be problems with the interpretation. For example, it is not entirely clear what constitutes a proper null hypothesis; for discussion, see Evans and Hatchwell (1993), Møller et al. (1993), and Sullivan et al. (1993). Also, in a comparative study of 63 bird species, Balmford et al. (1993) concluded that the pattern of asymmetry in long tails was better understood as the outcome of natural than of sexual selection.

Most models of condition-dependent ornaments assume that there is a trade-off between survival and ornament size. Although there is some evidence (chapter 10), such a trade-off has been difficult to demonstrate. Even if increasing ornament size raises the risk of mortality, large differences among males might permit those in best condition both to develop larger ornaments and to survive better than males in poorer condition (e.g., Zeh and Zeh 1988). There seems to be a parallel situation as regards several life history traits, such as clutch size and survival. Controlled experiments rather than observational studies often seem necessary to avoid misleading conclusions that could otherwise arise from individual variation in condition or environment (e.g., Partridge and Sibly 1991).

If the expression of sex traits increases with the overall phenotypic and genetic condition of the male, a preference for supernormal traits should be favored in females; the larger the trait, the better the condition of the male (Andersson 1982a; Weldon and Burghardt 1984). In four experimental studies of birds where males have long ornamental tails, the males with longest or even supernormal tails had highest mating success (section 13.6).

In sum, there is much evidence that individual variation in secondary sex traits depends in part on phenotypic condition, but the extent to which condition has a heritable genetic basis is not known.

3.7 Tests of Indicator Mechanisms

Like Fisherian models, genetic indicator models are difficult to test critically in a fashion that in principle makes it possible to refute the model. In one of the first thorough attempts at a test, Boake (1985, 1986) examined female choice of mate in the red flour beetle *Triboleum castaneum*. Males varied in pheromonal attractiveness and in developmental time. Male attractiveness and progeny fitness were not correlated, however, so no indicator mechanism seemed to be involved in this case. Other negative evidence was presented by Moore and Moore (1988), Payne and Westneat (1988), and Eberhard (1993).

A main line of work aims to clarify phenotypic and genetic consequences for offspring of mate choice by their mother. In an early study of female rejection of genetically unfit males, Maynard Smith (1956) found that female fruit flies *Drosophila subobscura* avoided mating with inbred males, who were incapable of performing the normal courtship "dance" (also see Steele and Partridge 1988). Female matings with outbred males produced many more viable offspring. Several later studies report that females allowed to choose their mates have offspring of higher phenotypic and perhaps also genetic quality than females with less scope for mate choice (e.g., Partridge 1980; Simmons 1987a,b; Crocker and Day 1987; Taylor et al. 1987; Woodward et al. 1988; Reynolds and Gross 1992; section 8.7 below). This is necessary but not sufficient for indicator mechanisms to apply.

A detailed study of extra-pair copulation in blue tit *Parus caeruleus* suggests that female choice may lead to genetic benefits (Kempenaers et al. 1992). DNA fingerprinting showed that one third of the nests in this socially monogamous species contained young sired by several males. Females in some pairs visited the territories of neighbor pairs and solicited copulations from the male. Some males were visited by many females, others by none. Females paired to unattractive mates often visited other males, and had several young sired by males other than their social mate, whereas females paired to attractive males rarely did so. Unattractive males were smaller and had lower survival, suggesting that they were of lower quality than attractive males. The mates of unattractive males did not, however, have lower weight or survival than other females. The offspring of attractive males had higher survival than those of unattractive males. Whether this holds also for the extra-pair young of attractive males is not clear.

These results are as expected if males differ in genetic quality, and if females seeking extra-pair copulations try to improve the genetic quality of their offspring. The possibility has not been entirely excluded, however, that quality differences among males are mainly phenotypic, perhaps being reflected in parental ability, and that females gain some nongenetic benefit from extra-pair copulations. For instance, females might then be better tolerated on the territory of the neighbor pair, or have higher chances of forming a new pair with the high-quality neighbor male, should his present mate die. The case for genetic benefits would become stronger if the extra-pair young of attractive males also have higher survival. This remains to be studied in the blue tit. Evidence for higher survival of offspring from attractive males comes, however, from a cross-fostering experiment in great tits *Parus major*, which also indicates that a plumage trait associated with male attractiveness, the black breast stripe, shows considerable heritability (Norris 1993; also see Lemel 1993).

A laboratory experiment in the guppy *Poecilia reticulata* suggests that a female preference for large males may lead to genetic advantages for offspring (Reynolds and Gross 1992). Large males sired sons and daughters with higher growth rates, and male size showed an estimated father-son heritability of 0.9. The higher growth rate of daughters led to higher reproductive output. The results are consistent with genetic indicator models, but other explanations are also possible (Reynolds and Gross 1992).

Genetic indicator models assume that females prefer males with the most well-developed sex ornaments, and predict that females choosing such mates will have offspring with higher than average viability. In addition, the higher viability must be shown to have a genetic basis, as in Norris' (1993) study of great tits. If the analysis is limited to phenotypic aspects, higher offspring quality might be a consequence of phenotypic benefits, such as nutrition, or male territory quality. For example, there is a direct reproductive advantage for females choosing the most ornamented males in house finch *Carpodacus mexicanus* (Hill 1990–1993; chapter 13 below) and great tit *Parus major* (Norris 1990a,b), and for female bicolor damselfish *Stegastes partitus* choosing the male with highest courtship display rate (Knapp and Kovach 1991). Other requirements for tests of indicator models are discussed, for example, by Partridge and Halliday (1984), Heisler et al. (1987), Wilkinson et al. (1987), Balmford and Read (1991), Kirkpatrick and Ryan (1991), and Petrie and Williams (1993).

Parasite-mediated Sexual Selection

One particular indicator process that has received much interest is the Hamilton-Zuk (1982) hypothesis, which suggests that genetic cycles in hosts and parasites maintain substantial fitness heritability (see revealing indicators, section 3.2 above). The idea was supported by a comparative analysis of plumage coloration in North American birds (Hamilton and Zuk 1982). The support from this and some other comparative studies may, however, be artifacts caused by phylogenetic relationships and confounding ecological variables. Comparative analyses may not be able to resolve the issue (for discussion, see Hamilton and Zuk 1989; Read and Harvey 1989a,b; Zuk 1989, 1991b; Hamilton 1990; Read 1990; Atkinson 1991; Harvey and Bradbury 1991; Harvey et al. 1991; Clayton et al. 1992; Folstad and Karter 1992). Comparative tests are reviewed in chapters 13 and 14.

More critical tests can be done within species (Harvey and Partridge 1982; Read 1988, 1990; Endler and Lyles 1989; Clayton 1991; Harvey and Bradbury 1991). In particular, detailed functional analyses are needed of the conspicuous trait and the preference, such as the study by Møller (e.g., 1988a, 1989a, 1990a, 1992a) of the swallow *Hirundo rustica*, and by Zuk et al. (1990a–d,1992) and Ligon et al. (1990) of red jungle fowl *Gallus gallus*. For the mechanism to work, the following conditions, among others, must be fulfilled (see, e.g., Read 1988, 1990; Endler and Lyles 1989; and Clayton 1991, for discussion of the criteria, methods, and problems of testing this hypothesis):

1. Host fitness decreases with increased parasite infection.
2. Ornament condition decreases with increased parasite burden.
3. There is heritable variation in resistance to parasites.
4. Female choice favors the most ornamented males.
5. Female choice favors the least parasitized males.

Tests of assumptions of the models of parasite-mediated sexual selection are summarized in table 3.7.1 (for a detailed review, see Clayton 1991). Several of the studies verify important assumptions. In three species, all five aspects are supported: guppy, pheasant, and swallow. There is evidence for heritable variation of resistance in several of the species in table 3.7.1, for example in guppies to *Gyrodactylus* monogeneans (Madhavi and Anderson 1985), pheasants to coccidians (Hillgarth 1990a), and swallows to mites (Møller 1990a) (also see Wakelin and Blackwell 1988). In the guppy *Poecilia reticulata*, females preferred males with relatively few parasites; such males displayed more than others, which may be the proximate reason why females preferred them (Kennedy et al. 1987; McMinn 1990). In the pheasant *Phasianus cholchicus*, there was a correlation between male display rate, coccidian parasite load, and female choice (Hillgarth 1990).

Male display or structural sex ornaments reflect degree of parasitism in 10 out of the 17 species reviewed in table 3.7.1. Females preferred males with the most pronounced sex traits in 12 of 13 cases. In 8 of the 13 species where the evidence is not conflicting, females preferred males with low parasite levels. Note that all but one of the negative results were based on observations; the better control of other factors that can be achieved by experiments may be necessary for adequate tests of these ideas (also see, e.g., Endler and Lyles 1989; Read 1990; Clayton 1991; Burley et al. 1991; Höglund et al. 1992a).

Support for a parasite-mediated mechanism of sexual selection comes from the study of swallows *Hirundo rustica* by Møller (1988a, 1990a). Cross-fostering tests and manipulation of the level of parasitism in a natural population suggested that host fitness is negatively affected by the bloodsucking mite *Ornithonyssus bursa*, as chick growth and adult orna-

Table 3.7.1

Tests of Assumptions of Hypotheses on Parasite-mediated Sexual Selection

		Assumption or Prediction Tested				
		Increased Parasite Load Reduces:		*Parasite Resistance Is Heritable*	*Females Prefer:*	
*Species**	*Secondary Sex Trait Studied*	*Host Fitness*	*Expression of Sex Trait*		*Males with Largest Sex Trait*	*Least Parasitized Males*
1 Cricket	Call, body size	p	n	—	n	N,p
2 Guppy	Display rate	p	p	p	p	p
3 Stickleback	Red color	P	P	—	P	P
4 Tree frog	Call	n	n	—	P	n
5 Toad	—	p	—	—	—	n
6 Lizard	Coloration	n	n	—	—	—
7 Lizard	Coloration	p	p	—	—	—
8 Pheasant	Display	P	P	P	p	P
9 Jungle fowl	Ornaments	P	P	—	p	P
10 Sage grouse	(a) —	—	P	—	P	P
	(b) Display	n	n	—	P	n
11 Rock dove	Display, plumage	P	P	—	p	P
12 Swallow	Tail ornament	P	P	P	P	p
13 Blackbird	Aggression	n	p	—	—	n
14 Grackle	Coloration	—	p,n	—	—	—
15 Zebra finch	Color	—	n	—	p	n
16 Parotia	Display	—	p	—	p	n
17 Bowerbird	Plumage, bower	—	n	—	p	p
18 Currasow	Bill knob	—	n	—	—	—

EVIDENCE: P positive, experiment; p positive, observation; N negative, experiment; n negative, observation; — not tested. (Only studies that manipulated the main independent variable are here labeled experimental.)

* SPECIES AND REFERENCES: 1 *Gryllus veletis* and *G. pennsylvanicus* (Zuk 1987a,b, 1988), 2 *Poecilia reticulata* (Kennedy et al. 1987; Madhavi and Anderson 1985; McMinn 1990), 3 *Gasterosteus aculeatus* (Milinski and Bakker 1990), 4 *Hyla versicolor* (Klump and Gerhardt 1987; Gerhardt and Doherty 1988; Hausfater et al. 1990), 5 *Scaphiophus couchii* (Tinsley 1990), 6 *Cnemidophorus arubensis* (Schall 1986), 7 *Sceloporus undulatus* (Ressel and Schall 1989), 8 *Phasianus cholchicus* (Hillgarth 1990a), 9 *Gallus gallus* (Zuk et al. 1990b,c), 10 *Centrocercus urophasianus* (a: Johnson and Boyce 1991, Spurrier et al. 1991; b: Gibson and Bradbury 1985, Gibson 1989, 1990), 11 *Columba livia* (D. H. Clayton 1990), 12 *Hirundo rustica* (Møller 1988a, 1990a, 1991f; Smith and Montgomerie 1991), 13 *Agelaius phoenicius* (Weatherhead 1990), 14 *Quiscalus quiscula* (Kirkpatrick et al. 1991), 15 *Poephilia guttata* (Burley et al. 1991; Burley and Coopersmith 1987); 16 *Parotia lawesii* (Pruett-Jones et al. 1990, 1991); 17 *Ptilonorhynchus violaceus* (Borgia 1985b, 1986; Borgia and Collis 1989, 1990); 18 *Crax daubentoni* and *C. rubra* (Buchholz 1991). (For details concerning the parasites involved, see the original sources.)

ment (tail) size is inversely related to parasite burden. The cross-fostering test, where half-broods were exchanged between nests, indicated that parasitism level is strongly heritable. The development of male tail length, which influences female choice of mate (Møller 1988a; Smith and Montgomerie 1991), reflects parasite load. For these reasons, female barn swallows favoring males with long tails should have offspring that inherit better than average resistance to the parasitic mites. This conclusion requires that no mites were transferred between nests during the exchange of chicks (Møller 1990a); otherwise, the effects might in part be environmental rather than genetic.

There was no relationship between male parasite level and display performance or mating success in several species, for example field cricket (Zuk 1987a), gray tree frog, spadefoot toad (Hausfater et al. 1990; Tinsley 1990), zebra finch (Burley et al. 1991), and the California population of sage grouse studied by Vehrencamp et al. (1989) and Gibson (1990). Their results differed from those of Boyce (1990) in Wyoming, where avian malaria was important. The long-term study of California sage grouse suggests that mainly other factors than parasites cause the variation in male success (section 7.6).

In satin bowerbirds, male mating success was negatively correlated with louse burden: the most parasitized males held no bower, a necessity for mating. But louse numbers were not correlated with male bower quality, although females probably respond to it when choosing mates (Borgia 1986; Borgia and Collis 1989). In Lawe's parotia, males with high parasite load did not mate, but there was no evidence that females chose mates in relation to traits that reflect parasite load (Pruett-Jones et al. 1990).

Several workers have suggested that transmission avoidance is an equally plausible explanation of the results as is the genetic Hamilton-Zuk mechanism (e.g., Gibson 1990; Hausfater et al. 1990; Pruett-Jones et al. 1990b; Buchholz 1991). In at least four species—rock dove, sage grouse, satin bowerbird, and Lawe's parotia—females may have avoided the most parasitized males simply by detecting ectoparasites or other signs of high parasite load, rather than by responding to secondary sex traits that reflect degree of parasitism (D. H. Clayton 1990; Boyce 1990; Borgia and Collis 1989, 1990; Pruett-Jones et al. 1990; also see Freeland 1976). Surprisingly, as Clayton (1991) emphasized, no experimental tests have been performed of alternatives such as the transmission avoidance hypothesis (also see Kirkpatrick and Ryan 1991). Yet another possibility is that the female preference for a male ornament has evolved for some other reason, and that its health-revealing nature is incidental (Read 1990).

In sum, several studies confirm important assumptions of the Hamilton-Zuk and other hypotheses of parasite-mediated sexual selection, but there is no conclusive evidence for any of the hypotheses as regards the selection of male secondary sex traits and female preferences. The evidence is compatible with several explanations, including those based on other sources of fitness heritability such as deleterious mutations, and spatial and temporal differences in selection coupled with dispersal (section 3.4). Deleterious mutations have been abundantly demonstrated, and there is also evidence of gene flow between areas with different selection pressures. These other possible sources of fitness heritability also need attention in tests of genetic indicator processes, which may involve many aspects other than host-parasite interactions.

Indicator models (based on viability advantages for offspring from ornamented males) and Fisherian models (based on self-reinforcing mating advantages for sons of ornamented males) have sometimes been treated as incompatible alternatives. The mechanisms are not mutually exclusive; they may work in concert, also together with other mechanisms, such as male contest competition. A major empirical problem is to estimate the relative importance of the various mechanisms, and to find out how much more a combination of them will explain, compared to single mechanisms (e.g., Andersson 1986a; Grafen 1990a; Iwasa et al. 1991; Balmford and Read 1991). This has not yet been done in any species.

3.8 Summary

Results from early genetic models of indicator ("handicap") traits were negative, but few-locus models now suggest that indicator mechanisms can play important roles in sexual selection. When coupled with a Fisherian self-reinforcing process, indicator mechanisms can eliminate the line of equilibria and lead to fixation of the male trait. Quantitative genetic models also suggest that indicator processes can be important, for example if there are deleterious mutations that reduce viability. Such mechanisms seem most likely to work if the preferred trait is condition-dependent, developing in proportion to the nutritional status and health of the male, as seems to be the case with many secondary sex traits. Another likely possibility is that indicators reflect direct material benefits to female or offspring.

Equilibrium states have received most interest in the models, but dynamic properties and rates of evolution may also be important for the distribution of sexually selected traits and preferences in nonequilibrium situations, which are probably common. A major empirical task is to estimate the relative roles of the various sexual selection mechanisms.

Many empirical studies have focused on the possibility of heritable variation in parasite resistance. Several predictions have been supported, but further critical testing is needed to distinguish between this and other possible explanations. As with Fisherian self-reinforcing selection, firm empirical evidence for genetic indicator processes is still lacking. There is better support for indicator mechanisms based on direct material benefits to the female or offspring, in cases where males provide parental care or other resources.

4 Empirical Methods

4.1 Introduction

The study of sexual selection over the last decades has become a highly diverse field. In addition to the scientific and esthetic fascination it offers, it leaves room for the most contrasting kinds of work. Approaches range from hardship-laden field studies of, for example, birds of paradise and bowerbirds in remote mountain forests of New Guinea, with blood-sucking leeches, headhunters, and other distractions (e.g., Beehler 1988; Pruett-Jones and Pruett-Jones 1990a; Diamond 1991b), through white-coat laboratory work clarifying parentage with the methods of molecular biology (e.g., Gowaty and Karlin 1984; Quinn et al. 1987; Burke 1989; Gibbs et al. 1990; Gyllensten et al. 1990; Westneat 1990; also see Gowaty and Gibbs 1993), to mathematical modeling of genetic mechanisms of sexual selection (chapters 2–3 above). Besides the fascinating unsolved problems, the many faces of the subject contribute to its attraction; there is important work to be done in sexual selection for biologists of all kinds, and much exciting dialogue and exchange of ideas and discoveries is taking place among them.

The most informative studies often combine widely different methods. For instance, good-old detailed field observations with binoculars enabled Davies (1983) to discover that competing male dunnocks *Prunella modularis* use a subtle display for stimulating a polyandrous female to eject sperm that she may have received from another male; the displaying male then mates with her. Observations also showed that a male feeds the young in proportion to his access to the female during the period of copulation. DNA fingerprinting analysis of parents and offspring verified that such access is a good predictor of paternity (Burke et al. 1989; Davies 1992). Controlled field experiments demonstrated that the mating system may shift from polyandry to polygyny depending on food conditions (Davies and Lundberg 1984); mathematical modeling has also clarified the system (Houston and Davies 1985; Davies and Houston 1986; also see section 8.4 below).

This chapter briefly describes methods of analysis of sexual selection; the next chapter presents some examples from a range of animals. Focus here is mainly on the methods used for analysis of selective mechanisms behind different kinds of secondary sex traits. A main goal of many studies is to identify these mechanisms and estimate their relative strength. Related questions concern the differences in selection between males and females, and the variation and trends in kind and degree of sex dimorphism among taxa. Detailed reviews of theory, quantitative methods, and results in the analysis of selection are found in Manly (1985) and Endler (1986a); its logical and conceptual basis is also treated by Sober (1984).

4.2 Observations and Experiments

Direct field observations usually give the first clue to the selective pressures acting upon a trait. As Darwin (1871) emphasized, the context in which a structure or behavior is normally used often provides strong evidence on its functions (also see section 13.1). For example, male gladiator frogs *Hyla rosenbergi* have a sharp bony spine on each hand (Kluge 1981). The spine is used as a dagger in fights between males over females and nests in the small ponds in which the female lays eggs. Many males are wounded after only a few nights on the territories, and some males are killed in fights. The spine therefore seems to be favored by sexual selection through male combat. A related, comparative question is why only gladiator frogs possess this weapon. Kluge (1981) suggested that part of the reason is unusually strong predation and sexual competition over breeding territories in gladiator frogs, which rarely become older than a year. Males that fail to defend a suitable nest will not get another chance in the next season.

For demonstration of sexual selection of a trait, it is important to show that variation in the trait leads to variation in mating success due to competition among rivals through contests, mate choice, or any other mechanism of sexual selection (see section 1.1). It is often clear in contest competition that the winner prevents access to the contested mate(s) for the loser. Also, in the other forms of competition over mates, successful mating by one individual often inevitably means that rivals will have lower success, for example if a female mates with only one male or if there is sperm precedence for, say, the first male to mate with her.

Even if a secondary sex trait is favored in contests between males, the trait is not necessarily directly sexually selected; this requires that victory in contests leads to higher success in competition over mates. This is sometimes the case (chapter 6), but natural selection might also be responsible if quality of territory or other factors are at stake that affect the production of offspring after mating has taken place. An indication that natural selection may also favor traits usually limited to males comes from birds in which females defend individual winter territories and sing during this period, but not during the breeding season. One example is the European robin *Erithacus rubecula*, in which adult females also develop a red breast, like males (e.g., Cramp et al. 1988; Schwabl 1992).

The selective pressures favoring a trait in most cases are unclear. There are often several plausible alternatives, and experiments may be necessary to show which mechanisms are at work. If a trait is sexually selected, competitive mating success may vary with the natural expression of the trait. Observational field studies have often produced such evidence. Observations may not reveal, however, whether a trait is directly sexually selected or indirectly selected through correlation with another trait. Multivariate techniques such as multiple regression can be clarifying (e.g., Lande and Arnold 1983; section 4.5 below), but there is a risk that some unknown or unmeasured correlated trait is the direct target of selection. In addition, it is sometimes hard to distinguish between cause and effect in observational studies. For example, is higher display rate in successful males the cause or the consequence of visits by attracted females? Controlled experiments are necessary to solve such problems.

With experiments, randomization reduces undesired correlations between the experimental variable (for instance, ornament size or display rate) and other factors (such as age, body size, condition, experience) (e.g., Box et al. 1978; James and McCullough 1985). This can make experiments more effective for clarifying causation, but they may not eliminate the problem of correlated unknown variables entirely; experiments can also create new artifacts. It is not possible to control for all conceivable factors, and not even the best experiment can guarantee a correct conclusion. This is the case whether the result confirms or refutes a hypothesis; Popper (1969) notwithstanding, refutations are also sometimes mistaken. In addition, observations and experiments in ecology have often involved pseudoreplication or other violations of the requirement that sample units be independent. One such violation is to treat several observations from the same individual as indenpendent sample units; another is treating related species as independent units in a comparative analysis. Such practice can render statistical significance levels meaningless (see, e.g., Hurlbert 1984; Harvey and Pagel 1991; and section 4.3 below). Other problems encountered in some experimental studies are discussed in section 14.5.

An experiment clarifying the importance of correlations among variables is the study of female choice in guppy *Poecilia reticulata* by Bischoff et al. (1985). Females prefer males with a large tail (figure 4.2.1). Display rate, which is also important in mate choice, increased with male tail size and changed with its experimental variation (Farr 1980; Bischoff et al. 1985; see Barnard 1990 for a similar case in a bird). By control of display rate through water temperature in the two different male chambers used in the test, the effects of the two variables could be separated. Had display rate not been controlled, the higher success of males with large tails might have

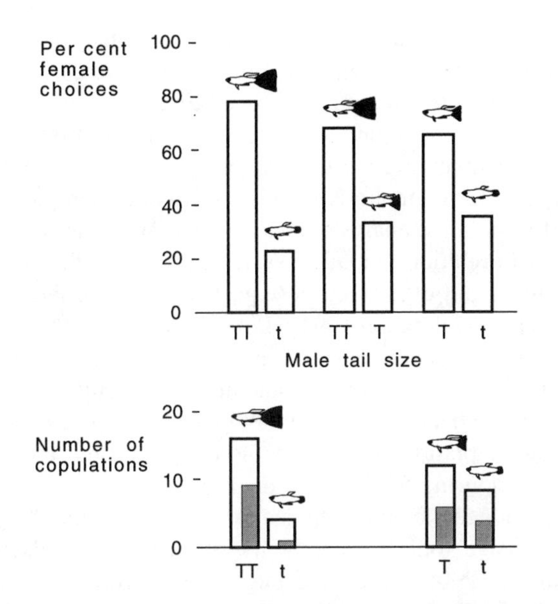

Figure 4.2.1 *Top*: Female guppies *Poecilia reticulata* offered a choice between two males with tails of different size preferred the male with larger tail. *Bottom*: When allowed free access to the two males, females mated more often with the male with larger tail (wide bars), which were more likely to sire offspring (narrow shaded bars). The three tail sizes used are: TT, large tail; T, surgically shortened tail; t, small tail. (After Bischoff et al. 1985)

been ascribed to tail size alone, but display rate in fact seems to explain more of the variation in male attractiveness (Bischoff et al. 1985).

Even though experiments have their own problems and cannot guarantee freedom from unknown correlated variables, they offer several advantages. One is that experimental randomization can break up correlations with other variables, recognized and sometimes also unrecognized. Multivariate methods based on observations permit control only of identified variables explicitly dealt with in the analysis. Another advantage is that experiments permit the range of variation of the independent variable to be expanded, making its effects stronger and easier to detect (e.g., Andersson 1982a; Wilkinson et al. 1987; Grafen 1988; Schluter 1988; Ryan 1990b). On the other hand, it is also desirable to make sure that the natural variation is large enough to create substantial selection effects.

4.3 Trends among Species: Comparative Approaches

Detailed observations and experiments are required to clarify present mechanisms of sexual selection and their strength within species. Other important questions concern patterns and trends in sex dimorphism among species. For example, why do some birds, fishes, reptiles, and butterflies, but not others, have more brightly colored males than females? Why are males larger than females in many mammals and birds, but females larger than males in many other animals? Why is sex dimorphism usually greater in polygamous than monogamous species? Such questions, whether they concern divergence among closely related species or convergence among distant ones, call for comparative analyses. Modern comparative methods offer powerful tools for analysis of taxonomic and other trends of secondary sex traits in relation to mating system, ecology, life history, phylogenetic constraints, and other factors (e.g., Ridley 1983; Pagel and Harvey 1988; Brooks and McLennan 1991; Harvey and Pagel 1991).

Since Darwin (1859), comparison among species has been a major tool for tracing the evolutionary history and adaptive significance of a trait. It was long assumed that comparative methods can only suggest but not test hypotheses on adaptation, and the methodology did not change much for about a century. During the last decade, however, exciting progress in the rigor of its phylogenetic and statistical aspects has made the comparative method an incisive tool for testing hypotheses (e.g., Felsenstein 1985, 1988; Maddison 1990; Harvey and Pagel 1991). Together with improved phylogenies based on, for example, molecular sequence data, DNA hybridization, and other molecular methods (e.g., Sibley and Ahlquist 1990), it will probably lead to a dramatic increase in the power of comparative studies for clarifying selection and evolution.

Sophisticated comparative analyses of secondary sex traits are already well under way; such traits were among the first to be explored with the new methods. Only a brief sketch is given here; progress in comparative methods has been so rapid that the first version of this chapter written a few years ago is now obsolete. For thorough reviews of modern comparative methods, see the books by Brooks and McLennan (1991) and Harvey and Pagel (1991); Williams (1992) also provides a stimulating discussion of the comparative method. The following brief survey gives some examples where the new methods have helped clarify sexual selection.

Qualitative Traits

A rigorous method for the comparative analysis of qualitative traits was pioneered by Gittleman (1981) and, in greater detail, by Ridley (1983). Owing to phylogenetic constraints, related species often cannot be regarded as independent sample units in statistical tests. Less problematic sample units are the different cases of independent evolution of the trait among the taxa under study. Such evolutionary shifts may characterize species within genera, whole genera within families, or higher (or lower) phylogenetic units (see e.g. Williams 1992). The analysis therefore will often involve different taxonomic categories. To estimate the minimum number of times that a trait has evolved, Gittleman (1981) and Ridley (1983) used available phylogenies for making cladistic out-group comparisons. They then tested the statistical association between the new trait and various ecological or behavioral preconditions. Such associations help identify conditions that promote evolution of the trait.

Ridley (1983) applied this technique to precopulatory mate guarding, a probably sexually selected behavior. The hypothesis under test—based in part on a game theory model by Grafen and Ridley (1983)—was that guarding will occur in species in which the female can be fertilized only during a brief, predictable period (e.g., Parker 1974). This idea had grown out of earlier attempts to explain mate guarding in a number of studies before Ridley's (1983) work. This seems no serious problem for Ridley's tests, as he applied them to many different arthropods and amphibia, most of which had probably not been in the minds of those who suggested the first versions of the hypothesis. But this can be a problem: if the source of inspiration for a hypothesis is knowledge of the biology of one or more taxa, a comparative test ideally should not involve those taxa. Otherwise, the test in part becomes a post hoc recalculation of what is already known, not an independent test of predictions.

Ridley (1983) inferred that there have been twenty separate evolutionary shifts to mate guarding, nineteen of which were correctly predicted. Mate guarding has been secondarily lost eleven times, ten of which were predicted. The conclusion seems clear: precopulatory mate guarding in arthropods and anurans usually evolves if the female is available for mating only during a short, predictable time.

Ridley (1983), like many others before him, remarked that causality is harder to infer with comparative methods than with experiments. By careful choice of species suitable for experiments, it may be possible to clarify in detail the selective mechanisms suggested by comparative analyses. In addition, a new comparative method developed by Maddison (1990) explicitly tests whether the evolution of a trait depends on the state of another variable. If this is the case, that other variable can be regarded as an independent causal factor in the evolution of the trait (also see Harvey and Pagel 1991).

Quantitative Traits

The approaches of Gittleman (1981), Ridley (1983), and Maddison (1990) are important advances in the comparative analysis of qualitative traits. The methods cannot, however, be applied to quantitative traits. Many characters, also among those that are often treated as distinct qualitative categories, show continuous variation, which may often contain crucial information. For instance, in comparisons involving different mating systems, species means and variances in numbers of mates per male and female may reveal interesting trends that do not turn up in cruder analyses based on a monogamy-polygamy dichotomy.

Over the last two decades, increasingly sophisticated methods have been developed for statistical comparative analysis of continuous traits (e.g., Felsenstein 1985, 1988; Harvey and Pagel 1991). Much of this work concerns secondary sex traits (e.g., Clutton-Brock and Harvey 1977, 1984; Clutton-Brock et al. 1977). A first step is to separate a variable into more than two classes. This has led to better understanding of secondary sex traits in relation to mating systems and other aspects of ecology. In primates, perhaps the most thoroughly studied group, male relative to female body and canine sizes are larger in polygynous than monogamous species. Moreover, they are larger in strongly polygynous forms, where a single male can have many mates, than in species with multimale troops, where a male has fewer mates (figure 4.3.1). These trends fit the idea that large relative canine and body sizes in males are sexually selected through con-

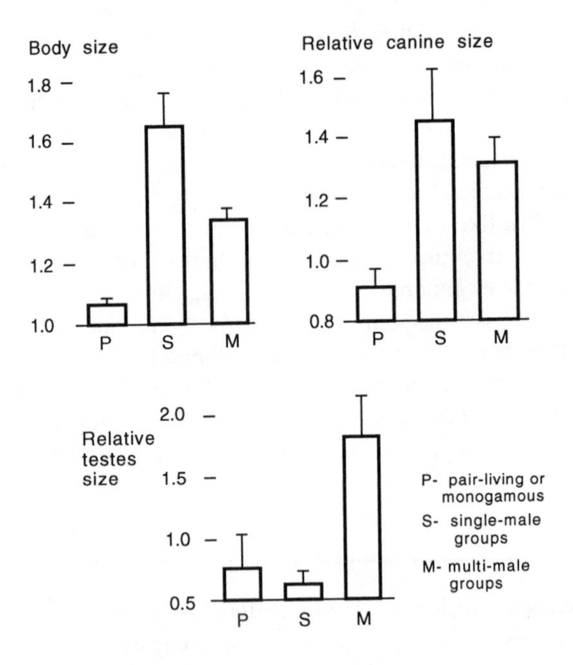

Figure 4.3.1 Sexual dimorphism in body and relative canine size (ratio of adult male : adult female), and male relative testes size, in primates with different mating systems. Variation bars = standard errors. (After Clutton-Brock and Harvey 1984)

test competition over females (Harvey et al. 1978; Clutton-Brock and Harvey 1984; Harvey and Pagel 1991). Testes size is discussed below

The independent variable (mating system) in this case was split into three categories, and the dependent variables were measured on continuous scales. To control for allometry and dietary causes of variation in male canine size, Harvey et al. (1978) calculated sex dimorphism in relation to the regression of female canine size on female body size. For each species they also used a correction factor based on the extent to which female tooth size differed from the expected regression value; this should compensate for the possible influence of dietary differences among species.

There are several alternative hypotheses for the sexual size dimorphism of primates. It might be related to food competition between the sexes; dimorphism could lead to separate diets in males and females and hence reduce competition (Selander 1972). If so, dimorphism should be largest in monogamous forms where the mates share foraging area. This is not so. On the contrary, in a partial regression analysis where the effect of body size was removed, dimorphism increased with the number of females per male in the breeding group (Clutton-Brock et al. 1977). In no case did the feeding apparatus appear to differ more than other traits between the sexes. The food niche hypothesis is therefore refuted as the main explanation for sexual dimorphism in primates; sexual selection appears to be a major cause (Clutton-Brock and Harvey 1977; Clutton-Brock et al. 1977; also see section 11.9).

A third, more gentlemanly hypothesis for the dimorphism is that larger canines is an adaptation for male defense of females and young against predators (e.g., DeVore and Hall 1965). In some troop-forming primates, males are more active defenders than females. Relative male canine size seems to increase with predator pressure, apart from any correlation with the breeding system. In the subsample of species with multimale troops, male relative canine size was larger in ground-living than in tree-living forms; predation risk is probably higher for ground-living primates. Body size can be ruled out as a cause in this case, as there is no correlation between relative canine size and body size in males among the species with multimale troops (Harvey et al. 1978). Larger canine weapons in males than females therefore seem to be maintained in part by sexual selection, in part by advantages in male defense against predators.

Such defense might also contribute to the larger body size in male than female primates. The ground-living primates in which predation risk appears to be highest show larger sexual size dimorphism than tree-living species. The reason could also be, however, that tree-living species are under stronger selection against large body size, if it reduces access to food at fine terminal twigs (Clutton-Brock et al. 1977). The issue is complicated further by a general tendency for sexual size dimorphism to increase with

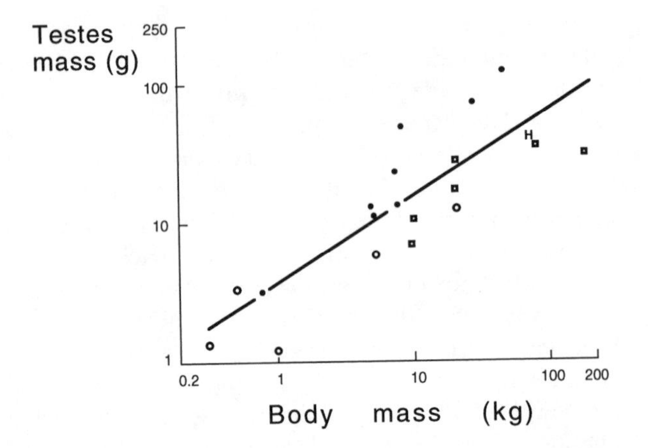

Figure 4.3.2 Testes mass versus body mass for primate genera in relation to the breeding system. Solid dots: multi-male groups; squares: single-male harems; open dots: monogamous; H: *Homo*. (After Harcourt et al. 1981. Closely related species within genera cannot be treated as independent units, so the line was estimated from points for genera, calculated from means for species.)

body size, and by phylogenetic effects (e.g., Cheverud et al. 1985; Gittleman and Kot 1991; Rogers and Mukherjee 1992; section 11.9 below).

Another example is sperm competition in primates. Parker (1970b, 1984) emphasized that competition among males can occur also after mating, if the female copulates with several males during her estrus period. This may lead to sperm competition among her mates, a possibility that has now been verified in a variety of animals and plants (for reviews of this rapidly growing field, see Smith 1984; Eberhard 1985; Willson 1990; Birkhead and Møller 1992). If the number of sperm per ejaculate increases with testes size, such competition should select for large testes (Short 1979; Møller 1988c). The largest relative testes size should then be found in the species with strongest sperm competition, for example where breeding groups contain several adult males that may copulate with a receptive female. This is indeed the case among primates (figure 4.3.2). Harcourt et al. (1981) and Harvey and Harcourt (1984) showed that testes size is larger in primates with multimale breeding groups than in species that are monogamous or have single-male harems (also see figure 4.3.1).

Comparative analyses of continuous traits present their own problems as regards choice of independent data points. There is now a variety of techniques for reducing such problems. Harvey and Pagel (1991) stress that the methods that make use of all the variation in the phylogeny are preferable (also see Williams 1992). Methods based on sets of statistically independent comparisons, across present-day taxa, or between ancestral and descendant phylogenetic nodes, emerge as the best available techniques.

A serious problem in many comparative studies is the accuracy of the phylogeny. It can usually only be estimated, not known with certainty, and the conclusions of the analysis may change with the phylogeny. The sensitivity of the conclusions to changes in the phylogeny therefore needs to be tested.

4.4 Measures of Sexual Selection

A long search for a suitable measure of the strength or intensity of sexual selection has been fuelled by a hope that it will help achieve more precise quantitative understanding of, for example, differences between the sexes and between species in the size of secondary sex traits. As several mechanisms may occur together, it is also important to estimate their relative strength. A number of measures have been suggested (table 4.4.1), but they all have various problems.[1] A major one is that they lack connection with the secondary sex traits that the theory seeks to explain. Rather than quantifying the strength of selection of particular traits, the measures in table 4.4.1 reflect the potential for sexual selection (Clutton-Brock 1988a).

Table 4.4.1

Some Proposed Measures of the Strength of Sexual Selection*

1	Relative parental investment of males and females (Trivers 1972; Sutherland 1985b)
2	Operational sex ratio (Emlen and Oring 1977; Westneat et al. 1990)
3	Relative time spent seeking mates (males : females; Sutherland 1985b)
4	Variance in male breeding success (Payne 1979, 1984)
5	Ratio of variance in breeding success between males and females (Ralls 1977; Payne 1979)
6	Variance in mating rate (McCauley 1983; Koenig and Albano 1986)
7	Ratio between relative variance in numbers of mates and relative variance in total breeding success (Wade and Arnold 1980)

*Modified from Sutherland 1985b and Clutton-Brock 1988a.

Several of the measures are based on some aspect of variance in success. Wade and Arnold (1980) and Arnold and Wade (1984a) suggested that the relative variance in mating success among males ($\sigma^2/\bar{X}^2$, where σ^2 is the variance in mating success, and $\bar{X}^2$ is the square of mean mating success) is useful in providing an upper limit to the strength of directional sexual selection. It has therefore been termed the "opportunity for sexual selection" (also see Wade 1979, 1987). But to identify and quantify actual

[1] For discussion, see, e.g., Wade 1979, 1987, Clutton-Brock 1983, 1988a, Sutherland 1985a,b, 1987, Endler 1986a, Koenig and Albano 1986, Downhower et al. 1987b, Partridge and Endler 1987, Grafen 1987, 1988, Wilkinson et al. 1987, Arak 1988a, Clutton-Brock and Parker 1992.

mechanisms of selection of particular traits, other approaches are required (Endler 1986a; also see below in this section).

The *operational sex ratio* is the ratio of adult males to females available for mating. It can sometimes be estimated with reasonable accuracy, and should be closely related to the strength of competition over mates (Emlen and Oring 1977; Sutherland 1985a; Ims 1988b; Clutton-Brock and Parker 1992; section 7.2 below). Westneat et al. (1990) modified the concept for situations where extra-pair copulations and sperm competition are important components of mating success, for example in some birds.

There are also other possible measures of selection, in particular the approach suggested by Lande and Arnold (1983), which deals explicitly with the selection of identified traits. Among their measures, most relevant here is probably the standardized selection gradient: the partial regression of relative fitness on the trait in question (Lande and Arnold 1983; Arnold and Wade 1984a,b). For traits under directional selection, the gradient indicates how much relative fitness will change with an increase in the size of the trait by a unit amount, if other traits are held constant. In contrast with the measures in table 4.4.1, the selection gradient therefore quantifies the effect of particular traits on fitness or some of its components. A trait can be said to be subject to stronger direct sexual selection, the higher its selection gradient with respect to competitive mating success. Different mechanisms of sexual selection may often influence the same trait; a major problem is then to estimate their relative strength, which can be done using selection gradients (Lande and Arnold 1983; Arnold and Wade 1984a,b; see section 4.5 below).

The most radical criticism of sexual selection measures comes from Grafen (1987, 1988), who questioned whether many of the quantitative measures are really of help in solving the problems of widest interest in sexual selection theory. Grafen suggested that most biologists working with sexual selection are interested in adaptation rather than selection in progress, for example in the reasons why red deer have big antlers rather than whether there are genes now changing in frequency that affect antler size. Measures such as those in table 4.4.1 that are not related to particular traits are not informative in this respect. In part because many of these questions are historical, there is no single quantity of central importance. Instead, Grafen (1987, 1988) emphasized experimentation as the most straightforward way to clarify selection mechanisms and adaptation.

While accepting its great value in many other respects, Grafen (1988) also suggested that the analysis of lifetime reproductive success and its variance (see the volumes edited by Clutton-Brock 1988b; Newton 1989) is rather irrelevant to the question of adaptive significance of a character. In itself, present variation in lifetime reproductive success has little to say about past and present selection of a trait. Besides not being related to any concrete trait, the variance in lifetime reproductive success often arises to a large degree through differences in life span, having little to do with sexual selection. On the other hand, data on lifetime success gathered separately for each morph or phenotype class can shed light on sexual and other selection. As the strength and direction of selection can vary among life history stages or age classes, partial short-term measures of success can be misleading (Clutton-Brock 1983, 1988a; Endler 1986a).

In the words of Koenig and Albano (1986), "it is critical to try to associate selective episodes with specific agents of selection and only then to propose that they represent sexual selection or natural selection. Otherwise, the potential for misunderstanding will almost certainly exceed the potential for sexual selection, whatever it may be." Grafen (1987, 1988) emphasized that variance measures such as the opportunity for sexual selection in general are less useful than trait-related approaches for understanding secondary sex traits (also see Clutton-Brock 1983, 1988a; Sutherland 1985a,b, 1987; Endler 1986a; Partridge and Endler 1987; Wade 1987; Wilkinson et al. 1987; Arak 1988a; D. Brown 1988; Howard 1988b; Lyons et al. 1989).

4.5 Selection of Quantitative, Correlated Traits

Theory

Probably most behavioral and morphological traits show polygenic inheritance; moreover, pleiotropy and linkage make many traits correlated. A theoretical framework for the analysis of selection and evolution of quantitative correlated traits was developed by Lande (1979), Lande and Arnold (1983), and Arnold and Wade (1984a,b) (see also Manly 1985 and Endler 1986a for review of these and related methods). The approach can be used to estimate the strength of direct and indirect selection upon a trait, and to predict how much it will change each generation, and at what size it will stabilize. The method is based on measurement of selection and heritability for an array of traits. Because traits are often correlated and vary together, genetic and phenotypic correlations must also be determined. Although it often meets practical problems, the method can sometimes be applied to field studies of suitable species, using multiple-regression techniques tailored to the analysis of selection.

Differences in survival and reproduction in a population depend directly on phenotypic differences among individuals, but only indirectly on genetic differences, to the extent that they have phenotypic effects. In this context, it is therefore convenient to define selection in relation to phenotypes, whose distribution it affects directly, and to separate phenotypic and genetic consequences of selection (Lande and Arnold 1983). For selection to have evolutionary effects in terms of genetic changes between generations, the phenotypic variation must in part be based on heritable genetic variation.

The directional selection differential, S, is the change in mean phenotypic value of a trait caused by selection working directly on the trait as well as indirectly via correlated traits (e.g., Bulmer 1980; Falconer 1989; the notation below mainly follows Lande and Arnold 1983, and Arnold and Wade 1984a,b). For a number of traits $z_1, \ldots z_n$, represented by a column vector **Z**, the directional selection differential **S** is the difference between the vectors of mean values of the traits after selection and before selection:

$$\mathbf{S} = \mathbf{Z}^* - \mathbf{Z}. \tag{4.5.1}$$

The selection differential measures the phenotypic change due to selection within a generation, or a suitably defined part of it. To keep the measure of selection purely phenotypic, the analysis should not cross generation borders, which would add effects of inheritance (Arnold and Wade 1984b).

The genetic evolution between two generations can be predicted from the selection differential together with the heritability of the trait. The difference in mean after selection compared to before selection will be carried over to the next generation only to the extent that the difference is heritable. For a single character without genetic or phenotypic correlation with other traits, the change ΔZ in character mean between two generations is simply the selection differential S multiplied by the heritability h^2 of the character (Falconer 1989) (figure 4.5.1). As the heritability is the ratio between the additive genetic variance and the total phenotypic variance of the trait, $h^2 = V_a/V_p$ (section 3.5), the change is given by

$$\Delta Z = V_a V_p^{-1} S. \tag{4.5.2}$$

In general, however, there are correlations between traits that must be taken into account. The genetic evolution $\mathbf{\Delta Z}$ of the mean phenotype between two generations is then given by an expression similar to equation 4.5.2, but generalized to n dimensions (number of correlated traits):

$$\mathbf{\Delta Z} = \mathbf{G}\mathbf{P}^{-1}\mathbf{S}. \tag{4.5.3}$$

G is the additive genetic variance-covariance matrix (Lande 1979; Lande and Arnold 1983). It describes the additive genetic variance of each trait z_i (the diagonal elements of the matrix), and the additive genetic covariances between all pairs of traits (off-diagonal elements). **P** is the matrix of the corresponding phenotypic variances and covariances. **G** and **P** in the multivariate case therefore correspond, respectively, to the additive genetic variance V_a and the total phenotypic variance V_p in the case of a single uncorrelated trait (see above in this section, and section 3.5). Equation

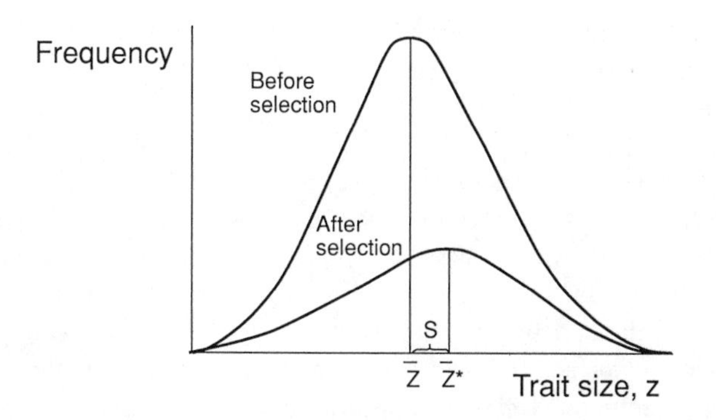

Figure 4.5.1 Distribution of trait size (Z) in a population before and after directional selection. The selection differential S is the change in mean trait size ($\bar{Z}^* - \bar{Z}$) arising during the selection episode. To the extent that the variation among individuals has additive genetic causes, it will result in an evolutionary response in the next generation, the mean trait size of which will then differ from $\bar{Z}$ by an amount $\Delta Z = h^2 S$, where h^2 is the heritability of the trait.

4.5.3 describes how the rate of evolution $\Delta\mathbf{Z}$ in character means depends both on directional selection, measured by the differential $\mathbf{S}$, and on phenotypic character variances and covariances $\mathbf{P}$ and their heritable components $\mathbf{G}$.

In sum, the change in a trait arises partly through direct selection of it, and partly indirectly through correlations with other traits that are also subject to direct selection. The selection differential $\mathbf{S}$ measures the effects of both these components: the total effect of directional selection on the phenotypic character.

The directional selection gradient, β, ideally should measure only the direct selection of a trait. It is the partial regression of relative fitness on the trait when all other characters are held constant. The selection gradients for all the traits together form a column vector $\boldsymbol{\beta}$, given by

$$\boldsymbol{\beta} = \mathbf{P}^{-1}\mathbf{S}. \tag{4.5.4}$$

The selection gradient predicts the change in relative fitness that results if the trait is altered by a unit amount while other traits are fixed (Lande and Arnold 1983). It offers a tool for measuring the strength of (direct) selection of a trait (section 4.4).

Together with the genetic variance-covariance matrix $\mathbf{G}$, the selection gradient $\boldsymbol{\beta}$ contains the information needed to predict the directional response to selection (provided that all correlated traits are included in the analysis). Both can be measured, albeit with various difficulties, in natural populations. The matrix $\mathbf{G}$ can be estimated from regressions between relatives for the same traits (this yields the genetic variances) and for different traits (covariances;

see Falconer 1989). The selection gradients β are estimated by the partial regressions of relative fitness on each trait in a multiple regression analysis. An example is discussed later in this section.

The previous analysis concerns *directional* selection. Lande and Arnold (1983) also developed a measure of *variance* selection (stabilizing or disruptive) of a trait near its population mean. Directional selection is estimated by the best linear approximation to the *slope* of the multidimensional selective surface of individual fitness as a function of the phenotypic traits. Variance selection is estimated by the best quadratic approximation to the *curvature* of the selective surface (for alternative methods, see, e.g., Mitchell-Olds and Shaw 1987; Schluter 1988; an application to sexual selection is found in Moore 1990a).

For a trait that changes with age, the analysis of selection in principle can be based on its growth trajectory. To take an example, antlers in red deer increase in size for several years (e.g., Clutton-Brock et al. 1982). A buck with large antlers at his peak reproductive age might not have reached that age if he had developed antlers of similar size already in his first year of life. If selection pressures vary with age and size, fitness should ideally be measured relative to the shape of the lifetime developmental trajectory for each trait. The trajectories might sometimes be possible to characterize by a few parameters, which in turn can be treated as individual traits contained in the vector **Z**, for example in equations 4.5.1–4.5.4 above (Lande 1982b; Atchley 1987).

There are many practical problems in the application of these methods. For example, if the population is age-structured with overlapping generations, it may be necessary to use fitness measures based on modified Euler-Lotka equations (Charlesworth 1980; Lande 1982b). Other difficulties are discussed by Lande and Arnold (1983), Arnold and Wade (1984a,b), Endler (1986a), Price and Boag (1987), Boag and van Noordwijk (1987), Mitchell-Olds and Shaw (1987), Wilkinson et al. (1987), D. Brown (1988), and Schluter (1988). One of the most serious problems is that selection may occur on unmeasured or even unrecognized traits correlated with those under study. The results might change markedly if such traits were included in the analysis. This uncertainty can never be eliminated entirely. To reduce it, it is desirable to measure many traits potentially subject to selection. But as the number of traits increases, so does the number of individuals needed for the multiple regression analysis, at a roughly quadratic rate. In addition, the presence of several strongly correlated traits can invalidate the regression analysis, which is based on a number of assumptions that need to be verified in each application (e.g., Draper and Smith 1981). Reducing the correlated traits to fewer principal components might solve the statistical problem, but at the price of obscuring the biological interpretation (Lande and Arnold 1983; Endler 1986a; Wilkinson et al. 1987). To translate a proverb from my own language: however you may turn, your rump is always behind you.

Arnold and Wade (1984a,b) presented a method of analysis of selection differentials and gradients based on multiplicative components of lifetime fitness, for instance survival from zygote to reproductive age, success in mating, and zygote production. Such or finer partitioning may be of help for identifying specific selection mechanisms even when lifetime fitness cannot be measured, but there are problems here too. For example, the method does not take into account covariation between different components of the life cycle (see D. Brown 1988 for a critical review of this and other methods of partitioning selection into different stages).

In spite of its problems, the approach of Lande and Arnold (1983), based on selection gradients and differentials, seems to me one of the best available methods for analyzing selection by using observational data. Applied with care and awareness of its problems and limitations, and combined with controlled experiments, the method may greatly help clarify selection mechanisms and their strength (also see Schluter 1988).

Example: Sexual Selection in Darwin Finches

Over the past two decades, Peter and Rosemary Grant and their coworkers have carried out extensive field research on *Geospiza* finches in the Galapagos islands, providing a wealth of insights concerning natural and sexual selection. Much of the work is summarized in Grant (1986) and Grant and Grant (1989). Several of these finches were subject to strong selection during periods of severe drought in 1976–1982. The resulting food shortage prevented successful breeding for several years. The worst drought cut the adult population of *G. fortis* on the island of Daphne Major to less than 1/5 of its size between June 1976 and December 1977 (Grant 1986 and references therein).

Price (1984a) examined selection of four characters: bill length, depth, width, and body mass, all of which are sex dimorphic, with males being largest. Price calculated selection differentials and gradients for these traits in males and females in relation to survival, mating success, and fertility in mating birds. Selection differentials were computed as the difference in sample mean of each character before and after the episode of selection. The gradients were estimated as partial regression coefficients of relative fitness on each character (normalized to unit variance).

The results for survival selection are given in table 4.5.1, based on birds breeding in 1976, many of which died during the 1977 drought. The positive differentials indicate that selection in total favored greater size in all four characters, apparently because hard seeds made up an increasing part of the food as the drought continued; only large finches could crack the hardest seeds. Selection gradients were, however, negative for bill width

Table 4.5.1

Selection Differentials and Selection Gradients Associated with Mortality in *Geospiza fortis*[a]

	Males		*Females*	
Trait	*Differential*	*Gradient ± s.e.*	*Differential*	*Gradient ± s.e.*
Weight	.039[c]	.38 ± .24	.030	.25 ± .32
Beak length	.018[b]	−.48 ± .27	.022	−.42 ± .46
Beak depth	.043[c]	.81 ± .36[b]	.038[b]	.76 ± .59
Beak width	.027[c]	−.24 ± .33	.023	−.21 ± .44

[a] From Price 1984a. [b] $P < .05$. [c] $P < .01$.

and bill length. Direct selection therefore seems to have favored a shorter, narrower bill; yet the total (including indirect) selection led to a larger mean bill width. This suggests that selection for greater size of some correlated trait(s) overrode direct selection for a narrower bill. Correlation analysis supports this conclusion: all four characters are highly positively correlated. This shows the importance of distinguishing between total selection (differentials, S) and direct selection (gradients, β): they can differ markedly, sometimes even working in opposite directions.

Sexual selection arose mainly from a heavily male-biased sex ratio of about 3:1, caused by sex differences in mortality during the drought. Roughly half the variance in production of young among males arose from differences in mating success, with many males remaining unmated. Sexual selection worked against small males. This difference in selection of males and females seems to explain at least part of the sex dimorphism. (Counteracting selection in some years favors small individuals because they survive better; small females may also breed earlier than large females; Downhower 1976; Price and Grant 1984; Price 1984a; Grant 1986).

Juvenile males have brown, female-like plumage, which darkens to black over 15 months or more. The sexual selection gradient was positive and statistically significant for territory size, plumage color, and bill depth (table 4.5.2, figure 4.5.2). Dark plumage and a large territory may be signs of high male parental quality.

Analyzing males born in the same year (1978), Price (1984b) separated the effect of male age from that of other traits, and found that the same three sexual selection gradients were still positive (table 4.5.2). All three traits seem to be favored directly by sexual selection; whether through male contests or female choice is not clear. It is possible, however, that one or more of the traits is favored by sexual selection not directly, but via some other correlated but unmeasured character.

A similar analysis of selection in the large cactus finch *G. conirostris* by

Table 4.5.2

Selection Gradients ($\beta \pm$ s.e.) Associated with Relative Mating Success in Male *Geospiza fortis*[a]

Trait	*Adult Males*	*1978 Males*
Beak depth	$.26 \pm .10$[b]	$.40 \pm .14$[b]
Territory area	$.30 \pm .10$[b]	$.24 \pm .14$
Plumage	$.24 \pm .10$[b]	$.26 \pm .14$

[a] From Price 1984b. [b] $P < .01$.

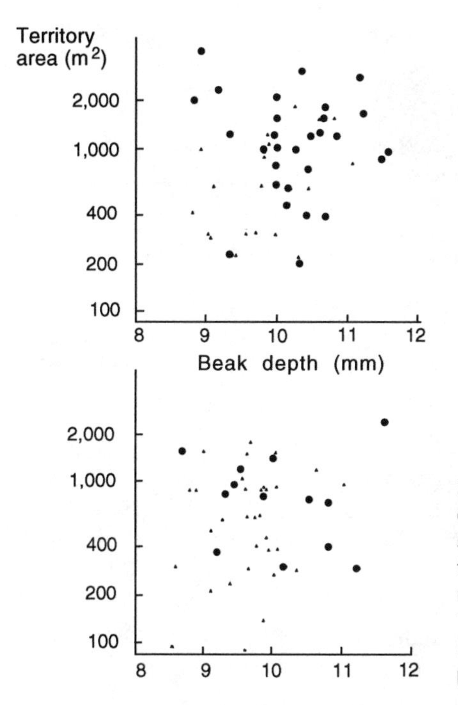

Figure 4.5.2 The contributions of beak depth and territory size to mating success in male medium ground finches, *Geospiza fortis*. Dots: mated males; triangles: unmated males. *Top*: Males in full adult plumage. *Bottom*: Males retaining some juvenile plumage. (From Price 1984b)

Grant (1985) and Grant and Grant (1987, 1989) gave partly different results. There was no evidence that females choose mate in relation to his body size, beak size, territory size, or territory quality. The only morphological trait correlated with mating success was plumage color, with black males being more successful than others. Male mating success was most consistently related to dark plumage and aspects of territorial behavior, both of which were associated with old age and breeding experience (figure 4.5.3). Such males also produced more fledglings than other breeding males. Females choosing males with plumage color and behavior that reflect age and experience may obtain mates with better than average parental

Figure 4.5.3 Categorization of male plumage coloration in studies of Darwin's finches (*Geospitza fortis* and *G. conirostris*). Males progress from female-like brown plumage in sequence over black 1 to black 5 (full adult plumage). (Drawing by P. Grant in T. D. Price 1984b)

behavior. This would lend a direct fitness advantage to female preferences for such male traits. There is also a possibility of genetic benefits, but, as in most other studies, this remains to be tested (Grant and Grant 1989).

To summarize, studies of Darwin's finches demonstrate direct and indirect sexual selection of several traits. Other analyses of sexual selection in animals, using selection gradients and differentials, are found, for example, in Gibson (1987), Conner (1988), Andersson (1989), Morris (1989), Hews (1990), and Moore (1990a). Similar analyses of selection in plants are found, for example, in Campbell (1991) and Johnston (1991a).

4.6 Summary

Field observations and controlled experiments have been the main methods for identifying mechanisms of sexual selection and estimating their strength. Comparative analyses, based on new phylogenetic and statistical methods, are becoming increasingly useful also for testing hypotheses. In addition, they offer possibilities for demonstration and explanation of taxonomic and other trends of secondary sex traits in relation to phylogeny, mating system, ecology, life history, and other aspects.

Among the many suggested measures of the strength of sexual selection, trait-related approaches such as selection gradients and differentials seem most useful. Multivariate versions of such methods can help clarify sexual and other selection among correlated traits in natural populations. An example from Darwin's finches is described.

5 Some Case Studies

Research in sexual selection during the 1980s has erupted in a flow of studies of organisms ranging from mammals to higher plants, providing an increasingly realistic picture of sexual selection as it occurs in nature. The main purpose of this chapter is to review some illuminating case studies that reflect the variety of problems, mating systems, traits, and animals that have been analyzed. The chapter describes in more detail than can be done later in the book the kinds of approaches and evidence on which the conclusions from empirical work are based. The subsequent chapter (6) reviews some broader patterns among studies of sexual selection.

Although they use quantitative statistical methods, the following studies aim mainly for qualitative demonstration of different mechanisms of sexual selection. Assessing their relative roles is more difficult and has rarely been attempted.

5.1 Causes of Sexual Selection in Male and Female Katydids

Most of the empirical work on sexual selection concerns its mechanisms and consequences. Fewer studies have tested the basic ideas on what determines the strength of competition over mates in the two sexes, such as the theory of sex differences developed by Bateman (1948), Williams (1966), Trivers (1972), and others (section 7.2 below). Much of this work on the determinants of sexual selection has been done in insects (e.g., Bateman 1948; Thornhill 1986a). Darryl Gwynne and his colleagues have studied orthopteran katydids or bush crickets, Tettigonidae, which offer unusual opportunities for testing the influence of relative parental investment on sexual selection in males and females. Their relative investment varies markedly within some of these species. Male katydids transfer sperm in a large, nutritious spermatophore to the female during copulation. The female eats it, and the nutrients enable her to lay more or larger eggs (e.g., Gwynne 1988a,b; Simmons 1990; also see figure 8.2.2). The food from the male provides females with a resource important for reproduction in times of food shortage. Changes in food abundance permit testing whether variation in relative parental investment of males and females affects the strength of mating competition in the two sexes, as theory suggests (Gwynne 1981, 1984a, 1985, 1990; Gwynne and Simmons 1990; Simmons and Bailey 1990; Simmons 1992).

Courtship role reversal and female fighting over males occurs in some populations of *Anabrus simplex* and *Metaballus litus*, two species with large spermatophores. Female competition is apparently restricted to areas where food shortage makes male spermatophores a scarce resource that limits female reproduction (Gwynne 1981, 1984a, 1985; also see Thornhill 1986a). This idea has been tested experimentally in *Requena verticalis*, an Australian species that is easily raised and mated in the laboratory. With abundant high-quality food, males can produce spermatophores rapidly, and many males are available for mating; this should eliminate the need for females to compete over males. With low-quality food, spermatophores take more time to produce, so fewer males are available for mating; females, to compensate for the low-quality food, should seek more copulations and nutrients from males via spermatophores. As predicted, when given access to members of the other sex, males mated less often if their diet was poor than if it was rich, and females showed the opposite trend (figure 5.1.1). In the wild, these differences should lead to male competition over females when high-quality food is abundant, and to female competition over males when it is rare.

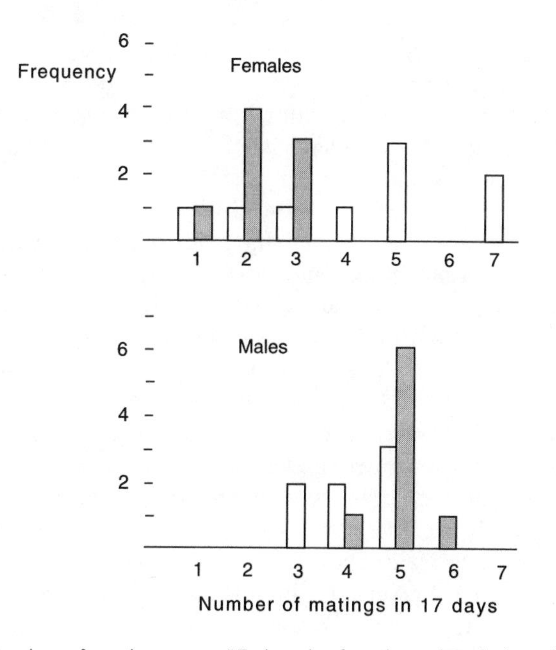

Figure 5.1.1 Number of matings over 17 days by female and male katydids *Requena verticalis* raised on low-quality (open bars) and high-quality (shaded bars) diets. (From Gwynne 1990)

These ideas were tested in the field in another Australian katydid, an undescribed zaprochiline species. Field observations by Simmons and Bailey (1990) showed that variation in sex roles was related to the food resources. During the early part of the reproductive season, this species feeds on pollen from the kangaroo paws *Anigozanthos manglesii;* later on the major food source is blackboy *Xanthorrhoea preissei*, a much richer pollen source. Females were lighter and less fecund, and males had smaller and more variable spermatophore glands during the early than during the late part of the season. Males were more critical in their choice of mate, and females were more competitive early in the season than during the late, food-rich period. These and other observations suggested that food resources influence the operational sex ratio (proportions of sexually active males to females; see section 7.2), and hence the strength of competition over mates in the two sexes (Simmons and Bailey 1990).

This conclusion was supported in a field experiment that excluded several alternative explanations, such as effects of population density or declining mating activity in aging females. At a study site in natural habitat, Gwynne and Simmons (1990) placed fiberglass-screened cages over eight clumps of flowering kangaroo paws for two weeks. Four randomly chosen cages were supplemented with extra food: bee-collected pollen offered on two honey-coated dried stalks of blackboy plants. The remaining four cages were controls and received only dried stalks. Twenty-four katydids of each sex were placed in each cage, and their behavior was observed during the evening period of calling and mating activity. The results were as predicted: in control cages, without extra food, males were much less active in courtship (fewer called). There were more female fights over males, which showed more critical mate choice, with males preferring large, fecund females. Females showed less critical mate choice than in cages with extra food. There, courtship roles were more traditional, with many calling males and more frequent female choice of mate. The opportunity for sexual selection in females (relative variance in female mating success) was higher in control than experimental cages.

Simmons (1992) took the analysis a step further by measuring the relative energy investment in eggs by males and females under two dietary regimes. When food was unlimited, females tended to invest more than males. When food was limited, however, males tended to invest more than females; this is the situation where females competed for males (see above). The result was therefore as predicted by theory (Williams 1966; Trivers 1972): when males show greater parental investment, the roles in sexual competition are reversed, with females competing for choosy males.

In sum, these experiments support the idea that the degree of competition over mates in the two sexes depends on their relative parental investment (Williams 1966; Trivers 1972; Thornhill 1986a; Gwynne and Simmons 1990; section 7.2 below). When food is abundant and male investment is of less importance for female reproduction in these orthopterans, males compete over females, which often reject courting males. When food is scarce and male investment is crucial for female reproduction, females compete more strongly for mates. Males then show less courtship and more critical choice of mate.

The result can also be viewed from the perspective of reproductive rates (Clutton-Brock and Vincent 1991; section 7.3 below). When nutritious food is scarce, there is a decline in male maximum reproductive rate, which is constrained by spermatophore production. Spermatophores then become a limiting resource for the reproduction of females, which therefore compete over males. When high-quality food is abundant, on the other hand, male spermatophore production is more rapid and no longer limits the reproduction of females, which therefore have less reason to compete over males.

5.2 Male Coloration and Female Choice in the Guppy

One of the most thoroughly studied species as regards sexual and natural selection is the guppy *Poecilia reticulata.* Its natural habitat is clear forest streams of northeastern South America and the nearby Caribbean islands. Males usually do not defend or provide any resource, but spend much time searching for receptive females and attempting to mate with them. Fertilization is internal. Females are ready to mate about once a month, but the proportion receptive at any given time is low, which creates a strongly male-biased operational sex ratio. In many populations, guppies live in dense groups along the edges of streams, with apparently strong mating competition and high display rate among males (e.g., Haskins et al. 1961; Liley 1966; Farr 1975; Endler 1978; Reznik and Endler 1982). Social organization, behavior, secondary sex coloration, and aspects of life history vary with water clarity and the types of predators present (e.g., Farr 1975; Liley and Seghers 1975; Endler 1978, 1980, 1983, 1992b; Breden et al. 1987; Reznik et al. 1990; Luyten and Liley 1991).

In a unique experimental study of evolution in the wild, Endler (1980), Reznik and Endler (1982), and Reznik et al. (1990) showed that coloration and life history traits can evolve rapidly in response to changes in the selective regime. The cichlid *Crenicichla alta* is a major predator of adult guppies. As predicted, transplanted guppies originating from streams with this predator evolved more conspicuous colors (see below in this section) and shifted their breeding to higher ages, and produced fewer but larger young. These changes took place over two years (10–15 generations) in a new environment, with predation mainly on juveniles by the killifish *Rivulus hartii*. After 30–60 generations (11 years), the life history changes were genetic, as shown by rearing guppies from the experimental and control waters in the laboratory for two generations, eliminating environmental effects (Reznik et al. 1990). This experiment shows that life history traits can undergo marked genetic evolution in surprisingly few generations.

Natural Selection of Male Color Patterns

Male guppies are highly polymorphic in coloration. They vary in numbers, size, position and reflectivity of color patches, which are controlled by many X- and Y-linked genes, expressed only in adult males (e.g., Haskins et al. 1961; Yamamoto 1975). There are two main types of color patches, formed by structural (mainly blue, iridescent, bronze) or pigment (mainly orange, red, black) colors. Some examples of the variation among males are shown in figure 5.2.1. Endler (1978, 1980, 1986b, 1987) tested the idea that the color pattern represents a balance between natural selection for crypsis by visually hunting predators, and sexual selection for conspicuousness.

The number and dangerousness of predator species increases downstream in many guppy waters. A first test of predation effects is therefore to see if the male spot pattern is related to the local fauna of predators. This is the case: the size and numbers of color spots decreased with increasing intensity of visual predation. This applied within each of two areas with different predator faunas, one of Caribbean and another of mainland origins (Endler 1978, 1983). To some extent this effect may be confounded by age-related differences in coloration, but at least some of the differences remain when this complication is considered (Endler 1986a).

Strong evidence comes from field and laboratory experiments. In the pioneering field transplantation test by Endler (1980; see above), the color patterns after two years had changed markedly in the predicted direction, converging on that of a population in a control stream where guppies coexisted with *R. hartii* but without *C. alta*.

In another test, Endler (1980, 1983) examined the evolution of color patterns over many generations in experimental populations derived from 18 sites in Trinidad and Venezuela. Ten compartments in an artificial greenhouse stream were used to study the evolution of spot patterns under different predation regimes. Four of the compartments were full-scale replicas of typical territories of *C. alta*, the main predator of adult guppies. Four other ponds were replicas of streams containing the less serious predator of mainly immature guppies, *R. hartii*. The two remaining compartments were free from predators. Whether a certain spot pattern is conspicuous or not depends on the background substrate. In each of the previous

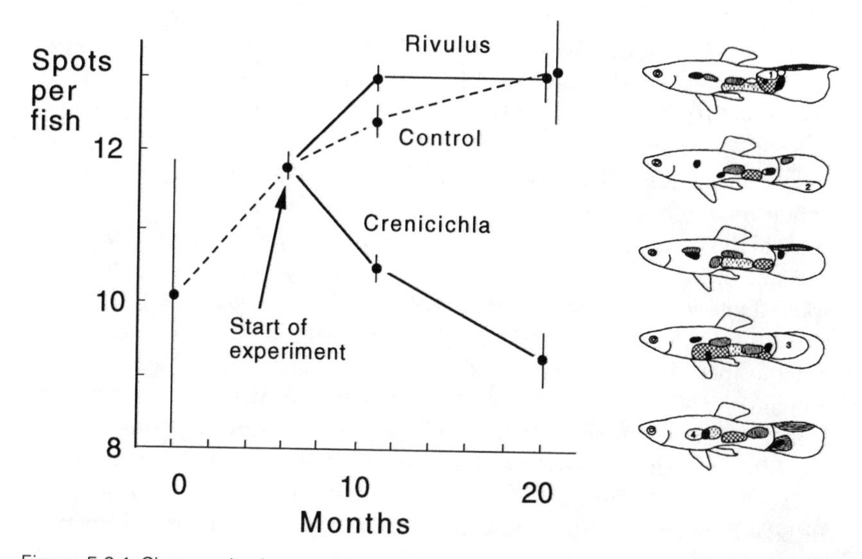

Figure 5.2.1 Changes in the number of color spots per male guppy *Poecilia reticulata* during the course of Endlers (1980) experiment, which spans about ten guppy generations. *Control* ponds had no predators; *Rivulus* ponds had six weak predators (*Rivulus hartii*); *Crenicichla* ponds had a dangerous predator (*Crenicichla alta*). Note the rapid change in spot numbers in *Crenicichla* ponds after predation started. Vertical lines are two standard errors. The guppies to the right illustrate some of the variation found in male coloration. (After Endler 1980; fishes from Houde 1988b. Color spots are as follows. Solid black: black; unshaded: white; vertical hatching: orange; horizontal hatching: yellow; cross-hatching: green; broken hatching: blue/purple; numbered spots: 1: cream; 2: brown; 3: black-and-yellow speckling; 4: brick red.)

three treatment categories, half the number of compartments had a substrate of fine gravel, the other half of coarse gravel, that is, a 3 × 2 factorial design, testing for the effects of background gravel size (fine and coarse) and predation intensity (high, low, and none).

The initial population of guppies was allowed six months to develop before predators were added. The predator ponds were then stocked with one *C. alta* or six *R. hartii* according to the treatment design. Two months (3–4 generations) later the guppy populations were censused, and all males were measured as to numbers, position, size, and colors of spots. A second identical set of measurements was taken again fourteen months after the predators were introduced (after 9–10 generations).

The number of spots had increased in the populations free from predation (figure 5.2.1) as well as in those with *R. hartii*, but decreased in those with the predators of adult guppies, *C. alta*. The trends were similar for spot size. The spot patterns in the two systems with predators converged toward the pattern characteristic for the corresponding predation regime in the wild. There was also a substrate effect: within predation treatments, populations on coarse gravel had larger color patches than populations on fine gravel (Endler 1980, 1983).

Both size and numbers of colored spots increase with age and body size, and the mean age and size of guppies should decrease with increasing predation on adults. For this reason, the reduced size and numbers of color spots in the treatments with high predation might be due to lower mean age and size of the guppies, rather than to selection by predators against color spots. Ontogenetic changes are not the whole explanation, however, because there was an effect of gravel size also within predation treatments, without there being significant differences in guppy size. Endler (1986a) emphasized that trait distributions should be measured at similar ontogenetic stages in studies of selection. An additional way of testing whether the differences in spot size and numbers reflect genetic evolution might be to breed offspring from the different categories in identical environments. If the differences persist (measured at the same ages), this is evidence for a genetic change.

Guppies show several features that should reduce predation risks associated with sexual advertisement (Endler 1987, 1991, 1992b; Reynolds 1993; Reynolds et al. 1993). Besides the adaptations in coloration and life history reviewed above, males reduce the cost of the bright colors by avoiding the most conspicuous courtship displays during high light intensity, when predation risk is greatest. Instead, they attempt less conspicuous "sneak" copulation tactics at such times. Compared to males from streams with lower predation risk, males from streams with many predators rely more on other courtship tactics than display (Farr 1975; Luytens and Liley 1985, 1991).

Sexual Selection of Male Color Patterns

In guppies under natural or near-natural conditions, sexual selection seems to occur mainly by female choice. Aggression among males is sometimes common in small experimental aquaria (e.g., Farr 1980), and may influence their mating success (Kodric-Brown 1992). It is, however, rare or absent in low-density populations in larger aquaria, and in artificial streams as well as in the wild (Farr 1975, 1980; Endler 1980, 1983; Houde 1988b). On the other hand, male courtship display is common both in nature and laboratory (e.g., Baerends et al. 1955; Liley 1966; Farr 1975, 1980, 1983; Endler 1980, 1983).

To examine the importance of male conspicuousness for female choice of mate, Endler (1983) tested the mating success of randomly drawn males from the *Rivulus* greenhouse pools. The results indicated that male color patterns that contrasted with the background were favored by female choice. Red and orange carotenoid color spots seemed to give an advantage in the initial stages of courtship, and blue structural colors in later stages. The brightness of carotenoid colors depends on diet: males with food rich in carotenoids have color spots several times brighter than those of their brothers fed a diet free from carotenoids (Kodric-Brown 1989). Bright carotenoid spots may be an indicator of feeding success and general vigor. Endler (1980, 1983) and Kodric-Brown (1985, 1989) suggested that the female preference for red and orange carotenoid colors has evolved through an indicator process based on heritable genetic variation, reflected in the foraging success of males; this suggestion remains to be tested. Nicoletto (1991) found that carotenoid pigmentation was correlated with male swimming performance in the laboratory, and therefore probably reflects male condition. There was no such correlation for tail size, another male ornament that influences female choice in the guppy (Bischoff et al. 1985; see section 4.2). An indication of the primary importance of carotenoid colors is that among more than 10,000 guppies scored for color, Endler (1983) never found one with only structural colors; some had only carotenoid colors, but most had both.

Kodric-Brown (1985) drew similar conclusions from a study of mate choice in three-chamber aquaria, with a male in each side chamber and a female in the center, recording the time she spent near each of the two males. The rate at which males perform the sigmoid courtship display is important for mating success (e.g., Farr 1980; Bischoff et al. 1985; section 4.2 above). Kodric-Brown (1985) therefore partly controlled for variation in male display rate by excluding tests where it differed by more than about 25% between the males. From these tests, a preference ranking with respect to male color patterns was calculated for the twenty females. The ranking correlated most strongly with male carotenoid colors, also when the effect of structural colors was held constant. Kodric-Brown (1985) concluded that female choice was based mainly on carotenoid pigments, even if structural colors also play a role. When male carotenoid colors were experimentally manipulated through diet, females also preferred the males with brightest carotenoid spots (Kodric-Brown 1989).

In another experiment, Houde (1987) found that females favor males with large total area of orange. There was, however, a leveling-off beyond a certain proportion of orange coloration. To make sure that females responded to orange color, Long and Houde (1989) tested female responses to orange under different colors of light. Females discriminated among males in relation to their orange coloration under white, blue, and green light. They did not do so when the light was orange, which largely eliminates the conspicuousness of orange color spots. The experiment also demonstrated the importance of interactions between ambient light color and the coloration of the fish.

The complex polymorphism in male color pattern may in part be a consequence of gene flow between guppy populations with different predator regimes and background substrates (Endler 1980, 1983). Frequency-dependent selection might also play a role. For example, males with rare types of color patterns had a mating advantage in the laboratory, which might contribute to the marked polymorphism among males. A female preference for rare males might evolve if it increases the chances that offspring will be heterozygous and better able to tolerate a variable environment (Farr 1980). The colorful gravel mosaic in natural guppy populations is also permissive of polymorphism, as there may be many equally cryptic color patterns on a complex background (Endler 1978, 1988a).

Genetic Correlation between Female Preference and Male Ornaments

Studies of different populations of guppies have provided some of the strongest evidence for joint evolution of female preference and male ornament. Preferences differ among Trinidad guppy populations; females from streams with high predation show lower preferences for orange spots than do females from streams with low predation, where males have more orange spotting (Breden and Stoner 1987; Houde 1988c; Stoner and Breden 1988; Houde and Endler 1990; also see Breden 1988; Endler 1988b). Estimating the copulation success of competing males by labeling their sperm with radioisotopes, Luyten and Liley (1991) found that males had highest success with females from their own population.

The degree of preference for orange color spots in females from seven populations was correlated with the amount of orange in males from the female's population (Houde and Endler 1990). The differences among populations remained after several generations in the laboratory, making it unlikely that they were only phenotypic. The results suggest that male carotenoid color spots and female preferences have evolved together and diverged genetically among populations (figure 5.2.2). Whether the joint evolution of trait and preference has occurred by Fisherian, indicator, or other processes, is not known (section 2.7). A strong heritability of the area of orange spots suggests that this aspect is unlikely to reflect variation in fitness and evolve by an indicator process (Houde 1992). The brightness of orange spots seems more likely to reflect condition (Kodric-Brown 1989; Nicoletto 1991).

To summarize, field and experimental studies of guppies show that conspicuous male coloration is favored by female choice and disfavored by predation. There is evidence of joint genetic evolution of female preference and male orange color spots.

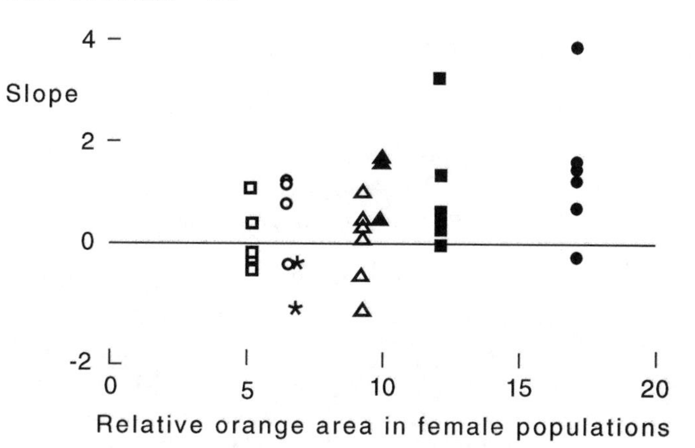

Figure 5.2.2 The degree of female preference for orange coloration in males differs among guppy populations. The preference is strongest in females from those among the seven study populations (represented with different types of symbols) where males have much orange. Each point gives the degree of preference (slope of regression line relating female response strength to total area of orange in male), for six females choosing among six males in one observation session. (After Houde and Endler 1990; see their paper for further details.)

5.3 Advertisement Calls in the Tungara Frog

Some of the most detailed studies of sexual signaling have been done in anurans. A combination of observations and experiments in field and laboratory have revealed a fascinating interplay between sexual and natural selection in the South American tungara frog *Physalaemus pustulosus* (e.g., Ryan 1980a, 1985; Ryan et al. 1982; Tuttle and Ryan 1981; Green 1990). Males gather and call from water pools visited by females, which select a mate by making physical contact with him. The pair builds a foam nest in which the female oviposits. The quality of the nest site is probably not important for female choice of mate, as the nest is built away from the male's calling site.

Males have two kinds of call, "whine" and "chuck" (figure 5.3.1a). Isolated males give only the whine. In response to vocal competition from other males they add chucks, which are more effective in attracting females. The whine appears to be the call used in species recognition, so chucks are not required for this function. The fundamental frequency of chucks decreases with increasing male size. Other aspects of the calls, such as repetition rate or intensity, were not correlated with male size, nor was any other measured aspect of male behavior (Ryan 1980, 1985; Rand and Ryan 1981).

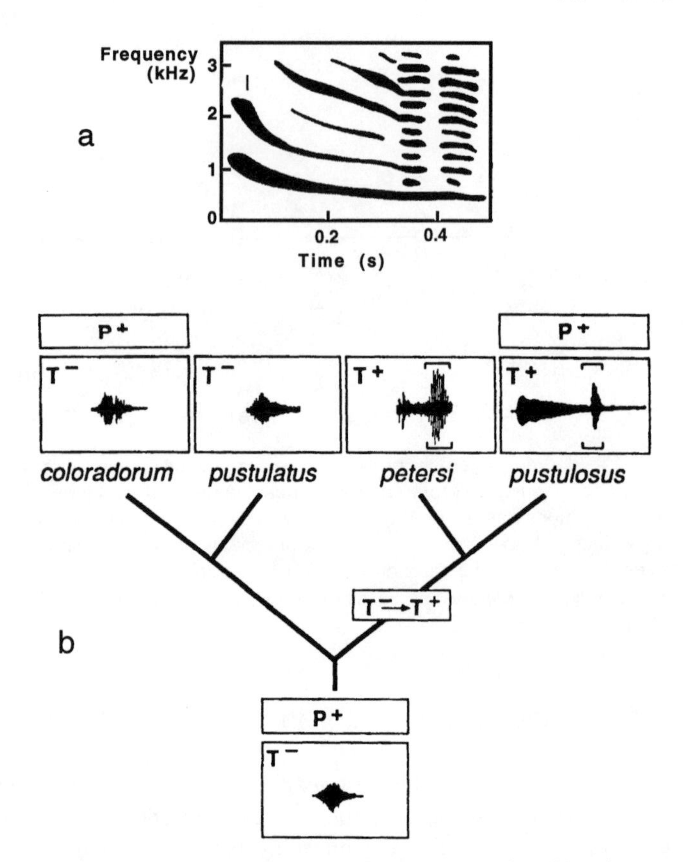

Figure 5.3.1 (a) Sonogram of advertisement call of the tungara frog *Physalaemus pustulosus* containing a whine and two chucks. (b) Estimated phylogeny of male chuck call (T+) and the female preference (P+) in *Physalaemus* frogs (T− = no male chuck). Only the closely related *P. pustulosus* and *P. petersi* add chucks to their calls (marked by square brackets in the oscillograms), so the chuck likely evolved in their immediate common ancestor. Females of *P. pustulosus* as well as *P. coloradorum* prefer calls with chucks, suggesting that the common ancestor of all four species had the preference. If this interpretation is correct, the preference evolved before the chuck itself. (From Kirkpatrick and Ryan 1991)

The proportion of unhatched eggs increased with the difference in size between male and female (she is usually largest). It should therefore be adaptive for most females to mate with a male of larger than average size. To see whether females discriminate between chuck calls of different frequencies, Ryan (1980, 1985) tested the response of females to calls with identical whine components but (synthesized) chucks of different fundamental frequencies. Each female was tested in an arena with two opposite loudspeakers sending different calls (see section 14.5 for a discussion of playback experiments). Females usually approached the speaker that gave the lower fundamental frequency in paired comparisons of 210 and 250 Hz, a response that should favor the larger of two males. In accordance, male mating success in the field increased with body size. Neurophysiological analyses showed that the frequency tuning of the female inner ear is biased toward low-frequency components of the chuck call, which may explain the preference for the low-frequency chucks from large males. Chuck calls also have a greater frequency range than whines, and stimulate both of the two sound receptors of the ear, the amphibian and the basilar papilla. This may be the proximate reason why calls that also contain chucks are most effective in attracting females (Ryan et al. 1990b; Ryan and Rand 1990).

Since chuck calls help attract females, one may ask why solitary males give no chucks, and what prevents chucks from being used more often in groups of males. One opposing selection pressure is bat predation: fringe-eared bats *Trachops cirrhosus* prey heavily upon male tungara frogs at chorus pools (Rand and Ryan 1981; Tuttle and Ryan 1981; Ryan et al. 1982). Loudspeaker tests showed that the bats respond to calls from edible frogs. The bats choose the recorded call of an edible over that of a poisonous species, and the call of a small over that of a very large species. Moreover, the bats were more attracted to calls that contained both chucks and whines than to whines only (Tuttle and Ryan 1981; Ryan et al. 1982). Bat predation therefore selects against the advertisement call of tungara frogs; predation works most strongly against chucks, the most effective mate attraction signal. This may explain why single males give no chucks, and why chucks are not more common than they are in choruses. Males respond to the presence of hunting bats in several ways, for example by ceasing to call, diving, and swimming away. In addition to bats, there are also other, less important predators of calling males (Ryan 1985).

How has the female preference for the chuck evolved? Ryan et al. (1990b) and Ryan and Rand (1990) suggested that Fisherian and indicator processes were refuted by the results, and that the female preference has not evolved together with the call. Instead, the male call may have been favored by selection because it happens to fit a preexisting bias in the female auditory system evolved under other selective forces (section 1.5). In support, Ryan (1990b) and Ryan et al. (1990b) pointed to the existence of a similar sensitivity to chuck calls in the related *P. coloradorum*, although males of this species do not produce chucks. The authors suggested that both species inherited the similar basilar papilla tuning from a common ancestor, and that only the ancestor of the tungara frog and its sister species *P. petersi* evolved a chuck call that takes advantage of the bias (figure 5.3.1b). An alternative interpretation is that the common ancestor of all four species in figure 5.3.1b had both the preference and the chuck call, which was then lost in the left branch, leading to *P. coloradorum* and *P. pustulatus* (Gardner 1990; also see Pomiankowski and Guilford 1990).

This interpretation would be supported if other closely related species also had "chuck" calls, but this does not seem to be the case (see Ryan 1985).

In sum, the female sensory characteristics favor male "chuck" calls in the tungara frog; bat predation selects against the same call. How the female preference evolved is debated; one possibility is that it arose before and independently of the favored male call.

5.4 Plumage Ornaments in the Red-winged Blackbird and Long-tailed Widowbird

Conspicuous colors and other plumage ornaments of birds have played a prominent role as likely examples of sexually selected traits since the birth of the theory. Experimental and other studies have now begun to reveal the mechanisms by which such traits are selected.

The North American red-winged blackbird *Agelaius phoeniceus* and the African long-tailed widowbird *Euplectes progne* are strikingly and similarly sex dimorphic in plumage coloration. Females are streaked brown (but see Whittingham et al. 1992); males are black except for a bright red wing epaulet, identical in the two species even in details such as the yellow lower margin, and the coverability of the epaulet by black scapular feathers. The black plumage makes the male conspicuous in the open habitat; both species have song flights that enhance the effect of the plumage.

Wing Epaulet

Males expose the epaulet during courtship and threat (see Orians and Christman 1968). Its function has been studied experimentally in the red-winged blackbird by dyeing the epaulet black in some territorial males, or by cutting away the red feathers (Peek 1972; Smith 1972; Morris 1975). Most of the altered males lost their territories, while control males kept theirs; there were many more intruders on the territories of the altered males (figure 5.4.1). Blackened males defended their territories by normal threat signals; the intruders avoided direct contact with the defender but did not leave (Peek 1972). Smith (1972) used a colorless solvent in painting the control males; among the over one hundred males, more than 90% of the controls kept their territories, but only one third of the dyed males did so. The two other studies gave similar results. As the territory is crucial for mating success (e.g., Searcy and Yasukawa 1983), the epaulet therefore seems to be favored by sexual selection through male contests over breeding territories (but see Eckert and Weatherhead 1987a,d). The epaulet may also be affected by natural selection, for example if territory quality influ-

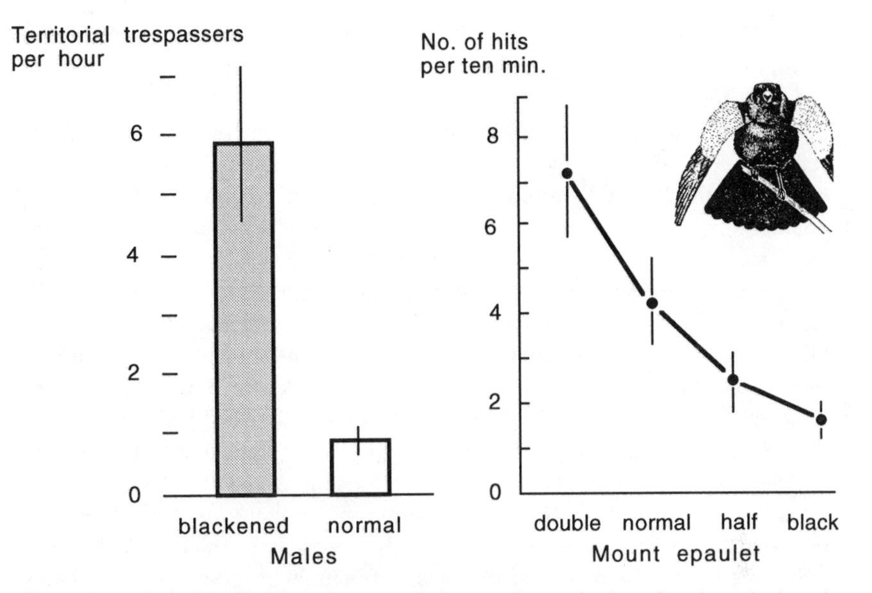

Figure 5.4.1 *Left*: Rates of intruder trespassing on the territories of male red-winged blackbirds *Agelaius phoeniceus* with blackened and normal epaulets during the premating period (data from Peek 1972, table I). *Right*: Number of hits on mounted red-winged blackbird males by territory owners in relation to the size of the red epaulet of the mount. Bars = s.e. (Data from Hansen and Rohwer 1986, table 2. Drawing of male in song posture with exposed epaulets by G. M. Christman in Orians and Christman 1968.)

ences the survival of offspring, or if the epaulet gives an advantage in contests over roost sites or other nonbreeding resources (Rohwer 1982; Weatherhead and Hoysak 1984).

Precisely how does the red epaulet help keep out intruders? Why did the intruders not leave the territories of blackened males? One possible reason is that the epaulet is an integral part of the threat signal. If no red epaulet is displayed, an intruder may not recognize the signal as threat. Second, an intruder might treat a defender without the red epaulet as another species (Peek 1972).

To test its function, Hansen and Rohwer (1986) changed the size of the epaulet. They used four categories of mounted redwing dummies to study the response of territory owners: (1) completely blackened males, the epaulet being painted with black hair dye; (2) males with half the epaulet blackened; (3) normal, unpainted males; (4) males with double-size epaulets, created by adding feathers alongside the normal epaulet. The species recognition hypothesis was refuted by tests where a blackened mounted red-winged blackbird was placed in redwing territories together with a mounted male of the all-black and sympatric Brewer's blackbird *Euphagus cyanocephalus*. Territory owners attacked the redwing much more, and clearly distinguished it from Brewer's blackbird, so the red epaulet is not required for species recognition in this case.

There was a clear trend in the response of territory owners to mounted red-winged blackbird males of the four categories. The response was weakest against the all black mount, and strongest against the mount with double epaulet size (figure 5.4.1). Larger epaulets clearly provoked more aggression from territory defenders. Why should this be so? The epaulet may be a "coverable badge" used for threatening opponents (Rohwer 1982; Hansen and Rohwer 1986). Field observations showed that an intruder usually conceals most of his epaulet and does not challenge the owner, should the site be occupied. A male ready to fight for the territory, on the other hand, exposes his epaulet. The owner therefore has more reason to attack a challenging intruder with fully exposed epaulets than one with covered epaulets. This may explain why owner attacks increased with the size of the intruder's epaulets. In accordance, floating males that took over the empty territories in a removal experiment first kept their epaulets covered, but began exposing them within minutes. After half an hour the new males acted as legitimate owners with fully exposed epaulets and song (Hansen and Rohwer 1986).

The blackened males in the experiments of Smith (1972) and Peek (1972) may have lost their territories in part because not showing the red epaulet, they failed to give the normal signal that the territory is occupied. Røskaft and Rohwer (1987) provided support for this explanation by replacing territory owners with mounted redwings with epaulets of different size: blackened, normal, and double size. Intruders avoided the mounts in proportion to the size of the red epaulet, as predicted by the coverable badge hypothesis. In groups of captive red-winged blackbirds, dominance status was also correlated with epaulet size (Eckert and Weatherhead 1987b). The red epaulet therefore is used as a threat signal; its size may be limited by the need for the epaulet to be coverable with black scapular feathers (Røskaft and Rohwer 1987). Clipping experiments suggested that also territory owners benefit from being able to conceal the red badge, since this facilitates trespassing on the territories of other males, and reduces unnecessary aggression with neighbors (Metz and Weatherhead 1992).

Besides in male contests, the epaulet might be favored by female choice of mate. Smith (1972) suggested that this is not the case, as dyed males that kept their territories also paired and bred successfully. Peek (1972) and Morris (1975) remarked, however, that such males had some difficulty in attracting females. The dyed males that kept their territories may also have been better than average competitors, so Smith's result does not completely rule out a role of the epaulet in mate choice.

Female choice in the red-winged blackbird may concern direct material benefits. In several populations, females favor males partly in relation to territory quality (e.g., Searcy 1979a; Lenington 1980; Yasukawa 1981c; Yasukawa and Searcy 1986; Searcy and Yasukawa 1989). For example, as suggested by the polygyny threshold model (section 8.4), male harem size is correlated with certain habitat features, which also correlate with female reproductive success (Lenington 1980; Picman 1980). Female red-winged blackbirds therefore seem to benefit by choosing a male in relation to aspects of his territory. Male parental care, such as nest defense and feeding of young, may also affect female choice. Additional studies are needed that control these factors independently and clarify their relative importance, which varies among populations (Yasukawa 1981c; Searcy and Yasukawa 1983; Eckert and Weatherhead 1987c).

According to the "unprofitable prey" hypothesis of Baker and Parker (1979; section 13.6 below), male red-winged blackbirds should display the epaulet to predators. This is not the case: when there is a predator nearby, the male conceals his epaulet, and there is no evidence that he uses it as a signal to predators. Field observations show that the male exposes his epaulet in territory defense and courtship (e.g., Orians and Christman 1968; Yasukawa 1978, Røskaft and Rohwer 1987).

In Cuban red-winged blackbirds, females are more similar to males in plumage coloration (black), song, and body size. Further, study of sex roles and mating systems are needed to clarify the reasons for reduced sexual dimorphism in this population.

Long Tail

The male long-tailed widowbird has one of the most remarkable ornaments among passerine birds: a tail that is often more than half a meter long whose function Andersson (1982a) tested by changing its length. Early in the breeding season, the territories of thirty-six males were mapped, and their numbers of nests, which females weave in 0.5–1 m high grass, were counted. In a randomized block experiment, the color-ringed males were partitioned among nine groups with four males in each, as similar as possible in territory quality and tail length (figure 5.4.2). The tail was cut to about 14 cm in one of the randomly drawn males within each group. Each removed feather was glued to the corresponding feather of another male, elongating his tail by 20–30 cm. The two other males were controls. One of them was ringed only; the tail feathers of the fourth male were cut off at the midpoint and glued back again.

Before the test, the number of nests on the territories of the four kinds of males was as shown in the upper part of figure 5.4.2; the number of new nests built on each territory after tail treatment is shown in the lower part. Using each male as his own control, by subtracting the number of nests

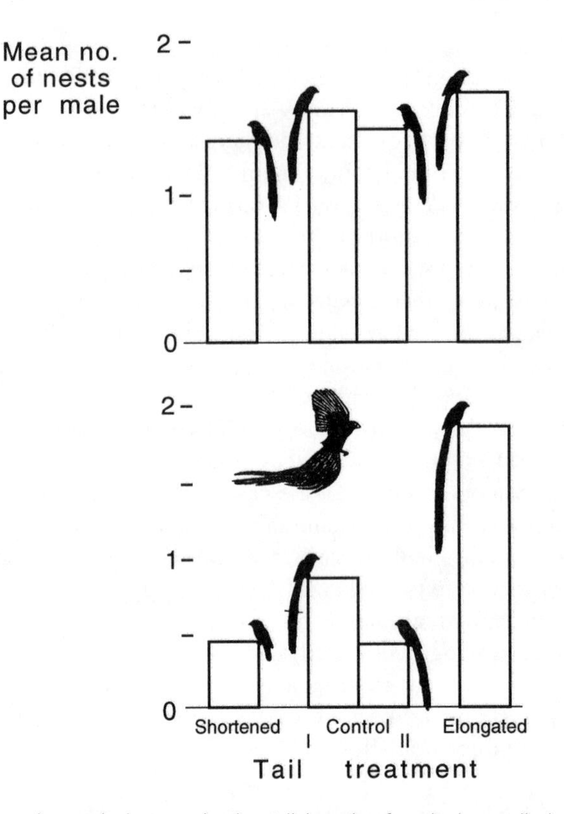

Figure 5.4.2 Experimental changes in the tail length of male long-tailed widowbirds *Euplectes progne* influence their attraction to females. The tail is expanded into a deep keel during the advertising song flight of the male, performed in the presence of females. Bars show the mean number of active nests per territory in each of the four groups of males, before the changes in tail length (*top*), and the number of new nests after the changes (*bottom*). Before the experiment, success was approximately equal in the four groups. After tail treatment, the attraction of females to the territory increased with the manipulated tail length of the male. (After Andersson 1982a)

present before treatment, reduces the influence of initial variation among males and territories. There was a trend in success, increasing from shortened to elongated males when the four groups were tested together. In pairwise tests, elongated males were more successful than both shortened males and control males of type II. The result indicates that the long tail is favored by sexual selection through female choice of mate. Females seem to prefer supernormal tails, as the elongated males had highest success. Two alternative explanations were refuted: shortened males did not become less active in courtship display, nor did they give up their territories more often than other males. The display of the tail also suggests that it is used in attracting females, not in contests among males. It is not expanded in the flight display used during territorial contests. During the advertising song flight, however, performed when females fly past or visit his territory, the male expands the tail into a deep keel while cruising over his territory in slow, conspicuous display flight (figure 5.4.2).

Why do females favor a long tail? A simple explanation is that the expanded tail enlarges the lateral surface of the male 2–3 times, and makes him visible from farther away in the open grassland. But this is probably not the whole explanation: females spend much time in the nesting area before breeding, and their home ranges are much larger than the male territories, so females can compare many males and their territories before settling to breed.

The function of the epaulet in the long-tailed widowbird has not been tested, but its use resembles that in the red-winged blackbird; the epaulet is displayed in threat and courtship on the territory. The two most conspicuous ornaments in the male long-tailed widowbird therefore may be favored in part by different mechanisms of sexual selection: the tail by female choice, the red epaulet mainly by male contest competition. Experiments and observations suggest that this is the case in the male malachite sunbird *Nectarinia johnstoni* (Evans and Hatchwell 1992a,b; section 13.6 below).

To summarize, the red epaulet in the red-winged blackbird is sexually selected by male contest competition over breeding territories, and perhaps also naturally selected by competition over nonbreeding resources. There is no clear evidence of female choice in relation to the epaulet. The tail of the male long-tailed widowbird is favored by female choice.

5.5 Sexual Size Dimorphism in Elephant Seals

Elephant seals (*Mirounga* sp.) are among the most sex dimorphic of all mammals. Adult males of the northern elephant seal *M. angustirostris* weigh more than three times as much as females; the sex difference is even larger in the southern species *M. leonina* (e.g., Deutsch et al. 1990). What are the reasons for this huge dimorphism in body size?

The answer is apparently found in the mating system and the forms of competition over mates: male contests and endurance rivalry. Female elephant seals gather on beaches during the three-month reproductive season to give birth and mate before returning to sea. The females form dense clusters, and dominant males can defend large groups. In this highly polygynous system, only large males are able to achieve top rank and high mating success. Small adult males and young males are forced by attacks to stay outside of breeding groups, obtaining few or no copulations (Bartholomew 1952, 1970; Laws 1956; Le Boeuf and Peterson 1969; Le Boeuf 1974; McCann 1981; Deutsch et al. 1990). Less than one third of the males resident on the beach copulate during a breeding season, and most matings are done by a few males. The pattern is similar in the southern elephant seal

(McCann 1981). The dominant bull interrupts copulations by others and chases away the intruder. When groups contain more than about fifty females this is not always possible, and other males obtain some matings, sometimes without being noticed by the dominant bull (Le Boeuf 1974; McCann 1981).

Many males die before reaching adulthood, and among those surviving, most never achieve high enough rank to mate. Those that do so, on the other hand, mate with many females; 8 males inseminated an estimated 348 females (Le Boeuf and Reiter 1988). Reproductive success is much more concentrated to high age (and large size) in males than females (figure 5.5.1).

Behavioral studies show that dominance greatly affects male mating success, and the highest-ranking males are among the largest present; large male body size therefore seems to be strongly favored by sexual contest competition (Bartholomew 1952, 1970; Laws 1956; Le Boeuf and Peterson 1969; Le Boeuf 1974; Sandegren 1976; Cox and Le Boeuf 1977; Cox 1981; McCann 1981; Le Boeuf and Reiter 1988; Deutsch et al. 1990). An-

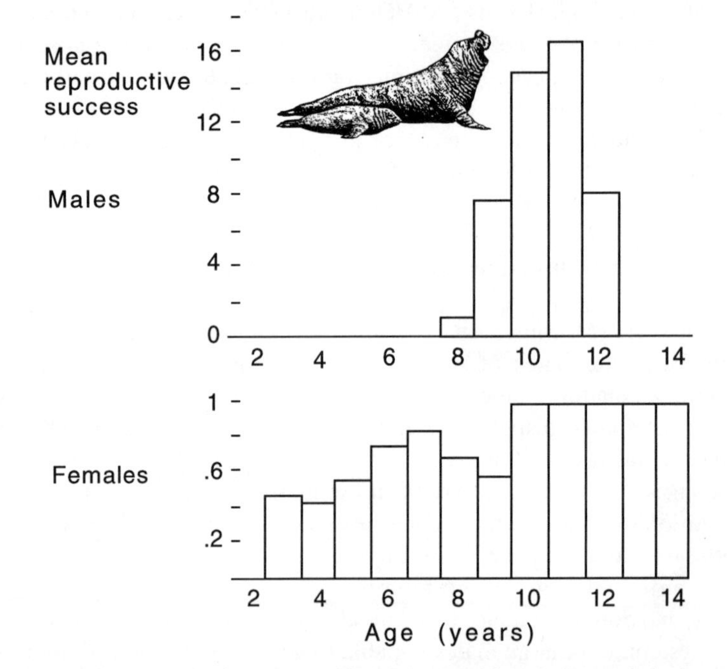

Figure 5.5.1 Mean reproductive success of male (top) and female (bottom) elephant seals *Mirounga angustirostris*. Success is much more concentrated to high age and large size in males than females. Success = estimated numbers of pups sired and weaned for males, and pups weaned for females. The values above ten years for females are extrapolated from cohorts. (After Le Boeuf and Reiter 1988. Drawing of elephant seals by Diane Breeze in Greenwood and Adams 1987)

other contributing factor may be the longer breeding season in elephant seals than in many other pinnipeds, favoring male endurance. In general, the capacity for energy storage should increase with body size more rapidly than metabolic costs (Calder 1984). Large size and stored energy in form of blubber enables a male to remain for a long period on the beach (91 days on average) and inseminate many females before going back to sea to feed (Bartholomew 1970; Deutsch et al. 1990). When doing so, a high-ranking male has lost on average 41% of his body mass, compared to 34% for low-ranking adult males. The relative mass loss of reproductive females is of similar magnitude, but the loss rate is much higher in lactating females, which stay on the beach only one third as long as males (Deutsch et al. 1990). The scaling of fasting ability with body size needs to be measured empirically to test whether large body size is advantageous in this respect, as seems likely.

In the northern elephant seal, females have some scope for mate choice. Females usually protest against being mounted, especially by subadult males, doing so by loud calls, evasive movements of the hindquarters, and other avoidance behavior. The calls alert nearby males, and the dominant bull usually discovers, attacks, and chases away other males that try to mate. Females protest less against dominant males. (Female elephants *Loxodonta africana* behave similarly; Poole 1989a). Cox and Le Boeuf (1977) and Cox (1981) suggested that the protest behavior raises the female's chances of mating with a male of superior genotype. Another possibility is that there are direct benefits to the female from seeking shelter near the top-ranking male and mating with him, for example a lower risk of being injured or killed during sexual harassment by peripheral males, as sometimes happens (Le Boeuf and Mesnick 1990). These alternatives remain to be tested.

To sum up, the strong sexual size dimorphism in elephant seals is apparently a consequence of sexual selection favoring large male size in contests over females, and perhaps also in endurance rivalry.

5.6 Secondary Sex Differences in Red Deer

With its large male antlers and sexual size dimorphism, stags being twice as heavy as hinds, red deer *Cervus elaphus* typify the sex differences that made Darwin develop his ideas on sexual selection. Red deer are the subject of a detailed and long-term field study in the Scottish island of Rhum. Clutton-Brock et al. (1982, 1988a) summarize 16 years of work, which covered the lifetime reproduction of several year classes of deer. The open environment of Rhum permits detailed records of behavior, yearly mating success, reproduction, and survival, providing estimates of lifetime repro-

Table 5.6.1
Relative Contribution (%) of Three Components of Lifetime Reproductive Success to Its Variation among Red Deer Hinds Reaching Breeding Age, and among Stags Reaching Five Years of Age[a]

Component	*Hinds*	*Stags*
Reproductive life span	27	7
Fecundity/mating success[b]	8	32
Offspring survival	57	20

[a] Modified from Clutton-Brock et al. (1988a).
[b] Fecundity for hinds, mating success for stags.

ductive success. (A DNA fingerprinting study verified that observations of harem membership and matings accurately identified male relative reproductive success, but underestimated the variance in true success; Pemberton et al. 1992).

Red deer are polygynous, and temporary groups of hinds visit attractive grazing areas defended by individual stags during the rutting season in October. Variation in lifetime reproductive success is larger in stags than hinds, and it is related to different factors in the two sexes. Variation among hinds depends on differences in life span and, particularly, offspring survival, but not on access to mates (table 5.6.1). Instead, female reproduction seems to vary mainly in relation to food resources and the size of the matrilineal group (daughters tend to remain with their mothers). This is suggested by variation in birth weight among years, by a reduction in fecundity that paralleled an increase in population density, and by lower reproductive success of hinds in large than in small groups. Much of the variation in success among hinds apparently had environmental origins. There is an advantage of large body size in hinds during dominance interactions over food; dominant hinds had much higher lifetime reproductive success than subordinates (Clutton-Brock et al. 1988a).

Variation in lifetime reproductive success among stags, on the other hand, depends mainly on mating success, but also on offspring survival (table 5.6.1). These results suggest that reproduction in hinds is limited mainly by food, and in stags by competition over females. Male success therefore depends strongly on the ability to win fights, which in turn increases with body size. Indirectly, male success also depends on the foraging resources of the stag's mother during his fetal and early juvenile growth, since adult body size is correlated with early growth rate.

Estimated fertilization success was closely related to the size of the harem held by a stag, and to the number of days he held it (Clutton-Brock et al. 1979). Gibson and Guinness (1980a,b) found a strong correlation in

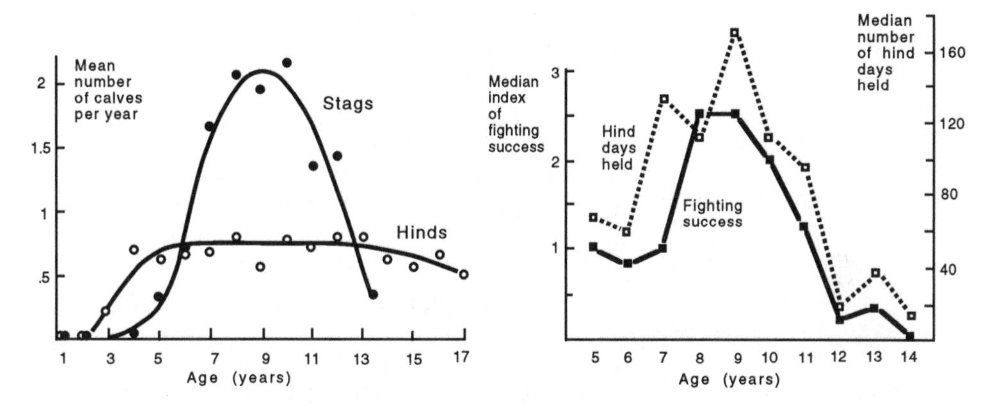

Figure 5.6.1 *Left*: Age-specific reproductive success in red deer *Cervus elaphus* hinds and stags. Success is more concentrated to prime age in stags than in hinds. *Right*: Median fighting success and hind-holding success in relation to age in stags that are over five years old. (After Clutton-Brock et al. 1982; drawing of deer by P. Barrett)

stags between estimated lifetime reproductive success and mean number of hind-days per year. Male reproductive success is strongly age-dependent and peaks at 7–11 years (figure 5.6.1). Older, senescent stags have low success. Breeding is more even among hinds, which reproduce well from 5 to at least 13 years.

What determines a stag's ability to hold a harem? Besides female avoidance of young stags, there is no evidence of female choice among the stags at Rhum (Gibson and Guinness 1980a,b; Clutton-Brock et al. 1982). (This does not mean that female choice is absent in this species; see Lincoln 1972; McComb 1987, 1991; section 14.3 below). Behavioral observations showed that harem size as well as timing and holding duration correlate with fighting success and dominance rank among stags, estimated from interactions during the rut (Clutton-Brock et al. 1979; Gibson and Guinness 1980b). Reanalyzing the data by the multivariate approach of Lande and Arnold (1983), Gibson (1987) reached partly different results. Selection gradients suggested that, as before, dominance was most important for mating success. Next came the duration of the rutting stage of the stag, putting a limit on the time during which he can defend a group of hinds. Male endurance rivalry therefore seems to play a role; perhaps it favors large body size through nutrient storage capacity. Rutting area was less important, in contrast with earlier conclusions.

A stag's reproduction therefore depends on his fighting ability and capacity of dominating other stags, which in turn is related to his age and size (fig. 5.6.1). Stags gain body mass until about 6 years of age; mass is then roughly stable until the age of 11, when it declines again (Mitchell et al. 1977). This weight curve roughly parallels that of fighting and reproductive success, suggesting that weight and condition, not mainly age and experience, are the most important variables.

As a consequence of sexual selection for large size and fighting ability, stags may have become energetically less well adapted and more likely to succumb during ecological stress. There are indications that hinds survive better than stags during such periods in this and other ungulates (e.g., Ahlén 1965; Owen-Smith 1993; reviewed by Clutton-Brock et al. 1982, 1985).

Fighting is costly, sometimes even lethal; an estimated 23% of the stags on Rhum show some sign of injury during the rut, and up to 6% are permanently injured. Fights during the rut are a common cause of death among stags in some European populations of red deer (Heptner et al. 1961; Müller-Using and Schloeth 1967). Males might reduce these risks by avoiding fights with clear superiors. It has been suggested that stags assess each other by antler size. Removal of antlers reduces fighting success and dominance rank, so antlers seem important for male competitive ability (Lincoln et al. 1970; Lincoln 1972). Antler mass (but not length) is correlated with fighting success, but it is also correlated with body mass (Clutton-Brock et al. 1982). In addition, there are better predictors of strength than antlers. The decline in condition and fighting ability during the rutting season is not reflected in antler size, but it is reflected in roaring rate and other display used in contests. Playback tests at Rhum showed that stags reply in proportion to the roaring rate of the perceived rival, apparently trying to outroar him—which succeeded against all but the most powerful, heavy playback equipment (Clutton-Brock and Albon 1979). Roaring rate in the tests was correlated with natural fighting success. In addition, high roaring rates seem to be favored by female choice (McComb 1988, 1991; section 14.3 below).

In sum, competition over mates is slight or absent in hinds, but strong among stags which, if successful, defend groups of females for a period during the rut. Contest competition over hinds favors large male body size; it probably also favors antlers as well as roaring and other threat display, but the evidence in these cases is weaker.

5.7 Summary

The case studies described here illustrate the variety of organisms, problems, approaches, and results of empirical work on sexual selection. Studies of katydids have clarified some of the factors that determine the strength of sexual selection in males and females. Work on guppies shows in great detail how conspicuous male colors are favored by female choice, and disfavored by predation; the tungara frog work does the same thing for male calls. Studies of red-winged blackbird and long-tailed widowbird show that male plumage ornaments in some cases are favored by male contests, in other cases by female choice. Large male body size and weaponry in elephant seals and red deer are favored by male contest competition, large body size probably also by endurance rivalry.

6 Empirical Studies of Sexually Selected Traits: Patterns

6.1 Introduction

There are now many quantitative demonstrations of sexual selection of various traits. This chapter reviews patterns that emerge from such studies as regards mechanisms of selection, traits selected, methods used, and groups studied. More than a century after the birth of the theory, the study of sexual selection has entered its own runaway phase, with dozens of new papers being published each month. Table 6.A (at the end of this chapter) summarizes all examples I found in the literature up to and including 1990, but my review no doubt has overlooked some relevant work. Yet the 232 studies listed should give at least a rough picture of some patterns in the empirical study of sexually selected traits.

The main criteria for the inclusion of a study in table 6.A are that it demonstrates a statistically significant relationship between some character and mating success, and that it identifies a mechanism of sexual selection. Most studies in the table have examined mating success in relation to quantitative variation in some trait(s). In a few cases where they supplement quantitative work, qualitative analyses based on presence versus absence of a trait have been included, but the table otherwise does not review such studies. Besides those listed, many other studies provide statistical evidence of sexual selection of a trait, but little or no evidence on the mechanism; yet other studies document variation in mating success without relating it to any particular trait.

Studies vary in their criteria for mating success; some do not follow success up to copulation. Least complete in this respect are the tests that measure female attraction to isolated male traits, for instance loudspeaker tests broadcasting natural or synthetic calls. Such experiments help reveal which aspects of the calls may be important for attracting mates, but they need support from field observations to verify that the conclusions also apply in nature.

Mating is far from synonymous with fertilization; sperm competition between males is common in many species (reviewed in Smith 1984; Eberhard 1985; Birkhead and Møller 1992). Genetic analysis of paternity is therefore desirable to ensure that mate attraction and copulation are reflected in production of offspring; few studies have yet done so (section

4.1). But even a measure of success based on fertilization is incomplete: what happens to offspring before or at adulthood may depend strongly on rearing conditions and phenotypic or genetic aspects of the parents (e.g., Clutton-Brock et al. 1988a). For these and other reasons, patterns of mating success in some of the examples in table 6.A are probably not reflected in reproductive success or genetic representation in future generations. For example, in the song sparrow *Melospiza melodia*, there was no correspondence between lab results on the stimulatory effect of male song repertoires, and male reproductive success in the field (Searcy 1984). In the Darwin finch *Geospiza fortis*, almost all young born in one year died due to a severe drought, so reproductive birds had no more surviving offspring than others that year (Price 1984b). For such reasons, field studies should follow reproductive success beyond the mating stage if possible. Unusually complete in this respect are, for example, studies of the great tit *Parus major* (McGregor et al. 1981), Darwin finches *Geospiza conirostris* (Grant and Grant 1987), and red deer *Cervus elaphus* (Clutton-Brock et al. 1988a), which measured recruitment of young into the breeding population.

A common difficulty is to separate the roles of mate choice and contests over mates. In some cases, experiments permit control of a mechanism such as contests, for example by isolating rivals out of reach or sensory contact with each other.

Indirect selection can lead to problems of interpretation: a trait correlated with mating success may be indirectly selected through correlation with some other, directly selected trait. For example, age and experience are not controlled in many studies; they might influence the outcome. The difficulty can in part be reduced by multivariate analysis and controlled experiments (chapter 4). Some of the studies in table 6.A have probably identified traits that are indirectly rather than directly favored by sexual selection. This does not mean that sexual selection is less important for such traits: indirect selection also affects their evolution. To the extent that the goal is to clarify mechanisms, however, direct and indirect effects should be separated. For example, the high mating success of large male wood frogs *Rana sylvatica* is perhaps not a direct effect of overall body size. The length of the arm, which is correlated with body size, may be the main directly selected trait that determines how well a male can hold on to the female when he is attacked by competitors (Howard and Kluge 1985). In accordance with this idea, males have longer arms than females of the same size. Although such problems may apply in particular to observational field studies, these on the other hand represent more natural conditions and may well provide more relevant evidence about sexual selection in the wild than do some of the better controlled laboratory experiments.

For a mechanism to qualify as *mate choice* in table 6.A, the behavior of the choosing sex must affect the mating success of members of the other sex in relation to their expression of a trait. In *contests*, some form of direct communication or physical contact between the competitors affects mating success. In *scrambles*, the order of contact with the potential mate influences success. In *endurance rivalry*, the length of time spent seeking, attracting, or defending mates influences mating success. Male contest competition is often identified by direct observation of fights or other interactions over mates or breeding territories. The clearest evidence of mate choice comes from studies that prevent contests by the design of the test. Loudspeaker and model experiments where no rival males are present offer the strictest control in this respect, albeit at the cost of lower realism: simple phonotactic responses to loudspeakers may not properly reflect mate choice under natural conditions (e.g., Gerhardt 1982, 1988; Butlin and Hewitt 1986). A combination of field observations and experiments can provide strong evidence on the mechanisms of selection (Endler 1986a; Searcy and Andersson 1986).

Mechanisms listed in table 6.A represent conclusions by the author(s) of the studies. The basis for these conclusions is sometimes weak, for example in some reports of female choice in relation to male size, where male contests were not excluded; whether they were so is indicated in the table.

6.2 Patterns

Table 6.A lists quantitative evidence of sexual selection of various traits in 186 species. Some of them have been analyzed in several respects and papers; numbers in the following summary are based on the units formed by the 232 rows of the table.

Four groups dominate: insects, fishes, amphibians and birds (table 6.2.1). As expected from their diversity and suitability for this kind of work (Thornhill and Alcock 1983), insects form the largest proportion of species. The high representation of birds and fishes is probably related to their often visually conspicuous sexual dimorphism and reproductive behavior. Among anurans, the simple calls are amenable to experimentation, and size-dependent mating success in some species can be readily quantified in the field; these aspects may explain the large number of anuran studies. Sexual selection has been examined at depth in only few reptiles. Many mammals depend on chemical communication not easily perceived by a human observer, which may partly explain their low representation. The very low number of lizards is more surprising, considering their importance in modern ecological work (e.g., Huey et al. 1983), and their often strong sexual dimorphism in size, morphology, coloration, and behavior (e.g., Stamps 1983).

Table 6.2.1
Methods of Study

Group	No. of Species	No. of Entries[a]	Field Obs. Only (FO)	Field Study (FO) and Experiment (FE or LE)	Lab Study Only (LO or LE)	Loud-speaker Test (LT)	Copulation Success (CS, SS or RS)[b]
Insects	55	63	19	24	20	7	49
Other arthropods[c]	7	7	2	4	1		4
Fishes	27	36	8	10	18		26
Amphibians	32	42	12	11	19	21	9
Reptiles	5	5	3	2			4
Birds	48	64	26	18	20	14	23
Mammals	12	15	11	3	1		10
Totals	186	232	81	72	79	42	125

[a] All sums, except no. of species, are calculated with the rows of table 6.A as units. See end of table 6.A for abbreviations.

[b] Amplexus in anurans is not included here, as pairs in amplexus may split before spawning, for example owing to struggles among males.

[c] Five crustaceans, one spider.

Types of Study

Among the 232 studies (rows) in table 6.A, roughly one third are based on field observations only, another third on field work and experiments (in field or lab), and the remaining third on laboratory work only (observations or experiments). The distribution is summarized in table 6.2.1. The highest proportion of laboratory studies is found in fishes, probably because they are less easily observed in the wild than most other groups, and in anurans, where loudspeaker tests of calls is the most common type of study.

Mating success was followed up to copulation or spawning in 125 of the 232 cases; the remainder used less direct measures. Only 14 of the studies in table 6.A followed success through to production of offspring.

Selection Mechanisms

One hundred sixty-seven of the studies concluded that female choice favors the selected character (table 6.2.2). Seventy-six of these studies had a design that prevented contests over mates (for example, loudspeaker playback of calls). Male choice of mate was recorded in 30 studies, male contests in 58, and female contests in only two. The pattern supports the traditional view that females are more choosy and males more competitive. Female contests may, however, not have been looked for to the same extent

Table 6.2.2
Selection Mechanisms

*Group**	*Female Choice (FCH*	*Female Choice, Contests Prevented (FCH, and CP or LT)*	*Female Contests (FCO)*	*Male Choice (MCH)*	*Male Contests (MCO)*	*Male Scramble or Endurance Rivalry*
Insects	42	22	1	9	19	1
Other arthropods	1			4	4	
Fishes	29	13		9	6	
Amphibians	27	20		5	11	7
Reptiles	3	1			3	
Birds	55	18	1	3	10	3
Mammals	10	2			5	3
Totals	167	76	2	30	58	14

* All sums are calculated with the the rows of Table 6.A as units.

as male contests, and are perhaps underrepresented. Male choice of mate is common among species in which females vary greatly in size and fecundity. Male scramble competition and endurance rivalry were found in four and ten cases, respectively. Seven of the endurance examples are from anurans.

Demonstration of one mechanism does not imply that others are absent, and these patterns may not reflect well the relative importance of the various mechanisms. Male contests over mates have long been well known, whereas mate choice has been controversial. Probably more studies have therefore looked for mate choice, making contests underrepresented. Such a bias seems even more likely against scrambles and endurance rivalry, which are less easily observed than conspicuous contests or courtship.

Selected Traits

Male song, advertisement calls, and other behavioral displays are the most common selected traits, with 81 cases (table 6.2.3). Sexual selection of body size is also frequent, with 51 cases in males, 27 in females. There are 24 cases of selection of male territories, and 18 of other material benefits, such as male nuptial food gifts in insects. Thirty-two studies found selection of male visual ornaments. Selection of male weapons and female visual ornaments was found in one study each (horn length in *Bolithotherus*, and coloration in *Nerophis*).

As with selection mechanisms, this pattern may be biased. It probably depends in part on the relative amenability of various traits to quantitative

Table 6.2.3

Characters Selected

*Group**	*Male Size*	*Female Size*	*Male Visual Ornament*	*Male Song, Display*	*Male Territory Size or Quality*	*Other Male Material Resource*	*Other Male Trait*	*Other Female Trait*
Insects	10	8	4	18		8	8	2
Other arthropods	4	4					1	
Fishes	14	8	5	2	6	7	3	2
Amphibians	12	5		26	2		3	1
Reptiles	3		1	1	2			
Birds	4	2	21	31	12	3	12	2
Mammals	4		1	3	2		8	
Totals	51	27	32	81	24	18	35	7

* All sums are calculated with the rows of Table 6.A as units.

measurement and testing. Body size can often be readily measured, and calls in many cases are easier to quantify, manipulate, and synthesize than are visual or chemical signals. This may partly explain why body size and calls dominate in the studies. In particular, there are many loudspeaker playback tests of responses to song or calls in frogs, insects, and birds (table 6.2.1).

Which mechanisms favor which traits? Closer study of table 6.A shows that large body size in males is usually selected by male contests (40 cases), sometimes by female choice or by both (29). Large body size in females, on the other hand, usually gives an advantage in male choice of mate (24 cases); contests favour large female size in only two cases.

Male visual ornaments are favored by female choice in 28 cases, by male contests in four, and by both in two cases. Male display, including song, is selected by female choice (72), sometimes by contests between males (9). Female choice also favors male territory size or quality (23) and other material resources of importance to the female (17).

Seven studies of anurans, two of mammals, and one of a lekking bird found correlations between mating success and the time spent at a lek or other mating site: endurance rivalry. Such a correlation is expected simply because the total number of females visiting the lek increases with time, providing more opportunities for mating (Gerhardt et al. 1987). In general, endurance rivalry should favor male energetic efficiency and stamina, in some species probably achieved by large male body size and fasting ability (sections 5.5, 11.9). Endurance rivalry may be a more important selective agent than often realized. For example, using multiple regression analysis,

Arak (1988c) found that the number of nights present at the breeding pond was the most important factor affecting male mating success in the natterjack toad *Bufo calamita*, accounting for 59% of the variance. Size accounted for an additional 9%.

Several categories of selected traits are reviewed in chapters 11 to 17.

In sum, massive empirical work in a variety of animals shows that sexual selection favors secondary sex traits in many species. Mate choice is clearly an important selection mechanism behind the evolution of many conspicuous secondary sex signals (also see Ryan and Keddy-Hector 1992). Contests over mates are important in some cases; male contest competition is probably underrepresented in studies of such traits.

Open Questions

Some of the most interesting questions about sexual selection still remain unanswered. To proceed further, more studies need to separate between direct and indirect selection of correlated traits, clarify selective mechanisms in more detail, and estimate their relative importance within and among species. Table 6.A concerns phenotypic selection during a limited part of one generation (chapter 4). In many studies, variation in the character selected is likely to be mainly phenotypic, for example body size in species with indeterminate growth. For sexual selection to have evolutionary consequences, the variation subject to selection must in part be genetic. This remains to be shown in most cases, even though it seems likely in the light of the heritabilities demonstrated for some secondary sex traits (section 2.7).

More tests are needed of genetic variation and responses to selection of secondary sex traits, such as the work on guppies by Endler (1980, 1983) and Houde and Endler (1990). It still largely remains to distinguish among the genetic mechanisms for the evolution of mate choice (Fisherian, indicator, sensory bias, etc.; sections 1.7, 2.3, 3.5). Comparative data suggest, however, that a Fisherian process is at least not the only mechanism of selection in mate choice. Ryan and Keddy-Hector (1992) found in their review of visual and acoustic traits that "if females prefer traits that deviate from the population mean, they usually prefer traits of greater quantity," which elicit greater sensory stimulation (also see Rowland 1989a,c). Such a pattern is not predicted from the Fisherian process alone, which is indeterminate as regards direction; reduced trait size is then as likely to evolve as is increased size. The pattern found by Ryan and Keddy-Hector (1992) therefore suggests that other processes are partly or entirely responsible. The authors emphasized, however, that preferences for average traits may often be important; there may be a bias in that researchers tend to study preferences for exaggerated traits.

Mating preferences and their variation and selection need to be clarified in much more detail. So far, emphasis has been on variation in the preferred trait; preferences have received less attention. Focus has been on the evolutionary, ultimate reasons for mate choice, but proximate mechanisms also need to be examined. Studies of preexisting bias in the female sensory system are now bringing a change in this direction (e.g., Ryan 1990b).

6.3 Summary

Many empirical studies of sexual selection have found a statistically significant relationship between mating success and the expression of some trait(s). Table 6.A reviews such results from 186 species, representing 232 demonstrations of sexual selection. Insects, fishes, anurans, and birds dominate among these studies. One third of them are based on field observations alone, one third also on experiments, and one third on laboratory work. Female choice is the most common mechanism (167 cases, of which 76 exclude any influence of male contests). Male choice was found in 30 cases, mainly of large, fecund females, and male contests in 58 cases. Male song or other display is the most common selected trait, followed by male body size, male visual ornaments, female body size, male territory, or other material resources. These patterns are probably biased in several ways. For example, contests, scrambles, and endurance rivalry over mates are likely under-represented; empirical as well as theoretical work on sexual selection has focused mainly on mate choice, the most controversial aspect. The other mechanisms also need more study. So does, in particular, the genetics of the processes, which remains largely unknown.

Table 6.A

Species in Which Sexual Selection Has Been Studied Quantitatively in Relation to Some Character(s)

Species	*Selective Mechanism*	*Character Selected*[a]	*Methods*[b]	*Reference*[c]
INSECTS				
Odonata				
Libellula luctuosa	FCH, MCO	M wing color; M size and wing color, respectively	FO,CS	Moore 1990a
Plathemis lydia	MCO	Color of M abdomen	FO,FE,CS	Jacobs 1955
Blattodea				
Nauphoeta cinerea	FCH	Odor of dominant M	LE,CP,CS	Breed et al. 1980, Moore & Breed 1986, Moore 1988, 1989
Orthoptera				
Acheta domesticus	FCH	Song of dominant M	LE,LT	Crankshaw 1979
Amphiacusta maya	MCO	M call	LE,CS	Boake & Capranica 1982, Boake 1984
Anabrus simplex	MCH	F size	FO,CS	Gwynne 1981, 1984a
Anurogryllus arboreus	FCH	M song perch height	FO	Walker 1983
Corthippus brunneus	FCH	M song syllable length	LO	Butlin et al. 1985
Gryllus bimaculatus	FCH, MCO	M size	LO,CS	L. W. Simmons 1986a,b, 1988a
Gryllus bimaculatus	FCH	Song of large M	LE,LT	L. W. Simmons 1988b
G. integer	FCH	M call bout length	LE,LT	Hedrick 1986, 1988
G. integer	FCH	M call loudness	FO,FE	Cade 1979
G. veletis	FCH	M age	FO,FE,CP	Zuk 1987a,b, 1988
G. pennsylvanicus	FCH	M age	FO,FE,CP	Zuk 1987a,b, 1988
Ligurotettix coquilletti	FCH	M song activity	FO,CS	Greenfield & Shelly 1985
Metaballus sp.	MCH	F size	FO	Gwynne 1985
Requena verticalis	FCH	M call rate, loudness, spectrum	FO,LE,LT, CS	Bailey et al. 1990, Schatral 1990
Scapteriscus acletus	FCH	M call loudness	FO,FE,CP	Forrest 1980
Telogryllus oceanicus	FCH, MCO	M song, fighting ability, respectively	LO,CS	Burk 1983
Tettigonia cantans	FCH	M song pitch, loudness	LE,LT	Latimer & Sippel 1987
Zaprochiline katydid	FCH	M song pitch	LE,LT	Gwynne & Bailey 1988
Thysanoptera				
Elaphrothrips tuberculatus	MCO	M size	FO,CS	Crespi 1986a
Holothrips carnyi	MCO	M size	LO,CS	Crespi 1986b, 1988

Table 6.A (*cont.*)

Species	*Selective Mechanism*	*Character Selected*[a]	*Methods*[b]	*Reference*[c]
INSECTS, Coleoptera				
Bolithotherus cornutus	MCO	M horn length	FO,CS	Brown & Bartalon 1986, Conner 1988, 1989
Brenthus anchorago	MCH, MCO	F size, M size, respectively	FO,FE,CP, CS	Johnson 1982, Johnson & Hubbell 1984
Chauliognathus pennsylvanicus	MCH	F color morph	FO	McLain 1982, 1988
Lytta magister	FCH, MCH	M size, F size, respectively	FO,CS	W. D. Brown 1990a,b
Monochamus scutellatus	FCH	M size	LE,CP,CS	Hughes & Hughes 1985
Nicrophorus spp (3)	MCO, FCO	M size, F size, respectively	FO,LE,CS	Otronen 1988a
Tetraopes tetraophtalmus	MCH, MCO	F size, M size, respectively	FO,LE,CP, CS	McCauley 1982, Lawrence 1986, McLain & Boromisa 1987
Triboleum castaneum	FCH	M pheromone	LE,CP	Boake 1985, 1986
Hymenoptera				
Centris pallida	MCO	M size	FO,CS	Alcock et al. 1977, Alcock 1984a
Pogonomyrmex spp.(2)	MCO	M size	FO,CS	Hölldobler 1976, Davidson 1982
Bittacus stigmaterus	FCH	Size of M nuptial prey	FO,CS	Thornhill 1978
Harpobittacus nigriceps	FCH	Size of M nuptial prey	FO,LE,CP, CS	Thornhill 1983, 1984b
H. australis	FCH	Size of M nuptial prey	FO,CS	Alcock 1979b
H. similis	FCH	Size of M nuptial prey	FE,CP,CS	Gwynne 1984c
Hylobittacus apicalis	FCH	Size of male nuptial prey	FO,CP,CS	Thornhill 1976b, 1980a, 1984a
Panorpa latipennis	FCH	M abdominal clamp organ	FE,LE,CS	Thornhill 1980b
Panorpa latipennis	FCH	M nuptial food gift	FE,LE,CS	Thornhill 1981
Panorpa latipennis	MCO	M size	FE,CS	Thornhill 1981
P. penicillata	FCH	M nuptial food gift	FO,LE,CS	Thornhill 1979
Lepidoptera				
Colias eurytheme	FCH	M genotype, flight activity	FO,CS,RS	Watt et al. 1986
Colias eurytheme	FCH	M uv-reflectance	FO,LE,CS, CP	Silberglied & Taylor 1978

Table 6.A (*cont.*)

Species	*Selective Mechanism*	*Character Selected*[a]	*Methods*[b]	*Reference*[c]
INSECTS, Lepidoptera (*cont.*)				
Danaus gilippus	FCH	M pheromone	FO,FE,LE, CS,CP	Brower et al. 1965, Myers & Brower 1969, Pliske & Eisner 1969
Ephestia elutella	FCH	M size	LE,CS,CP	Phelan & Baker 1986
Grapholita molesta	FCH	M pheromone	LO,LE,CS, CP	Baker & Cardé 1979, Baker et al. 1981
Papilio glaucus	FCH	M color morph	LE, CS	Krebs & West 1988
Phryganida californica	FCH	M wing length	FO,CS	Mason 1969
Pieris occidentalis	FCH	M color pattern	FO,FE,CP, CS	Wiernasz 1989
Pieris protodice	MCH	F size and age	FE,CP	Rutowski 1982b
Diptera				
Anastrepha suspensa	FCH, MCO	M size	LE,CS	Burk & Webb 1983, Burk 1984
Chironomus plumosus	FCH?	M small size	FO,CS	Neems et al. 1990
Drosophila melanogaster	FCH	Unmated, fertile M	LE,CS	Markow et al. 1978
Drosophila melanogaster	FCH	M courtship song	LE,LT,SC, CS,CP	Bennet-Clark & Ewing 1967, 1969, Kyriacou & Hall 1982
Drosophila melanogaster	MS	M size	FO,LE,LO, CS	Partridge & Farquhar 1983, Partridge et al. 1987a,b, Wilkinson 1987
D. pseudoobscura	MCO	M size	FO,LO,CS	Partridge et al. 1987a
D. subobscura	FCH	M size, nuptial food gift	LE,CS,CP	Steele 1986b
D. subobscura	FCH	M agility, small size	LO,LE,CP	Steele & Partridge 1988
Dryomyza anilis	MCO	M size	FO,CS	Otronen 1984a,b
Plecia nearctica	MCO	M size	FO,CS	Thornhill 1980e, Hieber & Cohen 1983
Plecia nearctica	MCH	F size	FO	Hieber & Cohen 1983
Scatophaga stercoraria	FCH, MCO	M size	FO,FE,CS	Parker 1970c, Borgia 1980–82, Sigurjonsdottir & Parker 1981
CRUSTACEANS				
Asellus aquaticus	MCH, MCO	F size, M size, respectively	FO,LO,LE, CS	Manning 1975, Ridley & Thompson 1979, Thompson & Manning 1981

Table 6.A (*cont.*)

Species	*Selective Mechanism*	*Character Selected*[a]	*Methods*[b]	*Reference*[c]
CRUSTACEANS (*cont.*)				
A. meridianus	MCH	F size	FO,LO,LE	Manning 1975
Gammarus pulex	MCH, MCO	F size, M size, respectively	FO,LO,LE, CS	Birkhead & Clarkson 1980, Ward 1983, 1984, 1988a, Elwood et al. 1987, Poulton & Thompson 1987, Dick & Elwood 1990
Paracerceis sculpta	MCO	M size	FO,LE	Shuster 1990b
Thermosphaeroma termophilum	MCH	F size	LE	Shuster 1981
Uca beebei	FCH	M burrow quality	FO,CS	Christy 1987, 1988
SPIDERS				
Linyphia litigiosa	MCO	M size	FO,CS	Watson 1990
FISHES				
Acanthemblemaria crockeri	FCH	M size	FO,LE,SS	Hastings 1988a
Cichlasoma nigrofasciatum	FCH	M size	LE,CP,SS	Noonan 1983, Keenleyside et al. 1985
Coralliozetus angelica	FCH, MCO	M size	FO,LE,SS	Hastings 1988b
Cottus bairdi	FCH, MCH	M size, F size, respectively	FO,LE,SS	Downhower & Brown 1980–83, Brown 1981
C. gobio	FCH	M guarding of eggs	FO,LE,SS, CP	Marconato & Bisazza 1986
C. gobio	FCH, MCO	M size	LE,CP	Bizassa & Marconato 1988
Cyprinodon pecosensis	FCH	M coloration, territory quality	FO,SS	Kodric-Brown 1983
C. macularis	MCH	F size	LE,SS	Loiselle 1982
Etheostoma nigrum	FCH	M defense of spawning site	FO,SS	Grant & Colgan 1983
Forsterygion varium	FCH	M size, territory quality	FO,FE,SS	Thompson 1986
Gasterosteus aculeatus	MCH	F size	LE,SS,CP	Rowland 1982a,1989c, Sargent et al. 1986
Gasterosteus aculeatus	FCH	M size	LE,CP	Rowland 1989a,b
Gasterosteus aculeatus	FCH	M coloration	LE	McPhail 1969, Semler 1971
Gasterosteus aculeatus	MCO	M coloration	LE	Rowland 1982b, Bakker & Sevenster 1983
Gasterosteus aculeatus	FCH	M guarding of eggs	LE,CP,SS	Ridley & Rechten 1981

Table 6.A (*cont.*)

Species	*Selective Mechanism*	*Character Selected*[a]	*Methods*[b]	*Reference*[c]
FISHES (*cont.*)				
Gasterosteus aculeatus	FCH	M nest site	LE,SS,RS	Sargent & Gebler 1980, Sargent 1982
Glyphidodontops cyaneus	MCO	M size	FO,SS	Thresher & Moyer 1983
Hyphessobrycon pulchripinnis	FCH	Unspawned M	LE,CP	Nakatsuru & Kramer 1982
Hypsypops rubicundus	FCH	M guarding of eggs	FO,FE,SS	Sikkel 1989
Malacoctenus hubbsi	FCH	M territory quality	FO,SS	Petersen 1988
M. macropus	FCH	M size, territory quality	FO,SS	Petersen 1988
Nerophis ophidion	MCH	F coloration	LE,SS	Berglund et al. 1986b
Nerophis ophidion	MCH	F size, nuptial skin folds	LE,CP	Rosenqvist 1990
Oncorhynchus kisutch	MCH	F size	FO	Sargent et al. 1986
O. nerca	FCH, MCH	M size, F size, respectively	FO,FE,SS	Foote 1988,1989, Foote & Larkin 1988
Ophioblennius atlanticus	FCH, MCH	M size, nest size, F size, respectively	FO,SS	Cote & Hunte 1989
Padogobius martensi	FCH, MCO	M nest size, M size, respectively	LE,SS	Bisazza et al. 1989
Pimephales promeles	FCH	M guarding of eggs	LE,RS	Sargent 1988, Unger & Sargent 1988
Poecilia reticulata	FCH	M coloration	FO,LE,CP, RS	Endler 1980, 1983, Kodric-Brown 1985, 1989, Houde 1987
Poecilia reticulata	FCH	M tail size, display rate	LE,CP,CS	Bischoff et al. 1985
Poecilia reticulata	FCH	M display rate, few parasites	LE,CP	Kennedy et al. 1987
Pseudolabrus celidotus	FCH	M territory quality	FO,FE,SS	Jones 1981
Stegastes rectifraenum	FCH	M territory quality	FO,FE	Hoelzer 1990
Syngnathus typhle	FCH, MCH	M size, F size, respectively	LE,SS,CP	Berglund et al. 1986b
Thalassoma bifasciatum	FCH, MCO	M size	FO,SS	Warner et al. 1975, Warner 1987, Hoffman et al. 1985
Xiphophorus helleri	FCH	M tail length	LO,LE,CP	Basolo 1990a,b
ANURANS				
Acris crepitans	FCH	M call pitch	FO,LE,LT, SC	Nevo & Capranica 1985, Ryan & Wilczynski 1988
Bufo americanus	FCH	M size, time calling	FO,AS	Gatz 1981a, Howard 1988a
Bufo americanus	MCO	M size	LE	Howard 1988a

Table 6.A (*cont.*)

Species	*Selective Mechanism*	*Character Selected*[a]	*Methods*[b]	*Reference*[c]
ANURANS (*cont.*)				
B. bufo	MCO	M size and call pitch	FO,LE,SS	Davies & Halliday 1977–79
B. bufo	FCH	M call loudness	LE,LT	Höglund & Robertson 1988
B. calamita	FCH, MER	M call loudness, pulse rr, time on lek	FO,SS,LE, LT,SC	Arak 1983a, 1988b,c
B. calamita	MCO	M size, call loudness, pitch	FO,FE,LT, SC	Arak 1983a
B. cognatus	FCH, MER	M time calling	FO,AS	Sullivan 1983a
B. gutteralis	MCO	M size	FO,AS	Telford & Van Sickle 1989
B. typhonius	MCO	M size	FO,SS	Wells 1979
B. woodhousei	FCH, MCO	M call rate	FO,AS,FE, LT	Sullivan 1982a, 1983b, 1987
B. woodhousei	MER	M time on lek	FO,AS	Woodward 1982a,b
Centrolenella fleischmanni	FCH, MER	M calling height, time calling	FO,SS	Greer & Wells 1980
Hyla cinerea	FCH	M call pitch	LE,LT,SC	Gerhardt 1978, 1981, 1982, 1987
H. chrysocelis	FCH, MER	M size, time calling	FO,AS	Goodwin & Roble 1983, Morris 1989
H. crucifer	FCH	M call loudness, rate, pitch	LE,LT	Forester & Czarnowsky 1985
H. crucifer	FCH	M call duration, pitch	LE,LT,SC	Doherty & Gerhardt 1984
H. ebraccata	FCH	M multinote calls; call rate	LE,LT,SC	Wells & Schwartz 1984, Wells & Bard 1987
H. gratiosa	FCH	M call pitch, pulse shape	LE,LT,SC	Gerhardt 1981
H.meridionalis	FCH	M call rate and duration	FE,LT	Schneider 1982
H. microcephala	FCH	M call rate, complexity, pulse rr	LE,LT,SC	Schwartz 1986, 1987b
H. rosenbergi	MER	M time on territory	FO,SS	Kluge 1981
H. regilla	FCH	M call rate and loudness	FO,LE,LT	Whitney & Krebs 1975a
H. versicolor	FCH	M call rate and duration	LE,LT,SC	Klump & Gerhardt 1987
H. versicolor	FCH	M call loudness	FE,LT	Fellers 1979
H. versicolor and *H. regilla*	FCH	M nonoverlapping calls	LE,LT,SC	Schwartz 1987a
Hyperolius marmoratus	MCO	M call loudness	FO,FE,LT	Telford 1985, Dyson & Passmore 1988a,b
Hyperolius marmoratus	FCH	M call loudness, M spacing	LE,LT	Telford 1985, Dyson & Passmore 1988a,b

Table 6.A (*cont.*)

Species	*Selective Mechanism*	*Character Selected*[a]	*Methods*[b]	*Reference*[c]
ANURANS (*cont.*)				
Litoria ewingi, verrauxi	FCH	M call pulse rr	LO,LT,SC	Loftus-Hills & Littlejohn 1971
Physalaemus pustulosus	FCH	M call pitch, complexity	LE,LT,SC	Ryan 1980a,1983,1985, Rand & Ryan 1981
Physalaemus pustulosus	FCH	M size	FO,SS	Ryan 1980a,1983,1985, Rand & Ryan 1981
Rana catesbeiana	FCH	M territory quality	FO,CS,RS	Howard 1978a,b, 1981b, 1983
Rana catesbeiana	MCO	M size	FO,RS	Emlen 1976, Howard 1978a, 1981a
R. clamitans	MCO, FCH	M size, territory quality	FO,SS	Wells 1977a,1978
R. sylvatica	MCO	M size	FO,FE,AS	Howard 1980, Howard & Kluge 1985, Berven 1981
R. sylvatica	MCH	F size	LE,AS	Berven 1981
Uperoleia laevigata	FCH, MCO	M size, call pitch	FO,LE,LT, SC,AS	Robertson 1986a,b, 1990
URODELES				
Desmognathus ochrophaeus	MCH	F size	LE,CS	Verrell 1989
Desmognathus ochrophaeus	FCH	M pheromone	LE,CP	Houck & Reagan 1990
Notophtalmus viridescens	MCH	F size and odor	LE,CP	Verrell 1982b; 1985
Triturus cristatus	MCH	F size	LE,CS	Malacarne 1984
T. vulgaris	MCH	F size	LE,CS	Verrell 1986
LIZARDS				
Anolis carolinensis	FCH	M dewlap color and display	FO,LE,CP	Greenberg & Noble 1944, Crews 1975, Sigmund 1983
A. garmani	FCH	M territory size	FO,CS	Trivers 1976
Sceloporus jarrovi	MCO	M size	FO,CS	Ruby 1978, 1981
Uta palmeri	FCH, MCO	M territory quality, M size and head depth, respectively	FO,LE,CS	Hews 1990
SNAKES				
Vipera berus	MCO	M size	FO,CS	Andrén 1986, Madsen 1988

Table 6.A (*cont.*)

Species	*Selective Mechanism*	*Character Selected*[a]	*Methods*[b]	*Reference*[c]
BIRDS, Nonpasserines				
Anas platyrhynchos	FCH	M size, display, plumage status	LO,LE	Holmberg et al. 1989
Centrocercus urophasianus	FCH	M position on lek	FO,CS	Wiley 1974
Centrocercus urophasianus	FCH, MER	M display rate, lek attendance, song structure	FO,CS	Gibson & Bradbury 1985, Gibson 1989
Circus cyaneus	FCH	M display, food provisioning	FO,RS	R. Simmons 1988a,b
Columba livia	FCH, MCH	Plumage color, reproductive experience	LE,CP	Burley 1977, 1981a, Burley & Moran 1979
Columba livia	FCH	M dominance	LE,CP	Burley 1977, 1981a, Burley & Moran 1979
Gallinago media	FCH	M display rate, lek position	FO	Höglund & Lundberg 1987
Gallinago media	FCH	M tail coloration	FE	Höglund & Robertson 1990, Höglund et al. 1990b
Gallinula chloropus	FCO	F size	FO	Petrie 1983a
Gallus gallus	FCH	M comb size, eye color	LE,CP,CS	Zuk et al. 1990a–d
Gallus gallus	MCO	M comb size	LE	Ligon et al. 1990
Perdix perdix	FCH	M vigilance behavior	LE,CP	Dahlgren 1990
Phasianus colchicus	FCH	M spur length	FO,FE	von Schantz et al. 1989, Göransson et al. 1990
Stercorarius parasiticus	FCH	M color morph, territory size	FO,RS	O'Donald 1983
Tetrao tetrix	FCH	M display duration, lek position	FO	Kruijt & de Vos 1988
Passerines				
Acrocephalus arundinaceus	FCH	M song length, repertoire size	FO,RS,LE, LT,EF	Catchpole et al. 1986, Catchpole1986
Acrocephalus arundinaceus	MCO	M song type	FO,LE,LT	Catchpole 1983, Catchpole et al.1986
A. schoenobaenus	FCH	M song repertoire size	FO,LE,LT, EF	Catchpole 1980, Catchpole et al. 1984
Agelaius phoeniceus	FCH	M territory quality	FO,RS	Searcy 1979a,c, Lenington 1980, Picman 1980, Yasukawa 1981c
Agelaius phoeniceus	MCO	M size	LO	Searcy 1979a,c, Eckert & Weatherhead 1987b

Table 6.A (*cont.*)

Species	*Selective Mechanism*	*Character Selected*[a]	*Methods*[b]	*Reference*[c]
BIRDS, Passerines (*cont.*)				
Agelaius phoeniceus	MCO	M song, repertoire size	FE	Peek 1972, Smith 1976, 1979, Yasukawa 1981a,b
Agelaius phoeniceus	FCH	M song repertoire size	LE,LT,EF	Searcy 1988
Agelaius phoeniceus	MCO	M red wing patch	FO,FE,LO	Peek 1972, Smith 1972, Hansen & Rohwer 1986, Røskaft & Rohwer 1987
Ammodramus maritumus	FCH, MCO	M song	FO,FE	M. V. McDonald 1989
Cacicus cela	MCO, MCH	M size, F nest-site quality, respectively	FO	Robinson 1986b
Calamospiza melanocorys	FCH	M territory quality	FO,FE	Pleszynska 1978
Carpodacus mexicanus	FCH	M coloration	FO,LE,CP	Hill 1990
Chiroxiphia linearis	FCH, MCO	M age, call, display, respectively	FO,CS	D. B. McDonald 1989a,b
Emberiza citrinella	FCH	M song dialect, repertoire size	LE,LT,EF	Baker et al. 1987a
Erithacus rubecula	FCH	M territory size	FO	Harper 1985
Euplectes progne	FCH	M tail size	FE	Andersson 1982a
E. jacksoni	FCH	M display, tail length	FO,CS	Andersson 1989
Ficedula hypoleuca	FCH, MS	M song rate, arrival date	FO,FE,LT	Eriksson & Wallin 1986; Gottlander 1987, Alatalo et al. 1990b
Ficedula hypoleuca	FCH	M territory size, quality	FO,FE	Askenmo 1984, Alatalo et al. 1986, Slagsvold 1986, Dale & Slagsvold 1990
Ficedula hypoleuca	FCH	M plumage color	FO	Røskaft & Järvi 1983, Järvi et al. 1987, Slagsvold & Lifjeld 1988
Geospiza conirostris	FCH	M song type, color, breeding experience	FO,RS	Grant 1985, Grant and Grant 1987
G. fortis	FCH	M size, color, territory size	FO	Price 1984b
Gymnorhinus cyanocephalus	FCH, MCO	M coloration, bill size, respectively	LO	Johnson 1988a
Gymnorhinus cyanocephalus	MCH	F mass, rank, bill depth	LO	Johnson 1988b
Hirundo rustica	FCH	M tail length	FE,FO,CS, RS	Møller 1988a, 1989a, 1990a,b

Table 6.A (*cont.*)

Species	*Selective Mechanism*	*Character Selected*[a]	*Methods*[b]	*Reference*[c]
BIRDS, Passerines (*cont.*)				
Melospiza melodia	FCH	M song repertoire size	LE,LT,EF	Searcy & Marler 1981, Searcy 1984
M. georgiana	FCH	M song repertoire size	LE,LT,EF	Searcy et al. 1982
Mimus polyglottus	FCH	M song repertoire size, territory quality	FO	Howard 1974
Molothrus ater	FCH	Song of dominant M	LO,LE,LT, CS	West et al. 1981
Parotia lawesii	FCH	M display	FO,CS	Pruett-Jones & Pruett-Jones 1990
Parus major	FCH	M song structure	FO,LE,LT, EF	McGregor & Krebs 1982, Baker et al. 1987
Parus major	MCO	M song repertoire size	FO,FE,LT	Krebs et al. 1978
Parus major	FCH	M song repertoire size	LE,LT,EF	Baker et al. 1986
Parus major	FCH	M breast stripe size	FO,RS	Norris 1990a,b
Passerculus sandwichensis	FCH	M song rate, territory size	FO	Reid and Weatherhead 1990
Phylloscopus trochilus	FCH, MS	M song rate, repertoire size, arrival date	FO	Järvi 1983, Radesäter et al. 1987, Radesäter and Jakobsson 1988, 1989
Ploceus cucullatus	FCH	M color, display, nest quality	LO,LE,CS	Collias & Victoria 1978, Collias et al. 1979
Poephilia guttata	FCH, MCH	Color of leg bands	LO,LE,RS	Burley 1981b, 1986a,c, 1988a
Prunella modularis	FCH	M territory size, quality	FO,FE,CS, RS	Davies & Lundberg 1984, Davies 1985, Davies & Houston 1986
Ptilonorhyncus violaceus	FCH	M bower quality	FE,FO,CS	Borgia 1985a,b
Ptilonorhyncus violaceus	FCH	M freedom from parasites	FO	Borgia & Collis 1989, 1990
Rupicola rupicola	FCH	M disruption of rival courtship	FO,CS	Trail 1985a,b, Trail & Koutnik 1986
Serinus canarius	FCH	M song repertoire size	LE,LT	Kroodsma 1976
Spiza americana	FCH	M territory size, quality	FO	Zimmerman 1966, 1971
Telmatodytes palustris	FCH	M territory size, quality	FO	Verner 1964, Verner & Engelsen 1970
Vidua chalybeata	FCH	M song rate	FO	Payne & Payne 1977, Payne 1983
V. regia	FCH	M display rate, tail length	LE,CP,EF	Barnard 1990

Table 6.A (*cont.*)

Species	*Selective Mechanism*	*Character Selected*[a]	*Methods*[b]	*Reference*[c]
BIRDS, Passerines (*cont.*)				
Xanthocephalus x.	FCH	M territory size	FO	Lightbody & Weatherhead 1987, 1988
Zonotrichia leucophrys	FCH	M song dialect	LE,LT,EF	Baker 1983
MAMMALS				
Antilocapra americana	FCH	M territory quality	FO,CS	Kitchen 1974
Cervus elaphus	MCO	M roaring rate	FO	Clutton-Brock & Albon 1979
Cervus elaphus	MCO	M size, fighting ability	FO,CS,RS	Clutton-Brock et al. 1979, 1988
Dama dama	FCH, MER	M antler size, time on lek	FO,CS	Clutton-Brock et al. 1989, Appollonio et al. 1989
Dama dama	FCH	M position on lek	FO,CS	Appollonio et al. 1990
Lemmus trimucronatus	FCH	Odor of dominant M	LO,CP	Huck et al. 1981, Huck & Banks 1982
Loxodonta africana	FCH, MCO	M musth, age, and size, respectively	FO,CS	Poole & Moss 1981, Moss 1983, Poole 1989a,b
Microtus ochrogaster	FCH	Dominant M	LE,CP,CS	Shapiro & Dewsbury 1986
Mirounga angustirostris	FCH, MCO	M size	FO,CS	Cox & Le Boeuf 1977, Le Boeuf & Reiter 1988
Mus musculus	FCH	Odor of viable M	LE	Lenington & Egid 1985
Ochotona princeps	FCH	M nest quality	FO	Brandt 1989
Oryctolagus cuniculus	FCH	Odor of resident M	LE	Reece-Engel 1988
Otaria byronia	MER	M time in breeding area	FO,CS	Campagna & Le Boeuf 1988
Spermophilus tridecemlineatus	MS	M mobility, range size	FO,CS	Schwagmeyer & Woontner 1985, 1986, Schwagmeyer 1988
Spermophilus tridecemlineatus	MCO	M size	FO,CS	Schwagmeyer & Brown 1983

CODES: F = female, M = male, FCH = female choice, FCO = female contest, MCH = male choice, MCO = male contest, MER = male endurance rivalry, MS = male scramble, FE = field experiment, FO = field observation, LE = laboratory experiment, LO = laboratory observation, CP = contests prevented or excluded by the design of the study, LT = loudspeaker test, SC = synthetic calls (i.e., not recorded natural calls), pulse rr = (call) pulse repetition rate, EF = estradiol-treated females (in tests of response to song), CS = mating success estimated from copulations, SS = mating success estimated from spawnings, AS = mating success estimated from amplexus (mating embrace in anurans), RS = mating success followed to reproductive success (production of offspring). Other studies use indirect evidence of mating success, such as attraction to the potential partner or to broadcast song or display, indicating readiness to copulate.

[a] "Call pitch" refers to the fundamental frequency of the call. Where body size is selected, large size is favored, unless otherwise stated.

[b] "Experiment" (FE or LE) refers to manipulative studies of the selected trait that include treatment and control groups. Other studies are listed as observational, even if involving manipulation.

[c] Where several studies are cited together, the information given concerns the studies in combination.

7 Sexual Selection in Relation to Mating System and Parental Roles

7.1 Mating Systems

Sexual selection is usually stronger in males than females. What are the reasons for sex differences in the strength of sexual selection, and for all the variation in this pattern among species? These and related problems are the main subject of this chapter.

It has long been clear that the expression of secondary sex traits depends on the mating system and parental roles. For example, Darwin (1871, I, 266) wrote "that some relation exists between polygamy and the development of secondary sexual characters appears nearly certain," and "with birds there often exists a close relation between polygamy and the development of strongly-marked sexual differences." The reasons for such relationships are also discussed here.

Mating system evolution and its dependence on sexual selection have been discussed from different angles in many reviews. Mating systems are dealt with here mainly from the reverse perspective, as one of the factors that influence the strength and consequences of sexual selection.[1] There are likely to be strong reciprocal relationships between mating systems and sexual selection (figure 7.2.2 below).

Although different mating systems have long been recognized in evolutionary biology, there is far from universal agreement on their classification, and the same terms are used in several senses. The number of mating partners per male and female is one of the main critera, but the exact meaning of terms such as monogamy, polygyny, and polyandry varies among students of different taxa and problems. Many ecologists, particularly ornithologists, have emphasized the nature and duration of the pair bond (e.g., Crook 1965; Selander 1965, 1972; Verner and Willson 1966, 1969; Lack 1968; Wittenberger 1979, 1981; Mock 1985; Brown 1987; Scott and Clut-

[1] Mating systems and their evolution in various taxonomic groups are reviewed in, e.g., Orians 1961, 1969, Crook 1965, Verner and Willson 1966, Lack 1968, Selander 1965, 1972, Wiley 1974, 1991, Bradbury and Vehrencamp 1977, Clutton-Brock and Harvey 1977, Emlen and Oring 1977, Kleiman 1977, Ralls 1977, Loiselle and Barlow 1978, Parker 1978, Borgia 1979, Bradbury 1981, 1985, Wittenberger 1981, Oring 1982, 1986, Gwynne and Morris 1983, Thornhill and Alcock 1983, Murray 1984, Payne 1984, Vehrencamp and Bradbury 1984, Gowaty 1985, Thornhill 1986a, Rubenstein and Wrangham 1986, Brown 1987, Ims 1987, Dunbar 1988, Clutton-Brock 1989, Westneat et al. 1990, Davies 1991, Duvall et al. 1992.

ton-Brock 1990; Westneat et al. 1990). For instance, Wittenberger (1981) defined *monogamy* as a mating system in which there is a prolonged bond and largely exclusive mating relationship between a male and a female, *polyandry* as the same relationship between a female and two or more males, and *polygyny* as the same relationship between a male and two or more females.

Others have found definitions based on the length and strength of pair bonds less useful, in part because bonds are harder to observe and quantify in other taxa, and a prolonged "bond" is difficult to define strictly (e.g., Ralls 1977; Thornhill and Alcock 1983; Murray 1984; Clutton-Brock 1989). Also within a category such as monogamy, there is much variation in the duration of bonds, and in the frequency of re-pairing, which may greatly influence the scope for sexual selection (Scott and Clutton-Brock 1990).

For some purposes in the study of sexual selection, mating systems have been classified in relation to the breeding sex ratio. With *monogamy*, equal numbers of females and males contribute gametes to zygotes; with *polyandry*, more males than females contribute; and with *polygyny*, more females than males contribute (e.g., Wiley 1974; Ralls 1977; Gowaty 1985). A similar classification was used by Thornhill and Alcock (1983): with *monogamy*, each individual has a single partner; with *polyandry*, some females mate with more than one male; and with *polygyny*, some males mate with more than one female per breeding season. In some animals, the mating system is perhaps best described as *polygynandrous*: some males as well as females have several mates per season. If prolonged bonding is taken as a requisite part of this system, it has been found so far only in communally breeding birds and mammals (see Brown 1987). If the bond is ignored, polygynandry occurs in many species; copulation with more than one member of the other sex probably occurs in some males and females in most animals that are predominantly monogamous, polygynous, or polyandrous. Social monogamy in many species also involves extra-pair copulations (e.g., Westneat et al. 1990; Birkhead and Møller 1992). These qualitative terms therefore are only rough labels for real mating patterns. In practice, those mating systems termed *polygamous* (i.e., polyandrous, polygynous, or polygynandrous) usually have at least several percent of non-monogamous relationships.

More exact descriptions of mating relationships can be given, for example, based on the mean and variance of the number of males that contribute genes to the offspring of the average female, and vice versa (e.g., Bateman 1948; Wade and Arnold 1980; Daly and Wilson 1983; Payne 1984). Such quantitative refinement holds potential to further clarify relationships between mating systems and sexual selection.

Mating systems are related to resources necessary for breeding, such as the spatial distribution of food, mates, and breeding sites, and the temporal distribution of mates (e.g., Orians 1969; Bradbury and Vehrencamp 1977; Emlen and Oring 1977). Because males, but usually not females, seem able to improve reproductive success by gaining several mates, male distribution should be influenced strongly by female dispersion, which in turn should be closely related to the distribution of resources needed for breeding. There is observational and some experimental evidence that this is the case. For example, in the gray-sided vole *Clethrionomys rufocanus*, experimental changes in female dispersion influences the dispersion of males but not vice versa (Ims 1988a); Warner (1990) obtained similar experimental results in a fish, the bluehead wrasse *Thalassoma bifasciatum*.

The mating system may change with conditions—for example, food abundance—also within species. An illuminating example is the dunnock *Prunella modularis*, where the system may shift from polyandry to polygyny (Davies and Lundberg 1984; section 8.4 below).

Two forms of polygyny have received particular attention in studies of sexual selection. In *resource defense polygyny*, a male defends sufficient resources, such as food or nest sites, to attract more than one female (Emlen and Oring 1977). This situation has been formalized in the polygyny threshold model of Verner and Willson (1966) and Orians (1969), discussed in section 8.4 below. In brief, it suggests that a female, when there are substantial quality differences between male territories, should choose an already mated male if his territory is sufficiently much better than those of unmated males.

In *lek polygyny*, males provide no material resources but genes, and are visited by females for copulation. This is perhaps the mating system with greatest within-season variance in male mating success; usually a few males on the lek obtain most matings, and the others few or none (e.g., Bradbury 1985; Bradbury et al. 1985; Trail and Adams 1989; Balmford 1991; Wiley 1991). In classical leks, such as that of the blackcock *Tetrao tetrix*, males are clustered on display arenas, but a similar distribution of mating success is also found in some species with dispersed males. Leks are known mainly from insects, frogs, birds, and mammals (e.g., Wells 1977b; Thornhill and Alcock 1983; Bradbury 1985; Clutton-Brock et al. 1988b; section 7.6 below).

In many taxa, sexual dimorphism in size, weaponry, or ornamentation increases with the degree of polygyny (e.g., Darwin 1871; Clutton-Brock et al. 1977, 1980; Ralls 1977; Harvey et al. 1978; Alexander et al. 1979; Shine 1979; Payne 1984; Björklund 1990, 1991a; chapters 11–13 below). Intuitively, it seems likely that the opportunity for sexual selection will increase with the degree of polygyny, and that this might explain such trends. Differences in the relative variance of male success between monogamous and polygynous species may, however, be smaller than expected. The variance in lifetime reproductive success among males may not be much larger among polygynous than monogamous species, if offspring survival to adulthood is included in the measure of success (Clutton-Brock 1988a).

More importantly, it is not the relative variance in mating success per se that determines the degree of dimorphism in sexually selected traits, but the relative effects of such traits on mating and reproductive success in males and females. A main reason that sex dimorphism tends to increase with polygyny is probably that the sex roles are more similar in monogamous than polygynous species. For example, many monogamous species have biparental care, whereas uniparental care (usually by the female) is more common in polygynous species. The sexes should therefore be subject to more contrasting selection pressures in many polygynous than in monogamous species (Clutton-Brock 1983, 1988a).

Ghiselin (1974) pointed out that, with the exception for dimorphism in body size, the most extreme secondary sex traits are found mainly among animals with parental care. They occur particularly in some species where males and females have markedly different sex roles because only one of the sexes provides care for the offspring, which imposes a number of selection pressures not present in the other sex. These plausible ideas need quantitative comparative testing in a range of animals with different mating systems.

7.2 Sexual Selection, Parental Investment, Sex Ratio, and Reproductive Rate

Male and Female Parental Investment

Why is sexual selection stronger in males than females? Darwin's (1871) review showed that males in higher animals usually have more elaborate weapons or other conspicuous sex traits than females, which often resemble juveniles. In addition, "it is the males that fight together and sedulously display their charms before the females," whereas a female "generally exerts some choice and accepts one male in preference to others." Darwin saw that these sex differences are fundamentally related to anisogamy (section 1.1 above): "The female has to expend much organic matter in the formation of her ova whereas the male expends more force in fierce contests with his rivals, in wandering about in search of the female, in exerting his voice, pouring out odoriferous secretions, etc. . . . On the whole the expenditure of matter and force by the two sexes is probably nearly equal, though effected in very different ways and at different times" (Darwin 1874, chap. 8, p. 224).

In a classical study where he placed equal numbers of fruit flies *Dro-*

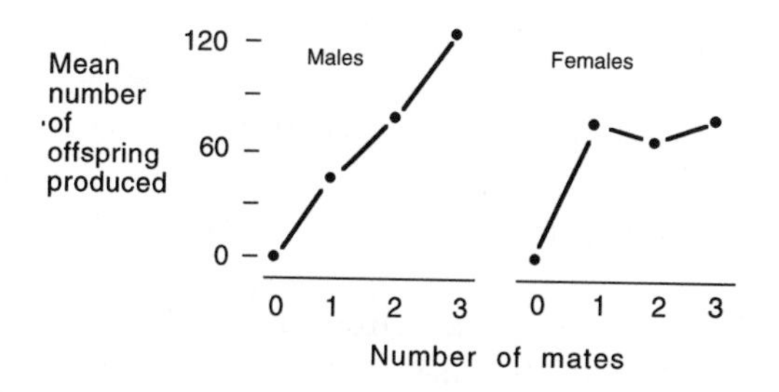

Figure 7.2.1 Bateman's (1948) demonstration that reproductive success in male *Drosophila* increases with his number of mates, whereas females do not increase their production of offspring by mating with several males. (After Daly and Wilson 1983)

sophila melanogaster of each sex together in bottles, Bateman (1948) showed that the fertility of a female fly is limited mainly by her ability to produce eggs, whereas male reproduction is limited by the number of females that he inseminates. The number of offspring fathered by a male therefore increases with his number of mates, whereas a female fruit fly probably has less to gain from mating with several males, at least as regards offspring numbers (figure 7.2.1). Traits that make it easier to get several mates should therefore be favored mainly in males. In the words of Williams (1966, pp.183–185): "A male can easily produce sperm in excess of what it would take to fertilize all the females that could conceivably be available. . . . Hence the development of the masculine emphasis on courtship and territoriality or other forms of conflict with competing males." On the other hand, pregnancy is energetically demanding, so "the traditional coyness of the female is thus easily attributed to adaptive mechanisms by which she can discriminate the ideal moment and circumstances for assuming the burdens of motherhood." In line with these ideas, males force copulation upon females in some species, whereas opposite cases are rare. Females often have various costs of forced copulation by males, including a risk of death (e.g., Thornhill 1980b; McKinney et al. 1983; Le Boeuf and Mesnick 1990). This is one of the conflicts between the sexes that arises from sexual selection (e.g., West-Eberhard et al. 1987).

Bateman (1948) showed that reproductive success varies more among males than females in *Drosophila*. Such a pattern is widespread (reviewed by Clutton-Brock 1988a). It has sometimes been taken as evidence for sexual selection among males, but Sutherland (1985a, 1987) and Hubbell and Johnson (1987) showed that the same pattern can arise even if mating is random. For example, one sex may contribute more toward the offspring than the other, and therefore will not be ready to mate again as soon as the other sex. Even if mating is random (i.e., all individuals of the same sex have equal chances of mating), this can lead to a sex difference in the variance of numbers of mates. Without further study of mating behavior, one cannot take sex differences in the variance of mate numbers and reproductive success as evidence for sexual selection (Sutherland 1985a; also see Wade 1979, 1987; Banks and Thompson 1985; Houck et al. 1985; Koenig and Albano 1986; Hafernik and Garrison 1986; Downhower et al. 1987b). The most direct way of measuring sexual selection is to quantify the effects of variation in phenotypic traits on success in competition over mates (section 4.5).

Williams (1966) pointed out that stronger sexual selection in males than females depends on sex differences not only in gamete size but also in relative contributions of resources and care for offspring. Trivers (1972, 1985) also emphasized that gametes are only part of the investment in offspring in many animals, and that parental behavior such as feeding of young and defense against predators is important. Trivers defined *parental investment* as any effort that raises offspring survival at the expense of the parent's ability to invest in other offspring. Present parental investment therefore reduces the residual reproductive value of the parent (its expected future contribution of offspring, discounted by the rate of population growth; e.g., Fisher 1930; Pianka 1978). The sex with largest parental investment, usually females, becomes a limiting resource for which members of the other sex compete.

Trivers's (1972) concept of parental investment helps clarify several aspects of sexual selection. For instance, in species where males and females have similar parental roles, both sexes may discriminate among and compete over potential mates (section 7.4). If males provide nutrition for females, this can enhance female fecundity and make males a limiting resource for which females compete (Thornhill 1976a,b, 1986a; Alexander and Borgia 1979; Gwynne 1984a, 1991; Thornhill and Gwynne 1986; Burley 1988a; Gwynne and Simmons 1990; Clutton-Brock and Vincent 1991). On the other hand, male mating effort that does not help the female produce offspring, such as displays and contests, will not be subject of competition among females.

Other reasons why males show larger mating effort and less parental effort than females may be that males have lower certainty of parentage, and therefore less control of resource flow to genetic offspring (Low 1978; Alexander and Borgia 1979; Gwynne 1984a). These theoretical developments shed light on several aspects that were not explained by Bateman's hypothesis based on gamete size. One example is reduced sexual selection in many monogamous species with similar sex roles; another is female competition over choosy males in species with reversed sex roles (Clutton-Brock 1991; Gwynne 1991). Unfortunately, parental investment is often difficult to measure quantitatively, but related sex differences in numbers of mates available for mating, and in reproductive rates, are more accessible (Clutton-Brock 1991; see below in this section).

Mate Control and the Operational Sex Ratio

Additional factors that affect the strength of sexual selection were identified by Emlen and Oring (1977), for instance the degree to which some members of one sex can monopolize access to the other sex. This *environmental potential for polygamy* depends on ecological factors. In species where mates can easily be monopolized, there can be strong polygamy and large variance in mating success. Especially important are the spatial distribution of resources, and the spatial and temporal distribution of mates. When food, breeding habitat, and other resources are clumped and readily defendable, it may be possible for a few individuals to monopolize access to many mates; there will then be strong polygamy and much scope for sexual selection. Temporal clumping of mates works in the opposite direction: if most potential mates are sexually receptive during the same brief period and each mating takes a long time, the potential for polygamy is low. On the other hand, a long breeding season leaves time for gaining several mates, and hence for stronger polygamy (Emlen and Oring 1977; Ims 1988b).

These aspects are reflected in the *operational sex ratio*: the ratio of fertilizable females to sexually active males at any given time (Emlen 1976; Emlen and Oring 1977). It depends on the degree of spatial and temporal clumping of the limiting sex, as well as on life history differences between the sexes. For example, in species where mortality is highest in males because of their sexually selected traits (sections 10.1–10.3 below), the adult sex ratio may become female-biased. On the other hand, in species where females, because of fecundity advantages, mature later and at a larger body size than males, the delay exposes females to mortality risk for a longer time, which reduces their numbers compared to mature males. Such life history differences will influence the operational sex ratio, and probably also the opportunity for sexual selection (Partridge and Endler 1987).

The operational sex ratio is a main determinant of the opportunity for sexual selection (e.g., Emlen and Oring 1977; Arak 1983b; Murray 1984; Partridge and Endler 1987; Ims 1988b; Westneat et al. 1990; Gwynne and Simmons 1990; Clutton-Brock and Parker 1992; also see section 10.5). For example, Höglund (1989a) showed experimentally in the toad *Bufo bufo* that, as predicted, when the sex ratio became more male biased, sexual selection apparently became stronger: larger males were then relatively more successful in gaining mates. For a particularly clear example, see the katydids reviewed in section 5.1 above.

There are exceptions, however: a more male-biased sex ratio does not always lead to stronger sexual selection for male traits favored in competition over mates. Arnqvist (1989, 1992a–c) found that female waterstriders

Gerris odontogaster show a general reluctance to mate, apparently owing to a conflict of interests between males and females. The reluctance is reduced as the sex ratio becomes increasingly male biased. Males with a well-developed abdominal organ used for grasping reluctant females therefore have reduced relative mating success as the sex ratio becomes more male biased. The relation between the sex ratio and the strength of sexual selection therefore need not be as simple as is often assumed. Other factors than the sex ratio, in this case female behavior, also influence the strength of sexual selection. Such factors may change with the sex ratio, sometimes making the relation between sexual selection and the sex ratio contrary to expectations (Arnqvist 1992b,c).

Differences in parental investment between the sexes is often a major influence on the operational sex ratio (box 7.2.7). In species where extra-pair fertilization and sperm competition is common, this aspect also needs to be included (Westneat et al. 1990).

These ideas shed light on the strong variation in mating systems and sexual selection among species; relative parental contributions alone do not account for all the variation. For example, among birds with little or no male parental care, species range from mainly monogamous, such as ducks (e.g., McKinney 1986), to extremely polygynous, such as the orange-rumped honeyguide *Indicator xanthonotus*, in which a few males are able to control the bees' nests to which reproductive females come to feed and mate (Cronin and Sherman 1977). Males in most ducks defend no external resource but the mate herself; as females do not flock during the breeding season, a male can rarely have more than one mate (Wishart 1983; McKinney 1986). Male-biased sex ratios also contribute to monogamy in ducks: when males were experimentally removed and females added to a local population, some of the remaining males were able to obtain two mates (Ohde et al. 1983; McKinney 1986).

Among mammals, males provide no parental care in over 95% of the species (Ridley 1978; Clutton-Brock 1989, 1991); those in which males do provide care tend to be monogamous. Apart from that, differences in parental care explain little of the variation in mating system and opportunity for sexual selection among mammals. The environmental potential for polygamy, especially the defendability of females by males, may explain much more of the variation. Food distribution and risk of predation determine female home range and group size, both of which influence the number of females that a male can monopolize (Jarman 1974; Bradbury and Vehrencamp 1977; Clutton-Brock 1989).

Ecological resource distribution for such reasons determines the environmental potential for polygyny, and hence the mating system, which in turn affects sexual selection in various ways (figure 7.2.2). There are causal pathways in both directions, however. Sexual selection influences male

BOX 7.2.7 WHAT DETERMINES THE OPERATIONAL SEX RATIO?

A formalization by Sutherland (1987) illustrates the role of some of the factors that determine the operational sex ratio. Consider a breeding period over which males and females can undertake one or more matings and associated activities, such as egg laying and parental care. Assume further that S is the absolute sex ratio of reproductive adults, T_f is the time devoted to breeding activities by a female, F_f is her number of reproductions during the breeding period, and H_f is the time required for each reproduction (egg formation, copulation, parental care, etc.). Corresponding symbols for males have index m (the time H_m required for each reproduction in males includes copulation, mate guarding, parental care, etc.). The operational sex ratio (males : females) is then

$$\mathrm{OSR} = S\,\frac{T_m - F_m H_m}{T_f - F_f H_f}.$$

In other words, the operational sex ratio is the product of the absolute sex ratio S and the ratio between the time remaining for obtaining mates in males and females, after the time spent mating, egg laying, etc., has been subtracted from the total time devoted to breeding. Such relationships are further clarified in models by Clutton-Brock and Parker (1992); also see Duvall et al. (1992).

The time available for obtaining mates may be spent searching for, attracting, or fighting over mates, depending on the biology of the species. If the available time is much smaller in one of the sexes, there may be many more members available for mating in the other sex, in which sexual selection is then likely to be strong. The operational sex ratio can vary among populations of the same species, and seasonally within a population. A detailed characterization of the operational sex ratio for a population should therefore be based on repeated measurements that reflect seasonal changes. Accurate measurements may be hard to obtain, for example if the population contains a fraction of nonterritorial males that easily escape census.

distribution in relation to resources. If, say, males that defend food territories have the highest mating success, resource defense polygyny is likely to evolve. If the most successful males do not defend resources, but advertise their presence by attractive display at "hotspots" particularly likely to be visited by females, a lek mating system with clustered males may evolve (e.g., Bradbury et al. 1985). Sexual selection is one of the major mechanisms that shape the mating system in relation to the resource distribution, and there are strong reciprocal causal links between mating system and sexual selection (figure 7.2.2). Sexual selection may also influence patterns of parental care (e.g., Maynard Smith 1977; Clutton-Brock 1991). For

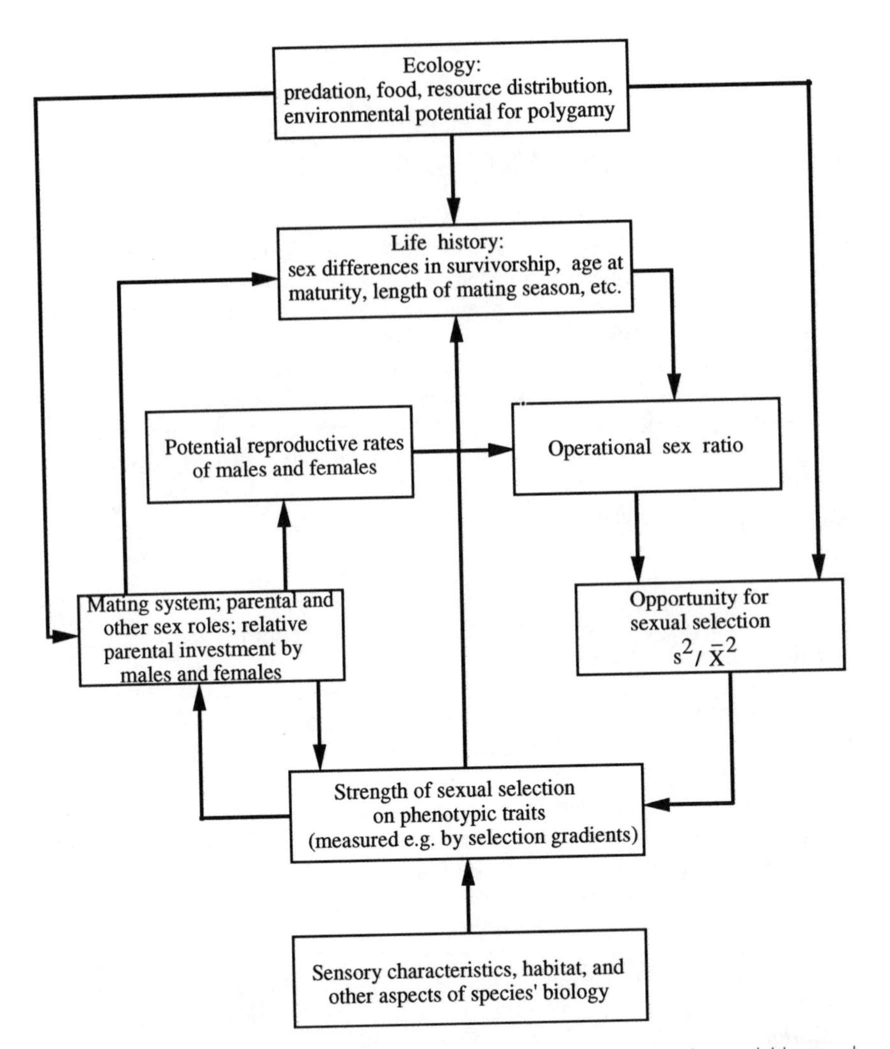

Figure 7.2.2 Suggested relationships among sexual selection and other variables, and their relations to ecology and life history. Phylogenetic constraints are important throughout the schema. Some possible causal paths have been left out for simplicity.

The boxes are connected by several closed loops, permitting feedback and mutual influences among sexual selection, mating system, parental and other sex roles, life history, ecology, and other variables. For example, if a change in the operational sex ratio leads to an increase in sexual selection, favoring males that spend more time on competition over mates and less time on parental care, the relative parental investment of males is reduced. This in turn can increase the operational sex ratio, amplifying the initial change. In other cases the feedback may be negative. For example, if the competitive ability of males is increased with age or size, sexual selection may favor higher age at maturity, reducing the number of sexually active males and hence the operational sex ratio. For these reasons mating system, parental and other sex roles, life history, and sexual selection will often be strongly interdependent.

The traits selected and the effects of sexual selection on secondary sex traits will depend on sensory characteristics, habitat, food and foraging behavior, predation risk, and other aspects of species' biology.

Table 7.2.1.a
Variance in Mating Success among Males of *Panorpa latipennis* in Relation to Resource Level*

Resource Level (No. of Crickets)	*Variance in Mating Success*
2	6.4
4	4.1
6	2.5

* Based on a total of 70 males and 140 matings for each resource level. After Thornhill 1986.

Table 7.2.1.b
Variance in Male Mating Success in Relation to Operational Sex Ratio in *Panorpa latipennis**

Males:Females per Cage	*Relative Variance in Success* ($s^2/\bar{X}$)	*% Males Unmated*
15:5	3.46	68
10:10	1.08	41
5:15	0.14	3

* After Thornhill 1986.

example, if there are few opportunities for obtaining additional females, a male may be more likely to remain with the family and help raise the offspring.

Thornhill (1986a) tested some of these ideas experimentally in natural habitats of the scorpionfly *Panorpa latipennis*, using cages with different sex ratios and food densities. Males defend dead arthropods or produce nutritious salivary masses, and attract females to the food by pheromones; some males instead attempt forced copulation. The relative variance in male mating success declined as food abundance (dead crickets) increased, and more males could present food to females (table 7.2.1.a). It also declined as the operational sex ratio became more female biased (table 7.2.1b), as suggested by Emlen and Oring (1977). There was an advantage of large male size in competition over mates. The advantage decreased when the sex ratio became more female biased, and some females had to accept small males in order to mate. Other experimental tests of what determines the opportunity for sexual selection have been done in orthopterans, corroborating many aspects of the theory (e.g., Gwynne and Simmons 1990; section 5.1 above).

In addition to ecological resource distribution and sex ratios, sperm competition can also influence mating tactics and mating systems (e.g., Parker 1970b; Smith 1984; Eberhard 1985; Westneat et al. 1990; Birkhead and Møller 1992). An example comes from *Spermophilus* ground squirrels. The relative importance of scrambles versus contest competition over females was studied in the thirteen-lined ground squirrel *S. tridecemlineatus* by Schwagmeyer and Brown (1983) and Schwagmeyer and Woontner (1985, 1986). Females in this species are dispersed at low densities. Male skill at locating females in estrus accounted for most of the variation in male mating success in three seasons ($r^2 = 0.83$, 0.82, and 0.62; fighting ability gave values of 0.05 or less). The first male to copulate with the female had an advantage and sired 75% of her offspring. Male mating success was not related to body size; in contrast with the situation in several related species, contests were mild and males were rarely injured (Schwagmeyer and Woontner 1986; Folz and Schwagmeyer 1989).

Idaho ground squirrels *S. brunneus* are also dispersed, but there seems to be no fertilization advantage for the first male. This may explain why males remain and guard females for several hours in this species, in contrast with the scrambling males of *S. tridecemlineatus* (Sherman 1989). In *S. beldingi*, females are clustered and there is a first-male advantage. Males fight over mating territories where females are clumped. The largest males have highest success; males do not guard females (Sherman and Morton 1984; Sherman 1989). The form of male competition over mates in *Spermophilus* ground squirrels therefore seems to depend on female dispersion as well as sperm competition (Schwagmeyer and Woontner 1988; Sherman 1989).

To sum up, ecological factors and the operational sex ratio can strongly influence the mating system, sex differences in parental care, and the form and opportunity for sexual selection. There are several reciprocal links among these variables, creating a complex network of mutual causal relationships (figure 7.2.2).

Potential Reproductive Rates of Males and Females

Several authors have remarked that there is sometimes no clear relation between sexual selection and parental investment or mating system (e.g., Vehrencamp and Bradbury 1984; Bradbury and Davies 1987; West-Eberhard et al. 1987). In particular, Trivers's concept of parental investment does not seem able to explain the variation in male and female competition for mates among species where males are responsible for parental care (e.g., Clutton-Brock and Parker 1992; Summers 1992b).

For example, in three-spined sticklebacks *Gasterosteus aculeatus* and some other fishes, the male cares alone for the offspring (Ridley 1978; Gross and Shine 1981; Wootton 1984). Yet he is more brightly colored and apparently subject to stronger sexual selection than the female. The relative amount of parental care does not determine sexual selection in these fishes (Williams 1966, 1975; Baylis 1981; Gross and Shine 1981; Gross and Sargent 1985; Clutton-Brock and Parker 1992). On the contrary, male parental care may in part be an indirect consequence of the particular form of sexual selection, where males attract females to spawning territories held for considerable periods, during which several females may be attracted. The male can then provide care for the offspring while he remains on the territory to attract more mates. (This may therefore be another example where sexual selection influences parental care, not only the other way around; figure 7.2.2). Territoriality is taxonomically more widespread than parental care in fishes, supporting the idea that male territoriality usually precedes the evolution of male parental care (Gross and Sargent 1985).

Williams (1975) suggested that paternal care in these fishes is not very demanding, but can be done at presumably little cost while the male defends his mating territory. In several species, females are even attracted by the presence of eggs in the nest, which may indicate lower risk of predation (e.g., Ridley and Rechten 1981; Unger and Sargent 1988). In the three-spined stickleback, males can guard up to six clutches at a time (Wootton 1984), and males seem to have a higher potential reproductive rate than females. Moreover, a male can probably care for many offspring without much loss in survivorship per offspring. Paternal care may not strongly reduce the male's future mating success or ability to invest in other offspring (Williams 1975; Baylis 1981; Gross and Shine 1981; Gross and Sargent 1985). Paternal care then is not a resource over which females have to compete in these fishes; hence it does not lead to a reversal of other sex attributes (Williams 1975). This may explain why males are more ornamented although they care alone for the offspring in some of these territorial fishes. To test this idea, the costs of paternal care in terms of reduced future reproduction (due, for example, to slower growth) need to be measured and compared to female costs, measured in the same currency.

Looking at sexual competitiveness from another angle, Clutton-Brock (1991) and Clutton-Brock and Vincent (1991) suggested that the sex with highest potential rate of offspring production should compete most strongly over mates (see also Williams 1975; Berglund et al. 1989). Among species with parental care, this idea agrees better with patterns of sexual selection than do sex differences in parental investment, or rate of gamete production. The latter aspect (Baylis 1981) does not explain why females are the most competitive sex, for example in some pipefishes Syngnathidae and in the mouth-brooding cardinal fish *Apogon notatus* (Berglund et al. 1989; Kuwamura 1985). In these as in most animals, the rate of gamete production is probably highest in males and gives no hint as to why females compete more strongly for mates in some species. Since, however, males in these fishes bear offspring in a limited body space, the potential rate of male reproduction seems to be limited by that space, and by the developmental time of offspring. Females seem to have higher potential rates of reproduction than males (Kuwamura 1985; Berglund et al. 1989; Clutton-Brock 1991; Clutton-Brock and Vincent 1991).

Potential reproductive rate may be easier to quantify than parental investment. The two approaches often make similar predictions concerning sexual selection, as the potential reproductive rate of a sex will depend on the parental investment it makes. Potential reproductive rate seems to have greater explanatory power, however, particularly as regards the variation among species with male uniparental care (Clutton-Brock and Parker 1992).

Comparing species from 23 genera of fishes, frogs, and birds, all with exclusive male parental care, Clutton-Brock and Vincent (1991) concluded

Table 7.2.2

Potential Reproductive Rate of the Sexes in Relation to Mating Competition, for Species in Which Only Males Provide Parental Care*

Maximum Observed Female Rate / *Maximum Observed Male Rate*	*Competition More Intense in Males*	*Competition More Intense in Females*
< 1	Fish: *Cottus* (2 spp.)	
	Oxylebius pictus	
	Chromis notata	
	Chrysiptera cyanea	
	Badis badis	
	Pimephales promelas	
	Etheostoma olmstedi	
	Gasterosteus aculeatus	
	Forsterygion varium	
	Frogs: *Alytes obstetricans*	
	Hyla rosenbergi	
	Eleutherodactylus coqui	
> 1	Fish: *Hippocampus* spp.	*Apogon notatus*
		Nerophis ophidion
		Syngnathus typhle
	Birds: *Rhea americana* ?	*Actitis macularia*
		Phalaropus (2 spp.)
		Eudromias morinellus
		Jacana (5 spp.)
		Rostratula bengalhensis
		Turnix sylvaticus

* From Clutton-Brock and Vincent 1991.

that sexual selection was strongest in the sex with the highest potential reproductive rate in all but two cases. Male rates were higher among most of the fishes (9 of 13 genera) and frogs (all of 3 genera), and female rates were higher among the birds (all of 7 genera) (table 7.2.2).

Clutton-Brock (1991) emphasized that the operational sex ratio tends to become biased when some individuals of the most competitive sex have several mates; this accentuates sexual selection. Sex role reversal and stronger competition for mates among females than among males should occur when females have higher potential rate of reproduction than males, and the operational sex ratio is female-biased. The ratio will also be influenced by other aspects, for example age at maturity, which in some species may be higher among males, in others among females.

The above ideas may partly explain an unusual combination of male uniparental care and sexual selection in some taxa other than fishes. In most birds, the sex most active in territory defense normally provides less parental care than the other sex. Puzzling exceptions occur among tinamous and ratites. In the ostrich and rheas, the male is probably the subject of stronger sexual selection than the female: he is larger, defends a territory, and displays for females (Bruning 1974; Handford and Mares 1985; Bertram 1992). Several females lay eggs in the same nest, which the male then takes care of alone (in the ostrich, one female also incubates). Part of the reason may be that the male can incubate more eggs than a single female can lay, and that chicks are well developed at hatching. The offspring of several females therefore may not need more care than a male can provide. In tinamous and rheas, females may sometimes be polyandrous, laying eggs for another male when the first mate incubates. Many aspects of the biology and mating system of ratites and tinamous are poorly understood (Handford and Mares 1985).

Game-theoretic modeling of reproductive tactics in the two sexes offers further potential insight into sex roles and sexual selection, for example whether contests or mate choice will predominate in a species. The games concern such aspects as parental investment, which sex will search for mates, and whether a certain mate should be accepted or not (e.g., Maynard Smith 1977, 1982; Parker 1979; Vehrencamp and Bradbury 1984; Houston and Davies 1985; Hammerstein and Parker 1987; reviewed by Clutton-Brock and Godfray 1991). The interests of male and female often differ in ways amenable to analysis by game theory, which seems likely to further clarify the evolution of parental care and mating systems, and the nature of sexual selection (e.g., Bradbury 1985; Bradbury and Davies 1987; West-Eberhard et al. 1987).

In addition to the factors discussed above, variation in quality among potential mates, for example in parental skills and investment, will also influence the strength of sexual selection, which should increase with such variation (e.g., Andersson 1982b; Halliday 1983; Petrie 1983a; Summers 1992b). Such effects are discussed in the next section.

7.3 Monogamy and Sexual Selection

Most studies of sexual selection concern polygynous animals with conspicuous secondary sex traits; monogamous species tend to be less sex dimorphic than related polygynous forms (e.g., Brown 1975; Clutton-Brock et al. 1977; Kleiman 1977). There can, however, be competition over mates also in monogamous species. With strict monogamy, no individual has more than one mate, but some will have none if the sex ratio is not unity. In addition, mates vary in fecundity or parental ability depending on age, condition, and other aspects. Such differences in quality among mates can lead to sexual selection in monogamous as well as polygynous species, and in females as well as males. Sexual selection for these reasons can be strong also in monogamous species (Darwin 1871; Fisher 1930; O'Donald 1980b; Mock 1985; Kirkpatrick et al. 1990; also see Darwin's finches, section 4.5). Moreover, in many species monogamy is not total; some males achieve higher reproductive success by polygynous matings or extrapair copulations (e.g., Gowaty 1985; Westneat et al. 1990; Birkhead and Møller 1992; Kempenaers et al. 1992).

Variation in Mate Quality

To explain sexual selection based on mate quality in monogamous animals, Darwin (1871) used birds as an example, suggesting that the females in best condition will be the first to breed in spring. Their good condition and long available season permit these females to raise more young than others. "The males, as we have seen, are generally ready to breed before the females; of the males the strongest, and with some species the best armed, drive away the weaker males; and the former would then unite with the more vigorous and best-nourished females, as these are the first to breed." Moreover, "the more vigorous females, which are the first to breed, will have the choice of many males; . . . they will select those which are vigorous and well armed, and in other respects the most attractive" (Darwin 1871, I, p. 261). This idea was supported and further elaborated by Fisher (1958, p. 153).

Lack (1968) dismissed the idea, arguing that the date of breeding in birds has evolved in relation to the ability of the female to produce eggs, and to the time when food is available for raising chicks. But Darwin's idea is compatible with Lack's (1968) argument, which applies to the mean start of breeding, whereas Darwin was concerned with its variation among females (Trivers 1972). O'Donald (e.g., 1980b, 1987) tested the logic of Darwin's (1871) and Fisher's (1958) hypothesis by few-locus genetic models, verifying that sexual selection in principle can work in monogamous species. So did Kirkpatrick et al. (1990) with quantitative genetic models (see section 2.3).

The clearest evidence for variation in the phenotypic quality of mates comes from animals with indeterminate growth, where larger females are more fecund (section 11.2). Males that mate with them sire more offspring than those mating with smaller females (e.g., Howard 1988b; figure 11.1.2 below). Such differences in fecundity might lead to strong sexual selection even in monogamous species. Also in polygynous species, successful males may not only get more mates, but mates in better condition; the yellow-rumped cacique *Cacicus cela* is an example (Robinson 1986a,b).

Sexual selection arising from differences in mate quality and time of breeding was studied by O'Donald (e.g., 1983b, 1987) in a monogamous seabird, the arctic skua *Stercorarius parasiticus*. This species is polymorphic in color. In the southern part of the range, where the study took place, the pale morph is selected against, and the morph ratio is apparently maintained by immigration of pale birds from more northerly areas where the pale:dark ratio increases, for unknown reasons. The study corroborated Darwin's (1871) idea that the females breeding earliest in the season tend to have highest fecundity (also see Price et al. 1988; Kirkpatrick et al. 1990). Moreover, the males with the largest territories mated earlier than others. Territory size therefore appears to be sexually selected. Males of the dark morph started their first breeding earlier in the season than pale males. Testing several models against dates of egg laying, O'Donald (1983b, 1987) concluded that sexual selection through female mating preferences for dark males explains their earlier mating. Direct behavioral observations might permit further tests of this idea.

In the monogamous moorhen *Gallinula chloropus*, females compete over mates and prefer fat or heavy males (Petrie 1983a). Such males seem to be efficient incubators, and have larger territories than others. Petrie (1983) found that large females were favored in competition over males in good condition. Also, males fight over mates and territories; large males were favored in fights, and had larger territories and higher reproductive success than small males (Gibbons 1989).

Burley (e.g., 1977, 1986c) demonstrated in laboratory experiments with zebra finches *Poephila guttata* that sexual selection can work in monogamous species. The attractiveness as mates was influenced by the color of the leg rings used for individual recognition (also see section 13.6). In a long-term test, birds with attractive ring colors had twice the reproductive success of birds with unattractive colors. This difference apparently arose in two ways: (1) attractive birds pair with high-quality mates, which are able to produce more offspring than others; (2) the mates of attractive birds make greater parental efforts than typical for their sex. As a consequence, offspring condition is improved, or the attractive mate can spend less parental effort, hence surviving better. There was evidence for both effects (Burley 1986c, 1988b). Sexual selection in another monogamous species, the pinyon jay *Gymnorhinus cyanocephalus*, was analyzed by Johnson (1988a,b), Johnson and Marzluff (1990), and Marzluff and Balda (1988).

Sexual selection of secondary sex ornaments in a monogamous species has been demonstrated experimentally in the swallow *Hirundo rustica*. The two outer tail feathers of the male are about 15% longer than those of the female. Møller (1988a) found that males with elongated tail streamers had higher reproductive success than males with shortened streamers; elongated males mated earlier in the season, and were more likely to have second clutches. A subsequent study (Møller 1991b) also showed that the most preferred males acquire mates of higher than average phenotypic quality, leading to yet another kind of reproductive advantage for such males. On average, they had more than twice as many fledging young as males with shortened tails. In addition, males with elongated tail performed more extra-pair copulations. Behavioral observations suggested that female choice rather than male contest competition was responsible for their higher success. Møller (1992a) also found that female choice favors symmetric as well as long tail streamers (see section 3.6 above).

In another experimental study of the same species in Canada, Smith and Montgomerie (1991) obtained results partly similar to Møller's (1988a) in Denmark. Elongated males in the Canadian population had a shorter prelaying period from the arrival until the female laid her first egg, indicating that such males mated before males with shortened streamers. In contrast with the Danish study, this mating advantage did not carry over into greater production of offspring. Among the possible reasons for this discrepancy are differences in the experimental procedure and less strong sexual selection in the Canadian swallows, where males help incubate the eggs and have shorter tail streamers than European swallows (Smith and Montgomerie 1991; Smith et al. 1991). DNA fingerprinting showed that elongated males fathered fewer than two thirds of the young in their nests, whereas shortened males fathered almost all their nestlings. Tail elongation probably hampered the ability of a male to guard his mate against extra-pair fertilizations. Among unmanipulated males, however, there was a negative correlation between male tail length and the proportion of illegitimate nestlings. The results therefore indicate that females prefer long-tailed males, which under natural conditions seem to be of higher quality than males with shorter tails (Smith et al. 1991).

Analysis of parasite levels suggested that the length of the tail streamers reflects phenotypic condition: their length was negatively correlated with degree of parasite infestation in adults (Møller 1990a), as expected if the development of secondary sex traits is condition-dependent (see section 3.5). These studies of swallows show that sex-dimorphic male ornaments can be favored by female choice also in monogamous species.

Skewed Sex Ratios

Sexual selection in monogamous species may often arise from skewed sex ratios. For example, in ducks, males outnumber females among adults, in some cases by more than 2:1 (e.g., Bellrose 1980). Such male-biased sex ratios may give rise to strong sexual selection in monogamous species, as demonstrated in Darwin's finches (section 4.5; Price 1984b; Grant 1986; Grant and Grant 1989). In *Geospiza fortis*, adult males outnumbered females by 2:1 or even 3:1 owing to higher female mortality. This skewed sex ratio created much opportunity for sexual selection in males, half their variance in offspring production being related to mating success (Price

1984b). Male-biased sex ratios and competition for mates may also be one of the reasons why most ducks form pairs already on the wintering ground (Lack 1968; McKinney 1986), and why males in many ducks are conspicuously colored. The evolution of sexual color dimorphism in monogamous ducks is most pronounced in species with serial monogamy, frequent remating, and markedly different sex roles, with only the female taking care of the offspring (Scott and Clutton-Brock 1990; section 13.6 below).

In sum, skewed sex ratios and differences in mate quality can give rise to strong sexual selection also in monogamous species.

7.4 Sexual Monomorphism

The sexes are similar in many monogamous species, for a number of possible reasons. One is that sexual selection is so weak compared to counteracting natural selection that no secondary sex differences have evolved. This seems possible, for example, in those mammals and birds where the sex roles are similar and adults do not differ markedly from juveniles in appearance (see Kleiman 1977 for mammals). In many monomorphic species, however, males are more active than females in display or territory defense, which suggests that males are subject to stronger sexual selection, but with mainly behavioral rather than morphological consequences.

A second possible reason is genetic correlations between the sexes, leading to secondary sex traits not only in the selected but in the other sex as well. Nonadaptive correlated expression is probably often important (e.g., Fisher 1930; Lande 1980; Lande and Arnold 1985), but there is no obvious reason why it should be more pronounced in monogamous than polygamous species. Other differences probably explain the pattern.

Third, Burley (1981c) suggested that selection for sexual indistinguishability may sometimes explain why males and females are similar. Individuals might benefit from concealing their sex most of the time, for example in monogamous, social animals with strong sexual competition. Communal and colonial birds appear to have similar sexes more often than do solitary species. Björklund (1984) suggested, hower, that monomorphism in monogamous birds is caused by similar natural selection of the two sexes, in combination with only weak sexual selection.

A fourth possibility is that competition over mates, or other social competition, is similar in the two sexes (e.g., Darwin 1871; West-Eberhard 1983; Trail 1990). This is likely in some species where males and females have similar parental roles, and the sex ratio is close to unity (Payne 1984; but see Trail 1985c). Even if competition over mates is strong (for example, because mates vary greatly in parental ability) this may hold for both sexes, leading to similar sexually selected traits in males and females. One possible example is the red-winged blackbird *Agelaius phoeniceus* in Cuba

(Whittingham et al. 1992). Others are monogamous birds in which both sexes change from juvenile to a more conspicuous adult plumage, such as the great crested grebe *Podiceps cristatus*, described in a classic study by Huxley (1914). He explained the crest by natural as opposed to sexual selection, arguing that the crest is developed and displayed similarly in males and females, and that the display occurs mainly after the mating period. Both points are questionable. There may be competition over mates in both sexes, which have similar parental roles. Later studies have shown that the crest, which is larger in males, is used in display also during pair formation, before egg laying (Cramp et al. 1977).

A recent experimental study of the crested auklet *Aethia cristatella* (figure 7.4.1) supports the idea that mutual sexual selection by mate choice may favor similar display structures in both males and females of a monogamous species. Manipulating the size of the ornamental crest in realistic models made from mounted skins, Jones and Hunter (1993) showed that males as well as females respond to accentuated models of the other sex

Figure 7.4.1 The crested auklet *Aethia cristatella* is a monogamous species in which both sexes have ornamental crest feathers, and mutual preferences for such feathers. (Jones and Hunter 1993; birds after photograph by J. Unosson)

with more frequent sexual displays (see also Jones and Montgomery 1991, 1992).

Noble (1936) found that the yellow underside of the tail in a woodpecker, the flicker *Colaptes auratus*, is displayed by both males and females, mainly in contests with individuals of the same sex. This suggests the bright coloration is favored by sexual contest competition in females as well as in males, an idea that might be tested by larger sample size in an experiment similar to Noble's (1936; also see Short 1972; W. S. Moore 1987). The possibility that sexual selection in monogamous species with unequal sex roles might favor different attributes in males and females is discussed in section 11.8 (birds of prey).

Comparing two lekking birds, the color dimorphic cock-of-the-rock *Rupicola rupicola* and the monomorphic capuchine bird *Perissocephalus tricolor*, Trail (1990) found that females were often aggressive toward one another in the latter species, but not in *Rupicola* (3% vs. 17% of female courtship visits were terminated by female interactions, respectively). Trail (1990) proposed that monomorphism in lekking birds is a result of intrasexual competition in females as well as in males.

It is still not clear why some monogamous species have adult display structures, whereas other monogamous as well as some lekking and other polygynous species lack such structures. One possibility is that sex dimorphism in some species is expressed in behavior rather than morphology (e.g., Höglund 1989b). Among bowerbirds, the species with most elaborate bowers often have the least conspicuous male plumage (Gilliard 1969). Trail (1990) pointed out, however, that some species have more elaborate behavioral as well as morphological dimorphic traits than others. Although different kinds of display traits in some taxa tend to replace one another, there is often no simple replacement of one kind of dimorphism by another (e.g., Shutler and Weatherhead 1990). For example, the male of the superb lyrebird *Menura novaehollandiae* has one of the most remarkable ornamental tails found in birds; he also builds a number of large display mounds on the territory, and he has a conspicuous, elaborate song with much mimicry (e.g., Rowley 1975; L. H. Smith 1982).

The reasons why some species mainly evolve behavioral secondary sex traits, others also morphological ornaments, remain to be clarified by comparative work.

7.5 Leks and Sexual Selection

For several reasons, lekking animals are of special interest in sexual selection theory (e.g., Bradbury and Gibson 1983; Balmford 1991; Wiley 1991). Since females get no direct material benefits from males on leks, such species may offer better prospects than others for identifying traits by which females choose mates. Mating success often varies greatly among the males on a lek. Males in many lekking species have conspicuous morphological ornaments that may be targets of female choice, but male contest competition may also be involved. As males in lekking species are free from any constraint imposed by parental behavior, and as a few males get most of the matings, sexual selection and dimorphism are expected to be stronger in lekking than in other related species (e.g., Darwin 1871; Payne 1984). The degree to which this is actually the case is debated.

In spite of much study, female preferences for male ornaments have rarely been convincingly demonstrated in lekking animals. Among the possible reasons are that (1) females tend to favor certain sites rather than certain males; (2) strong sexual selection depletes genetic variation in the preferred traits; (3) the variation in male mating success also depends on other aspects that vary more than ornaments, especially male attendance and behavior at the lek; (4) females often copy others in their choice of mate. These factors can make female preferences for male ornaments hard to identify (Gibson et al. 1991; Balmford 1991; Wiley 1991). It is also possible that (5) male ornaments in lekking species are favored not by female choice, but mainly by contests among males. This alternative has rarely been tested in detail.

About one quarter of all lekking birds are sexually monomorphic. Trail (1991) suggested that monomorphism in lekking species is favored by strong sexual and other social selection in both sexes (see section 7.4). Höglund (1989b) made a phylogenetic analysis (based on Ridley's outgroup comparison method, section 4.3 above) of size and plumage dimorphism in relation to lekking in ten families of birds (Tetraonidae, Otididae, Scolopacidae, Psittacidae, Cotingidae, Pipridae, Tyrannidae, Pycnonotidae, Paradisidae, and Ploceidae). He concluded that neither plumage nor size dimorphism was correlated with lekking, which did not precede dimorphism phylogenetically more often than the other way around. Males are larger than females particularly in species displaying on the ground; aerial or arboreal display seems to select for small agile males (Höglund 1989b; also see Payne 1984, and chapter 11).

Ridley's (1983) comparative method requires that data be classified in discrete categories, leading to loss of information for continuous traits such as size. To use more of the information and make a stronger test, Oakes (1992) reanalyzed Höglund's (1989b) data using Felsenstein's (1985) method of matched pairwise comparisons. It avoids the problems of statistical dependence that affect some directional comparative methods (see Harvey and Pagel 1991). With this more sensitive method, Oakes (1992) found that sexual size dimorphism in wing length is greater in lekking than in closely related nonlekking birds, also when body size is controlled for.

Greater sexual dimorphism is expected in lekking than in other species if the roles of males and females differ more strongly, and if the opportunity for sexual selection is larger, in lekkers (section 7.2). Male lifetime reproductive success may, however, not show larger variance in lekking than in other polygynous species. Even if mating success on a lek is highly skewed in favor of a few males in a given year, mating success is often strongly age-dependent (e.g., Kruijt and de Vos 1988a). The lifetime variance among males may be much smaller than the yearly variance (Clutton-Brock 1983, 1988a). In addition, sexual selection may be stronger than previously expected also in monogamous species (section 7.3).

Copying the mate choice of other females is common in some lekking and other polygynous animals. Copying may be adaptive when sampling and choice of mates is costly, for example because males are difficult to assess (e.g., Boyd and Richerson 1985; Losey et al. 1986; Reynolds and Gross 1990; Gibson et al. 1991; Gibson and Höglund 1992; Pruett-Jones 1992). There is observational evidence of copying, for example in isopods, fish, deer, and grouse (Marconato and Bisazza 1986; Clutton-Brock et al. 1989; Gibson et al. 1991; Shuster and Wade 1991b; Balmford et al. 1992), and experimental evidence from the guppy *Poecilia reticulata* (Dugatkin 1992; Dugatkin and Godin 1992). Copying among females leads to higher variance in male mating success, and hence formally to greater opportunity for sexual selection (Wade and Pruett-Jones 1990).

Copying might under some circumstances lead to stronger sexual selection by increasing the correlation between male mating success and his secondary sex traits, but this is not a necessary consequence (Gibson et al. 1991). One reason is that with copying, fewer females are assessing the males independently. As the number of independent choices is reduced, chance may have a greater influence on the outcome, and correlations between male trait size and mating success may be weaker than if all females independently assessed and chose mates. Female copying (and site fidelity) might then lead to less pronounced male sex traits than if all females were choosing males independently. Copying among females can also lead to problems in the analysis of mate choice, as females then cannot be treated as statistically independent units.

Although females obtain no material resources from males on leks, and the heritability of the father's fitness is probably low (section 3.4), female choice is often conspicuous in lekking species. This has been called the "lek paradox" (Borgia 1979; Taylor and Williams 1982; Williams 1992). Kirkpatrick and Ryan (1991) extended the term lek paradox to all species in which females seem to receive little more from males than their sperm.

Some such species, for example guppies and swordtails, do not seem to be lekking in the traditional sense of the word, but the problem highlighted by the lek paradox, if there is one, applies also to other than lekking species in which males provide nothing but sperm to females.

The paradox may be more apparent than real, however. First, it is doubtful whether females choose more carefully among males in lekking than in other species: female site fidelity and copying on leks suggests that this may not be the case. Second, it has been suggested that female choice of certain males on leks may lead to reduced risks of predation, harassment by undesired males, mating disruption, infertile matings, or contraction of diseases or parasites (e.g., Wrangham 1980; Clutton-Brock et al. 1988a; Borgia and Collis 1990; Reynolds and Gross 1990; Gibson et al. 1991; Kirkpatrick and Ryan 1991; Wiley 1991). These possibilities largely remain to be studied.

Female choice of spawning site rather than male has been well demonstrated in a coral reef fish, the bluehead wrasse *Thalassoma bifasciatum* (Warner 1987, 1988a, 1990). In lekking animals, however, the importance of male lek position seems to vary among species. Some studies have concluded that there is a position effect, usually favoring central males (e.g., Kruijt et al. 1972; Kruijt and de Vos 1988; Wiley 1973; Höglund and Lundberg 1987; Clutton-Brock et al. 1988a; Gosling and Petrie 1990; Appollonio et al. 1992; Balmford et al. 1992a); other results are negative (e.g., Foster 1977; Gibson and Bradbury 1985; Andersson 1989; Clutton-Brock et al. 1989). Where central males are favored, the *position hypothesis* suggests that females for some reason prefer to mate at central sites, for which males therefore compete. The *attractiveness hypothesis* on the other hand suggests that some males are more attractive than others on the lek, which cluster around attractive males and attempt to mate with females visiting them (Waltz 1982; Arak 1983b, 1984; Beehler and Foster 1988; Höglund and Robertson 1990). In the first case therefore, some males are successful because they occupy a central, favored site; in the second, successful males become central because they are attractive. The two mechanisms might work together, if the most attractive males also get the most attractive sites.

Male mating success in lekking species may depend on a variety of factors, such as male lek attendance, dominance, and female preferences for particular territories and male phenotypes. Female choice may be based on several different male cues (Bradbury and Gibson 1983; Balmford 1991; Wiley 1991). As discussed above, females may also copy each other. After a thorough long-term study of sage grouse (see below in this section), Gibson et al. (1991) felt that all these factors probably play a role. A challenge for future studies of lekking species is to estimate their relative importance. The following brief review illustrates the variety of results obtained.

An Australian Frog

Males of *Uperoleia laevigata* (formerly *U. rugosa*) defend small territories in aggregations near breeding ponds (Robertson 1984, 1986a,b, 1990). Males have three kinds of calls: (1) a long-distance advertisement call; (2) an encounter call used in contests between males; (3) a courtship call given when the male perceives a female nearby. The fundamental frequency of a frog call tends to decrease with increasing male size (section 14.2). In play-back experiments with advertisement calls from males differing in size, 17 males that were at least as large as the played-back male attacked the loudspeaker, whereas only 3 out of 22 lighter males attacked. This suggests that the advertisement call in large males is favored by sexual contest competition. Encounters between two calling males were observed 220 times during 480 hours. Some fights were violent: two males had broken blood vessels where they had been clasped, and another male died an hour after losing a vigourous fight (Robertson 1986b). Calling and holding a territory is energetically costly: territorial males lost up to 30% mass per month. Some gave up calling, became silent satellites near calling males and put on mass, and then regained a territory and resumed calling.

A gravid female spends one or more nights moving slowly among the males, apparently listening to several before approaching and touching one of them. The male then clasps the female, and she carries him to the pond for mating and egg laying. There is no parental care. Robertson (1986a) tested female responses to male advertisement calls in arenas with two loudspeakers, one broadcasting calls of a light male, the other of a heavy male. Of 18 females tested, 10 preferred the call of the heavier male; 8 were not attracted to either call; none preferred the call of the lighter male. In another similar experiment with synthetic calls, the females chose between a call with frequency corresponding to a male 70% the mass of the female (the most preferred mass) and calls with 2%–10% higher or lower frequency. Twenty-two out of 29 females preferred the call representing a male 70% of her own mass (Robertson 1990). A potential problem in this and many other similar studies is that the test used a sample of females but calls that represent only a single male (or a single synthetic call) of each category (see section 14.5).

Males in amplexus pairs usually weigh about 70% as much as the female (figure 7.5.1). In laboratory tests, egg-hatching success declined drastically if the male was not 65%–80% the mass of the female. In addition, a too heavy male may drown the female (Robertson 1990). This is one of the few studies that have identified a direct phenotypic benefit of female choice in a lekking species.

Robertson (1986a) separated the mates in twelve pairs found in amplexus in the field and put the male back in the chorus. He usually resumed

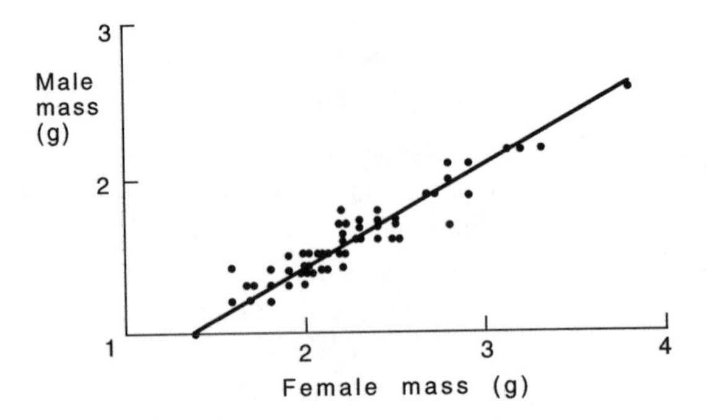

Figure 7.5.1 Body mass of male and female in mating pairs (amplexus) of the Australian frog *Uperoleia laevigata.* Females pair with males that are about 70% their own mass. (After Robertson 1990)

calling within 3 minutes; the female was then released in the chorus at roughly equal distance from about five males. Each of the twelve females consistently returned to her previous mate. Repeated trials with three or four separations confirmed that she returned to the original mate, even if he was moved from his original calling site. These and other observations suggest that females after separation returned to their previous mate in response to some aspect of his call. Together with the loudspeaker tests and observations of natural mating events, this is evidence for female choice of mate in relation to his call. It is not clear, however, to which aspects of the call the females responded in the field. The dominant frequency may be important, but due to large statistical scatter it is a poor predictor of male size; the 95% confidence limits cover a large portion of the male size distribution. The experiments and field observations suggest that female choice is much more accurate than the confidence limits would suggest. Female choice is therefore probably based on more cues than the dominant frequency of the call (Robertson 1986a; Gerhardt 1988).

Tetraonids

In sage grouse *Centrocercus urophasianus*, male mating success is related to his lek attendance and display rate, and to an acoustic component of the strut display (Gibson and Bradbury 1985; figure 7.5.2 and 10.4.1). When the natural display of males was supplemented in a controlled experiment with loudspeaker playback of a successful male, females tended to approach males with playback support, not only on days of playback, but also the day after, suggesting that some females remembered and returned to the site of the attractive display (Gibson 1989; figure 7.5.2).

Display is energetically costly, so metabolic constraints may limit male lek attendance and mating success (Bradbury et al. 1989a,b; Vehrencamp et al. 1989; section 10.4 below). There was no evidence that the relative

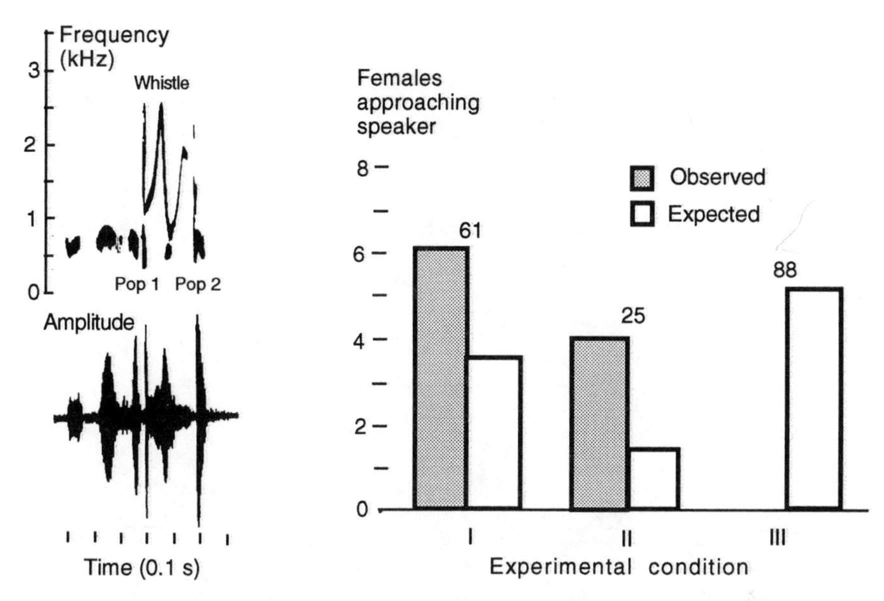

Figure 7.5.2 *Left*: Sonogram (top) and oscillogram of the last part of a typical strut display of the sage grouse, *Centrocercus urophasianus*. Three coo notes are followed by the first pop, a frequency-modulated whistle, and the second pop. *Right*: Observed and expected numbers of females approaching the speaker's location under three conditions in Gibson's (1989) experiment: I, playback days; II, non-playback days preceded by a playback day; III, non-playback days preceded by a non-playback day. (Numbers of female lek days are given above bars for each condition.) (After Gibson 1989; Gibson et al. 1991)

position on the lek, or the size of a male's territory, was important (Wiley 1973 found a position effect). Females tended to return to lek territories where they mated the previous year, and there was copying of mate choice among females. Both these latter results suggest that female assessment and comparison of males may be costly (Gibson et al. 1991). Estimating the various costs of female choice, Gibson and Bachman (1992) concluded, however, that it is low. For example, it should reduce annual survival by only 0.1%. The results suggested that even very small indirect (genetic) or direct benefits might suffice to favor female choice in this species.

THE BLACK GROUSE *Tetrao tetrix* is one of the classical lek species that, through the English hunter and naturalist Lloyd (1867), has supplied the local Scandinavian word "lek" (play) for this mating system. The cocks defend small territories on arenas in open ground, permitting detailed observations in field studies (e.g., Hjorth 1970; Kruijt et al. 1972; de Vos 1979, 1983; Kruijt and de Vos 1988; Alatalo et al. 1991, 1992). During the mating season, the female visits a number of displaying males and copulates with one of them. Females favor central cocks in the densest part of the lek, but at least one solitary male in the study of Kruijt and de Vos

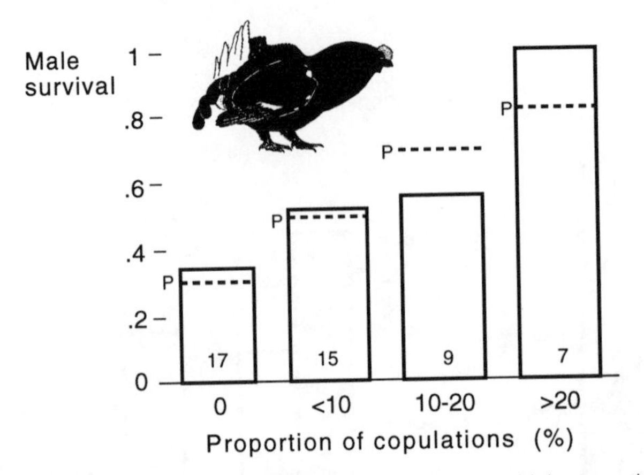

Figure 7.5.3 Among male black grouse *Tetrao tetrix*, those with highest mating success also had higher survival than others. Sample sizes are given in the bars; hatched lines indicate predictions from a logistic regression model. (After Alatalo et al. 1991; bird from Hjorth 1970)

(1988) also had high success. Competition for central sites is strong; mainly males in their prime age (3–5 yr) occupy such sites. Older males seem to lose physical condition (de Vos 1983; Kruijt and de Vos 1988). Cocks with high mating success tend to produce the "rokooing" song more rapidly than less successful males. Song rate (which, like the ability to hold a central territory, probably depends on male physical condition) therefore may influence female choice (Kruijt and de Vos 1988).

Alatalo et al. (1991) showed by experiments and observations that dominant males have highest mating success, and are able to keep other males farther away from visiting females. Dominant, successful males had intact tail ornaments, whereas many other males had tail damages, usually inflicted in fights. Preferred males had higher survival than others (figure 7.5.3). Choosing dominant males, females will therefore have their offspring fathered by males with high viability. To the extent that viability is heritable, this might lead to higher offspring survival. It is also possible that females have some direct benefit of choosing viable males (see above in this section).

Sexual selection in lekking peacock is reviewed in section 13.6 below.

Birds of Paradise

With their magnificent male plumages and conspicuous, even bizarre displays (e.g., Cooper and Forshaw 1977; and figure 7.5.4), birds of paradise seem an epitome of sexual selection. Owing to their mainly New Guinean forest distribution, however, they are hard to study and have only recently been subject of quantitative field work.

Figure 7.5.4 Sexual dimorphism in three birds of paradise. *Top*: Wilson's, *Diphyllodes respublica*. *Middle*: Superb, *Lophorina superba*. *Bottom*: Twelve-wired bird of paradise, *Seleucidis melanoleuca*. (From Jägerskiöld 1908)

In Lawe's parotia *Parotia lawesii*, males attract females to display courts on the ground. Males are clustered, but with considerable variation in dispersion (Pruett-Jones and Pruett-Jones 1990a). Male mating success is correlated with courtship behavior, especially with the probability and rate of display to visiting females. Females visit several different males before mating. There was no evidence that contests or disruptions affect male mating success, which was apparently not site related. Female choice, not male contests, seems to be the main agent of sexual selection in this species (Pruett-Jones and Pruett-Jones 1990a).

In the classically lekking *Paradisaea decora*, male dominance and interactions on the lek seem important for male mating success. LeCroy et al. (1980) and LeCroy (1981) concluded that female choice plays no role, and that dominance contests among males determine who mates. There were no quantitative observations, however, and in the closely related lesser bird of paradise *P. minor*, field observations suggest that female choice influences male mating success (Beehler 1983). In a larger study of the national bird of Papua New Guinea, the raggiana bird of paradise *P. raggiana*, Beehler (1988) found that (1) copulation did not occur without prior female solicitation, (2) any solicitated male could copulate with the female. Females therefore can choose among males on the lek, but it is not known which male traits make for success. Aggression might also influence male chances of mating; possibly, contests partly determine access to the lek.

In sum, female choice influences male mating success in at least some birds of paradise. Male contests are probably also important, but this has not been convincingly shown. The detailed mechanisms favoring the magnificent plumage ornaments of birds of paradise remain to be demonstrated.

Bowerbirds

Males in most of the eighteen species of bowerbirds (Ptilonorhynchidae, apparently less close relatives of birds of paradise than previously thought; Sibley and Ahlquist 1990) attract females by building an elaborate, decorated structure, the bower (figure 7.5.5). The most extreme species, *Amblyornis ornatus*, shows local, probably cultural differences between areas, its bowers ranging from high maypoles to a stick hut up to 2 m in diameter. Bowers are often decorated with conspicuous flowers, colored fruits, snail shells, leaves, butterfly wings, and beetle elytrae. Using colored plastic chips, Diamond (1986a,b, 1987, 1988) found that certain colors are more attractive than others to males, which steal decorations from each other. Construction and maintenance of the bower probably reflect male quality, age, and dominance; a nicely decorated bower may signify a dominant male in good condition, free from genetic or phenotypic disorders, and therefore a favorable mate (Borgia et al. 1985; Diamond 1988).

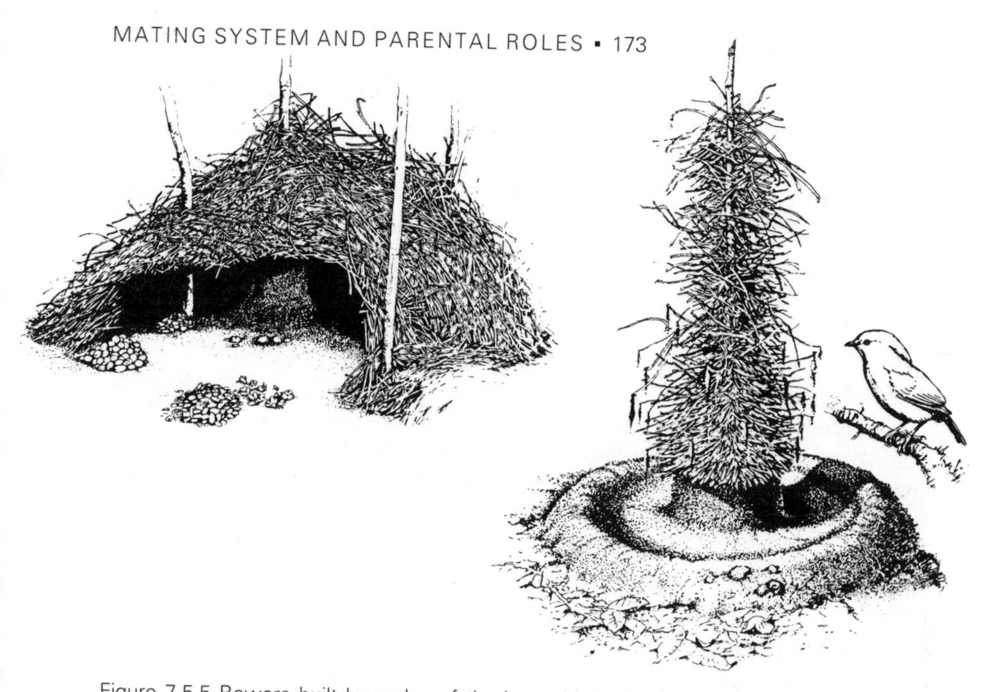

Figure 7.5.5 Bowers built by males of the bowerbirds *Amblyornis inornatus* (*left*) and *A. macgregoriae* (*right*). (Drawings by W. T. Cooper in Cooper and Forshaw 1977)

Several factors may explain the transfer of display to a bower (Diamond 1988). Males provide no parental care; bowerbirds are large and dominate most other species at rich fruiting food trees, so enough food is easily obtained. Males can therefore devote much time and energy to bower building and display. This can be done on the forest floor because New Guinea and Australia are relatively free from specialized bird-hunting mammals, such as small felids. Bower construction probably evolved from the building of simpler, undecorated courts. A broad comparative analysis of court and bower building might provide clues to its evolution. Bowerlike structures are also found in several other birds, for example Jackson's widowbird *Euplectes jacksoni*, lyrebirds (*Menura*), and even in some cichlid fishes (e.g., Lill 1979; McKaye et al. 1990; Andersson 1991).

Male mating success was studied in detail in the satin bowerbird *Ptilonorhynchus violaceus* by Borgia (1985b). Using automatic cameras at twenty-two bowers, he showed that males with highly decorated bowers obtained most matings. Blue feathers, snail shells, and yellow leaves were particularly important. Males often destroy bowers of other males, and steal attractive blue feathers. The males that destroyed most bowers were dominant in aggression at feeding sites; the males with most feathers on their bowers stole more often than they were stolen from. This suggests that the number of blue feathers and some other decorations reflect male dominance. Mating at a highly decorated bower, a female is therefore likely to have offspring fathered by a dominant male (Borgia 1985a; Borgia and Gore 1986). The mating success of males was inversely related to their number of ectoparasites. Testing predictions from several hypotheses, Borgia and Collis (1989, 1990) concluded that the results agreed best with female avoidance of males with ectoparasites that might be transmitted to her (also see section 3.5).

Cotingids and Manakins

In a South American equivalent of birds of paradise, the cock-of-the-rock *Rupicola rupicola*, males fight over lek territories and increase their success by disrupting the courtship of others. A female is, however, relatively free to choose her mate and often returns for a second mating with the same female (Trail 1985a,b). Analyzing the sampling sequences of individual females, Trail and Adams (1989) compared the observations with proposed models of female choice tactics (e.g., Janetos 1980; Parker 1983b; Wittenberger 1983; Real 1990). Females did not make simple threshold decisions or sequential comparisons, but usually sampled a subset of males before mating, following a "pool comparison" tactic.

Manakins (Pipridae) have been the subject of several studies of sexual selection (e.g., Lill 1974, 1976). In *Chiroxiphia* manakins, two or more males form long-lasting teams that cooperate in courtship display (Snow 1971; Foster 1977, 1981; McDonald 1989a,b). Age-related dominance seems to determine male mating success within groups; it is highly skewed, with the top male performing almost all copulations. Other males remain in the group and display probably because they occasionally copulate, and may rise to primary status if the top male disappears (Foster 1977, 1983; Wiley and Rabenold 1984; McDonald 1989b). Female home ranges in the long-tailed manakin *C. linearis* include a number of male display teams, permitting choice among 5–8 top males. The frequency of male "toledo" calls was correlated with female visitation rate. Given a female visit, male mating success was correlated with a component of the dual-male dance display (D. B. McDonald 1989a). In swallow-tailed manakins *C. caudata*, with leks of 4–10 males, the dominant male performs all matings, and females may not be able to choose a mate within leks, only among them (Foster 1981, 1983).

Other Birds

The great snipe *Gallinago media* is another lekking species with vocal as well as visual male display (e.g., Lemnell 1978; Höglund and Lundberg 1987; Höglund and Robertson 1990). Male mating success at two leks was correlated with display rate and proximity to the lek center. Females seemed free to choose among males. As in some other lekking birds (re-

Figure 7.5.6 Displaying great snipe *Gallinago media*. (Drawing by A. Roos in Höglund et al. 1990)

viewed by Wiley 1991), central males tended to be older, and differed in some other respects from peripheral males. Those without a permanent lek territory were smaller and weighed less than others (Höglund and Lundberg 1987). Toward the end of a display sequence, the male fans and twists his tail with its conspicuous white outer feathers (figure 7.5.6). Central, old males had more white on the tail than fringe males. In an experiment, males with whiter tails received more matings than others, but the highly skewed mating success complicates statistical interpretation. Tail color had no effect in male contests (Höglund et al. 1990).

Höglund and Robertson (1990) tested the position and attractiveness hypotheses (see above) by experimental removal of males. When five dominant males were removed, none of the territories was reoccupied, contrary to what is expected if position is important. When five subordinate males were removed, four of the territories were quickly occupied again. These results accord better with the attraction than with the position hypothesis.

Jackson's widowbird *Euplectes jacksoni* is the only lekking ploceid (Craig 1980). Like the long-tailed widowbird *E. progne* (section 5.4 above), it is strongly sex dimorphic. Males cut down the grass around a central tuft at individual dance rings on leks in open grassland, displaying by repeated, up to 1-m-high jumps (Van Someren 1946; Andersson 1989, 1992). Male mating success at four leks was correlated with jump rate and, as in the long-tailed widowbird, tail length. The jump display, which ends when a female lands at the dance ring, seems to function in long-range attraction, whereas other traits, such as the tail and the central tuft, may play a role at close range (Andersson 1989, 1991). Tail length is correlated with male body condition. There was no evidence that contests influence mating success among established males, but fights were common during arena formation, and disappearing males were rapidly replaced. In addition to female choice, contest competition over dance rings therefore may influence male mating chances. Established males were larger than males without known lek territories (Andersson 1993).

Fallow Deer

Leks are known from eight mammals, five of which are ungulates (Clutton-Brock et al. 1988b). Among fallow deer *Dama dama* in a 4-square-kilometer park in southern England, most bucks defend small territories, 15–20 m in diameter, on a traditional mating ground. Male mating success was apparently determined by female choice. In an experiment where the most successful bucks were forced to change site after their lek territories were covered with black polythene, these males remained most successful in spite of moving up to several hundred meters across the lek (Clutton-Brock et al. 1989). Territory position therefore was not the main determinant of male mating success. (On a lek in Italy, on the other hand, there was correlational evidence of position effects on male mating success; Appolonio et al. 1989, 1992). Male mating success was correlated with antler and body size, and with fighting success. There was also an effect of female copying: the number of does attracted to a buck increased with the number of does already present on his territory, at a faster rate than expected if females choose males independently of each other (Clutton-Brock et al. 1988b, 1989).

In sum, studies of a variety of lekking vertebrates have shown that contests often determine male access to lek territories. Females choose among males on the lek, and mating success is skewed toward one or a few males. It is less clear what criteria females use in their choice of mate. Some studies have found correlations with aspects of display, and a few with morphological traits. How male morphological ornaments are selected remains to be clarified in most lekking species: it is usually not known whether male contests, female choice, or both are responsible. Male lek attendance correlates with mating success in a number of species, and so does display rate, dominance, and position on the lek (also see table 6.A). Female mate choice is sometimes influenced by fidelity to previous mating sites, and by choices made by other females. In spite of much progress through a number of painstaking experiments and long-term field studies, much remains to be learned about the mechanisms of sexual selection and their consequences in lekking animals. In particular, the morphological sex ornaments in such species remain poorly understood. As with other mating systems, almost nothing is known about the genetics of the processes.

7.6 Sex Role Reversal

Animals in which males show greater parental effort than females are of special interest for the theory of sexual selection; the exception tests the rule. Whatever the reason for larger male than female contributions to offspring, higher potential reproductive rate in females, and female-biased operational sex ratio: when they are at hand, the theory predicts (1) stronger female than male competition for mates, (2) more critical choice of mate by males, (3) higher variance in female than male mating success, and (4) more pronounced female secondary sex traits.

Most animals in which males contribute parental care are polygynous, some are monogamous, and a few are polyandrous. These latter species often show complete sex-role reversal, females competing strongly for males (reviewed by Ridley 1978; Trivers 1985). Conspicuous role reversal is found in certain insects (e.g., Smith 1980; Svensson and Pettersson 1987, 1988), pipefishes and seahorses (e.g., Fiedler 1955; Berglund et al. 1986a,b), amphibia (Ridley 1978; Wells 1980; Summers 1989, 1992a,b), and birds such as jacanas (Jacanidae), some plovers (Charadriidae), sandpipers (Scolopacidae), and button-quail (Turnicidae) (e.g., Jenni 1974; Erckmann 1981, 1983; Lenington 1984; Oring 1986; Emlen et al. 1989). More limited courtship reversal is found, for example, in some orthopterans (e.g., Gwynne 1991; section 5.1 above). The reasons for sex-role reversal in these animals are still obscure and debated.[2] Oring (1986) concluded that there are polyandrous birds "whose life histories are so different that one is hard pressed to imagine common environmental circumstances that may have led to the selection of the system."

A crucial question is why males accept to provide most or all parental care in many of these species. Two often suggested reasons are that (1) males are better able to care for the clutch than are females after egg-

[2] See reviews, e.g., by Jenni 1974, Maynard Smith 1977, Ridley 1978, Wittenberger 1981, Erckmann 1981, 1983, Oring 1982, 1986, Petrie 1983b, Lenington 1984, Murray 1984, Thornhill and Gwynne 1986, Clutton-Brock 1991, Clutton-Brock and Vincent 1991, Gwynne 1991.

laying; (2) males are unlikely to obtain secondary mates, whereas females can do so (Oring 1986). Critical tests of these ideas require that a variety of normal and role-reversed species be compared in depth. In most cases, the theory is supported by field studies, but there are exceptions. Also, cases exist where females are more active in courtship than males, even though there is no known paternal contribution of the kind usually coupled with courtship role reversal (Hatziolos and Caldwell 1983). The following examples represent some of the most thoroughly studied species with sex role reversal.

Giant Waterbugs

Most of the insects known to have only male parental care are giant waterbugs (Belostomatinae). The biology of one species, *Abedus herberti*, was studied in detail by R. L. Smith (1979a,b, 1980). Females deposit their eggs on the back of males which carry, aerate, and protect the eggs for about three weeks until they hatch. The clutch may weigh twice as much as the male and it cuts his swimming speed to half, probably making him less able to catch prey and more vulnerable to predation. Males or, more precisely, male back space, is a limiting resource for a gravid female, which cannot lay eggs until a male without a brood accepts her (also see Kruse 1990). A female will not mate with a male that already carries a full set of eggs. Male parental care, not fertilization, therefore is the main limiting resource sought by females (Williams 1975).

Females can copulate with several males, but a male may not accept the eggs unless he has fertilized them. Probably for this reason, bouts of copulation alternate with egg laying. This should assure male paternity, as the last copulating male fertilizes most of the eggs (Smith 1979a,b).

As predicted, the courtship roles are reversed in some respects. The female approaches the male, which announces readiness to mate by a special display. He largely decides whether mating will take place, and sometimes declines her invitation to mate (also see katydids, section 5.1). Females lack courtship display, so the roles are not completely reversed. R. L. Smith (1979a,b) suggested this has to do with large investment in the eggs by females. The time it takes for a female *A. herberti* to produce a clutch is similar to the time during which the male broods them, and the physiological cost of making the eggs may be similar to the cost of brooding. In *Belostoma flumineum*, however, females can rapidly produce a second partial clutch, which apparently makes males backspace a limiting resource for female reproduction early in the season (Kruse 1990). Until these and other costs and benefits have been measured, it is not clear how much sex role reversal to expect. Phylogenetic constraints might also influence the sex roles in giant waterbugs.

Pipefishes

In some pipefishes and seahorses (Syngnathidae), females transfer the eggs to males, which brood the embryos on the ventral body surface and supply nutrients to the offspring through the epitel (Kronester-Frei 1975; Haresign and Shumway 1981; Berglund et al. 1986a). In some species, males even have a brood pouch into which the female inserts her penis during copulation (e.g., Gronell 1984; Berglund et al. 1986a,b). Females can produce broods more rapidly than males can raise them, so males are likely to be a limiting resource for female reproduction in these species (Berglund and Rosenquist 1990). As expected, females are more ornamented than males in some pipefishes (e.g., Fiedler 1954; Berglund et al. 1986b). Differences among species offer an opportunity for testing whether variation in male parental effort is accompanied by the predicted variation in relative sexual selection of males and females.

Comparing two species, *Syngnathus typhle* and *Nerophis ophidion*, Berglund et al. (1986a,b, 1989) measured the energy contents of adults and zygotes, and the respiration of zygotes. Males and females are colored similarly in *S.typhle*; they spend roughly equal energy on offspring. Males in *N. ophidion* spend less than females. Surprisingly, in spite of their larger parental effort, female *N. ophidion* have bright nuptial colors and other conspicuous secondary sex traits, and are larger than males. The male has no special brood pouch but carries the offspring on his ventral side and invests little energy in them. There was no correlation between male body size and brood size, but female fecundity increases with body size. In *S. typhle*, where males have a brood pouch, large males can carry more offspring, so large size is favored by fecundity advantages in both sexes. These differences between the species may explain why females are larger than males in *N. ophidion* but not in *S. typhle*. The reason for conspicuous nuptial colors in female *N. ophidion* rather than in the other species is, however, not understood.

In mate choice experiments with *S. typhle*, where males and females could choose between two mates of different size, both sexes preferred the larger mate (figure 7.6.1). In female *N. ophidion*, fecundity increases not only with body size but also with the area of blue nuptial colors. When female size was held constant, males discriminated among females and preferred those with the largest area of blue coloration, which therefore is favored by male choice of mate. Males also preferred females with large body size and large ornamental skin folds. There was no direct aggression among females, but dominance relationships may prevent some females from developing ornamental skin folds (Berglund et al. 1986b; Rosenqvist 1990). Females also reduce their reproduction and grow faster instead in the presence of a large as compared to a small female (Berglund 1991).

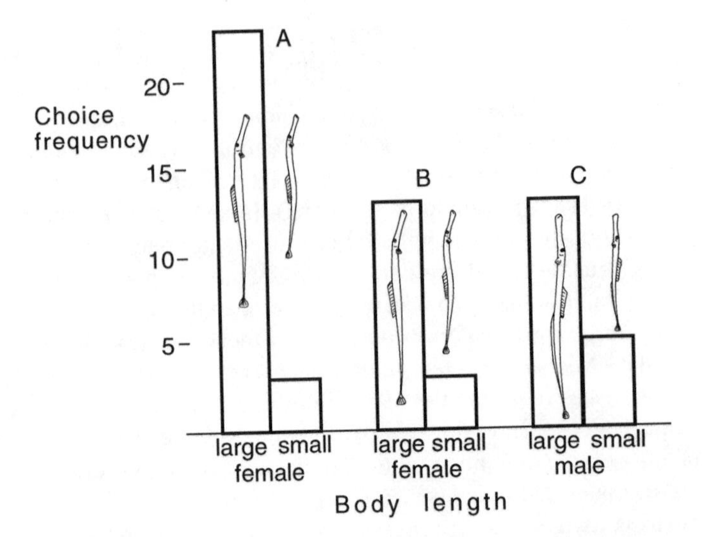

Figure 7.6.1 Results of mate choice experiments in *Syngnathus typhle*. In (A), one male chose between two females of different size. In (B), the two females were isolated from each other by plastic enclosures. In (C), one female chose between two males of different size. The larger individual was chosen most often in all three tests. (After Berglund et al. 1986b)

In both species, females can produce broods almost twice as rapidly as males can rear them. Males apparently do not invest more energy than females in offspring. The results suggest that the potential rate of reproduction better explains the sex differences in sexual selection than does parental energy investment (Berglund et al. 1989; also see Clutton-Brock and Vincent 1991). Males, however, have larger costs of reproduction in terms of mortality or reduced growth (Svensson 1988). Such costs reduce the expected future reproduction, and therefore should be included in parental investment in Trivers's (1972) sense. Whether males will then turn out to show greater parental investment than females remains to be tested.

Spotted Sandpipers

The spotted sandpiper *Actitis macularia* has been studied by L. Oring and his colleagues for two decades at Leech Lake, Minnesota (summarized by Oring and Lank 1986; Oring 1986; Oring et al. 1991a,b). Males in this population carry out most of the incubation and care for the chicks, which gather their own food. In lower-density populations, females also care for chicks. As in other waders, clutch size is limited at four eggs, probably because their large size prevents incubation of more eggs. In the food-rich environment and long breeding season at Leech Lake, females lay up to five clutches in succession for different males, whereas a male never raises more than one clutch per season. Food is abundant, and the main limit to female reproduction is the availability of mates. As most males incubate or care for broods, there is usually a strongly female-biased operational sex ratio (Lank et al. 1985).

These conditions create strong sexual selection in females. Contrary to the usual pattern among birds, females arrive before males on the breeding ground and compete by contests over territories and mates. The number of males attracted increases with female age, and with the size of the territory and length of beach that she controls (Oring et al. 1991b). Fights among females are often severe: 10% of the breeding females have visible injuries, some being crippled for life (Maxson and Oring 1980; Lank et al. 1985). Spotted sandpipers differ from most other birds (Greenwood 1980; Greenwood and Harvey 1982) also in another respect: they show female-biased philopatry. Females are more likely than males to return to their birth site and breed there. Perhaps this has to do with advantages of site familiarity for the sex that defends territories, in this case females (Oring and Lank 1982, 1984). A reproductive advantage of site familiarity has been demonstrated in male collared flycatchers *Ficedula albicollis* (Pärt 1991).

In contrast with the situation in Bateman's (1948) study of *Drosophila* (see figure 7.2.1 above), the number of offspring produced by a female spotted sandpiper increases markedly with her number of mates (figure 7.6.2, also see Oring et al. 1991b). This is the pattern expected for a species with stronger female than male sexual selection, evident from the behavioral differences between the sexes reviewed above. The spotted sandpiper therefore corroborates predictions from parental investment theory (section 7.2): the sex with largest parental effort and lowest potential reproductive

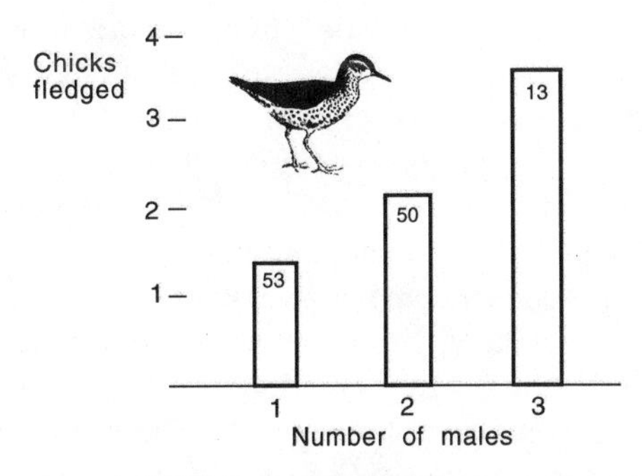

Figure 7.6.2 The number of chicks fledged for female spotted sandpipers *Actitis macularia* increases with the number of males monopolized by the female. Sample sizes are given in the bars. (After Oring and Lank 1986)

rate, in this case the male, is a limiting resource for the reproduction of the other sex, leading to stronger competition over mates in females.

Further studies that quantify relative parental effort and potential reproductive rate in the two sexes may permit more detailed tests of the theory. The reasons for parental sex role reversal in some birds and other animals are not well understood, and call for more theoretical and field work.

7.7 Summary

The relative strength of sexual selection in males and females depends on sex differences in parental effort, such as differences in gamete size, and in provisioning of material resources and care to the offspring. The sex that invests most heavily in offspring will usually have a lower maximum rate of reproduction, and becomes a limiting resource for reproduction in the other sex, which therefore competes more strongly over mates. In most animals, parental effort is greatest in females; they have lower maximum reproductive rates than males, which are subject to stronger sexual selection. Among animals in which males care alone for the offspring, sexual selection tends to be strongest in the sex with highest maximum reproductive rate. In some cases this is the female sex, in other cases the male.

The strength of sexual selection depends on the mating system and the operational sex ratio, which is related to the relative rates of reproduction of the sexes, and to other aspects such as sex differences in mortality. These aspects in turn depend on sex differences in parental effort and competition over mates, creating several likely feedback loops in the evolution of sexually selected traits. For these reasons such traits, mating systems, parental roles, and life history differences between the sexes will be strongly interdependent (figure 7.2.2).

Dimorphism in secondary sex traits tends to increase with polygyny. A likely reason is that males and females have less similar sex roles and are subject to more different selection pressures in polygynous than in monogamous species. Theory suggests and empirical studies of birds verify that sexual selection can have marked effects also in monogamous species, based on skewed sex ratios, and on quality differences among individuals. Both males and females may compete over mates in species where individuals vary markedly in fecundity or other aspects of quality.

Leks seem to offer favorable conditions for sexual selection by mate choice uninfluenced by material resources, but there may be other direct or indirect benefits of female choice. Few studies have demonstrated female preferences for male morphological ornaments. Male lek attendance, display rate, other aspects of display and, in some cases, position on the lek, are more often related to male mating success. Site fidelity and copying of other females influences female choice in some lekking species. Male contest competition may often determine access to lek territories.

Sexual selection is stronger in females than males in some insects, fishes, anurans, and birds. As predicted, females seem to make less parental effort and have higher potential rate of reproduction than males in such species. The reasons for reversed parental roles are not well understood.

8 Benefits of Mate Choice

Why do females, and often also males, choose mates, although such choice is likely to have costs? What benefits, besides the advantage of mating with a partner of the same species, favor discrimination among potential mates? Advantages of choice among conspecific mates are reviewed here, and discrimination against other species is covered in the next chapter.

Most genetic models of mate choice focus on heritable aspects of mating success and viability, but several kinds of nongenetic benefits can also favor mate choice (e.g., Williams 1975; Borgia 1979; Searcy 1982; Halliday 1983; Thornhill and Alcock 1983; Reynolds and Gross 1990; Kirkpatrick and Ryan 1991). Focus here is mainly on direct effects on offspring production, for example through fecundity advantages, food, parental care, or a good territory. In all the points listed in table 8.1.1, the benefit will ultimately lead to greater production of surviving offspring for the choosing sex. Some evidence for possible genetic benefits of mate choice is reviewed in section 8.7, but empirical results more directly related to Fisherian and genetic indicator mechanisms were discussed in sections 2.7 and 3.7.

8.1 Fertilization Ability and Fecundity

Female Choice

Differences in fecundity or fertility among potential mates may favor mate choice in many animals. If males vary in fertilization ability, for example owing to differences in sperm supply, female choice of the most fertile males should reduce the risk of producing infertile eggs (Halliday 1978; M. B. Williams 1978; G. C. Williams 1992). In the frog *Uperoleia laevigata*, females prefer males of a size that leads to high fertilization success (Robertson 1990; section 7.6 above). In addition, if there is sperm competition, females mating with males with highest fertility might give birth to sons with higher success in sperm competition (Eberhard 1985).

Experiments with aquarium stocks of lemon tetra (*Hyphessobrychon pulchripinnis*) indicate that females discriminate against males with reduced sperm supply. Male fertilization ability declines after several successive matings; females preferred males that had not spawned recently (Na-

Table 8.1.1

Possible Benefits of Mate Choice

Discrimination among potential mates can be advantageous because it leads to mating with a partner
(1) of the correct species (chapter 9)
(2) with better fertilization ability or fecundity (section 8.1)
(3) that provides more food to its partner (8.2)
(4) with better parental ability (8.3)
(5) offering better breeding territories or other defended resources (8.4)
(6) associated with lower risks of predation, harassment, or other hazards (8.5)
(7) that better complements the partner (8.6)
(8) that produces offspring of higher heritable viability or other quality (8.7)

katsuru and Kramer 1982). A reduction of male courtship success after mating also occurs in *Drosophila* (Markow et al. 1978). In smooth newts *Triturus vulgaris* and checkered white butterflies *Pieris protodice*, male display rate is correlated with sperm supply (Halliday 1976; Halliday and Houston 1978; Rutowski 1979). Male newts and frogs may deplete their sperm supply, but it is not clear whether females choose mates in relation to display rate in these species (Verrell 1982, 1987; Malacarne and Cortassa 1983; Gibbons and McCarthy 1986; Halliday 1987). Sperm depletion has been found in a variety of animals (reviewed by Dewsbury 1982; Birkhead and Møller 1992). It is not known, however, how often sperm are in sufficiently short supply to justify female discrimination in favor of more fertile males.

Multiple Matings by Females

Multiple matings by females occur in many species. Paternity studies of animals regarded as monogamous have shown that a proportion of the offspring have been sired by a male other than the apparent father, and extra-pair fertilizations are more common than previously thought (reviewed by Gowaty 1985; Bradbury and Davies 1987; Westneat et al. 1990). Many explanations have been suggested for multiple matings by females, for example reduced risk of infertility, benefits of material contributions from several males, genetic benefits for offspring, and female mating tendency being a correlated response to sexually selected multiple matings in males (e.g., Parker 1970c, 1992; Davies 1983; Halliday and Arnold 1987; Ridley 1988; Sherman and Westneat 1988; Westneat et al. 1990; Møller 1992b; also see section 8.6 below).

Two recent studies of reptiles demonstrate female reproductive advantages of copulating with several males. In a small, inbred population of adders *Vipera berus*, the proportion of viable offspring increased with the number of mates and matings of the females. Madsen et al. (1992) concluded that the males most successful in sperm competition and egg fertilization also produce the most viable offspring. In an isolated population of sand lizard *Lacerta agilis*, the proportion of viable eggs also increased with the number of mates of the female (Olsson 1992a). Both these studies suggest that females in inbred populations may give rise to more successful offspring if mating with several males. Possible reasons for such an effect are discussed by Parker (1992). Additional studies in other populations and species are needed to test how general this result may be.

Male Choice of Females

There is a potential advantage to male choice of mate in species where females differ markedly in fecundity, related for example to body size (section 11.2). To the extent that mating takes time, or depletes male sperm or other resources, and reduces males' chances of fertilizing other females, males should favor the most fecund females as mates (e.g., Parker 1970b; Trivers 1972; Rutowski 1982a; Forsberg 1987). As expected, males discriminate among females in many species where females differ markedly in size; 24 cases are listed in table 6.A. Examples range from arthropods such as crustaceans, beetles, orthopterans, butterflies, and flies, to fishes, frogs, newts, and lizards. For instance, fecundity increases with body size in the milkweed beetle *Tetraopes tetraophtalmus* (Price and Willson 1976). Males fight over females, but they also show mate choice. The scope for choice should depend on the operational sex ratio, and seems to be more critical when the proportion of males is low (McCauley 1982; Lawrence 1986; McLain and Boromisa 1987).

A particularly clear example of male choice of fecund females comes from the Mormon cricket *Anabrus simplex*, where males are mounted by females before copulation. In two-thirds of 45 cases observed, the male pulled away from the female without transferring his nutritious spermatophore. Males can probably assess the mass of mounting females; those accepted for mating weighed on average 18% more than the rejected females. This corresponds to a fecundity advantage of about 50% (Gwynne 1981).

Males prefer large, fecund females in several fishes (e.g., Downhower and Brown 1980, 1981; Rowland 1982a; Sargent et al. 1986; Cote and Hunte 1989). So do male salamanders *Desmognathus ochrophaeus* (Verrell 1989) and sand lizards *Lacerta agilis*; large females in the latter species produce up to three times as many eggs as small females (Olsson 1993a). Male choice of mate is probably more common than previously assumed, at least in species where females differ markedly in fecundity; male choice has attracted less attention than female choice.

Sperm competition can also favor male choice (Parker 1970). In the thirteen-lined ground squirrel *Spermophilus tridecemlineatus*, males avoid copulating with an already mated female if the time since her previous mating exceeds a certain limit, beyond which the sperm of the second male has poor fertilization success. Schwagmeyer and Parker (1990) modified an approach from foraging theory (e.g., Stephens and Krebs 1986) to predict the time beyond which a second male should leave and search for another mate. He should do so if the expected reproductive gain per unit time from staying and mating is lower than his average gain rate from continued search. Males left as predicted in 54 out of 73 cases.

8.2 Nutrition Provided by the Male

In many birds and some insects, courting males offer nutrition to females. The offerings among insects are of several kinds: prey, seminal nutrients transferred together with sperm, somatic gifts such as specialized male organs eaten by the female during copulation, and suicidal food transfer, the male himself being eaten (e.g., Thornhill 1976a; Rutowski 1982a; Thornhill and Alcock 1983; Zeh and Smith 1985; Simmons and Parker 1989).

Prey Gifts

Food gathered and offered to the mate by male birds was long viewed mainly as a means of strengthening the pair bond (e.g., Lack 1940; reviewed by S. M. Smith 1980). It is now clear that such courtship feeding (nuptial feeding in entomological parlance) can also enhance female fecundity in birds and insects (e.g., Nisbet 1973, 1977; Thornhill 1976a,b, 1983; S. M. Smith 1980; Gwynne 1984b, 1988b; Steele 1986a,b; Carlsson 1989; Simmons 1988b, 1990).

As in many other birds, the male common tern *Sterna hirundo* feeds his mate before she lays eggs. Male courtship feeding is correlated with his later rate of feeding the young (Nisbet 1973; Wiggins and Morris 1986). Females with mates that provide much food tend to lay earlier and produce larger eggs or clutches than others. The ability of the male to feed the female should therefore influence her production of offspring. It is not clear, however, whether female terns choose mates based on variation in their courtship feeding. In the red-billed gull *Larus novaehollandie scopulinus*, the likelihood of copulation after courtship increases if the male feeds the female (Tasker and Mills 1981).

Some insect females choose a mate in relation to his courtship feeding, and reproduce better as a result (reviewed by Thornhill and Alcock 1983). This has been shown particularly clearly by Thornhill (e.g., 1976b, 1980a, 1981) in the hangingfly *Hylobittacus apicalis*, where females allow the male mating time and sperm transfer in proportion to the size of his food

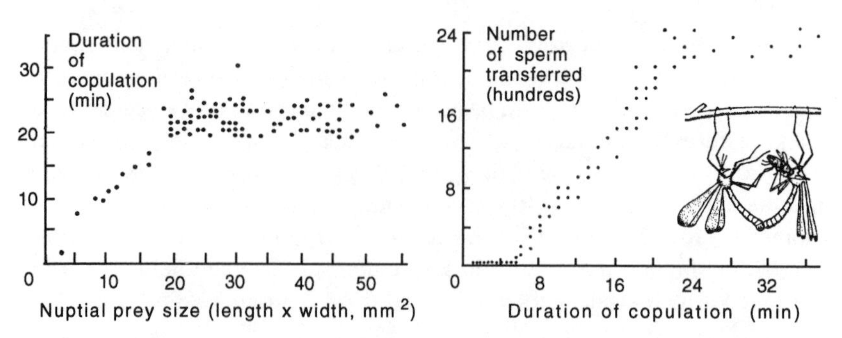

Figure 8.2.1 Female hangingflies *Hylobittacus apicalis* mate with males for a time that increases with the size of the prey that he offers her (*left*). The number of sperm transferred increases with the duration of the copulation (*right*). (After Thornhill 1976b, 1980. Drawing of hangingflies by T. Prentiss. Copyright © 1980 by Scientific American, Inc.)

gift. During copulation, the female feeds on the nuptial prey presented by the male. Females favor males offering large prey: matings in which the male presents a small prey are brief, little or no sperm is transferred, and the female may soon mate again with another male (figure 8.2.1). If the prey is large, copulation lasts longer, more sperm is transferred, and the female accepts no new male before she has laid a clutch of eggs. Males that offer large gifts are therefore likely to sire more offspring. This creates competition among males for presenting large prey, which is not an easy task; less than 10% of the males are carrying prey at any one time, and some prey are too small for mating to result in sperm transfer.

Females benefit directly in at least two ways from the nuptial gift. First, hunting is risky, and many scorpionflies are killed in spider webs (section 10.1). Males forage and fly much more than females, and although the initial sex ratio is close to 1, more than twice as many males as females die in spider's webs. A female should increase her survival by favoring males with large gifts, reducing the need for her to hunt. Second, females seem to gain fecundity by accepting large gifts and rejecting males with small prey. Such females lay more eggs per unit time than do less discriminating females (Thornhill 1976b, 1980a). In *Hylobittacus nigriceps*, the rate at which a female lays eggs increases with the size of her male and his nuptial prey. This effect favors large males presenting large prey, in addition to other advantages that such males may have in sexual competition (Thornhill 1983).

Nutrients Synthesized by the Male

Males in some insects produce seminal fluids that provide females with extra nutrition before egg laying (e.g., Markow 1988). Sperm possibly have a secondary, nutritive function in some species (e.g., Sivinski 1980; Thornhill and Alcock 1983).

In grasshoppers and other Orthoptera, females obtain sperm and nutrients from the male in a spermatophore applied to her genital opening (e.g., Sakaluk and Cade 1980; Gwynne 1984b, 1988a,b; Sakaluk 1986; Butlin et al. 1987; Simmons 1988b; Wedell 1993; also see section 5.1 above, and figure 8.2.2). In *Corthippus brunneus*, proteins or other nutrients are transferred internally to the female from the spermatophore; some of the nutrients end up in the eggs. When maintained on a restricted diet, multiply mated females laid more eggs and did so at a faster rate than females mating only once, suggesting a fecundity effect of nutrients from the spermatophore (Butlin et al. 1987). There is such an effect on egg size in several species (e.g., Gwynne 1988b; L. W. Simmons 1988b).

In some crickets and bush crickets, females eat the spermatophylax part of the large spermatophore (reviewed by Thornhill and Alcock 1983; Quinn and Sakaluk 1986). Male tettigonids form a spermatophylax that in some species can make up 30%–40% of the male's body mass (Gwynne 1982, 1984b, 1988a,b). Feeding on the spermatophylax enables females to lay more as well as larger eggs (figure 8.2.2, also see Simmons 1990). The spermatophylax in some species is an important limiting resource, and females compete for access to males, showing sex role reversal in several respects (reviewed in section 5.1).

The nutrition provided by the male at least in part represents a mating effort that raises his chances of fertilizing the eggs, for example by making the female less likely to end copulation or eat the spermatophore before all sperm are transferred (Thornhill 1976a; Alexander and Borgia 1979). In several species, the spermatophore is about the size needed to keep the female eating until full insemination is achieved (Thornhill 1976a; Sakaluk 1984, 1986a; Wedell and Arak 1989). In the bush cricket *Requena verticalis*, however, the spermatophylax is twice as large as needed for full in-

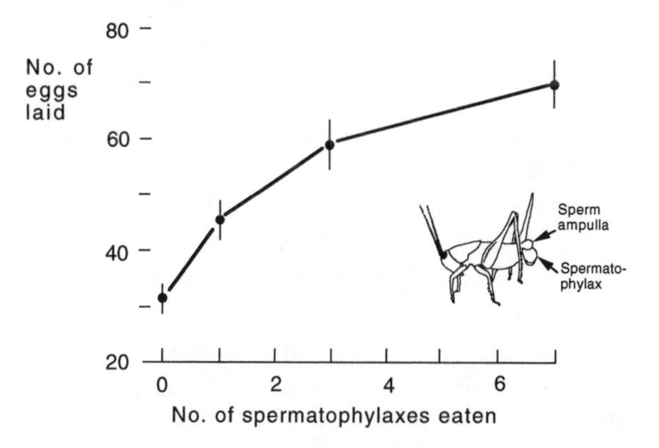

Figure 8.2.2 The number of eggs laid by a female bush cricket *Requena verticalis* increases with the number of male spermatophylaxes she has eaten. Vertical lines = s.e. The drawing shows a female with a spermatophore (sperm ampulla and spermatophylax). (Data from Gwynne 1984)

semination. It may partly represent male parental effort, as nutrients from the spermatophylax influence egg size and numbers (Gwynne 1986, 1988a; also see Butlin et al. 1987). The relative importance of mating versus parental investment in courtship feeding is still debated (see below in this section).

Also in butterflies, males transfer a nutritious spermatophore to the female. Protein and carbohydrates from the spermatophore seem to play a role in egg production (e.g., Boggs and Gilbert 1979; Boggs and Watt 1981; Marshall 1982; Rutowski et al. 1987; Svärd 1985; Oberhauser 1989). In an experiment with the alfalfa butterfly *Colias eurytheme*, female egg production and survival were higher in a group receiving large than in one receiving small spermatophores (Rutowski et al. 1987).

Some other studies of butterflies found no effect of the spermatophore on the quantity or quality of eggs produced (e.g., Jones et al. 1986). Svärd and Wiklund (1988) suggested that female fecundity is increased by male-derived nutrients only in harsh environments, and that males may produce the spermatophore (consisting of 85% water) to delay or prevent remating by the female. The relative role of spermatophores in sperm competition and sexual selection, as compared to parental investment, needs to be assessed in a wider range of species and circumstances.

Also in certain cockroaches, Blattodea, spermatophore nutrients are transferred to the eggs (Mullins and Keil 1980). In *Xestoblatta hamata*, females feed on urate glands of the male, which can make up 20% of his body mass; the urates enhance female fecundity (Schal and Bell 1982). Females feed on specialized male glands or tissue in several other insects. In certain Orthoptera, the female eats parts of the male body such as leg spurs and fleshy hind wings that appear to have been modified for this function (e.g., Morris 1979; reviewed by Thornhill and Alcock 1983).

Mating Cannibalism

The ultimate male transfer of nutrients occurs in some spiders, scorpions, mantids, diptera, and a few other species, where the female devours her spouse (e.g., Polis 1981; Thornhill and Alcock 1983; Buskirk et al. 1984; Elgar 1991, 1992). This lopsided wedding dinner may increase the fecundity of the female; theoretically, it might raise the fitness of the male as well (Parker 1979; Buskirk et al. 1984). In some cases, the male may have no chance of finding additional females, nor of contributing parental care. If the fecundity of his mate increases if she eats him, selection might then favor the male offering himself as a meal after copulation.

While plausible, this idea is not supported by tests of some of its predictions. For example, if a male really gains fitness by being eaten, he should willingly offer himself as a meal (Gould 1984). The male seems far from willing, however. Male mantids, spiders, and scorpions that run a risk of being eaten approach the female slowly and cautiously, and move away quickly after copulation (e.g., Liske and Davis 1987; Birkhead et al. 1988; Polis and Sisson 1990; Elgar 1992; possible exceptions are reviewed by Elgar 1992). Contrary to what has been suggested (Roeder 1935), successful copulation does not depend on female decapitation of the male mantid, and he seems at pains to avoid this outcome (Liske and Davis 1987).

In orb-web spiders Araneidae, where sexual cannibalism is common, several male traits may reduce the risk of cannibalism. For example, the male waits until the larger female has captured prey before courting her; or he weaves a thread that permits him to mate without entering the central portion of the female's web. Males also differ morphologically from females in ways that should reduce the risk (e.g., Elgar et al. 1990; Elgar 1991). Extreme sexual size dimorphism in spiders occurs mainly in orb-web forms; sexual cannibalism is one suggested explanation (see section 11.2). Within Araneinae, males use their legs to manipulate the female. Relative male leg length is largest in the forms where males are smallest compared to females (figure 8.2.3). From a comparative analysis, Elgar et al. (1990) concluded that sexual cannibalism is responsible for this pattern, and that long legs in these spiders help males avoid being eaten by the female.

In many cases, sexual cannibalism is clearly maladaptive to the male, as the female eats him before they have mated. Such cannibalism might lead to nutritional advantages and higher fecundity of the female, if her chances are high of mating soon with another male (Elgar and Nash 1988; Newman and Elgar 1991).

If adaptive also in males, sexual cannibalism is probably a form of male parental effort that increases the fecundity of his mate. Male mating effort

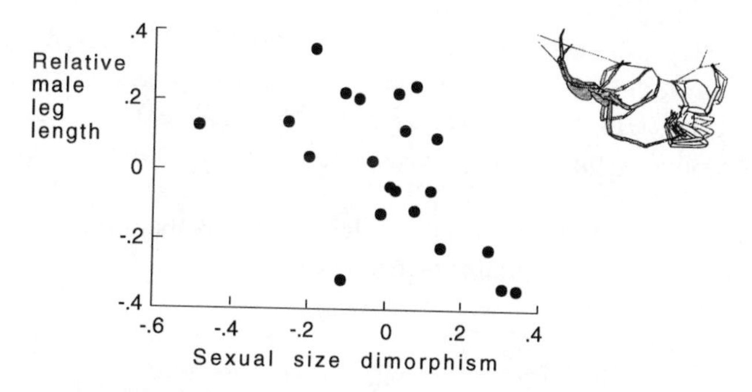

Figure 8.2.3 Relative male leg length is negatively correlated with sexual size dimorphism among genera of Araneinae spiders, which construct mating threads for courtship (shown for *Araneus diadematus*; female to the right). Large negative values represent species where males are much smaller than females. (After Elgar et al. 1990; spiders by D. R. Nash in Elgar and Nash 1988)

is possibly involved in ceratopogonid midges (Simmons and Parker 1989; Elgar 1992). Whether sexual cannibalism is sometimes adaptive for the victim is, however, still doubtful.

Food Gifts: Male Mating or Parental Effort?

In sum, there is potential for female choice based on nutrients offered by the male before or during mating in many insects and birds. These nutrients in some species improve female fecundity or survival, as well as male mating success. The degree to which nutritional gifts are favored mainly via male mating success and sexual selection, or by natural selection raising the number of offspring produced by the male's mate, is debated. Wickler (1985) suggested that the nutrients are unlikely to be built into the eggs quickly enough to permit a male to fertilize the eggs that he helps nourish; later males may fertilize them. He therefore argued that the nuptial gift is a sexually selected mating effort. In several species, however, the nutrients are built into the eggs within 24–48 hours, and there is evidence that the male in some species does fertilize the eggs to which he contributes nutrition (Gwynne 1986, 1988a; Sakaluk 1986a,b; Parker and Simmons 1989; Simmons 1990). Both mating and reproductive effort may therefore be involved.

The factors favoring nuptial gifts during their evolutionary origin may differ from the selective pressures that now maintain them (Gwynne 1988a). Simmons and Parker (1989) suggested that seminal and somatic gifts likely arose as small, sexually selected offerings that took on a parental investment function when they became larger. Prey offerings were probably more substantial from their origin, and more likely had naturally selected effects on offspring production from the start. Sexual selection may, however, also have contributed.

A next step to clarify the selection of courtship offerings would be to quantify the strength of sexual and natural selection in systems where both can be estimated in the wild. This might be done, for example, by manipulating the value of the gift and measuring its effects on mating success and mate fecundity (see, e.g., Thornhill 1986a; Gwynne and Simmons 1990). A quantitative phylogenetic analysis of the conditions that favor evolution of male food offerings to females is also needed.

8.3 Parental Ability of the Mate

Potential mates vary in parental ability in many species. For example, female mottled sculpins *Cottus bairdi* prefer large males, which are better egg guardians than small males (Downhower and Brown 1980, 1981;

Brown 1981). In several fishes, females are more willing to spawn in the nest of a male if it already contains eggs (e.g., Noonan 1983; Constantz 1985; Marconato and Bisazza 1986; Unger and Sargent 1988; Sikkel 1989; but see Jamieson and Colgan 1989). Such a nest may be safer than an empty nest (e.g., Rohwer 1978; Ridley and Rechten 1981; Sargent 1988).

Also in several birds, female choice is based in part on the quality of male parental care (e.g., Burley 1981a). Female moorhens *Gallinula chloropus* compete for fat males with good energy reserves (Petrie 1983a). Such males incubate more than others, perhaps enabling females to produce more clutches per season. In some populations of red-winged blackbird *Agelaius phoeniceus*, experienced males feed their young more than others do, and tend to have larger territories and harems. Male parental ability influences the number and weight of fledglings produced (Yasukawa 1981c; Muldal et al. 1986). The high mating success of experienced males possibly arises because females prefer the best territories (Searcy 1979a,c; Yasukawa 1981c).

8.4 Territories and Other Defended Resources

Whether a male is chosen or not may depend on a combination of factors in addition to his behavior and morphology, such as the availability of nest sites, food, or other resources on his territory. For example, in the dragonfly *Plathemis lydia*, males defending the best oviposition sites get most matings (Campanella and Wolf 1974). In *Anthidium* bees, the male defends flowers used for food, and permits a female to feed only if she mates with him. Males attract females in proportion to the number of flowers defended (Alcock et al. 1977; Severinghaus et al. 1981). In species with male resource defense and territoriality, contests may determine male access to resources, among which females can then choose (Searcy and Yasukawa 1983). Male contest competition and female choice may therefore work in combination. If male dominance is important for the males' mating success, and dominance is heritable to some degree, females should have more successful offspring if mating with dominant males (e.g., Borgia 1979; Borgia and Collis 1989).

Male mating success is related to aspects of his territory in many fishes (table 6.A). For example, female wrasse *Pseudolabrus celidotus* prefer males with territories in deep water, probably with reduced egg predation (Jones 1981). Also in some frogs (e.g., *Rana clamitans* and *R. catesbeiana*), males that defend the best oviposition sites attract most females (Wells 1977a; Howard 1978a,b; figure 8.4.1).

In many birds, male mating success increases with territory size or quality (table 6.A). Some examples are long-billed marsh wren *Telmatodytes*

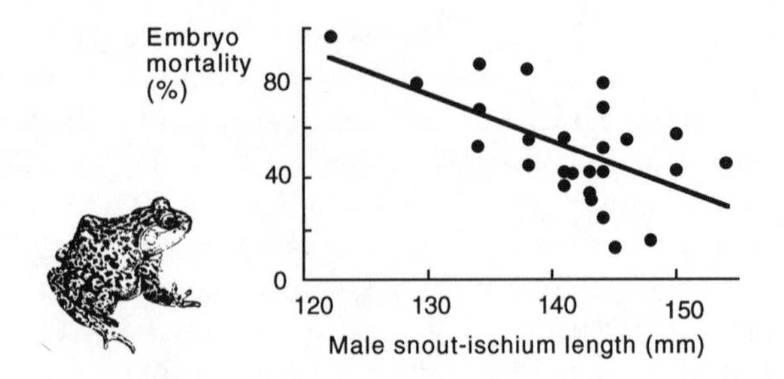

Figure 8.4.1 Male bullfrogs *Rana catesbeiana* defend territories where females lay their eggs. The largest males gain the best territories, with low embryo mortality. (After Howard 1978a)

palustris (Verner 1964; but see Leonard and Picman 1988), arctic skua *Stercorarius parasiticus* (O'Donald 1983), dunnock *Prunella modularis* (Davies and Lundberg 1984), medium groundfinch *Geospiza fortis* (Price 1984b), European robin *Erithacus rubecula* (Harper 1985), yellow-headed blackbird *X. xanthocephalus* (Lightbody and Weatherhead 1987, 1988), and pied flycatcher *Ficedula hypoleuca* (Alatalo et al. 1986; Dale and Slagsvold 1990). This pattern need not imply that females compare males or territories; random female settlement in suitable habitat could favor males with large territories (Harper 1985; Lightbody and Weatherhead 1987, 1988). In the monogamous robin, chick survival is better in large territories, resulting in higher offspring production (Harper 1985).

Male red-winged blackbirds *Agelaius phoeniceus* with territories that contain the most suitable vegetation for nesting attract most females (Holm 1973; also see Searcy 1979a; Yasukawa 1981c). Other similar examples are dickcissel *Spiza americana* (Zimmerman 1966, 1971), lark bunting *Calamospiza melanocorys* (Pleszynska 1978; see below), and long-tailed widowbird *Euplectes progne* (M. Andersson, in prep.). Male red-winged blackbirds provided with extra food on their territories attracted more females (Ewald and Rohwer 1982; Wimberger 1988).

In several birds, for example the village weaver *Ploceus cucullatus*, females choose a mate in relation to the quality of the nest that he builds (Collias and Victoria 1978; Collias and Collias 1984). Variation in the quality of nest cavities defended by male pied flycatchers *Ficedula hypoleuca* is more important for female choice than is variation among the males themselves (Alatalo et al. 1984, 1986; Askenmo 1984; Slagsvold 1986; Lundberg and Alatolo 1992; also see section 8.8 below).

A mammalian example of sexually selected defense of a feeding territory is the pronghorn antelope *Antilocapra americana*, where males with the best forage on their territories attract more does and have higher mating success than others (Kitchen 1974).

Natural selection can also favor male territoriality and traits that help make territory defense successful. This is the case, for example, if males with better territories obtain not only more or better mates, but also have offspring with higher survival as a consequence of territory quality. Such benefits need to be distinguished from mating advantages to clarify the relative roles of sexual versus natural selection of territoriality, and of weapons and signals used in territory defense. The outcome might vary widely depending on the species.

The Polygyny Threshold Model

To explain female choice in polygynous birds with breeding territories, Verner (1964), Verner and Willson (1966), and Orians (1969) developed a model based on variation in territory quality among males (also see, e.g., Wittenberger 1979; Davies 1989, 1991; Searcy and Yasukawa 1989; Bensch and Hasselquist 1991). Females should respond to the total breeding situation, which includes, among other things, the quality of the male as well as his territory, and the presence of other females (Emlen 1957). The first few females that arrive in the nesting area are assumed to pair monogamously with males in the best territories. A later female can either pair monogamously in one of the remaining territories, or become second mate of an already paired male in a better territory. In the latter case, the female will have lower success than if that male were previously unmated. A female is assumed to choose the alternative that maximizes her reproductive success, given the number of females already on the territories (figure 8.4.2). If differences among territories are large, some males may get many mates, others none.

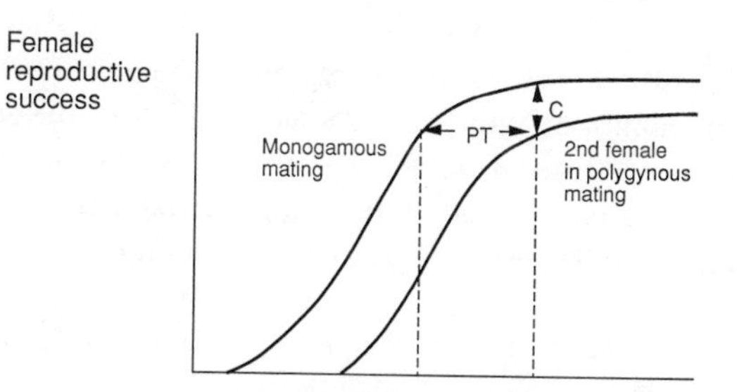

Figure 8.4.2 The polygyny threshold model. The curves represent the fitness (reproduction) of females mated monogamously, or as second females of a previously mated male. *C* is the cost, for a given breeding situation, of settling with an already mated male. *PT* is the polygyny threshold: the gain in quality of breeding situation required to justify settling on the territory of a previously mated male, rather than with an unmated male. (After Orians 1969)

This model has played an important role and stimulated much field work on resource defense polygyny. For example, in the lark bunting *Calamospiza melanocorys*, Pleszczynska (1978) found that males attracting more than one female had territories with more nest cover than others, an important resource in the open grassland. Experimental addition of suitable shading cover made males more successful at attracting females, and also raised female breeding success. Secondary females on average produced at least as many fledglings as did monogamous pairs at the same time of the breeding season. Another important prediction was not upheld, however; several late females paired monogamously, although they were expected to nest bigamously on better territories (Pleszczynska 1978).

A possible human example comes from the Kipsigis people of Kenya. In a recently settled area, there was a strong correlation between a man's land ownership and his number of wives. Men were most preferred (by women, or by parents on their behalf) if they had large plots of land, and if they were bachelors or married to few wives. Females had reproductive costs of polygyny: fewer children survived per female if there were many co-wives (Borgerhoff Mulder 1990, 1991).

Testing the polygyny threshold model in detail is difficult for many reasons, as discussed by Wittenberger (1979), Vehrencamp and Bradbury (1984), Davies (1989, 1991), and Searcy and Yasukawa (1989). It is not easy to identify the critical resource, check whether the assumptions apply, and measure the relevant fitness components. For instance, female survivorship might differ between polygamous and monogamous matings, if monogamous males provide more parental care. Female errors in judging the quality of the breeding situation will lead to a mismatch between predictions and observations. Owing to such problems, many tests of the polygyny threshold model are hard to interpret. Lightbody and Weatherhead (1988), Davies (1989, 1991), and Searcy and Yasukawa (1989) pointed out that the model cannot explain mate choice in all cases of territorial polygyny, and emphasized the need for a multiple-hypotheses approach, with tests among alternatives. Some of the main alternative explanations of how female settlement patterns can lead to polygyny in birds are listed in table 8.4.1.

Contrary to the assumption of the polygyny threshold model, there seems to be no cost to females of polygyny in some populations of the red-winged blackbird *Agelaius phoeniceus* (Searcy and Yasukawa 1989). In the pied flycatcher *Ficedula hypoleuca* there is a cost, but the reasons that females mate polygynously is debated. The "deception hypothesis" (Alatalo et al. 1981, 1990d; Lundberg and Alatalo 1992; also see Temrin 1991) suggests that secondary females in this polyterritorial species do not know that the male is already mated, as his territories are often several

Table 8.4.1

Hypotheses for Territorial Polygyny through Female Settlement Patterns in Birds*

I. There is no cost of polygyny to females.
   1. There is no benefit either, and females settle at random in suitable habitat (Lightbody and Weatherhead 1988).
   2. Females benefit because reproductive success increases with the number of females nesting nearby, for example owing to joint nest defense (Altmann et al. 1977).

II. Polygyny has a cost for females.
   1. Females accept the cost because
      (a) unmated males are rare and/or hard to find (Leonard and Picman 1987; Stenmark et al. 1988);
      (b) mated and unmated males are hard to distinguish (Alatalo et al. 1981).
   2. Females choose polygyny because the cost is compensated by benefits: the Polygyny Threshold Model.
      Accepting polygyny gives access to
      (a) a better breeding situation (food, nest sites, male parental care, etc.; Verner and Willson 1966; Orians 1969;
      (b) a male of higher genetic quality (Weatherhead and Robertson 1979).

* Modified from Searcy and Yasukawa 1989 and Davies 1991.

hundred meters apart in the forest. Stenmark et al. (1988) countered that a mated male advertises his secondary territory less and is absent more than unmated males, perhaps enabling a female to recognize his mating status. Slagsvold et al. (1992) found that, at least for distances of up to about 100 m, primary females may visit their mate's second site and show aggression toward the secondary female. These authors suggested that secondary females are prevented from settling close to the primary female because of her aggressive behavior.

Another experiment suggests that mated and unmated pied flycatcher males do not differ in behavior when a new female is present, and there was little evidence of female aggression (Searcy et al. 1991). Females did not discriminate between the two categories of males even when they could, and did hear and visit both (Alatalo et al. 1990d). Females choosing a mated male had lower success than those choosing an unmated male. This experiment suggests that secondary females were not aware of the mating status of the male; they did not discriminate against mated males, although reproductive success was reduced as a consequence.

In the likewise polyterritorial wood warbler *Phylloscopus sibilatrix*, secondary females had at least as high reproductive success as monogamous females, so the deception hypothesis does not seem to apply in this species (Temrin 1989, 1991).

The remarkably flexible mating system of the dunnock *Prunella modularis* does not fit easily into the framework of the polygyny threshold model. The variable mating system illustrates the importance of conflicts within and between the sexes (Davies 1989, 1992; section 4.1 above). In the same population, individuals may be mated in monogamous, polyandrous, polygynandrous, or polygynous relationships; some remain unmated. Much of the variation seems to depend on the distribution of food. When food was added, female home ranges shrunk, and males could more easily monopolize females; the mating system therefore shifted from polyandry to monogamy or polygynandry (Davies and Lundberg 1984).

In this and several other species, the polygynous status of some males cannot be fully understood by considering only female choice of mate and territory, as in the polygyny threshold model. There are strong conflicts of interest between the sexes. Polygyny in the dunnock is likely when female range size is small, and two females have similar fighting ability but differ much in reproductive timing. A male can then prevent access by other males to the females, and make it difficult for a female to drive away her rival (which would lead to greater reproduction for the remaining female). When, on the other hand, the territory is large and vegetation dense, and two males have similar fighting ability, females often gain higher success by mating polyandrously with two males. The number of mates obtained by male and female dunnocks is therefore determined in a more complex way than suggested by the polygyny threshold model (Davies 1989).

In sum, the polygyny threshold model represents an important step toward understanding territorial polygyny, but other aspects than those dealt with in the model also play a role in many cases; there are several plausible alternative explanations.

8.5 Risk Reduction

Mate choice can reduce risks associated with mating in several ways. For example, some males and territories may offer better protection than others from predators. Choice of a dominant male reduces the risk that the mating is interrupted by other competing males, which might injure the female or delay her breeding (e.g., Trivers 1972; Borgia 1979; Partridge and Halliday 1984; Trail 1985; Trail and Adams 1989; Clutton-Brock et al. 1988a; also see section 7.5). In animals ranging from insects to mammals, females are sometimes killed by males attempting to mate, often in struggles between competing males (reviewed by Le Boeuf and Mesnick 1990). Female dung flies *Scatophaga stercoraria* prefer large males, which are less likely to be attacked by competitors while mating (Borgia 1980, 1981; Sigurjonsdottir and Parker 1981). A preference for males in good condition could also help females avoid mates with contagious disease or parasites (e.g., Freeland 1976; Borgia and Collis 1989; section 3.5 above).

8.6 Mate Complementarity

In long-lived birds, pairs with low reproductive success are more likely than others to split (reviewed by Rowley 1983). One or both mates may have poor parental ability, but low success may also arise from incompatibility between the mates, for example poor synchrony of incubation (Coulson and Thomas 1983). Divorce may be adaptive in raising future reproductive success (Rowley 1983). It need not imply mate rejection; compared to successful birds, failed breeders return the next season at less consistent times, making it more likely that one member of the pair takes a new mate before the old partner returns (Coulson and Thomas 1983; Rowley 1983). In the great tit *Parus major*, however, experimental manipulation of fledgling production led to higher divorce rate and longer dispersal of females in pairs with reduced success, compared to pairs with increased success. In this species, divorce therefore seems to involve active avoidance of the previous mate or territory (Lindén 1991). There is evidence that females have higher reproductive success after divorce (Källander 1983).

Whether poor success in these cases has a genetic or purely phenotypic basis is not known. Bateson (e.g., 1980, 1983) and Shields (1982) emphasized the importance of genetic complementarity, and suggested that mild inbreeding preserves evolved favorable gene complexes. There might be an optimal degree of outbreeding that maximizes the number of successful offspring of the parents (also see Price and Waser 1979; Partridge 1983; Charlesworth and Charlesworth 1987; Rørvik et al. 1990). In laboratory studies of quail *Coturnix coturnix* (Bateson 1980), zebra finch *Poephilia guttata* (Slater and Clements 1981), and white-footed mouse *Peromyscus leucopus* (Keane 1990), females seemed to prefer distantly related males over non-kin. Early learning may influence the preference (ten Cate and Mug 1984; also see McGregor and Krebs 1982; Barnard and Fitzsimons 1988). In mice, males as well as females prefer to mate with partners with major histocompatability complex haplotypes that differ from their own, a preference that may also have to do with benefits of avoiding mating with close relatives (Yamazaki et al. 1976, 1988; Egid and Brown 1989).

There is now plenty of evidence for inbreeding depression (Partridge 1983; Charlesworth and Charlesworth 1987). In animals, inbreeding can be avoided in two main ways: (1) dispersal of one sex from its natal area (e.g.,

Greenwood 1980; Cockburn et al. 1985; Motro 1991; but other aspects than inbreeding can sometimes explain such patterns—see Hammerstein and Parker 1987); (2) kin recognition and avoidance of mating with close relatives (e.g., Crozier 1987; Wilson 1987).

8.7 Mate Choice and Offspring Quality

In several insects, female choice of mate seems to influence the quality of her offspring. These results are of interest in relation to indicator models of sexual selection (chapter 3). The studies reviewed below have not, however, examined female choice in relation to male secondary sex traits.

Flies

In an experiment with *Drosophila melanogaster*, Partridge (1980) mated some females to males drawn randomly from a population cage, and allowed others to mate in the cage, which contained many males. In competition with larvae from another stock, the offspring of females mating in the cage had about 2% higher larval survival than offspring of females mated to a randomly drawn mate. Whether the fitness difference was heritable or purely phenotypic is not known; the result can have several explanations (Partridge 1980). Female choice could be responsible, but direct competition among males in the cage might also determine with whom a female mated.

Schaeffer et al. (1984) failed to repeat this result, but in a larger similar study Taylor et al. (1987) found a strong correlation (0.76–0.92) between male mating success and larval competitive success in their offspring. The latter authors concluded that sexual selection in *D. melanogaster* results in offspring of higher quality, but the study leaves open whether direct male competition or female choice produced the result. If female choice is responsible, the choice cues remain to be identified. Further work on the genetic mechanisms involved is also needed; heterosis effects is one possibility.

In another dipteran, the seaweed fly *Coelopa frigida*, females given a randomly drawn male by the experimenter had lower fertility and gave rise to offspring with lower viability than did females mating in a population cage (Crocker and Day 1987). This species is polymorphic for several chromosome inversions that include over 25% of the genome. There is negative assortative mating with respect to this inversion, and the offspring from heterozygous matings have higher viability (Butlin et al. 1984; Day and Butlin 1987; Day et al. 1987; Crocker and Day 1987). Mate choice is probably involved, since mating is nonrandom in relation to the chromosome inversion. Direct observations showed that females reject some males, and there seems to be genetic variation in the female preference (Engelhard et al. 1989).

Grasshoppers

A grasshopper has been the subject of a similar study by Simmons (1987b). The offspring of *Gryllus bimaculatus* females that were allowed to choose mates survived better and developed faster than those of females given a male by the experimenter. Even if male interactions may play a role, females probably had some scope for mate choice; they could mate repeatedly and remove the spermatophores of some males before insemination was completed (Simmons 1986b).

In this and the previous experiments, the possibility was not excluded that females housed with many males produced larger or better nourished eggs. This might be the case, for example, if females in the population cage were stimulated to a greater reproductive investment, or if they received seminal nutrients from several males. The greater egg production by females in the mate choice category of Simmons's (1987b) study suggests this possibility. Tests of differences in nutrition and egg size in females of the different categories therefore seem desirable. Similar precautions often need to be taken in studies of genetic and other benefits of mate choice, as direct material as well as indirect genetic benefits might occur together in many cases. Estimating their relative importance is a major problem.

Sulfur Butterflies

Female choice in relation to male genotype has been demonstrated in natural populations of sulfur (*Colias*) butterflies. A long-term study has revealed genetic differences in flight capacity, survival, and mating success, related to genetic polymorphism at the phosphoglucose isomerase (PGI) enzyme locus. Glycolysis, and hence a number of physiological abilities, including flight, is influenced by the PGI genotype (Watt 1983; Watt et al. 1985, 1986).

Examining several populations of *Colias eurytheme* and *C. philodice eriphyle*, Watt et al. (1985, 1986) discovered a mating advantage for males of the PGI genotypes with most favorable enzyme-kinetic properties, which differ among areas in relation to temperature. Compared to their proportion in samples of males flying along with females, such males were overrepresented in mating couples. The difference does not seem to arise simply from an advantage for the best flyers in scrambles over females, but at least partly from female choice. The number of times a female had mated was judged by the number of spermatophores in her reproductive tract.

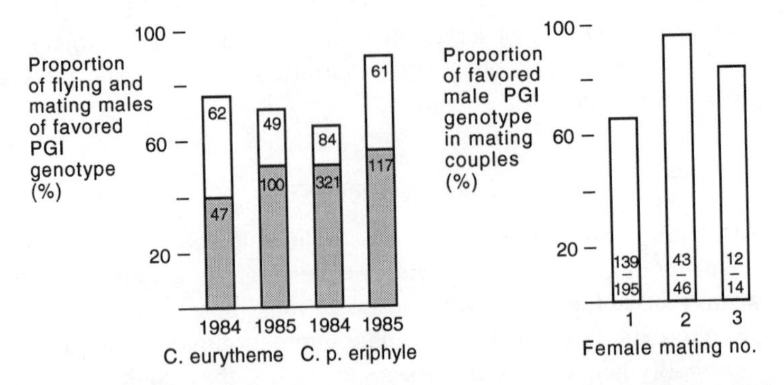

Figure 8.7.1 *Left*: In four samples of *Colias* butterflies (*C. eurytheme* and *C. philodice eriphyle*), physiologically advantageous genotypes at the PGI locus are overrepresented among males in mating couples (entire bars) as compared to males flying together with females (shaded parts of bars). *Right*: The advantageous PGI genotypes are more strongly overrepresented among older, more discriminating females compared to young females mating for the first time. Sample sizes are given in each bar. (Data from Watt et al. 1986)

Genetic and other analyses suggested that the frequency of favorable PGI genotypes was highest among males mated with old, discriminating females that had mated several times before (figure 8.7.1). Assortative mating does not explain the result; an age-dependent choice mechanism in females seems to favor males with efficient PGI genotypes, which have an advantage with old females.

These results suggest female choice of male genotype. It remains to find out exactly which PGI-related male trait leads to higher mating success. Since the PGI genotype probably affects general functions such as foraging ability (Watt 1983), other traits besides flight might also be influenced. The kinetically most favorable PGI genotypes are mainly heterozygotes, so their success is perhaps an outcome of heterosis (Watt 1983; Watt et al. 1985). For this reason, there may be little effect on the fitness of progeny from females mating with favorable PGI males; as far as is known, heterozygote advantages are not inherited by offspring (Partridge 1983). In a temporally fluctuating environment, however, a mating preference for heterozygotes can be favored under certain circumstances (B. Charlesworth 1988). How and why females have come to favor such males remains an open question. Watt et al. (1986) suggested that female choice might favor particular male homozygotes at other loci; this remains to be shown (also see Rutowski 1984; Sappington and Taylor 1990).

Frogs

Several laboratory studies of anurans suggest that there is male genetic variation of likely importance for fitness (e.g., Berven 1987; Travis et al.

1987). In the spadefoot toad *Scaphiopus multiplicatus*, maternal half-sibs sired by large males grew faster in the lab after metamorphosis than did their half-sibs sired by small males. In *S. couchi*, there was a survival advantage for the half-sibs with large fathers (Woodward 1986, 1987). In the spring peeper *Hyla crucifer*, males larger than average sired offspring that grew faster than offspring of smaller males (Woodward et al. 1988). Laboratory-spawned offspring from males found mated in the field had higher growth rates than offspring from males found unmated. As mating is external and males are not known to transfer nutrition to the offspring, Woodward et al. (1988) and Mitchell (1990) concluded that male genotype may influence offspring performance.

There is evidence for similar effects also in *Bufo woodhousei*. Large males produced offspring that were 10% heavier at metamorphosis than offspring of small males. Large males have a mating advantage in this species (Woodward 1982a,b), but it is not known whether male contests or female choice are responsible, nor whether the paternal effects lead to fitness advantages for females mating with large males. The studies verify one assumption of indicator models (chapter 3): paternal effects can be predicted from male phenotype, in this case size. Also in the salmon *Oncorhynchus gorbuscha*, the offspring from large males grow faster and are larger at maturity (2 yr) than those from small males (Beacham and Murray 1988). For similar results in the guppy, see Reynolds and Gross (1992), and section 3.7 above.

8.8 Direct Benefits Indicated by Male Traits

Indicator mechanisms (chapter 3) have mainly focused on possible genetic benefits from choice of mates with well-developed secondary sex traits. As yet, there is no fully convincing empirical demonstration of genetic benefits in such a process. Several models suggest, however, that indicator processes based on material benefits is a plausible possibility (Heywood 1989; Hoelzer 1989; Grafen 1990a). There is also empirical evidence for direct material benefits from female choice in relation to male traits. For example, in the red-winged blackbird *Agelaius phoeniceus*, experienced males tend to feed offspring more and have better territories than other males (Searcy 1979a; Yasukawa 1981c). Courtship time and song repertoire size increase with male age (Yasukawa et al. 1980; Searcy and Yasukawa 1981b), so female preferences for intense male courtship and large song repertoires may be favored by direct benefits (Eckert and Weatherhead 1987b; Searcy 1988).

Female birds choosing males with high song rate seem likely to obtain direct benefits (also see section 14.3). The time a male can spend singing should be related to the amount of food in his territory. In some cases, it might be easier for an arriving female to estimate the food situation by the singing rate of the male, who already knows where food is most plentiful, than to search and sample the food abundance herself. In the stonechat *Saxicola torquata*, male parental feeding as well as nest defense is correlated with his song rate early in the season (Greig-Smith 1982), perhaps as a consequence of variation in food among territories, or quality among males.

Evidence that male song rate reflects food abundance comes from observations as well as provisioning experiments. Temperature influences food abundance for insect-feeding species, as insects become more active at high temperatures. There is a relation between song rate and temperature in several birds, for instance in the Carolina wren *Thryothorus ludovicianus* (Strain and Mumme 1988) and the willow warbler *Phylloscopus trochilis* (Radesäter et al. 1987; Radesäter and Jakobsson 1988). The latter species shows an inverse relationship between the proportion of time spent singing versus feeding (Arvidsson and Neergaard 1991).

A relationship between song rate and experimentally added food has been demonstrated in red-winged blackbirds *Agelaius phoeniceus* (Searcy 1979d), dunnocks *Prunella modularis* (Davies and Lundberg 1984), and pied flycatchers *Ficedula hypoleuca*. In the flycatcher, males with experimentally increased song rate obtained mates sooner than others (Gottlander 1987; Alatalo et al. 1990b). Females choosing the males with highest song rate should usually end up in the territories with most food, and with the males in best condition (Gottlander 1987; Radesäter et al. 1987; Alatalo et al. 1990b). In the willow warbler, females pairing with early-arriving males laid eggs sooner after mating than did females pairing with late-arriving males. A removal experiment suggested that this was because early arriving males were in better condition and had territories richer in food (Arvidsson and Neergaard 1991). Female settlement may also in part be directly influenced by food abundance in the territory. In these territorial species, females may therefore have direct material benefits from mate choice in relation to male song, but the detailed mechanisms remain to be demonstrated.

In the bicolor damselfish *Stegastes partitus*, with male uniparental care of the eggs, females favor males with high rates of courtship display (Knapp and Kovach 1991). The eggs guarded by such males survived better than others, apparently because males with high display rate had better energy reserves and were less prone to cannibalize eggs.

Male coloration may also reflect parental ability. Female cactus finches *Geospiza conirostris* prefer experienced males in fully black plumage. Females mated to black males contribute more offspring than others to the breeding population; a main reason may be superior parental care by such males (Grant and Grant 1987). Female house finches *Carpodacus mexicanus* choose a mate in relation to his carotenoid coloration, which is correlated with his rate of food delivery to the family, leading to higher breeding success for females mated to colorful males (Hill 1990, 1991; see section 13.6 below for details).

Male great tits *Parus major* have larger black breast stripes than females. In a three-year study, females paired to males with large stripes laid larger clutches than others; in one year they also began laying earlier. No significant correlation was found between male stripe size and territory quality, but males with large stripes were more likely than others to defend their brood, and had heavier fledglings that grew faster. These results suggest that the size of the black breast stripe reflects male parental ability (Norris 1990a,b; also see Norris 1993, and section 3.7 above).

In the pheasant *Phasianus colchicus*, females mating with long-spurred males had higher reproductive success than females mating with other males (von Schantz et al. 1989; Göranson et al. 1990; see section 12.3 below).

Choice of males with well-developed secondary sex traits might also lead to other direct benefits compared to random mating. Such males are likely to be in good nutritional condition and health. They should therefore be less likely than others to carry contagious disease or parasites, and more likely to be dominant and able to protect the female from harassing or competing males that might injure her (section 8.5). Such advantages remain to be demonstrated.

Choice of a male of preferred phenotype will not always result in phenotypic benefits for the female. In a laboratory experiment with zebra finches *Poephilia guttata*, preferred male phenotypes provided less parental care than others (Burley 1988b; see section 13.6).

8.9 Summary

Female choice of mate is widespread in relation to direct phenotypic benefits such as prey offered by the male, nutrients transferred via spermatophores, and male parental care. Females favor males in relation to territory quality in many species. Male mate choice is common in animals where females differ much in fecundity, mainly in relation to body size.

Experimental prevention of female choice and male contests usually leads to lower offspring success in several insects; the mechanisms are not known. Female choice of fit male genotypes has been found in butterflies, but it is not clear whether there is any heritable advantage. Female choice in relation to male secondary sex traits sometimes leads to direct benefits. For example, male song rate may be related to food abundance in the territory. In some birds and fishes, male coloration or courtship display is correlated with parental ability.

Direct benefits are probably often important in the evolution of mating preferences for ornamented males. Such benefits seem less likely, however, to fully explain preferences for the most extreme ornaments, which are found mainly in polygynous species where males provide females with little or no material resources. But even in such species, there might be direct benefits of mate choice, such as reduced risk of injury or disease transmission to females mating with the most ornamented males. A main problem is to estimate the relative importance of the possible direct and indirect mechanisms influencing mating preferences; several mechanism may occur together.

9 Species Recognition, Sexual Selection, and Speciation

9.1 Introduction

Selective advantages of species recognition have long been suggested to play a role in the evolution of secondary sex traits and mating preferences (section 1.5). Sexual selection may be influenced by mate recognition mechanisms, and sexual selection in turn can affect the evolution of such mechanisms, a debated issue in speciation theory (e.g., Paterson 1978, 1985; Templeton 1981, 1989; Coyne and Orr 1989; Endler 1989; Ryan 1990a). For example, divergent sexual selection among populations might increase their differences in mate recognition traits, reducing hybridization. Conversely, Fisherian and indicator processes might start with traits and preferences that are favored initially because they reduce the risk of hybridization (section 1.5). Such traits do not differ fundamentally from traits influencing choice among conspecific mates. Species recognition traits are a subset of those involved in choice of a suitable mate, one crucial aspect of which is species identity.

Species isolation was overemphasized as an explanation for secondary sex traits and mate choice during the modern synthesis of evolution, with its emphasis on speciation and related problems (e.g., Dobzhansky 1937; Huxley 1942; Mayr 1963; for critical reviews, see Paterson 1978, 1985; Thornhill and Alcock 1983; West-Eberhard 1983; Butlin 1989; Templeton 1989). Work on sexual selection may even have been delayed by a mistaken view that mate choice and secondary sex signals in general were explained by a species isolation function. It is now clear that such an explanation often is not sufficient, but it took several decades before renewed interest in sexual selection turned up important problems that had been swept under the species isolation carpet (e.g., Lloyd 1979; Thornhill and Alcock 1983; West-Eberhard 1983; Endler 1989; Ryan 1990a).

Divergence of Species Recognition Traits

There are two main alternative explanations for the divergence of mating preferences and preferred traits in speciation. The hypothesis of allopatric divergence suggests that isolating characters evolve by chance while two diverging, geographically separated lineages accumulate different mutations under partly different selective regimes. As a side effect, the mate recognition systems may diverge so much that the two forms do not interbreed if they come into secondary contact (e.g., Mayr 1963; Templeton 1981; Nei et al. 1983). This explanation has some empirical support (Mayr 1963, but see Parulid warblers in 9.2 below). For example, in *Epicauta* meloid beetles, pairs of allopatric species have probably not been in contact, yet they show species-specific courtship displays, and males do not court females of the other species (Pinto 1980). Nevo and Capranica (1985) drew similar conclusions from a study of *Acris* cricket frogs. During the 1980s, species recognition was often explained as an incidental side effect of divergence in allopatry.

Several other interpretations are, however, possible (e.g., Endler 1977, 1989), including the idea of sympatric divergence. It suggests that isolating traits evolve particularly in areas of secondary overlap between two forms. Individuals that resemble the other form, or fail to discriminate against it, are likely to mismate. If the hybrid matings produce less viable or fertile offspring, selection disfavors such individuals (e.g., Fisher 1930; Dobzhansky 1940). The result can be reinforcement or reproductive character displacement, where differences between the two forms become stronger in areas of contact and overlap (Brown and Wilson 1956; Grant 1972). *Reinforcement* is now often used as a label for the mechanism proposed by Dobzhansky (1940): there is gene flow between the populations, which diverge owing to selection against hybrid offspring (e.g., Butlin 1989; Endler 1989). *Reproductive character displacement* then refers to the further evolved situation of interactions between distinct species without gene flow. Displacement reduces wastage of gametes on infertile matings, or reduces signal interference between species and makes the signals more effective (Butlin 1989; Otte 1989).

Skeptics have pointed out that reproductive character displacement has rarely been demonstrated, in spite of much search for it.[1] Several of the few studies of such aspects of mating preferences have, however, produced positive evidence (reviewed by Gerhardt 1993). It is also debated whether species recognition mechanisms that diverge in sympatry do so mainly by reinforcement or character displacement. Templeton (1981), Paterson (1985), Butlin (1989), and Sanderson (1989) presented theoretical arguments that reinforcement is unlikely to be common. Coyne and Orr (1989) and Otte (1989) drew a different conclusion from extensive comparative studies of fruit flies and Hawaiian crickets (see section 9.2 below).

Sexual Selection

Besides sympatry-allopatry and reinforcement-character displacement, the roles of natural versus sexual selection in the divergence of species is also

[1] For criticism, see, e.g., Paterson 1978, 1982, Templeton 1981, Payne 1983, Thornhill and

debated. Ecological differences were stressed by the architects of the evolutionary synthesis (e.g., Dobzhanskhy 1940; Mayr 1963). Following Darwin (1871), Gulick (1888), and Fisher (1930), later authors emphasize that also sexual selection can make traits diverge among species.[2]

West-Eberhard (1983) gave detailed reasons why sexual selection should make secondary sex signals particularly labile (section 1.7). They may diverge rapidly in isolated populations; as an incidental consequence, this can lead to species-specific signals and speciation. Some genetic models suggest that this is possible even if there is gene flow among populations, that is, without geographic isolation (Lande 1982a; Wu 1985). Many authors have suggested that female choice favors species-specific, conspicuous traits in males because they make it easier for females to choose mates of their own species. Alternatively, such male traits might evolve because they reduce mistaken aggressive interactions with similar species (Butcher and Rohwer 1989; Otte 1989).

Enquist and Arak (1993) used models of neural networks to explore the evolution of male trait and female preference in the context of mate recognition. They found that such recognition systems will have inevitable biases that impose selection on the signal. The system can evolve to a state where females prefer extreme "supernormal" male traits; the precise course of evolution may be highly unpredictable.

Sexual selection may have been crucial in some of the most remarkable taxonomic divergences known: the profuse speciations resulting in large species flocks of, for example, Hawaiian fruit flies, crickets, and several other taxa, African Great Lakes cichlids, and Lake Baikal amphipods (e.g., Templeton 1979; Dominey 1984a; Kaneshiro and Boake 1987; Otte 1989). Probably in part owing to rich opportunities for geographic speciation in the volcanic Hawaiian archipelago, with large altitudinal and habitat differences and with dispersal barriers in form of lava flows within islands, Hawaiian fruit flies have speciated explosively. They comprise an estimated 800–900 species and contain about one-fourth of the world's described fruit flies. A contributing reason for the species richness is probably the absence of many other taxa on the islands, leaving more "empty niches" for new species to occupy. As in several other taxa, the most remarkable differences among fruit flies in many cases concern mating system and

---

Alcock 1983, West-Eberhard 1983, 1984, Eberhard 1985, Spencer et al. 1986, Butlin 1987, 1989, Butcher and Rohwer 1989. Reviews of reproductive character displacement are found in, e.g., Otte 1974, Walker 1974, Ferguson 1977, Rohwer et al. 1980, Becker 1982.

[2] E.g., Trivers 1972, Alexander and Borgia 1979, Templeton 1979, 1981, Lande 1981, 1982a, Payne 1983, Thornhill and Alcock 1983, West-Eberhard 1983, 1984, Dominey 1984a, Kiester et al. 1984, Eberhard 1985, Henry 1985, Carson 1986, Kaneshiro and Boake 1987, Ahearn and Templeton 1989, Butcher and Rohwer 1989, Carson et al. 1989, Endler 1989, Otte 1989, Otte and Endler 1989, Claridge 1990, Ryan 1990a.

Figure 9.1.1 Fruit flies have speciated massively in the Hawaiian islands. The *top left* figure shows a species, *D. heteroneura*, that has diverged markedly from the more usual head shape represented by *D. silvestris* (*bottom*). The head is displayed in the fighting stance of male *D. heteroneura* (*right*). (From Val 1977 and Spieth 1981)

secondary sex traits, suggesting that sexual selection has been important in the speciations (e.g., Spieth 1974, 1981; Carson 1978; Templeton 1979; Kaneshiro 1983, 1988; Kaneshiro and Boake 1987; Hoy et al. 1988). One of the Hawaiian speciations, between *Drosophila silvestris* and *D. heteroneura* (figure 9.1.1), has been examined in detail. Ahearn and Templeton (1989) and Carson et al. (1989) suggested that the two species have diverged by sexual selection in allopatry after a founder event, and that sexual selection maintains their distinctness in spite of hybridization.

Even if sexual selection has played a role in these massive speciations, ecological divergence is also a likely primary reason in many cases. This applies in particular to recently colonized areas such as some of the African Great Lakes and the Hawaiian archipelago, with rich opportunities for ecological divergence, which might secondarily lead to differences in mating system and sexual selection. The roles of ecological divergence and sexual selection in these speciations therefore are hard to disentangle.

Fisher (1930) proposed that hybrids may be less successful in competition over mates. Such a disadvantage can provide a starting point for runaway sexual selection that makes two forms diverge further, leading to sex traits that are much more extreme than required for species recognition. A genetic model by Lande (1982a) suggested that there need be no reproductive character displacement in the zone of overlap (see section 2.6). Owing to the runaway process, trait and preference spread rapidly into areas of allopatry, and there arose no cline reversal of the kind associated with character displacement in sympatry. Even if selection favors species recognition in sympatry, gene flow to the nonoverlapping parts of the distribution areas might make two forms differ no more in sympatry than allopatry (e.g., Walker 1974; Lande 1982a; Otte 1989). Evolution of species recognition by sexual selection therefore need not be associated with reproductive character displacement; lack of displacement is not strong evidence against selection of species recognition in sympatry. In addition, most attention has been directed to signals, but perhaps the main divergence occurs in preferences, which is harder to demonstrate (Waage 1979; Gerhardt 1982, 1993; Ratcliffe and Grant 1983; Otte 1989). More work is needed on preferences in this and other respects.

Scarcity of well-documented examples of reproductive character displacement for these and other reasons need not imply that selection against mismatings with other species is rare and unimportant. Such selection does not necessarily lead to character displacement, and even if it occurs, it may be hard to demonstrate. The following conditions, among others, should ideally be verified (Walker 1974; Waage 1979; Littlejohn 1981).

1. The character is important for premating isolation, and displacement reduces the likelihood of mismatings.
2. Sympatry has arisen from previous allopatry.
3. The displacement is unique to sympatry and does not result from extension of a geographical trend present also in allopatry.
4. The divergence arises from interactions between the two species, and not from unique aspects of the environment in the area of sympatry.

Several new results on mating preferences are hard to explain by species recognition arguments. For example, male traits in some species are not matched by corresponding species-specific female preferences. In some cases, characteristics of the sensory system even make females prefer males of another species, if available (e.g., Basolo 1990a,b; reviewed by Ryan 1990a,b). This need not mean, however, that selection for species recognition is generally unimportant in the evolution of mating preferences and preferred traits. To the extent that mechanisms favored by reduced mismatings will also influence choice among conspecific mates, they can play a role in sexual selection in several ways (e.g., Otte 1979, 1989; Searcy and Andersson 1986). For example, they might make choice among conspecific mates more discriminating, and they might trigger the evolution of extreme traits and preferences by a Fisherian process.

9.2 Selection for Species Recognition

Results from several taxa now suggest that selection for avoidance of matings or other mistaken interactions with other species has sometimes been important in the evolution of secondary sex traits or mating preferences. Some of the evidence for traits that may also be sexually selected is reviewed below.

INSECTS

In spite of much research, reproductive character displacement has been reported for few insects, and an early review by Walker (1974) found no convincing case. Later research has pointed to likely cases in fruit flies, damselflies, and crickets, suggesting that selection for species recognition plays an important role in the evolution of courtship signals and mating preferences in some insects.

DROSOPHILA MOJAVENSIS AND *D. ARIZONENSIS*

These closely related species show stronger reproductive isolation in their Sonoran area of sympatry than in allopatry (Wasserman and Koepfer 1977; also see Koepfer 1987). Experimental tests in the laboratory indicate that *D. mojavensis* has become more reluctant to mate with *D. arizonensis* in sympatry, suggesting reproductive character displacement. Hybridization in the wild seems to be rare or nonexistent. It is not known, however, which characters are involved in species recognition, and alternative explanations are difficult to exclude (see Butlin 1987).

SPECIATION AND REPRODUCTIVE ISOLATION IN DROSOPHILA

In a comparative study of 119 pairs of closely related fruit flies, Coyne and Orr (1989) examined the evolution of reproductive isolation. Geographic ranges, mating discrimination, and strength of hybrid sterility and inviability had previously been studied in these species pairs, and genetic distances estimated by protein electrophoresis. If there is a roughly constant "molecular clock" rate in these cases, the genetic distance between species should change linearly with time (e.g., Nei 1987), and reflect the divergence time for each pair. This makes it possible to compare the rates of evolution of prezygotic species isolation (mating discrimination) and postzygotic isolation (hybrid sterility or inviability) in sympatric and allopatric pairs of species. As expected, the rates were similar in allopatric pairs; in allopatry, both forms of isolation are likely to arise as unselected side effects of genetic divergence between the two species. Prezygotic isolation cannot be selected for as such in allopatry, and postzygotic isolation, which leads to wastage of gametes, is not normally favored by selection. It is therefore also expected that postzygotic isolation evolves at similar rates in both sympatric and allopatric species; this prediction was borne out by the results.

The controversial issue, whether prezygotic isolation is enhanced by selection in sympatry, can be tested by comparing the rate of evolution of prezygotic isolation in sympatric and allopatric species pairs. If there is such selection, isolation should evolve faster in sympatry than allopatry.

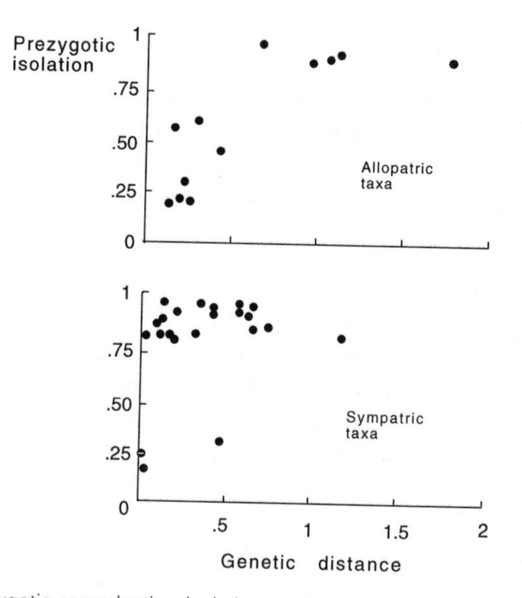

Figure 9.2.1 Prezygotic reproductive isolation evolves more rapidly among sympatric pairs of *Drosophila* species (*bottom*) than among allopatric species pairs (*top*). The genetic distance between two species is assumed to reflect the time since their divergence from a common ancestor. (After Coyne and Orr 1989)

This is the case. The results show that mating discrimination and postzygotic isolation evolve at comparable rates in allopatry, whereas mating discrimination evolves more rapidly in sympatry (figure 9.2.1). The difference was verified by several statistical analyses, some of which were not based on the assumption of a constant molecular clock rate. For example, significantly more sympatric than allopatric species pairs have greater pre- than postzygotic isolation (Coyne and Orr 1989; also see Coyne 1989). Some form of selection of the mate recognition system therefore seems to be at work in sympatry, making specific mate recognition evolve more quickly there than in allopatry. Whether the receptor side, the signal side, or both are involved in the changes is not known.

Several possible alternative explanations were refuted. For example, species pairs that happen already to have prezygotic isolation may be more likely than others to coexist stably if they come into secondary contact. This would imply that the observed pattern has arisen in part owing to fusions or extinctions after contact between populations separated by little mating discrimination (Templeton 1981). This and other alternatives were, however, contradicted by other results (see Coyne and Orr 1989).

Mating preferences or preferred traits in many of these fruit flies therefore seem to have been shaped in part by selection against mismatings with other sympatric species. The selective mechanism might perhaps involve natural selection against hybrids, as suggested by Dobzhansky (1937). An alternative possibility is reduced mating success for hybrid offspring in competition over mates with the pure strains, that is, a sexual selection disadvantage for hybrid offspring (Coyne and Orr 1989). Hybrid sterility or inviability is yet another possibility. The alternatives need to be tested further to clarify the causes of this remarkable pattern.

CALOPTERYX DAMSELFLIES

Perhaps the strongest example of reproductive character displacement among insects is the damselflies *Calopteryx maculata* and *C. aequabilis* (Waage 1975, 1979). Males of *C. maculata* have uniformly black wings, and females have translucent brown wings (figure 9.2.2). In male *C. aequa-*

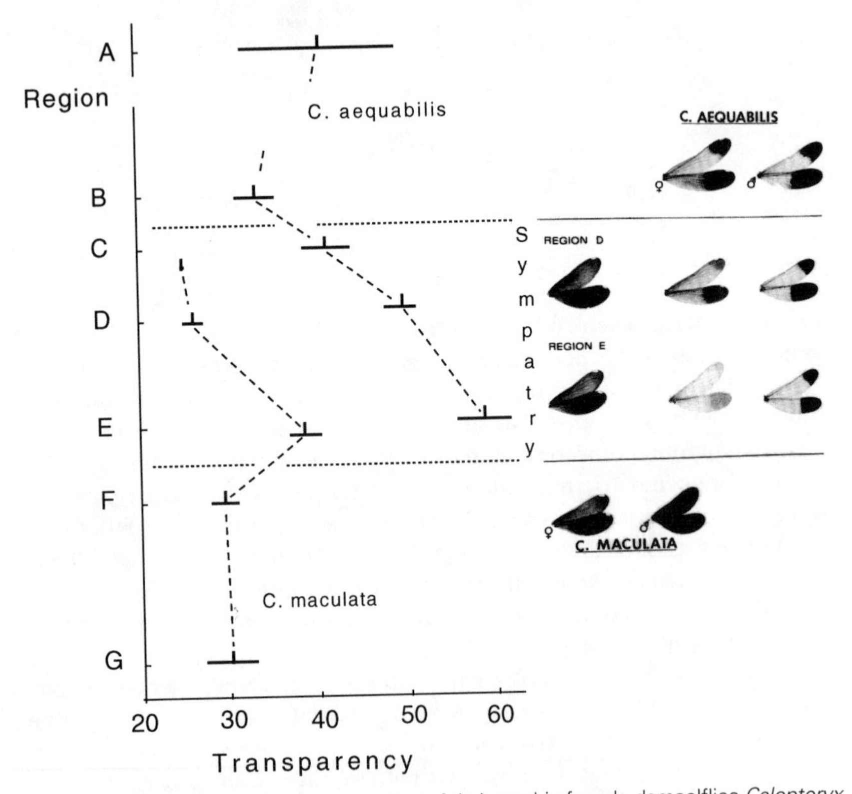

Figure 9.2.2 Wing coloration (transparency of dark area) in female damselflies *Calopteryx aequabilis* shows evidence of reproductive character displacement in regions of sympatry (C, D, E) with *C. maculata* (vertical lines = means; horizontal lines = s.e.). The photograph shows wing pigmentation in two regions of allopatry (B, F) and two of sympatry (D,E). Only females are shown for *C. maculata* in region D and E; males are all black. (From Waage 1979)

bilis only the distal 20%–40% of the wing is black; females often have similar dark wing tips. In field experiments, *C. maculata* males discriminated between females of their own and the other species on the basis of wing pigmentation. Discrimination ability appears to be enhanced in sympatry (Waage 1975). Along a transect across an area of sympatry and extending into regions of allopatry for both species, wing pigmentation in female *C. aequabilis* varied in a fashion consistent with reproductive character displacement (figure 9.2.2). Matings between the two species sometimes occur in the wild, but they do not seem to produce viable hybrid offspring (Waage 1975).

Waage (1979) concluded that this is one of the few verified cases of reproductive character displacement. Remarkably, it concerns a female structural trait, not a male trait. West-Eberhard (1984) suggested the reason may be that females are less strongly sexually selected than males, in which signals perhaps diverge so fast that specificity has arisen incidentally before the diverging populations come into secondary contact (Walker 1974).

Another possibility is that the sex roles in species recognition are less fixed at female discrimination among signaling males than we have come to expect. Female signaling occurs in many insects (Thornhill and Alcock 1983), and male choice may play a role in species isolation for many of them (e.g., de Winter and Rollenhagen 1990). For example, von Schilcher and Dow (1977) concluded that males, not females, are responsible for the termination of courtship between heterospecific couples in *Drosphila* fruit flies.

CONOCEPHALUM NIGROPLEURUM

This North American katydid (bush cricket) discriminates against the song of a closely related sympatric species, *C. brevipennis* (Morris and Fullard 1983). In loudspeaker tests, female *C. nigropleurum* did not, however, prefer the conspecific song over the rather different song of a third species, *C. attenuatus*, with which there is no risk of mismatings as it does not occur in the same locations. On the other hand, female *C. nigropleurum* fail to respond to the song of *C. brevipennis*, which does occur in the same locations. The evidence suggests that females avoid mismatings by recognizing and ignoring the song of the sympatric *C. brevipennis*. This mechanism has probably not evolved in allopatry, as *C. nigropleurum* females do not prefer the song of their own species over that of *C. attenuatus*. Discrimination against *C. brevipennis* suggests that a risk of mismating with this species in sympatry has influenced the evolution of mating preferences in *C. nigropleurum* (Gwynne and Morris 1986).

Another possible case of reproductive character displacement in a ground cricket, *Allonemobius fasciatus*, was described by Benedix and

Howard (1991). Some populations of this species showed evidence of song displacement in sympatry with the related species *A. socius*.

SPECIATION IN HAWAIIAN CRICKETS

Not only fruit flies have radiated enormously in Hawaii: the islands have at least twice as many cricket species as the entire continental United States. They apparently derive from four original colonizing species (Otte 1989). Analysis of the song of many of these flightless crickets in relation to their distribution suggested that the evolution of song properties, especially pulse repetition rate, is often influenced by interactions among species. For example, the differences in pulse rate between sympatric *Laupala* species that share the same habitat is much greater than expected if the songs were independent of each other (Otte 1989). Several parallel clines and other patterns suggest that character displacement in pulse rate is common. Also, song variation is much greater within isolated species than in species sympatric with others. In the isolated *Prognathogryllus waikemoi*, song varies as much as it does in an entire genus of crickets.

None of these patterns is predicted by sexual selection or other intraspecific selective agents: interactions between species are required. The results indicate that song in Hawaiian crickets is often influenced by the presence of sympatric species, favoring divergent mate recognition systems. Otte (1989) suggested that both reinforcement and character displacement may be involved.

SPECIES RECOGNITION AND SCENT EMITTING ORGANS IN MALE MOTHS

A comparative study of the evolution of species-specific courtship signals in five families of moths was done by Phelan and Baker (1987). In some but not all species of each family, males carry scent-emitting organs: hair pencils, brush organs, wing folds, or other structures (figure 9.2.3). Their presentation to the female during courtship is crucial for copulation success (e.g., Birch 1974; Baker and Cardé 1979; Löfstedt et al. 1989; also see Cardé 1986, and section 15.1 below). They are probably also important for reproductive isolation (e.g., Phelan and Baker 1990b).

To find out whether male scent organs diverge in sympatry or allopatry, Phelan and Baker (1987) compared their occurrence with the ecological distribution of closely related species. Two moths of the same genus should be more likely to mismate if they share a common host plant than if not. The hypothesis of sympatric divergence, but not that of allopatric divergence, therefore predicts that species sharing host plants with congeners will evolve isolating scent organs more often than species not sharing hosts. The result was clear: within each of the five families, male scent

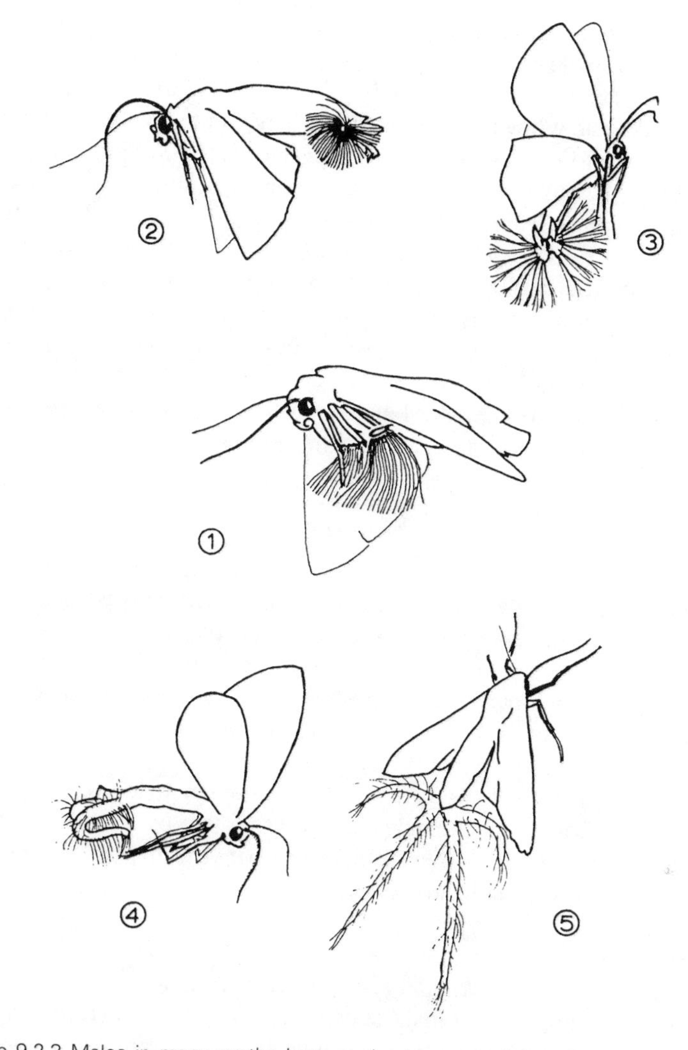

Figure 9.2.3 Males in many moths have evolved scent-emitting organs presented for the female during courtship. The species are: (1) *Apaema monoglypha*, (2) *Plusia gamma*, (3) *Bapta temerata*, (4) *Xanthorhoe ferrugata*, and (5) *Creatonotus gangis* (1–2 are noctuids, 3–4 geometrids, and 5 an arctid moth). (From Birch 1974)

organs were overrepresented among species that shared host plants. If each family is monophyletic, the correlation pattern has evolved at least five times independently in the direction predicted ($p = 1/32$, one-tailed sign test). (Male scent organs occurred in 53% of 396 species that shared host plants, compared to 28% of 419 species without host-plant sharing.) Disadvantages of mismatings therefore have probably been important in the evolution of the male trait. Other evidence points to the same conclusion (Phelan and Baker 1987; but see Birch 1974). The evolution of male scent organs is associated with elaborate courtship behavior that helps deliver the male pheromone to the female (Phelan and Baker 1990a). The female receptor side of the story has not been the subject of similar comparative analysis; it may have been affected as well.

Although the results point to sympatric divergence, there is no sign of reproductive character displacement. One interpretation is that sexual selection by a Fisherian process governs the sympatric divergence, and that it spreads rapidly to areas of allopatry, preventing character displacement (Phelan and Baker 1987; see section 9.1 above). In this scenario, a main disadvantage of hybrids is reduced mating success, a possibility that might be studied by behavioral tests.

Anurans

Litoria (Hyla) ewingi–L. verrauxi

These tree frogs overlap in a large zone of southeastern Australia. Their calls, especially pulse repetition rates and numbers of pulses, differ more in sympatry than allopatry. The differences are large enough to influence female responses and probably reduce mismatings (Littlejohn and Loftus-Hills 1968). Pulse repetition rates in this and several other similar cases differ between species by a factor 2 or more (e.g., Littlejohn 1965; Loftus-Hills and Littlejohn 1971; Foquette 1975; Straughan 1975). Playback of synthetic calls showed that females can discriminate by pulse repetition rate alone, the call property that differs most between the two species in sympatry (Loftus-Hills and Littlejohn 1971).

Pseudacris frogs

The chorus frogs *P. nigrita* and *P. triseriata ferarium* are sympatric in a small part of the southern United States. They show high interfertility and hybrid viability in laboratory tests, but hybridization is rare in the populations studied by Foquette (1975), even though the two species often reproduce at the same ponds and times. Females discriminate among males in relation to calls (Michaud 1962). The two species are more similar in allopatry than sympatry, and the slowest calls of *P. t. ferarium* at all sympatric sites are at least twice as fast as the fastest calls of *P. nigrita*. Geographical variation in pulse repetition rate suggests reproductive character displacement of the call of *P. t. ferarium* (Foquette 1975).

Character displacement and reproductive isolation is no necessary outcome of sympatry in these two species. Farther west, they hybridize massively in a narrow zone where the calls are similar and there is no evidence of displacement (Gartside 1980). This is in line with the idea that hybridization is prevented by the call divergence in the populations studied by Foquette (1975).

OTHER ANURANS

Other possible examples of character displacement in frog calls are weaker than those discussed above, but other kinds of evidence from, among others, *Hyla* treefrogs suggest that the risk of mismating with other species has influenced the evolution of female preferences and male calls (e.g., Blair 1964; Gerhardt 1982, 1988, 1991, 1993; Schwartz 1987b). For example, the three species *H. cinerea*, *H. gratiosa*, and *H. squirella* often breed in the same places. Hybrids of the first two species occur, and female *H. cinerea* sometimes respond to calls of the other species (Oldham and Gerhardt 1975; Gerhardt 1982). In loudspeaker tests, female *H. cinerea* preferred calls with a low frequency peak of 900 Hz, typical of their own species, over calls closer to those of the other two species (700 or 1000 Hz, respectively, or more extreme values; Gerhardt 1982).

These results and other aspects of female responses, for example in relation to pulse repetition rate in *H. versicolor* and *H. chrysoscelis*, may be explained by a risk of mismatings (e.g., Gerhardt 1982). Female *H. versicolor* prefer synthetic calls with a pulse repetition rate typical of their own species over calls with a higher pulse rate, typical of the sympatric *H. chrysoscelis*. Pulse shape and other call properties also differ, and females favor the shape typical of their own species. The preference is very strong, and persists even if the conspecific calls are played only half as often as the calls of the other species, or with eight times lower sound level (Klump and Gerhardt 1987; Gerhardt and Doherty 1988). This may have to do with a hybrid disadvantage: female *H. versicolor* that mate with *H. chrysoscelis* males have sterile offspring. Experiments by Gerhardt (1993) indicate that female preferences in *H. chrysoscelis* have undergone reproductive character displacement in sympatry with *H. versicolor* (figure 9.2.4).

LIZARDS

Coloration may be important for species recognition in some lizards. One possible example is the color of the dewlap in the *Anolis brevirostris* complex of Haiti, where clinal variation in dewlap color may have arisen from selection for species recognition, but other interpretations are also possible (Webster and Burns 1973). Dewlap color seems to play a role in species recognition in several anoles (e.g., Williams and Rand 1977; Fitch and Hillis 1984). This need not be the primary reason for dewlap color divergence, which may have occurred in allopatry for other reasons (West-Eberhard 1983).

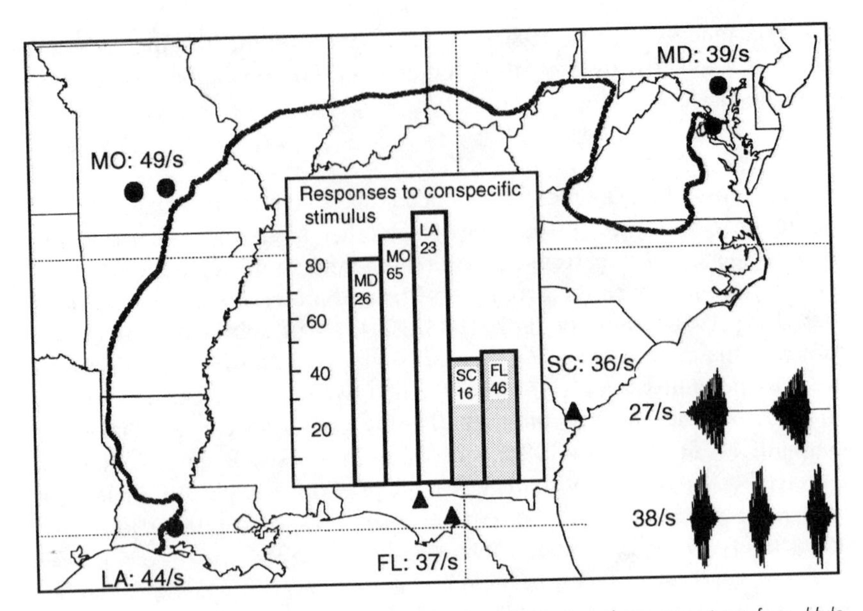

Figure 9.2.4 Female mating preferences in the North American gray tree frog *Hyla chrysoscelis* show evidence of reproductive character displacement in sympatry with *H. versicolor*. The bold line shows the approximate limits of the distribution of *H. versicolor*. Only *H. chrysoscelis* occurs in most of the southeastern United States. Dots indicate sample sites in sympatry, triangles in remote sites of allopatry. Numbers indicate mean pulse rate of the male call at each site. The oscillograms show alternative synthetic stimuli with 27 and 38 pulses/second, respectively, presented to females in choice experiments. The histograms give the responses of female gray tree frogs to synthetic calls with a pulse rate typical of a conspecific male. Females from populations sympatric (open bars; numbers = sample sizes) with *H. versicolor* show stronger responses to such stimuli than do females from allopatry areas (shaded bars), suggesting that female responses in sympatry have been selected for avoidance of mismatings. (Modified from Gerhardt 1993)

Birds

Reproductive character displacement has been inferred in a handful of birds, but the support is weaker than in most cases reviewed above. Grant's (1975) detailed study of *Sitta* nuthatches is an exception: eye-stripe size in *S. neumayer* has apparently diverged from that of *S. tephronata* in some areas of sympatry (also see Rohwer et al. 1980; Butcher and Rohwer 1989).

A possible example involving a conspicuous sex trait is the red shoulder patch of the red-winged blackbird *Agelaius phoeniceus*. In most areas of sympatry with the tricolored blackbird *A. tricolor*, the shoulder patch of the red-wing differs from the patch in other populations, in a manner that makes it deviate more from the tricolored species. The evidence can, however, be interpreted in several ways, and does not necessarily imply character displacement (Hardy and Dickermann 1965; Orians and Christman 1968; Butcher and Rohwer 1989). Another possible example is the reduced plumage darkness of male pied flycatchers *Ficedula hypoleuca* in sympatry with *F. albicollis* (see section 13.6).

Song is a potential species-isolating character in many birds (e.g., Payne 1973; reviewed by Becker 1982; also see section 9.3), for example in the pied flycatcher *Ficedula hypoleuca* in the Baltic islands of Öland and Gotland (Alatalo et al. 1990a). Studying hybridization in relation to song structure, Eriksson (1991) found that male pied flycatchers singing a song similar to that of the collared flycatcher *F. albicollis* were more likely to end up in a hybrid pair than were males singing the typical pied flycatcher song.

In several cases, laboratory tests have shown that females discriminate against song that deviates from the norm of the population or species. For example, estradiol-treated female great tits *Parus major* respond with more copulation solicitation displays to songs from their local than from more distant populations (Baker et al. 1987b; also see McGregor and Krebs 1982; and section 14.3 below). This preference appears to depend on early female experience of local song.

Female brown-headed cowbirds *Molothrus ater* prefer the song of their own over that of other subspecies, and female song sparrows *Melospiza melodia* strongly prefer the conspecific song over that of the sympatric congener *M. georgiana* (King et al. 1980; Searcy and Marler 1981). Females in some species may also respond in relation to the degree of familiarity of the song dialect (e.g., Baker 1983), and to the degree of similarity of the song to that of neighboring males (Grant 1984). Males seem to discriminate more strongly against the song of closely related species in sympatry than allopatry in North American parulid warblers (Gill and Murray 1972) and European firecrests *Regulus ignicapillus* (Becker 1982).

Hybridization occurs in several bird families with distinctive and conspicuous male plumages, for example ducks and geese Anseriformes, grouse and partridges Galliformes, and hummingbirds Trochiliformes (Grant and Grant 1992). In Southern Hemisphere ducks, which often lack conspicuous male coloration, hybridization seems to be even more common than in most northern ducks (Milstein 1979). If so, this is in line with the idea that male plumage plays a role in species recognition in ducks.

Among North American passerines, hybridization is common in zones of secondary contact between some species with distinctively colored males (Rohwer et al. 1980). Parulid warblers offer many examples, reviewed by Morse (1989). Intergeneric hybrids between dissimilar parulids are surprisingly common compared to intrageneric hybrids. Apparently, allopatric morphological divergence has not led to effective species isolation between members of different genera, in spite of their different appearances. This evidence does not seem to fit the hypothesis of allopatric divergence of mate recognition traits (section 9.1 above). Parkes (1978) suggested that selection for species recognition among sympatric congeners has reduced hybridization among them compared to allopatric noncongeners. Hybridization seems to be most likely in areas where one of the species involved is rare (Morse 1989; also see Sawyer and Hartl 1981). The circumstances and consequences of hybridization in these and other birds, which is more important than previously thought, need further study (Grant and Grant 1992; also see Dowling and DeMarais 1993 for fishes).

Females should be more discriminating than males in relation to male signals. A female that accepts courtship from another species may waste efforts on hybrid offspring with low fitness. A territorial male that responds to the song of another species makes less serious mistakes such as wasting display on a noncompetitor. Too narrow responses, on the other hand, could lead to the more serious mistake of accepting another conspecific male on the territory. In accordance, female red-winged blackbirds *Agelaius phoeniceus* are more discriminating than males, and fail to respond to abnormal songs that do evoke male responses (Searcy and Brenowitz 1988; Searcy 1990).

Experiments in Darwin's finches *Geospiza* suggest that recognition in some species has been enhanced by selection of mating preferences in sympatry. In populations sympatric with another species, males discriminate more against mounts of females of the other species than do males from populations where the other species does not occur. This indicates that male mating preferences have become more acute in sympatry (Ratcliffe and Grant 1983). Divergence in at least one of the distinguishing characters of these species, bill morphology, may have arisen mainly as a consequence of ecological character displacement, perhaps with sexual selection playing an additional role (Grant 1986). Yet, hybridization is common among some Darwin's finches, and hybrid offspring sometimes have high fitness (Grant and Grant 1992).

Plants

Reproductive character displacement also occurs in plants, for example with respect to flower color (reviewed by Levin 1978; Waser 1983). In *Phlox pilosa*, the normal form with pink corolla has been partly replaced by a white form in several areas of close contact with the congener *P. glaberrima*. The white form is less affected by pollen transfer from *P. glaberrima* than is the pink form. The improved isolation reduces pollen wastage and halves the proportion of hybrid seed from 4% to 2% (Levin and Kerster 1967; Levin and Schaal 1970).

There is another example from the same genus. *P. drummondi* and *P. cuspidata* both have pink flowers in allopatry. In sympatry, *P. drummondi* has red flowers. Experiments in sympatry with *P. cuspidata* showed that the pink morph of *drummondi* suffers several times higher hybridization than the red morph; hybrids have reduced seed set (Levin 1985).

A third example of character displacement in floral traits is found in Central American populations of *Dalechampia scandens* (Armbruster 1985). Character displacement in these and other similar cases should reduce competition over pollinators and gametic wastage on hybrid matings (Levin 1978; Waser 1983).

9.3 Sexual Selection, Species Divergence, and Speciation

Signals favored by sexual selection may play important roles in species recognition, speciation, and species divergence (e.g., Darwin 1871; Lande 1981; Thornhill and Alcock 1983; West-Eberhard 1983, 1984; Kaneshiro and Boake 1987). For example, in the butterfly *Pieris occidentalis*, field experiments show that females choose mates in relation to their forewing melanin pattern. Females favor males with much melanin, which reduces the risk of mismating with the sympatric sister species *P. protodice* (Wiernasz 1989; Wiernasz and Kingsolver 1992). *P. protodice* males painted to resemble males of *P. occidentalis* had much higher mating success with female *P. occidentalis* than had control male *P. protodice*. The results suggest that the lack of hybrids in nature is a result of female choice, based on the same melanin preferences as in female *P. occidentalis* choosing among males of their own species. Wiernasz and Kingsolver (1992) suggested that the divergence in color pattern of the two species may have evolved by sexual selection through female choice of male melanin pattern.

Recent evidence from widely different taxa suggests that mechanisms of mate recognition, which often seem to be based on sexually selected traits, may be crucial in speciation. In particular, rates of speciation may depend on sensory abilities, and on the scope for distinct variation of traits used in mate recognition. Capacity for production and detection of signal variations may be crucial. For example, the most species-rich lineages of anurans have inner ears that can detect a wide range of frequencies (figure 9.3.1). Ryan (1986, 1990b) suggested that the mating calls in such lineages can diverge over a wider frequency range, making speciation more likely than in lineages with a less complex inner ear.

Speciation in birds has been particularly rapid among passerines, which form 5274 out of 9021 species recognized by Bock and Farrand (1980). “Key adaptations” have been suggested to explain their rapid radiation, but the explanations are problematic (Raikow 1986, 1988). Small size seems to be one facilitating factor (Kochmer and Wagner 1988), but not the only

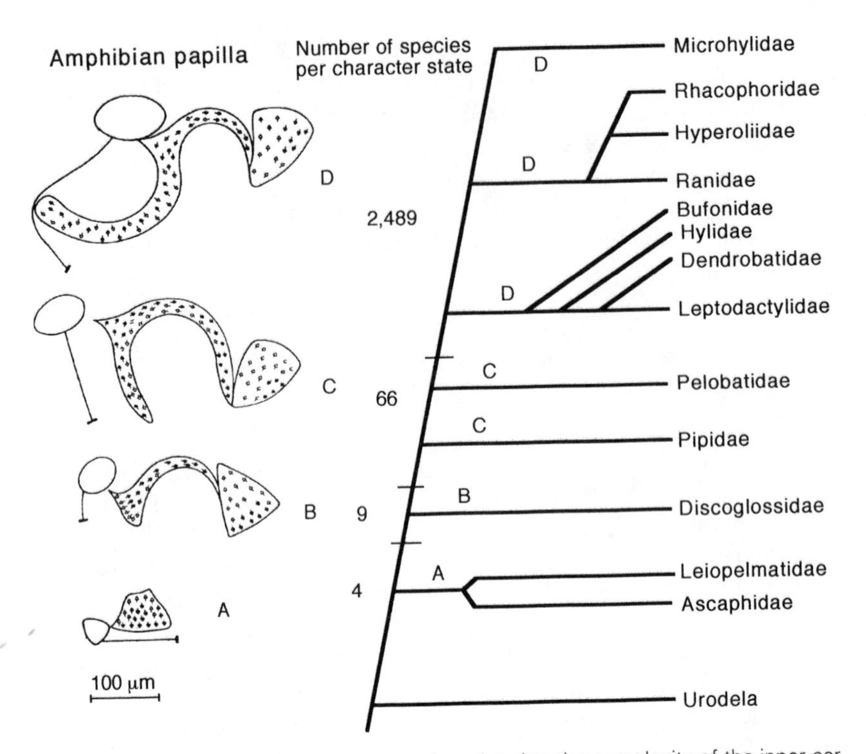

Figure 9.3.1 Speciation in frogs appears to be related to the complexity of the inner ear. The most species-rich lineages in the estimated phylogeny (*right*) have the most complex amphibian papillas (*left*), capable of detecting a wide range of frequencies. Ryan (1986) suggested that sensory ability may have influenced speciation rates in frogs. (After Lewis 1981; Ryan 1986)

one. Raikow (1986) pointed out that most passerines belong to either of three groups that probably independently have evolved advanced syringes, capable of diverse sounds: Oscines, Furnari, and Tyranni (table 9.3.1) (see Raikow 1987). The other four groups, with simpler syrinx and sound production, contain in total only 46 species.

This pattern suggests that advanced syrinx and sound production in males may have been crucial for the flush of speciation in three groups of passerines. A possible reason is that an advanced organ for sound production makes it possible for song to diverge rapidly among species, leading to reproductive isolation (Raikow 1986; also see Vermeij 1988; Fitzpatrick 1988; Gill 1990). For example, song has diverged greatly within several groups of morphologically rather uniform warblers, such as *Phylloscopus* and *Acrocephalus*. This idea might be explored, for example, by comparative studies of acoustic insects or other groups that permit independent tests (section 4.3). Groups with advanced production and detection of variable sounds are predicted to show higher speciation rates than related but vo-

Table 9.3.1

Species Numbers of Passerine Birds in Relation to Syrinx Function*

		Number of Species
Taxa with advanced syringes		
Oscines		4,177
Furnarii		541
Tyranni		510
	Total	5,228
Taxa with simpler syringes		
Acanthisittidae (Xenicidae)		4
Pittidae		24
Philepittidae		4
Eurylaimidae		14
	Total	46

* Based on Raikow (1986, 1987).

cally less advanced taxa. A hint that this may be the case comes from the anuran example above, and from the profuse speciations among Hawaiian fruit flies and crickets, both of which have acoustic courtship (section 9.2 above; see also Hoy et al. 1988; Otte 1989).

Possible similar cases involving male visual signals are ducks, pheasants, birds of paradise, African cichlids, Lake Baikal amphipods, and, again, Hawaiian fruit flies (West-Eberhard 1983; Dominey 1984a). The potentially important roles of signal divergence and sexual selection in speciation should be testable with comparative methods (section 4.3). Other mechanisms, such as ecological divergence, also seem likely to play primary roles in rapid speciations (section 9.1).

A similar mechanism has perhaps contributed in the profuse speciations of angiosperm plants (e.g., Stebbins 1981; Crepet 1984; Kiester et al. 1984; Willson 1991; Eriksson and Bremer 1992; Pellmyr 1992; also see section 17.8). Most angiosperms depend on insects for pollination, mediated by visual and scent stimuli and pollinator rewards that may in part be sexually selected (chapter 17). Angiosperms would therefore seem to have more scope for evolution of reproductive isolation by prezygotic mechanisms than do wind- or water-pollinated plants. Visual and olfactory discrimination by pollinators should permit much more accurate transfer of pollen to plants of the same species than does pollen transport by wind or water. In addition, slight changes in flower color, shape, scent, or other features perceived by pollinators might suffice for evolution of reproductive isolation and speciation. The mediation of pollination by acute animal senses may therefore partly explain the rapid speciation and radiation that apparently have taken place among angiosperms. Testing several hypotheses by comparative data, Eriksson and Bremer (1992) concluded that animal pollination has been a major factor in the diversification of angiosperms. Life form has also been important, herbs speciating more rapidly than shrubs and trees.

Fisherian runaway selection might increase speciation rates in animals by making populations diverge rapidly in mating preferences and preferred traits (Lande 1982a; West-Eberhard 1983; section 2.6 above). On the other hand, although Fisherian processes usually seem unlikely in plant-pollinator systems (section 17.8 below), many angiosperms may have speciated rapidly through divergence in pollinator attractants. Fisherian runaway selection has perhaps not been of crucial importance in the rapid speciations among animals either; further comparisons with plants might be clarifying.

9.4 Summary

There is now considerable evidence that selection for species recognition sometimes plays a role in the evolution of secondary sex traits or mating preferences in several taxa. This is suggested for example by comparative studies of reproductive isolation in fruit flies and moths, and by reproductive character displacement in some insects, frogs, and flowering plants.

Mating discrimination against closely related species in some cases probably evolve incidentally in allopatry, but in other cases sympatric interactions are at work, affecting the evolution of secondary sex traits and mating preferences. Among the possible factors responsible are parental sterility barriers or offspring inviability. There can also be disadvantages for hybrid offspring in form of low ecological efficiency, low success in competition over mates, and reduced or no production of offspring. Interference with the signals of related species may also favor divergence. The relative roles of these alternatives need to be assessed in a number of taxa before a clear general picture of the evolution of species recognition can be drawn. This holds also for the relative role of species recognition advantages in the evolution of mating preferences and secondary sex traits.

Sexual selection and divergence of secondary sex traits together with ecological divergence may have played an important role in the profuse speciations among some taxa, such as Hawaiian fruit flies and crickets, African cichlids, passerine birds, and angiosperm plants. Advanced ability of producing and detecting variable acoustic signals, used in male contests or attraction of mates, is one of the factors that may have facilitated rapid speciations in certain insects, anurans, and birds.

10 Constraints

The theory of sexual selection predicts that sexually selected traits eventually are prevented from further exaggeration by opposing natural selection (section 1.5). Such opposing selection can take many forms, from dramatic predation to subtle reduction in growth rates. It can arise at various stages of the life cycle, from early development in form of higher juvenile mortality, to reduced future reproduction owing to present mating effort. Sexual selection can also be self-limiting: greater development of a secondary sex trait may increase mating success in one way but reduce it in another.

This chapter first reviews evidence for the classic idea that conspicuous secondary sex traits raise the risk of predation. Other constraints are then discussed, such as energetic costs, starvation risk, life history trade-offs, genetic correlations, and sensory bias. In spite of encouraging progress, constraints on sexually selected traits in general are poorly known, and need much more study (also see Price et al. 1987). Although many mechanisms of counteracting selection have been identified, there is still no quantitative empirical demonstration of a balance between sexual and natural selection of a secondary sex trait.

10.1 Predation

Results from a variety of animals show that male secondary sex traits sometimes raise the risk of predation. Suggestive evidence comes from cases where adult males suffer higher predation than females; the most convincing cases are those in which the risk is clearly related to the expression of a male trait.

There are many reasons why searching, signaling, and contesting for mates may increase the risk of predation (reviewed by Burk 1982; Magnhagen 1991). The need for a signaling male to make himself conspicuous to females or rivals will often make him easier to find for predators and parasites as well. Sexual display and contests can make males less wary and easier to catch.

Male Bias in the Prey of Predators

Analyses of the prey taken by certain predators suggest higher predation on adult males than females in some species. The reasons in many cases may have little to do with male morphological sex traits or signals; sex differences in exposure to predators often seem to be responsible. For example, male brown lemmings *Lemmus trimucronatus* are more common than females as prey of snowy owls *Nyctea scandiaca*, and the proportion of males in a lemming cohort declines with time (Pitelka 1957). Male lemmings have larger home ranges and spend more time active outside of burrows than females, apparently to secure as many matings as possible (Banks et al. 1975; also see Ims 1987, 1988a). It therefore seems that sexually selected extensive movements raise the risk of predation in male lemmings, as in several insects (see below in this section). Similar male mating tactics are found in many other animals, so this cause of male mortality may be common. For example, Jormalainen and Tuomi (1989) suggested that higher predation on the more active and mobile males of the isopod *Idothea baltica* explain the change from male-biased to female-biased sex ratio over the reproductive season.

Predation by peregrine falcons *Falco peregrinus* also falls more often on conspicuous males than on females of some species. Waterfowl taken by Alaskan peregrines consisted almost entirely of drakes, and 75% of the sixty-four ruffs *Philomachus pugnax* found in peregrine nests in Lapland were males (Cade 1960; Lindberg 1983). Such results need not depend on the conspicuous coloration of males, however. The adult sex ratio in the prey population is often not known, and the traits that cause higher predation are not easy to identify. The reason for higher predation on males might well be greater exposure to predators due to (in part sexually selected) behavioral differences from females, which may be difficult to discover while they are incubating. The rapid moult to inconspicuous plumage in male ducks and ruffs after the mating season suggests, however, that the conspicuous nuptial plumage has a cost (Andersson 1983).

Other kinds of predation and other mortality may fall more heavily on females of the same species because of their parental behavior (e.g., Promislow et al. 1992). For example, severe predation by carnivores on nesting female ducks often leads to a strongly male-biased sex ratio (e.g., Johnson and Sargeant 1977; Bellrose 1980). As a consequence, there may be strong sexual selection among males also in these monogamous species (sections 7.3, 13.6).

Visual Signals

An increased risk of predation on visually signaling males has been demonstrated in several fishes. In one of the first experimental tests, male three-spine sticklebacks *Gasterosteus aculeatus*, with a red throat, ran at least twice as high risk of predatory attack in aquaria with cutthroat trout *Salmo clarki* as did males with a dull throat (Moodie 1972). Males develop a red throat during the breeding season only in some populations; the color seems to be under genetic control (McPhail 1969). Differences in risk of predation were previously thought to explain the variation in male coloration, but other causes now seem more likely, for instance differences in water visibility (e.g., Reimchen 1989; section 13.3 below).

The firetail *Notobranchius guentheri* is a cyprinodontid African killifish that lives in shallow temporary waters; its eggs survive periods of drought in diapause until the next wet season. Like most other "annual" killifishes (e.g., Van Ramshorst and Van Den Nieuwenhuisen 1978), the firetail is strikingly sex dimorphic. Females are dull brown, whereas males have red and blue flanks, yellow head and abdomen, and carmine caudal fin. Haas (1976) concluded from tests in aquaria that males in turbid water are discovered more rapidly than females by conspecifics, and that females in choice tests approached the male with the reddest caudal fin for spawning. Haas also staged hunting by a heron in a pond with equal numbers of male and female firetails. The heron took males more rapidly than females, but the difference was not tested statistically. Another complication was that males are somewhat larger and perhaps also for this reason more conspicuous than females. Strong evidence that predation selects against bright colors in male fish comes from the detailed studies of Trinidad guppies by Endler (e.g., 1980, 1983, 1987), reviewed in sections 5.2 and 13.1.

A remarkable form of predation on males occurs in fireflies, subject to a series of fascinating studies by J. E. Lloyd (e.g., 1975, 1979, 1980, reviewed 1984; figure 10.1.1). Males flying in twilight or darkness emit a species-specific pattern of light flashes, which are answered by perching receptive females, enabling males to find them. Other, predatory fireflies have broken into this communication system: female reply patterns of, for example, *Photinus* fireflies are mimicked by females of some *Photuris* fireflies, which attract *Photinus* males not for a mate but for a meal. In certain *Photuris* species, males may search for females of their own species by giving the same signals as males of several *Photinus* species preyed upon by *Photuris* females (J. E. Lloyd 1980; figure 10.1.1). Lloyd's work has revealed a fascinating world of male and female mating and predation tactics by flash signals in these little animals, "with brains the size of the comma in this sentence" (Lloyd 1979).

Promislow et al. (1992) present comparative evidence for mortality costs of male bright plumage in birds. For negative evidence from a lizard, see Olsson (1993c).

Acoustic Signals

Predation on calling males has been observed in many insects (reviewed by Burk 1982; Thornhill and Alcock 1983). For example, male short-tailed crickets *Anurogryllus celerenictus* are located through their song and eaten by little blue herons *Florida coerulea* (Bell 1979). Singing male field

Figure 10.1.1 Male *Photuris* fireflies produce species-specific patterns of light flashes when cruising in the treetops. They can also descend and mimic the flickering and glowing flash patterns of males of other sympatric forms (*Photinus* and *Pyractomena* species). (From Lloyd 1980. Copyright 1980 by the AAAS)

crickets *Gryllus integer* are parasitized by tachinid flies. Observations and loudspeaker experiments showed that flies as well as female crickets locate males through their calls (Cade 1975, 1979). The flies deposit larvae on or near the cricket; the larvae eat and kill the host in a few days. The calling behavior of the cricket shows partly genetic variations. Some males do not call, but wait near calling males and try to mate with arriving females (see

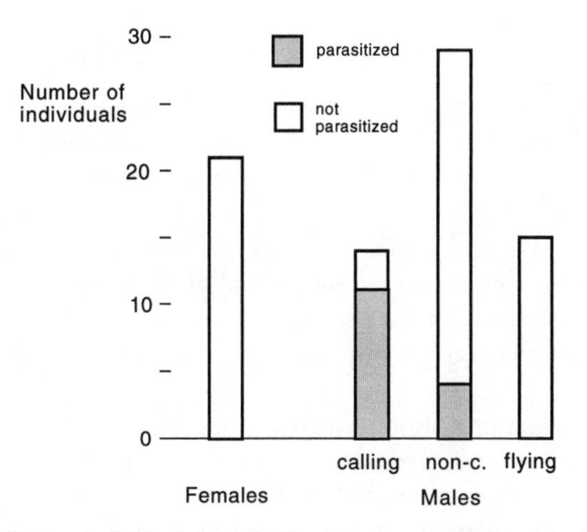

Figure 10.1.2 Calling male field crickets *Gryllus integer* run a higher risk of being parasitized by tachinid flies than do females and noncalling males. (Data from Cade 1979)

section 16.2). Calling males are subject to much higher risk of parasitism than are silent males or females (Cade 1979; figure 10.1.2). Whereas calling attracts females, it also raises the risk of being parasitized, which reduces the reproductive life of the male. In another cricket, *Gryllodes supplicans*, the male call attracts predatory geckos (Sakaluk and Belwood 1984).

A parasitic sarcophagid fly, *Colcondamyia auditrix*, also locates its host, the cicada *Okanagana rimosa*, by the song of the male (Soper et al. 1976). Parasitized males lose their singing ability, and therefore probably also the ability to attract females. In some years, 14%–19% of the cicadas were parasitized. A vertebrate example is the predation by bats and several other predators on calling male tungara frogs *Physalaemus pustulosus*, reviewed in section 5.3.

Pheromones

Tachinid flies also parasitize chemically signaling males of the stinkbug *Nezara viridula*. The flies are attracted by the male pheromone; a higher proportion of males than females are parasitized (Mitchell and Mau 1971; Todd and Lewis 1976; Harris and Todd 1980). Additional cases are reviewed by Thornhill and Alcock (1983).

Fireflies are not the only insects with signals mimicked by a predator. Mimicry of the pheromones of female moths enables bolas spiders *Mastophora dizzydeani* to attract and catch males of several noctuid moths searching for females. The spider also produces a silken thread ending in a sticky ball. When a moth approaches, the spider tries to catch it by swinging the ball (Eberhard 1977, 1980a).

Other Sexually Selected Behavior

The studies of scorpionflies by Thornhill (see section 8.2) show that male mating efforts raise the risk of predation. *Bittacus* scorpionflies are caught in spider webs during hunting flights, whereas *Panorpa* species in addition are captured when feeding on dead prey in spider webs (Thornhill 1978). Spider predation is usually more than twice as high on males as on females, apparently because males forage for nuptial gifts. An exception is *Bittacus strigosus*: both sexes hunt for prey in this species, and there is no nuptial feeding, nor any marked sex difference in predation by spiders (Thornhill 1978, 1980a).

The risk of spider predation also differs between the sexes in the ticktock cicada *Cicadetta quadricincta*. Males fly much more than females in order to locate mates, and are more likely to be captured in spider webs (Gwynne 1987). The sex ratio among emerging cicadas was close to unity, but in a sample of seventy-two cicadas caught in spider webs, males were more than five times as common as females. Males run a higher risk of being killed in spider webs also in the butterfly *Euphydras editha*; Moore (1987) found 116 males but only 15 females dead in webs. A possible reason is that males sometimes mistake dead butterflies in webs for newly hatched females and approach them to mate. Moore (1987) placed dead butterflies as baits for males in fifty webs, and used fifty empty but otherwise identically treated webs as controls. As predicted, a higher proportion of males than females was killed in the webs containing a dead butterfly, but not in the empty webs (table 10.1.1).

Swarming males may run high risks of predation (reviewed by Burk 1982; Thornhill and Alcock 1983). For example, aggregations of male syrphid flies attract not only females, but also *Oxybelus* sphecid wasps that specialize on this prey (Peckham and Hook 1980; Burk 1982). On the other hand, aggregations of males can sometimes provide antipredatory advantages, for example because of more rapid detection of predators or dilution of the risk per male (e.g., Sweeney and Vannote 1982). In choruses of

Table 10.1.1

Number of *Euphydras editha* of Each Sex Found in Spider Webs with and without Butterfly Baits*

Sex	*Empty Webs*	*Baited Webs*	*Totals*
Males	1	16	17
Females	4	1	5

* Males are more likely than females to occur in baited webs ($p < 0.01$, Fisher exact test). Data from S. D. Moore 1987.

tungara frogs *Physalaemus pustulosus*, the risk of predation per male decreased with increasing chorus size (Ryan et al. 1981). There was no indication that the chance of mating per male decreased; if anything, the tendency was toward an increase, so males seem to benefit from large chorus size.

Mate guarding can lead to higher predation. For example, slow-moving *Gammarus* pairs in amplexus may be subject to high predation by fish (Ward 1986). In some cases, however, copulation may reduce predation on males, as copulation lowers their locomotory activity (reviewed by Gwynne 1989). Male territory defense and contests can increase predation and parasitism. Territorial male digger wasps *Philantus bicinctus* are killed more often than females by robber flies (Asilidae). This is apparently a consequence of higher male exposure to the flies during territorial patrolling and defense (Gwynne and O'Neill 1980).

Among the bullfrogs *Rana catesbeiana* studied in a Michigan pond by Emlen (1976) and Howard (1981b), territorial males were subject to size-dependent predation. Large males are favored in contests, and are more likely than small males to obtain territories and females. Calling, fighting, and mating is costly, however: snapping turtles *Chelydra serpentina* apparently are attracted to such commotion, and they killed a number of large territorial males. Predation was lower in females and small males (Howard 1981a). Turtle predation therefore selects against large male size in bullfrogs. Owen-Smith (1993) presented similar evidence for kudu antelopes *Tragelaphus strepsiceros* subject to lion predation.

Selander (1965) suggested that male body size in blackbirds (Icteridae) is set by a balance between sexual selection for larger size and survival selection for smaller size. This suggestion is supported by some results for red-winged blackbirds *Agelaius phoeniceus*, but contradicted by others. Likewise, the suggestion that there is a mortality cost of large male size during the early growth period remains an interesting idea, but there is little support for it in the red-winged blackbird at present (reviewed by Searcy and Yasukawa 1994).

In addition to the traits favored by competition over mates, mating preferences and mate choice can also have costs (reviewed, e.g., by Kirkpatrick 1987b; Iwasa et al. 1991), with profound effects on sexual selection; see chapters 2 and 3 above.

Although sexually selected traits will often increase the risk of predation, this is not always the case. For example, sexually selected male weapons may confer some protection against predators. The enlarged claw of male fiddler crabs *Uca* apparently makes males less likely than females to be eaten by herons (Bildstein et al. 1989). The claw does not seem to reduce male foraging efficiency (Caravello and Cameron 1987), and it is not known what prevents further increase in its size.

Evolutionary Consequences of Predation on Males

Increased risk of predation may be an unavoidable cost for sexually competing males in some species, limiting the evolution of their sexually selected traits. The risk may often be reduced, however, by defenses such as structural or chemical weapons, spatial, temporal, or other adjustment of signaling, and use of alternative tactics (Burk 1982; Magnhagen 1991). In some insects, male signals are less conspicuous than in related species subject to lower predator pressure. For example, in certain neotropical forest katydids (Tettigonidae), song and other behavior seems to be adapted for reducing the risk of predation by bats. Such katydids have much shorter song than species in areas without these bats (Morris 1980; Belwood and Morris 1987). Predation or parasitism may also favor alternative male mating tactics, such as silent satellite behavior in crickets and frogs (see above).

Predation may influence the form and strength of sexual selection in several ways (e.g., Endler 1983, 1986a; Partridge and Endler 1987). For example, elaborate courtship, mate choice, and mate guarding may be selected against if the risk of predation is high, perhaps instead favoring male contests. The sensory mode and kind of secondary sex traits favored by sexual selection may also depend on the senses used by major predators, and on the sensory conditions during courtship, which are related to ecological aspects such as habitat, light conditions, and breeding season. The guppy *Poecilia reticulata* provides several examples of local adaptations that reduce predation associated with sexual advertisement (e.g., Endler 1987, 1991, 1992b; see section 5.2 above).

Synchronous swarming may reduce the risk of predation in some insects, for females as well as signaling or searching males. Mass occurrence of prey during a brief period should satiate local predators and reduce the risk per individual (e.g., Edmunds 1974). Mayflies (Ephemeroptera) provide likely examples. Sweeney and Vannote (1982) rejected the alternative that synchronization is favored largely because it raises the chances of finding mates: three parthenogenetic species showed as strong synchronization as did twelve sympatric sexual species.

The mass swarming of North American periodical cicadas (*Magicicada*) may also be an antipredator adaptation (Lloyd and Dybas 1966). These species require 13 or 17 years to develop as subterranean larvae to a reproductive adult stage of a few weeks. There are three partly sympatric species of each kind (Alexander and Moore 1962). The main population in an area emerges as adults during a few weeks every 13 or 17 years, at a density that may exceed ten animals per square meter. (Since 13 and 17 are prime numbers, no multiple of a shorter life cycle, save for an annual one, will enable a predator to hit the next cicada year with its own offspring.) The calling of hundreds or thousands of males produces a sound so intense that it has been suggested to repel predators (Smith and Langley 1978). Whether this is correct or not, it seems likely that the mass emergence every 13 or 17 years satiates predators and reduces the risk of predation per cicada. The synchronous emergence of three sympatric species during the same year supports the ecological explanation based on predation; there is no known long-term cycle in the food resources of the cicadas (Lloyd and Dybas 1966).

10.2 Other Mortality in Sexually Competing Males

Secondary sex traits can reduce survival in many ways other than through predation. Male relative to female total mortality rate increases with sexual size dimorphism among mammals and birds, and with male plumage brightness among birds (Promislow 1992; Promislow et al. 1992).

One probably common effect of sexual selection for larger or more ornamented males is increased juvenile mortality (Clutton-Brock et al. 1985; Weatherhead and Teather 1991). Faster or longer growth and higher nutritional requirements can make growing males more prone to starve than females. In a review of results from ten bird and nine mammal species, Clutton-Brock et al. (1985) concluded that the male bias in nestling or juvenile mortality apparently increases with the degree of sexual dimorphism (male:female size) (figure 10.2.1).

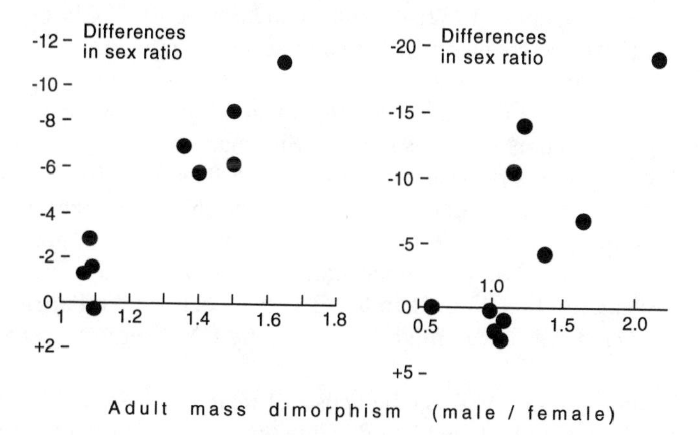

Figure 10.2.1 The differences in sex ratio (% males) between birth and the end of the first year of life in ungulates (*left*), and between hatching and fledging in birds (*right*), increases with the degree of sexual size dimorphism (male : female body mass). These patterns suggest that the male bias in nestling or juvenile mortality increases with the sexual size dimorphism. (After Clutton-Brock et al. 1985)

There are several alternatives to this sexual selection hypothesis for higher juvenile mortality in male than female mammals. One is the expression of deleterious recessive genes from the X chromosome of the heterogametic males. This explanation cannot, however, account for the likewise higher mortality in juvenile males of several birds, where females are the heterogametic sex (Clutton-Brock et al. 1985). Another explanation was proposed by Trivers and Willard (1973). Owing to advantages in competition over mates, the reproductive success of sons probably increases more strongly with large size than does the success of daughters. Females in nutritional condition below the average might therefore bias their parental investment toward daughters. This hypothesis predicts higher male mortality early in the period of parental investment, when important resources can still be saved if a son dies. The sexual selection hypothesis predicts higher juvenile mortality in males whenever there is food shortage, also after juveniles are no longer fed by their parents. This pattern is found in several ungulates and birds, supporting the sexual selection hypothesis for higher male than female juvenile mortality, even if other factors may also be involved (Clutton-Brock et al. 1985; Weatherhead and Teather 1991; and references therein).

In several lekking species, mortality increases at the age when males become sexually active and engage in energy-demanding and risky sexual competition (reviewed by Wiley 1991). Sexual competition can lead to higher male mortality through fights over mates. Such fights are often damaging and even lethal in many species, from insects (reviewed by Hamilton 1979; Thornhill and Alcock 1983) to large vertebrates such as lions and ungulates (e.g., Schaller 1972; Wilkinson and Shank 1977; Clutton-Brock et al. 1982; also see chapter 12). Lethal or seriously injuring fights among rival males are common in species with specialized weapons such as fig wasps (Hamilton 1979). They occur also in some species without specialized weapons, such as ground squirrels (Sherman and Morton 1984), and even in such symbols of peaceful idyll as butterflies (e.g., Eff 1962). The risk of injury and death is an important cost of contest competition over mates in some animals. Male fighting may be most severe in polygynous species where several females are at stake, and in forms where males have only one chance of breeding (Hamilton 1979; Alcock 1984b; Partridge and Endler 1987). These ideas might be tested by quantitative comparative studies.

Marler and Moore (1988) studied costs of male aggression by testosterone implants in side-blotched lizards *Uta stansburiana*. Free-living males treated with testosterone were more aggressive, active, and conspicuous, and lost more weight and survived less well than controls. The possible benefits of greater success in contests therefore were counteracted by reduced survival. This may be a common effect of increased testosterone levels, which not only stimulate the development of many sexually selected male traits but probably also reduce immuno-competence and make males more susceptible to pathogens and parasites (Folstad and Karter 1992).

10.3 Other Costs of Sexually Selected Traits

While increasing mating success in one way, a larger sexually selected trait may also tend to reduce it in other ways. A possible example is large male body size in the red-winged blackbird *Agelaius phoeniceus*, which is advantageous in male contests, at least in groups of captive males (Searcy 1979b; Eckert and Weatherhead 1987b). Large males seem, however, to have energy problems. They spend more time foraging, and have less time for defending the territory and displaying to attract females. (Also in sage grouse *Centrocercus urophasianus*, small males display at higher rates than large males; Gibson and Bradbury 1985). Searcy (1979d) suggested that body size in male red-winged blackbirds may be stabilized by these opposing effects on competitive mating success; a disadvantage in terms of natural selection is not necessarily required (also see section 10.1).

Male swallows *Hirundo rustica* experimentally provided with longer tail streamers had higher mating success than males with shortened tails (Møller 1988a, 1989a; also see section 7.3). Their foraging ability was impaired, however, and after moult these males grew shorter tails than they had before manipulation. The mating advantage from an elongated tail in one year was therefore counteracted by reduced mating success the following year, so sexual selection may be self-limiting, as Searcy (1979d) suggested. In addition, among the males with elongated tail in Møller's (1989a) experiment, those with short initial tail, and hence with greatest relative increase in tail length, seemed to survive less well than those with long initial tail.

Another example is mate guarding in amphipods. Precopulatory mate guarding in *Gammarus lawrencianus* reduced male growth almost by half (figure 10.3.1; Robinson and Doyle 1985). The male cannot feed while holding the female, which he may do for a number of days or weeks. Reduced growth lowers future mating and reproductive success, because large males are most likely to mate with large, fecund females. Reduced growth also occurs, for example, in sexually signaling male anurans (see below). As several of these examples show, a secondary sex trait can be favored by one component of sexual selection, and counteracted by another component. In practice, many sexually selected traits are probably also limited by natural selection.

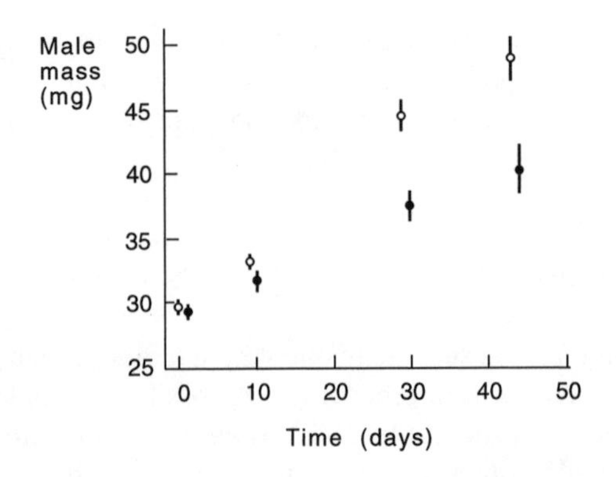

Figure 10.3.1 Growth is faster in free-swimming (open circles) than in mate-guarding (dots) males of the amphipod *Gammarus lawrencianus* (measured over 43 days). (From Robinson and Doyle 1985)

10.4 Physiological Constraints

Physiological constraints may limit courtship and other sexually selected behavior, for example in various forms of endurance rivalry. There is evidence from a variety of species that sexual display is energetically costly (reviewed by Halliday 1987; Ryan 1988a, but see Borgia 1993). Some birds perform song flights that are likely to be highly energy demanding, for example the skylark *Alauda arvensis* (Møller 1991e). The display flight of the flappet lark *Mirafra rufocinnamomea* requires an estimated power output 16 times the basal metabolic rate (Norberg 1991). Song raises the instantaneous rate of energy expenditure an estimated 18 times above resting metabolism in male bladder cicadas *Cystosoma saundersii*, about 21 times in the gray tree frog *Hyla versicolor*, and 14–17 times above the basal metabolic rate in sage grouse *Centrocercus urophasianus* (MacNally and Young 1981; Taigen and Wells 1985; Wells and Taigen 1986; Vehrencamp et al. 1989).

Song seems to be one of the most energy-demanding activities in animals. Some frogs may approach the maximum sustainable rate of energy expenditure when calling. In other species, song is less likely to be limited in the short term by physiological constraints (Taigen and Wells 1985; Wells and Taigen 1986; Ryan 1985, 1988; Sullivan and Walsberg 1985; Klump and Gerhardt 1987; Prestwich et al. 1989). Over longer periods, on the other hand, energy consumption may put limits to behavior used in competition over mates (Halliday 1987).

One indication of such a constraint is the weight loss of males competing for females in a variety of animals, for example frogs and toads (*Rana clamitans*, Wells 1978; *Bufo bufo* and *B. calamita*, Arak 1983b; *Uperoleia rugosa*, Robertson 1986b; *Rana virgatipes*, Given 1988), urodeles (*Triturus vulgaris*, Verrell and Halliday 1985), mammals (*Cervus elaphus*, Clutton-Brock et al. 1982; *Halichoerus grypus*, Anderson and Fedak 1985; *Mirounga angustirostris*, Deutsch et al. 1990), and birds (*Tetrao tetrix*, Angelstam 1984; *Centrocercus urophasianus*, Vehrencamp et al. 1989; *Anthus spinoletta*, Askenmo et al. 1992). Such weight loss, which in some cases exceeds 20%, is usually a consequence of reduced feeding by males that display and compete for females on the mating ground. For example, in the Australian frog *Uperoleia rugosa*, males lost up to 30% of their mass during periods of territorial calling, and regained mass during periods of silent satellite behavior (Robertson 1986b). Male red deer *Cervus elaphus* reduced the time spent feeding by over 90% when holding a harem, and lost up to 20% of their body mass during the rutting period. Deteriorating body condition is apparently one of the factors that put an end to a stag's rut (Clutton-Brock et al. 1982; section 5.6 above). In northern elephant seal *Mirounga angustirostris*, males lose over one third of their body mass during the breeding season. The mass loss is largest in dominant, successful bulls, which apparently must pay for their success by more rapid loss of energy reserves (Deutsch et al. 1990; also see section 5.5, and Anderson and Fedak 1985).

Although mass loss in many of the cases cited is likely an unavoidable consequence of sexual competition, mass loss in reproductive animals may sometimes be adaptive, for example by reducing the energy costs of some reproductive activities, such as transport of food to nestlings (Norberg 1981).

The energetic cost of sexual display was studied in detail in sage grouse *Centrocercus urophasianus* by Vehrencamp et al. (1989; also see section 7.5). Energy expenditure was measured with doubly labeled water technique in eighteen adult males and related to their display rate, body condition, mass loss, blood parasites, hematocrit level, foraging behavior, and other variables. As expected, daily energy expenditure increased with display rate (figure 10.4.1) and time spent on the lek. Males that attended leks were in better condition (heavier in relation to size) than others. Vehrencamp et al. (1989) concluded that the most vigorous males were probably displaying at the maximum rate that can be sustained over several hours. The daily energy expenditure was four times the basal metabolism, which may be close to the maximum sustainable rate in homeoterms (Drent and Daan 1980; but see Bryant and Tatner 1991).

Unexpectedly, the mass loss per day decreased with increasing display activity, so the most active males seemed best able to recover their energy losses. Such males foraged farther from the lek than others. The evidence therefore suggests that males vary in their capacity for energetically costly lek attendance and display, important determinants of mating success

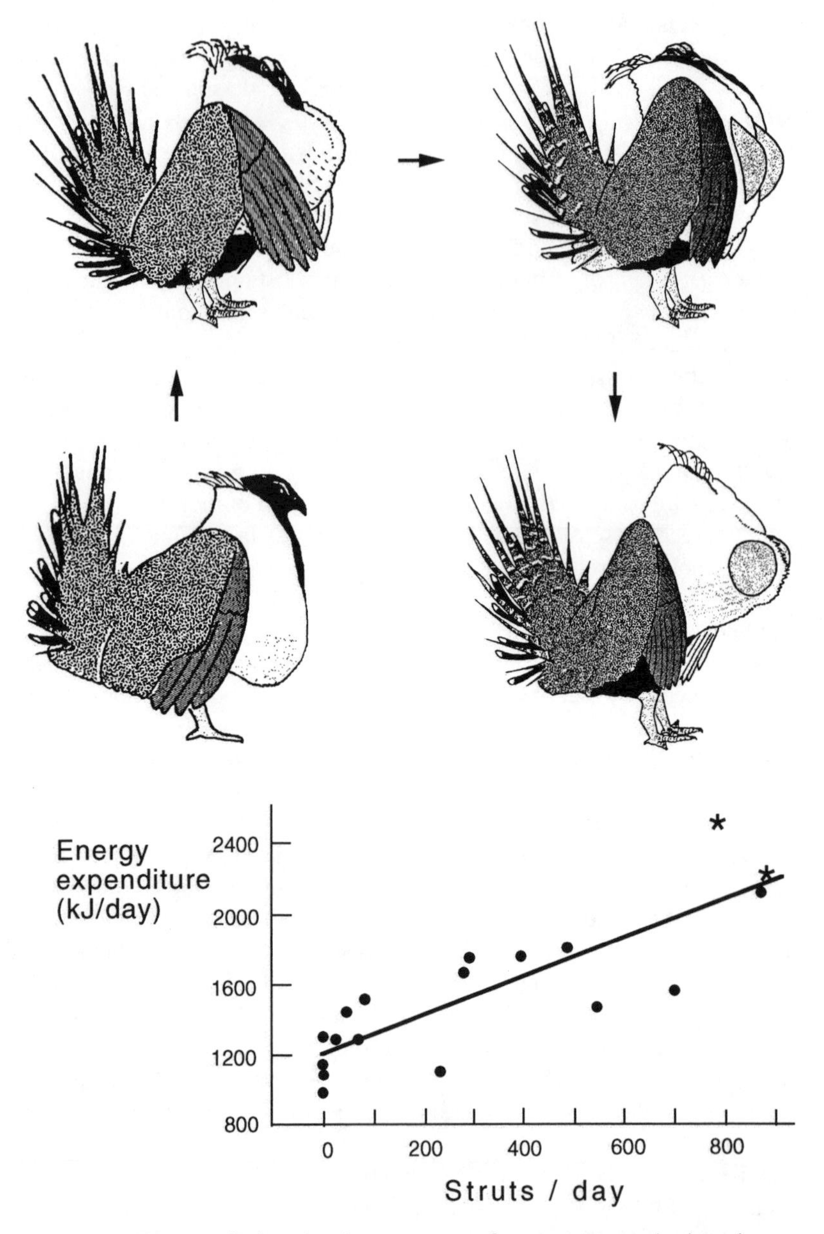

Figure 10.4.1 The strut display of male sage grouse *Centrocercus urophasianus* is energetically demanding (the drawings show four phases of the strut). The daily energy expenditure increases with the number of struts performed per day, and is roughly twice as high in the most active and successful males as in nondisplaying males. Stars indicate the only two males known to have mated. (After Vehrencamp et al. 1989; drawings from Hjorth 1970)

(Gibson and Bradbury 1985). The reasons for the variation among males are not known. There was no evidence that it was caused by blood parasites or diseases as reflected by hematocrit levels. The relationship between display rate, energy expenditure, and foraging distance suggests that foraging differences are involved, but their role is not clear. Vehrencamp et al. (1989) discuss possible interpretations that might be tested by further field work.

In another tetraonid, the black grouse *Tetrao tetrix*, adult cocks lost about 7% of their body weight during the peak lekking period, when subadult nonterritorial males lost no weight (Angelstam 1984). In great snipe *Gallinago media*, males lost almost 7% of their body mass during a night of display activity, and seem to display at nearly maximum sustainable load (Höglund et al. 1992b).

These results suggest that energetic constraints limit the proportion of the breeding season during which a male can compete over females. The length of time a male can spend competing for mates is an important determinant of his mating success in a variety of species (table 6.A). In several cases, such endurance rivalry seems to be a major cause of variation in success among males.

10.5 Sexual Selection in Relation to Life History

Reproductive Effort

Sexually selected traits represent part of the effort an animal expends in order to reproduce. Development of a sex trait at one stage of the life cycle is likely to affect not only current but also future reproduction, as several examples above show. Many sexually selected traits may therefore follow similar life history patterns as do other aspects of reproductive effort, and may be subject to similar constraints (Williams 1966; Andersson 1982b; Partridge and Endler 1987). Such aspects have received little study, but there are some interesting patterns. The degree of iteroparity (lifetime number of reproductions) may be important for the effects of sexual selection. Life history theory predicts that, among species, the reproductive effort per season should decrease with increased iteroparity, especially in species with continued growth in body size and fecundity. The largest effort should occur in semelparous forms. As they breed only once, their entire reproduction is then at stake, and there is no reason to withhold any effort that could be spent on reproduction (e.g., Williams 1966; Stearns 1976, 1992; Charlesworth 1980; Roff 1992). There is some evidence that secondary sex traits are more strongly developed in such animals than in related iteroparous forms (Williams 1966; Partridge and Endler 1987).

For example, among salmonid fishes, the most extreme reproductive efforts and secondary sex traits are found in semelparous salmons that make one long upstream migration from the Pacific to the spawning area (Williams 1966). The digestive system is resorbed and all effort is devoted to the single reproduction; the fish then dies. Males develop bright red colors and a hooked lower jaw, useless for feeding but an effective weapon in fights over females (figure 16.2.4). Annual fishes in temporary waters also show more extreme male secondary sex traits than their iteroparous relatives (Williams 1966; Partridge and Endler 1987). In annual species, males have long fins and bright colors and are very aggressive against each other (e.g., Van Ramshorst and Van Den Nieuwenhuizen 1978). Field studies and comparative analyses of sex roles, mating system, and sexual selection in these species are desirable.

There are other possible examples. Male gladiator frogs *Hyla rosenbergi* in contrast with most other amphibia carry an efficient weapon, a thumb spine, and use it in fights, sometimes with lethal effect. The gladiator frog apparently is semelparous (Kluge 1981; section 4.2 above). In dasyurid marsupials, male size relative to female size increases with reduced iteroparity. The extremes are found among species such as *Antechinus swainsonii*, where males are semelparous and up to four times as heavy as females (Lee and Cockburn 1985). The larger mass should enable males to maintain reproductive activity based on stored resources for a longer time before succumbing (also see chapter 11). Although more studies are needed, these cases suggest that greater reproductive effort, including expression of sexually selected traits, is favored by semelparity (Williams 1966; Partridge and Endler 1987).

Another prediction from life history theory is that iteroparous species should often increase their yearly reproductive effort with age, at least over the first few breeding seasons (e.g., Williams 1966; Gadgil and Bossert 1970; Charlesworth 1980; Stearns 1992). This is often the case with secondary sex traits. For example, male lyrebirds *Menura novaehollandiae* develop full adult plumage first at 6–8 years of age, although they are capable of breeding earlier (Smith 1982). Antlers in deer likewise increase in size for several years after the stag has reached sexual maturity (e.g., Hyvärinen et al. 1977). There are many other examples of gradual development of male secondary sex traits after maturity. Whereas life history theory predicts such an increase with age, it may occur also for other reasons, one of which is continued allometric growth after the first reproduction. Another is improved foraging ability with experience (e.g., Marchetti and Price 1989; Wunderle 1991). The nutritional status of a male may then increase with age, permitting full development of secondary sex traits first some years after sexual maturity (see also section 13.6).

Like costs of reproductive effort in general, the costs of sexually selected traits are difficult to measure (e.g., Partridge and Sibly 1991). Natural relationships among the size of sexually selected traits, mating success, predation, survival, growth rate, and other aspects are of little help, as differences in condition and other uncontrolled variation among individuals may affect many fitness components in the same way (Bell and Koufopanou 1986; Partridge and Harvey 1985; Reznik 1985; Clutton-Brock 1988a; Partridge 1989a). Even if there are costs of developing secondary sex traits, individuals in good condition may therefore show the largest traits as well as the most rapid growth, highest survival, and highest mating success. Experiments are needed to disentangle such relationships. Even when feasible, however, experiments can be problematic, since manipulation of a sex trait usually does not reflect all the changes in developmental and other costs that would result from corresponding natural changes in the expression of the trait. Ideally, its entire ontogenetic trajectory should be changed. Perhaps this can sometimes be done by hormonal experiments, as many secondary sex traits are under hormonal control. For some traits, such as bright nuptial colors in fishes and birds, the developmental cost is perhaps negligible compared to the increased risk of predation or social competition that the trait entails.

Another possibility might be genetic manipulations, but in addition to being ethically questionable, the effects of a genetic change cannot usually be limited to the trait under study. Quantifying the costs of secondary sex traits and other aspects of reproductive effort for these reasons is replete with problems. In addition, cause and effect in the interactions among sexual selection and life history patterns are often difficult to identify (e.g., Wiley 1974; Partridge and Endler 1987; Partridge 1989a,b). Much more work is required to place sexual selection in a general life history perspective, and to explain quantitatively the expression of secondary sex traits by their most important lifetime costs and benefits.

Sexual Bimaturism

Females in many polygynous animals mature and start to reproduce at a younger age than males (e.g., Selander 1965, 1972; Wiley 1974, 1991; Charlesworth 1980; Murray 1984; Manning 1985). Such sexual bimaturism is correlated with polygyny in, for example, North American blackbirds, grouse, ungulates, and pinnipeds (e.g., Selander 1965; Wiley 1974; Jarman 1983; Bryden 1972). A likely reason is strong sexual competition among males in polygynous species, where large size is important for male mating success (sections 11.8–11.9). By deferring reproduction, a male may gain in both respects while avoiding risky competition with older and larger males, against which he would have little chance of success (Wiley 1974, 1991; Wittenberger 1979).

All else being equal, deferred reproduction in males will reduce the ratio of breeding males to females and increase the degree of polygyny. As a consequence, there is an increase in the maximum possible variance in mating success, and in the potential reproductive benefits from traits that raise a male's mating success. The evolution of delayed reproduction in males may therefore be self-reinforcing (Wiley 1974). Many species with marked bimaturism show large sexual size dimorphism. Bimaturism may be necessary for evolution of strong dimorphism in species where the rate of body growth even in the smallest sex is close to the limits set by ecology or physiology. In some dimorphic polygynous species, however, males grow more rapidly than females, apparently at the expense of survival, as juvenile mortality is higher in males than females (section 10.2).

10.6 Genetic Constraints

Owing to limited genetic variation, and to genetic correlations among traits within and between the sexes, there are likely to be strong genetic constraints on the evolution of many secondary sex traits (sections 2.5, 4.5). The importance of such constraints is underscored by the role of phylogeny found in several comparative studies (e.g., Clutton-Brock and Harvey 1977; Cheverud et al. 1985; Gittleman and Kot 1991; also see section 11.9). Genetic evolution and adaptive divergence among species can occur only as suitable genetic variation arises, which may take a long time. For the same reason, genetic correlations among traits and between the sexes will often constrain the evolution of secondary sex traits (e.g., Price et al. 1987). Selection pressures differ among traits and between the sexes. A selective change in one trait (or sex) will therefore often be opposed by counteracting selection of another, genetically correlated trait. Negative genetic correlations have been found for components of fitness in *Drosophila* (e.g., Rose and Charlesworth 1981; Rose 1984; Wilkinson 1987).

The effects of genetic constraints on the evolution of sex-dimorphic traits can be profound, as suggested by studies of selection in the wild (Price 1984a,b; Grant 1986), by artificial selection in the laboratory (Shaklee et al. 1952; Frankham 1966; Eisen and Hanrahan 1972), and by theoretical models of quantitative traits (section 2.5; Lande 1980, 1987; Lande and Arnold 1983, 1985). In particular, when a trait in males and females is strongly correlated genetically (e.g., van Noordwijk et al. 1980; Price 1984a) and shows similar variation, sexual dimorphism in that trait may evolve much more slowly than the average phenotype (Lande 1980, 1987; Rogers and Mukherjee 1992).

The relative importance of genetic constraints versus selection in morphological evolution is debated (e.g., Gould 1989; Eberhard and Gutierrez 1991; for examples, see sections 11.9 and 12.2 below).

10.7 Sensory Bias and Constraints

It is becoming clear that secondary sex ornaments and other signals are often constrained by sensory or neural characteristics, which may bias the evolution of such traits in various directions (e.g., West-Eberhard 1979, 1984; Ryan 1985, 1990b; Kirkpatrick 1987b; Endler and MacLellan 1988; Endler 1992a,b; section 1.5 above; see Reeve and Sherman 1993 for critical discussion). In some cases, a male trait is favored by selection apparently because it fits preexisting features of the female sensory system (reviewed by Ryan 1990b; Ryan and Keddy-Hector 1992). There are several likely examples, such as male calls in tungara frogs *Physalaemus pustulosus* (Ryan et al. 1990; see section 5.3 above), male-constructed nest pillars in fiddler crabs *Uca beebei* (Christy 1988), and male display motion patterns in *Neumania* water mites and *Anolis* lizards (Proctor 1991, Fleishman 1992). Female swordtails *Xiphophorus* may even prefer males of another species, if given a coice (Ryan and Wagner 1987). Females prefer long tails not only in the swordtail *X. helleri* but also in playtfish *X. maculatus*, as Basolo (1990a,b) demonstrated by attaching artificial tail swords to platyfish males, which lack long tails. The most parsimonious phylogenetic explanation is that the female sensory system had the properties that favor a long tail before male swordtails evolved this trait (see section 5.3 for a similar case in the tungara frog).

Large male song repertoires are favored by female responses in several birds. The mechanisms leading to the evolution of such responses are debated (e.g., Searcy 1992; chapter 14 below). In a study of the common grackle *Quiscalus quiscula*, Searcy (1992) found that females favor artificial repertoires of four song types even though each male sings only one song type. The result suggests that none of the traditional hypotheses explains the preference. It may instead be a consequence of a common property of the nervous system: habituation to specific stimuli. Searcy (1992) presented evidence that male song repertories in birds may evolve because they are favored by this "bias" in the neural response system; there may also be further coevolution of repertoire size and female preferences.

Endler (e.g., 1978, 1989, 1992b) stressed that there is a complex network of evolutionary interactions among signal properties, signaling behavior, sensory system, habitat, foraging, and avoidance of predation. These aspects should evolve together because each influences selection acting on the others. As a result, there should be correlations among all these aspects, an idea that has been applied in detail to guppies by Endler (1992b; section 5.2 above); similar studies of a variety of species are needed (also see Fleishman 1992).

10.8 Summary

The evolution of sexually selected traits is constrained by a variety of factors that may balance sexual selection and perhaps lead to an equilibrium development of such traits. Predation selects against signaling or searching for mates in a variety of insects and vertebrates. Contests over territories and mates can also lead to higher risk of predation, and to mortality in fights between males. Sexually selected larger body size in some species makes males more prone to starve during juvenile growth than females. Sexual selection can also be self-limiting, if greater expression of a trait increases mating success in one way but reduces it in another. Secondary sex traits, in particular male display behavior, may also be limited by energetic and physiological constraints. In some cases males approach short-term limits to energy expenditure; in other cases energetic costs over longer periods limit the time during which a male is able to compete over mates. Life history aspects, such as number of reproductions in a lifetime, and demographic differences between males and females, also influence the strength and consequences of sexual selection. Genetic correlations among traits and between the sexes can strongly constrain the evolution of secondary sex traits. Preexisting sensory characteristics may often bias the evolution of secondary sex traits in various ways; several likely examples have recently been discovered.

For these reasons sexually selected traits are limited in a number of ways, but it is not known in detail for any such trait how the various benefits, costs, and phylogenetic constraints together determine its size. A balance between sexual and natural selection remains a plausible idea, not a firmly established empirical result. Much more detailed quantitative work is required to make it so.

11 Sexual Size Dimorphism

11.1 Introduction

Body size is the most conspicuous difference between males and females in many animals. There are dramatic extremes, ranging from sea elephants with three-ton males, more than five times the mass of females, to endoparasitic barnacles in which the shriveled male, attached inside the female, has been mistaken for the testes of a hermaphrodite. Not surprisingly, sexual size dimorphism has prompted a variety of explanations and tests.

The aim of this chapter is to review the empirical evidence for and against these hypotheses, and to summarize the extent to which sexual size dimorphism is understood in various groups. In addition to sexual selection, this chapter therefore deals with a number of other selection agents that may influence males and females differently (so do the remainder of the chapters, for the same reason). The resulting picture of sex dimorphism will be complex, and there is much poorly understood variation among taxa. If anything, the results confirm that variation is a striking feature of the organismal world, and that similar patterns can arise for very different reasons.

Darwin (1871, II, p. 260) suggested that females are larger than males in most animals, particularly in species that grow throughout life, because egg production increases with body size. He also suggested that males are larger than females in most mammals because large size gives an advantage in contests over mates. There is now quantitative support for these ideas in a variety of species (see below), but many other factors also contribute to the dimorphism (e.g., Selander 1972; Ghiselin 1974; Ralls 1976, 1977; Endler 1983; Clutton-Brock and Harvey 1983; Arak 1988a; Hedrick and Temeles 1989; Shine 1989).

As contest success will depend not mainly on absolute size, but on relative size compared with the other members of the population, an evolutionary arms race of escalation in body size or weapons seems possible. Selection of large body size through contest competition has been modeled, for example, by Haig and Rose (1980), Maynard Smith (1982), Parker (1983a), B. Charlesworth (1984), and Maynard Smith and Brown (1986). Depending on genetic and other assumptions, the outcome can be a stable

BOX 11.1.1 A MODEL OF CONTEST COMPETITION AND BODY SIZE

Based on a game theory model of size effects in contest competition, Maynard Smith and Brown (1986) explored the consequences of a range of assumptions (also see Parker 1983a; B. Charlesworth 1984). In the simplest case, size is genetically determined in asexual haploid individuals that grow to adult size, then reproduce. There are two adult size classes, *S* and *L*, of sizes $m_1 < m_2$, respectively. The probability $s(m)$ of surviving to adult size m is assumed to be a decreasing function of m. Large body size therefore reduces survival (for example because size exceeds the ecological optimum), whereas it is advantageous in contests. There is also a cost c of fighting, which occurs between individuals of similar size. With some additional assumptions, for example about the breeding success of *S* and *L* individuals, the dynamics of the game can be determined. With a large cost of fighting, polymorphism becomes likely. Of greater interest is, however, the more realistic case with a number of adult size classes, $m_1 < m_2 < \ldots < m_n$. As the population evolves large mean body size (favored by high contest success), small individuals have much higher survival to adult size. Simulations showed that these effects can lead to population cycles in body size. The likelihood of a stable polymorphic size distribution increases with the cost of fighting, c. The conclusions were verified in an analytic version of the same model, assuming continuous distribution of body size.

Parker (1983a) found in another game theory model that environmental variation in size can lead to a pure ESS in body size. Adding environmental variance in their model, Maynard Smith and Brown (1986) reached the same

monomorphic size, a stable size distribution, size cycles, or even an indefinite size increase (box 11.1.1). The models show that stability is promoted by environmental variance in body size, by mortality that increases with size, and by costs of contests with individuals of similar size. Owing to sexual selection by male contests, the average body size becomes larger than that favored by natural selection. Sexual selection is one possible reason for the trend toward larger size in many fossil mammalian lineages, perhaps accelerating their extinction (e.g., Ghiselin 1974; Lande 1980; Parker 1983a; Maynard Smith and Brown 1986; see box 11.1.1 and section 12.2; for alternative explanations, see Peters 1983).

To explain the degree of sexual size dimorphism of a species, the variety of selective factors affecting the sizes of male and female needs to be considered (e.g., Ralls 1976). An example of this approach is Price's (1984a) study of size dimorphism in Darwin's ground finch *Geospiza fortis*. For example, sexual selection favors large male size, and reproductive (fertility) selection favors small females, which are energetically more efficient and can start breeding more quickly when the fluctuating conditions be-

BOX 11.1.1 CONT.

conclusion: if sufficiently large, such variation leads to a stable, genetically monomorphic population in which all size variation is environmentally determined.

To further increase realism, Maynard Smith and Brown (1986) assumed diploid sexual inheritance in one-locus and polygenic models. As in the asexual case, a stable distribution was favored by environmental variance, high cost of contests, and increasing mortality with body size. With diploid sex, there were no size cycles, but in some cases there was indefinite increase in body size. The reason appears to be that, when mean body size has become so large that a very small individual would do better, there is no way that such a rare individual can reproduce its kind in the model. Owing to the diploid sexual inheritance, its offspring will also inherit alleles for larger body size from the other parent, and so will loose the benefits of very small size. Maynard Smith and Brown (1986) suggested that a "macromutation" might then invade the population. One kind of mutation with sufficiently large effect could perhaps be a dominant mutation for reduced production of growth hormone, leading to dwarf size.

The stabilizing factors in the model seem plausible: environmental variance, increasing mortality with large body size, and a cost of fighting with rivals of similar or larger size. Should we therefore conclude that evolutionary instability in body size is unlikely to occur? Maynard Smith and Brown (1986) pointed out that, according to the model, this would also imply that species with frequency-dependent size competition should not start breeding until reaching senescence, which is clearly not the case in nature. Perhaps evolutionary instability is more common than we realize: the fossil record suggests that most mammalian lineages increase in body size over time. Since mammals on average do not increase in size, this implies that extinction is more common among large species, and speciation in small species. Maynard Smith and Brown (1986) suggested that there has perhaps been long-term selection for larger males (as evident from the sexual size dimorphism) in many species, increasing the risk of extinction (also see section 12.2).

Parker (1983a) dealt specifically with male contest competition over females in some of his models, finding that the greater the number of females in the contested group, the higher the ESS investment in male armament. He suggested that the correlation in primates between sexual size dimorphism and female group size is a case in point (Clutton-Brock and Harvey 1977; see sections 4.3, 7.1, and 11.9 for discussion). Parker also predicted that the ESS investment level will be highest if most males are relatively small and poorly armed, as in some long-lived vertebrates where male body size increases throughout life.

Similar results as for body size in most cases should also apply for horns and other weapons used in contests over mates (chapter 12).

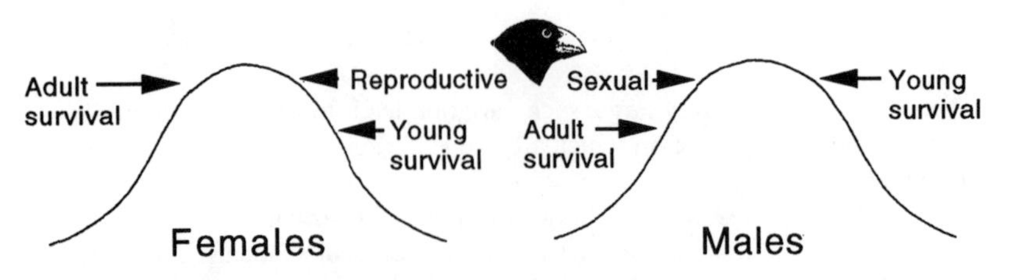

Figure 11.1.1 Body size is subject to many partly opposing selection pressures, several of which may differ between males and females, and between age classes. For example, in the medium ground finch *Geospiza fortis*, survival selection favors small size in young birds but large size in adults, which is also favored by sexual selection in males. Reproductive selection favors smaller size in females. For clarity, male and female size distributions are separated in the figure, but in practice they overlap, males on average being about 5% larger than females. (After Grant 1986. Copyright © 1986 by Princeton University Press)

come favorable (also see Downhower 1976; Grant 1986). These and other selection pressures are indicated in figure 11.1.1.

In many lower vertebrates and invertebrates, for example the amphipod *Gammarus pulex* and the hermit crab *Clibanarius digueti*, large male size is favored by sexual selection, and large female size by fecundity advantages. The advantages seem to increase more strongly with size in males, which are larger than females in these species (Ward 1988a; Harvey 1990). In wood frogs *Rana sylvatica*, reproductive success increases more strongly with body size in females, which are larger than males in this species (Howard and Kluge 1985; Howard 1988b). In bullfrogs *R. catesbeiana*, the advantages increase at similar rates in males and females, which in accordance are of similar size (figure 11.1.2).

The relative merits of a certain body size depend not only on present reproductive and other functions, but on life history aspects such as time to maturation, and survival in relation to body size throughout the ontogeny.

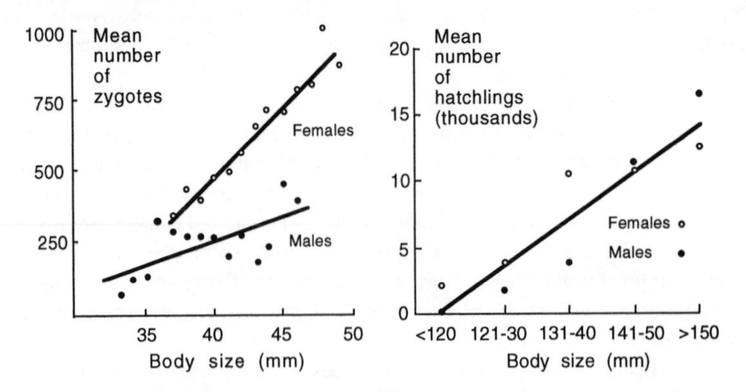

Figure 11.1.2 *Left*: In woodfrogs *Rana sylvatica*, reproductive success increases more strongly with size in females than males. *Right*: In bullfrogs *Rana catesbeiana*, reproductive success increases similarly with size in males and females. (After Howard 1988)

Although such a perspective has not often been adopted in studies of sexual size dimorphism, its explanation may require life history considerations (e.g., Murray 1984; Halliday and Verrell 1986; Roff 1986; Arak 1988a; Shine 1988; Kozlowski 1989; Charnov and Berrigan 1991; Shine and Schwarzkopf 1992). Quantitative understanding of sexual size dimorphism requires knowledge of both the fitnesses associated with a range of growth trajectories and adult sizes for males and females, and the reasons for any sex differences in these respects.

Genetic correlations between males and females are also relevant; together with other phylogenetic constraints, they may partly explain the pattern of sexual size dimorphism among species (e.g., Lande 1980; Payne 1984; Lande and Arnold 1985). Owing to its complexity, many aspects of sexual size dimorphism are still poorly understood, and there is no general consensus on the explanation of several of its trends. Some of the factors that may give rise to sexual size dimorphism are listed in table 11.2.1. In addition, there are several possible ecological reasons for sex differences in body size, including foraging competition between the sexes, and independent adaptations to foraging in males and females (e.g., Darwin 1871; Selander 1966; Slatkin 1984; Shine 1989; section 1.2 above).

Sexual size differences can arise in many different ways, and at different stages of the life cycle. In some cases they may be genetically determined, in others, environmental effects may be stronger. For example, energy that is used for growth in females may be used in display, contests, or search for females by males in some species, reducing their growth compared to females (Ghiselin 1974). Adult size differences can also reflect differences in age structure between the sexes, for example if growth continues after maturation, and one sex survives better than the other (e.g., Howard 1981a; Halliday and Verrell 1986). Differences in size might arise mainly during the juvenile ontogeny, after maturation, or both. Comparing lower vertebrates, Shine (1990) found that the sex that matured at a larger size in about 90% of the cases also attained a larger average size. This suggests that mainly differences in juvenile growth rate or age at maturity determine adult size dimorphism. Continued growth after maturation may affect the degree of dimorphism, but probably rarely changes its direction.

11.2 The Direction and Degree of Sexual Size Dimorphism

Many selective factors can influence sexual size dimorphism (table 11.2.1). Some of these factors are reviewed below for each sex. In both sexes, some factors favor larger size than the present one, and others favor smaller size.The net balance will depend on many aspects of the biology of the species, leading to an enormous variation in sexual size dimorphism

Table 11.2.1

Some Selective Factors that May Influence Direction and Degree of Sexual Size Dimorphism

Female advantages of large size: Higher fecundity; better parental care; male preferences for large females; dominance in contests over resources, or over males in role-reversed species

Female advantages of small size: Earlier maturation, with shorter generation time and more rapid reproduction as conditions become favorable; more effective shunting of resources into offspring production

Male advantages of large size: Dominance in contests over females or resources when strength is crucial; better performance in endurance rivalry; female preferences for large males; higher success in sperm competition

Male advantages of small size: Dominance in contests over resources when maneuverability rather than strength is crucial; earlier maturation, with more rapid reproduction and shorter generation time; higher success in scrambles; more surplus energy available in searching for mates; female preferences for small males

among organisms. The main factors are reviewed in this section; their relative roles in various groups are discussed in subsequent taxonomic sections. The relative importance of the various factors are, however, far from well understood in most cases.

Female Advantages of Large Size

Females are larger than males in most animals. Although there are many exceptions, this is the rule among invertebrates, and also in many lower as well as some higher vertebrates, including the blue whale *Balaenoptera musculus*. The largest existing animal therefore is likely to be a female (Ralls 1976).

The most common reason for larger female size is not sexual selection; fecundity advantages may explain why females are larger than males in most animals, higher vertebrates being the main exception (e.g., Darwin 1871; Williams 1966; Ghiselin 1974). Egg production increases with body size in many species.[1] Such an increase is often expected simply because larger females have more internal space for eggs (e.g., Williams 1966). In addition, when female reproduction is based on energy reserves accumulated over a long period, large female size may confer a fecundity advan-

[1] Female fecundity has been shown to increase with body size in a variety of animals. See, for *insects*: Thornhill and Alcock 1983, Honěk 1993; *other invertebrates*: e.g., Ridley and Thompson 1979, Birkhead and Clarkson 1980, Greenspan 1980, Shuster 1981; *fishes*: e.g., Perrone 1978, Berglund et al. 1986; Loiselle 1982 provides additional references; *amphibians*: e.g., Tilley 1968, Verrell 1982a, Gibbons and McCarthy 1986, Howard 1988a; Woolbright 1983 gives more references; *reptiles*: e.g., Fitch 1970, Wiewandt 1982; *birds*: e.g., Bryant 1988.

tage because the capacity for storing energy reserves increases more rapidly with body size than do metabolic costs (e.g., Downhower 1976; Calder 1984; Millar and Hickling 1990, 1992).

The fecundity advantage has been widely accepted as an explanation for larger female than male size in many invertebrates, and in lower vertebrates that grow throughout their life. Shine (1988) pointed out, however, that there are problems with this idea. Whereas clutch size in many species is correlated with female size, this need not mean that larger females will have higher lifetime reproduction or fitness, as their growth to larger size takes time and energy. A life history perspective is required that also takes into account survival and fecundity in relation to body size, as well as age at maturity (section 11.1). Until this has been done, the fecundity advantage is not a satisfactory explanation, even though much empirical evidence is in line with it (but not all; Shine 1988).

In animals with advanced parental care, such as most birds and mammals, female reproductive success may be limited less by egg numbers (Trivers 1972) than by ability of raising the offspring to independence. Offspring success does not necessarily increase with female size, but there is evidence that it does so in red deer *Cervus elaphus* (Clutton-Brock et al. 1988a). For mammals in which females are larger than males, Ralls (1976) suggested that large females are better mothers (see section 11.9 below). Female contest competition over food or some other resource may also favor larger females (e.g., Trivers 1972; Clutton-Brock et al. 1988a). There is a strong advantage of large body size in competition over spawning sites among female coho salmon *Oncorhynchus kisutch* (van den Berghe and Gross 1989; also see Foote 1990). Large body size is favored in female (and male) contests over food required for raising offspring in burying beetles *Nicrophorus spp.* (Otronen 1988a), and in female competition over mates in the moorhen *Gallinula chloropus* (Petrie 1983a).

Yet another advantage is that males may prefer large, fecund females as mates; there are 24 such cases in table 6.A. Large, preferred females may then obtain better territories or other resources than smaller females (e.g., Downhower and Brown 1980; Grant and Colgan 1983; Gwynne 1984b; Berglund et al. 1986b; Cote and Hunte 1989).

In some species with reversed sex roles, females are larger than males, apparently in part owing to advantages in contest competition over mates (sections 7.6 and 11.8).

Female Advantages of Small Size

Small females may be able to shunt more resources into reproduction (e.g., Wiley 1974), or start breeding more rapidly in a fluctuating environment when food becomes abundant (Downhower 1976; Price 1984a; Grant and Grant 1989). Langston et al. (1990) concluded that reproductive energetics favor small size, whereas competition over the breeding situation favors large size in female red-winged blackbirds *Agelaius phoeniceus*. Smaller size may also permit reproduction at an earlier age and a faster rate, for example in weasels *Mustela nivalis*, where females are half the size of males (e.g., Erlinge 1979; King and Moors 1979; Ralls and Harvey 1985). Unless it overly reduces clutch size, earlier reproduction, with shorter generation time and higher survival to maturity, can strongly increase fitness under certain circumstances (e.g., Lewontin 1965; Charlesworth 1980; Stearns 1992).

Male Advantages of Large Size

Larger male than female size, typical of most birds, mammals, and some reptiles, amphibia, beetles, crustacea, and certain other groups, have usually been explained by a size advantage in contests over females (e.g., Darwin 1871; Ghiselin 1974). Studies of many different animals show that larger male size often gives an advantage in dominance contests and fights over females. Forty-eight such cases are listed in table 6.A. Only few studies show clearly, however, that dominance also leads to higher production of offspring (reviewed by Dewsbury 1982a; Clutton-Brock 1988a; Dunbar 1988).

Sequestering of females and forced copulation may also favor large male size (e.g., Ghiselin 1974; see turtles, section 11.7 below). So may sperm competition, if large size leads to higher sperm production. Another advantage is female preferences for large males, for example because they provide better resources than smaller males. There are 29 such cases in table 6.A (also see Cote and Hunte 1989). Large size may also give an advantage in endurance rivalry, permitting a male to spend longer time at a breeding site and compete for females (e.g., Bartholomew 1970; and sections 5.5–5.6). The reason is the same as for females that base their reproduction on stored reserves (see above in this section): maximum storage capacity should increase with body size more rapidly than do metabolic costs (Calder 1984). This seems likely to be a common advantage of large male body size in endurance rivalry.

Male Advantages of Small Size

As in females, a small male can mature and start breeding earlier by finishing growth at a younger age than a larger male (Wallace 1867; cited by Darwin 1871). When there is scramble competition to inseminate females, early-maturing males are likely to find females before later males. Such situations may lead to protandry, males emerging before females (Wiklund and Fagerström 1977; Fagerström and Wiklund 1982; Bulmer 1983; Iwasa et al. 1983). Early-maturing or protandrous males are usually smaller than females (e.g., Ghiselin 1974; Singer 1982; for an exception, see Gunnarson and Johnson 1990).

Other things being equal, reduced size increases agility (Andersson and Norberg 1981). Therefore, sexual contest competition in three-dimensional habitats (air, water), where maneuverability would seem especially important for success, may select for smaller male than female size. In addition, large body size and strength may be less advantageous when there is no substrate against which the animal can take firm support in a fight. Food provisioning by the male for the family may also favor smaller male size (e.g., Ghiselin 1974; Mosher and Matray 1974; Ralls 1976), for reasons discussed in section 11.8. Small males may also be favored in some cases where courtship takes place on the ground, for example in *Drosophila subobscura*, where small males are better than large males at tracking the female during the courtship dance (Steele and Partridge 1988; also see Maynard Smith 1956).

In species where males are larger than females, small males may be less likely to starve during food shortage (see section 10.2, and Clutton-Brock et al. 1985).

Dwarf Males

Extreme dimorphism, with dwarf males many times smaller than females, occurs in, for example, marine invertebrates and algae, and some vertebrates. Ghiselin (1974) suggested that the main factors favoring dwarf males are (1) low population density and low adult motility, which makes it difficult to find a mate, and (2) a premium on long life in the female but not the male. His ability of finding mates is then likely to be much more important than his fighting prowess. Some examples are deep-sea anglerfishes that live at low density, sessile low-density animals such as certain barnacles, and endoparasites where the male must not only find a host, but one harboring a female of his own species. In these cases, the chances of mating with several females are low, so the most successful male tactic may be to remain with a female once found. In some species, the male even becomes physically attached to her: dwarf males that are little more than a supply of sperm nourished by the female occur in several marine invertebrates and Ceratioid anglerfish (Berthelsen 1951; Ghiselin 1974; Pietsch 1976; Gould 1983) (figure 11.2.1).

The relative importance of sexual versus natural selection in the evolution of dwarf males is difficult to estimate (Darwin 1871; Ghiselin 1974). Even if there were no competition over mates, a population density so low that some males risk never meeting a female should favor reduced male size, if this raises his search efficiency. So should an advantage of early

Figure 11.2.1 Dwarf males in deep-sea Ceratioid anglerfish. *Top*: The male is attached to the female in *Ceratias holboelli* and (below) *Neoceratias spinifer. Third from top*: At the metamorphis stage, the male *Cryptopsaras couesi* (*right*) is adapted for efficient search through a more streamlined body shape and better developed sensory organs than in females (*left*). *Bottom*: In some species, such as *Edriolychnus schmidti*, the attached male obtains nutrition from the female through integration with her vascular system. (From Berthelsen 1951)

reproduction in more quickly maturing small males. There is, however, probably also scramble competition over first finding a female, which puts further premium on male motility and rapid maturation. The relative roles of competitive versus noncompetitive mating advantages are hard to distinguish (also see section 1.2).

Male anglerfish illustrate dramatically the effects of selection for finding and remaining with a mate. Mature females are probably hard to locate. The male in some species apparently finds his spouse by olfactory cues; in the darkness at 2000–3000 m depth, he carries the largest nasal organ relative to body size of any vertebrate (Berthelsen 1951; Jumper and Baird 1991 provide a theoretical analysis of mate location by olfaction in deep-sea fishes). In other species, he probably locates the female through the light emitted by the luminiscent lure at the tip of her elongated first dorsal fin ray, used to attract prey. Males in these species have enormous eyes (figure 11.2.1). The small males are adapted for efficient search also through a streamlined body shape, and a large tail fin for propulsion (figure 11.2.1; Berthelsen 1951). Fast travel is not compatible with the foraging technique and body shape of the female.

After metamorphosis to the search-adapted morphology, the male in some species never feeds normally again, but upon finding a female attaches himself with the jaws to her body wall, obtaining nutrition from her in return for his masculine services. In certain species, the vascular systems of the mates are integrated, the male depending entirely on nutrition from the blood stream of the female (Berthelsen 1951; Pietsch 1976). The attachment should eliminate the risk that the mates lose contact in the dark depth. Reproductive competition among males may sometimes occur in anglerfish: two or more males have been found attached to the same female in several species. Such examples are rare, but no known mechanism prevents other males from attaching to a female that already has a mate (Pietsch 1976). Aggressive competition and defense of the female, favoring larger male size, seems impossible when he is permanently attached to her.

In specialized endoparasites belonging to several marine phyla, males are so reduced to essentials that some species were once thought to be hermaphroditic (e.g., Yanagimachi 1961). The "testes" were later found to be the degenerated male that once found his way into the body of the female (see Ghiselin 1974; Gould 1983). Dwarf males are found in many parasitic species. They do not occur in parasites where both sexes are motile, but only in those were the adult female is sedentary, indicating that natural or sexual selection for locating mates favors small male relative to female body size in these forms (Ghiselin 1974).

Several other factors may lead to smaller males than females. In some male-haploid organisms (mainly Hymenoptera), females supply sons with fewer resources than daughters. This is probably in the genetic interest of the mother, but not necessarily of the son (section 11.4). Parental manipulation may therefore lead to smaller males than females in these species (Ghiselin 1974; Alcock et al. 1977; Cowan 1981). It is not known whether this is the reason for dwarf males in other haploid species, especially rotifers. Another factor is sharing of food: when the male obtains his nutrition from the female or her foraging area, as in endoparasites and certain sessile animals, reduced male size may be favored if this leaves more food for his mate to turn into offspring.

In certain barnacles, filter feeding may favor low female motility, small male size, and search for females. The only terrestrial taxon where dwarf males are common is spiders, many of which are filter feeders with sedentary females. With few exceptions, the female is larger than the male (Vollrath 1980a; Foelix 1982). The smallest males are found in spiders that spin webs or sit on flowers and ambush prey, sit-and-wait predators that are analogs of barnacles and anglerfish. In extreme cases such as tropical *Nephila* spiders, the male is only a few per cent the mass of the female (Bristowe 1929; Vollrath 1980b). As in anglerfish, the male matures at a younger age than the female, and after his last molt he turns vagabond and searches for mates, in some species catching prey no more; he may even be small enough to travel airborne on his own thread (Foelix 1982). Earlier male maturation in spiders is not always associated with smaller male than female size, however. In the sheetweb spider *Pityohyphantes phrygianus*, adult males are larger than females in spite of emerging earlier in the season (Gunnarson and Johnsson 1990).

Another factor favoring small male size in spiders, the risk of being eaten by the female, was suggested by Darwin (1871). If the male is sufficiently small, the risk that the female will treat him as prey may be reduced. This idea was supported by Elgar (1991, 1992) in comparative studies of orb-weaving spiders (also see Newman and Elgar 1991). There remains, however, much variation in the degree of size dimorphism that apparently cannot be explained by differences in the risk of sexual cannibalism (Vollrath 1980b; Elgar 1991).

Although males in some spiders are much smaller than females, they often fight over mates, the largest male usually winning (e.g., Bristow 1929; Foelix 1982; Vollrath 1980b). It is not clear why, in spite of this male size advantage, females are larger. A possible explanation, which may also apply to other animals with sedentary females and dwarf males, was presented by Vollrath and Parker (1992). Owing to the contrasting life-styles of adult males and females in these species, males will have much higher mortality. This tends to make the operational sex ratio female-biased, reducing the importance of contest competition over mates among males, as most males will not have rivals over the females they have found. Large size therefore becomes less important for male mating success. The fitness of a male is then increased by earlier maturation at smaller size. As earlier maturation means higher chances of survival to adulthood for males, it in turn reduces the female bias in adult sex ratio, and increases competition among males, finally making further size reduction disadvantageous as the opposing effects balance each other. This game theory model was supported by field data on mortality and maturation patterns in the orb web-building spider *Nephila clavipes*.

Spiders seem to offer good possibilities for clarifying the evolution of dwarf males. Some insects with scramble competition are probably also suitable for tests of ideas on the evolution of dwarf males (Darwin 1871; Ghiselin 1974; Vollrath and Parker 1992).

11.3 Taxonomic Review

Sexual size dimorphism has been analyzed by quantitative comparative methods in several vertebrate groups. Such studies largely remain to be done, however, in fishes and invertebrates, which offer much opportunity for testing hypotheses in this field (e.g., Thornhill and Alcock 1983).

Measuring dimorphism is problematic in forms that grow throughout life, such as many invertebrates and lower vertebrates. Owing to sex differences in growth rate, the relative dimorphism (male:female size) may change with age. In some cases it may approach an asymptotic value, in other cases not. Other problems are that males and females may differ in survivorship, or in the age at which they mature and become available for sampling in the breeding area. Many polygynous vertebrates show sexual bimaturism, males starting to reproduce later in life than females (section 10.5 above). The situation is reversed in some invertebrates and lower vertebrates. Measures of dimorphism may therefore be based on different mean ages for the sexes (Howard 1981a; Halliday and Verrell 1986). For these and other reasons, many of the results discussed below are tentative; especially the conclusions for animals that grow throughout life may have to be modified as more accurate data become available, based on individuals of known age. Some reported values of size dimorphism are likely to partly reflect sex differences in age rather than in age-specific size.

The most common measures of size dimorphism are the length of the body or some suitable part of it, or body mass. For some higher vertebrates, especially primates, sex dimorphism has often been calculated in relation to body mass. Measures that vary less in relation to nutritional status may be more suitable, such as the length of skeletal parts. The length of feathers in birds, for instance wing length, also varies more than skeletal parts (Freeman and Jackson 1990).

11.4 Arthropods and Other Invertebrates

Females are larger than males in many insects, probably owing to a fecundity advantage for large females (Honěk 1993). A mating advantage for small males in scrambles over females might also contribute, as it is a common form of sexual competition in insects. But males are also the smallest sex in some species in which they struggle over females (e.g., *Drosophila melanogaster*; Partridge and Farquhar 1983; and the spiders discussed in section 11.2).

In some haplo-diploid parasitic wasps, a female can influence the size dimorphism among her offspring by laying male and female eggs on hosts of different sizes. Males are usually smaller than females. Charnov (1979a, 1982; also see Ghiselin 1969) suggested that female fitness increases more strongly with size than does male fitness; mothers should therefore reserve the largest hosts for female eggs. A test in *Lariophagus distinguendus* suggests that large body size is indeed more important for female than male fitness (van den Assem et al. 1989).

Larger males than females are found, for example, in horned beetles (e.g., Darwin 1871; Otte and Stayman 1979). Male contests seems to be the main selective factor responsible: in beetles where males have much larger horns or mandibles than females and use them in fights, males also tend to have larger body size than females (e.g., Otte and Stayman 1979; section 12.4 below).

Larger male than female size is perhaps not always caused by sexual selection, but convincing exceptions are hard to find. Adams and Greenwood (1983) suggested that superior swimming of amphipod *Gammarus pulex* pairs in which males are larger than females explains their sexual size dimorphism. The male carries the female in mate-guarding precopula before she molts and copulation takes place. Sexual selection may also be involved, however, because large males are more likely to enter precopula with large, fecund females, and large males sometimes displace smaller rivals from females (Ward 1983, 1984; Elwood et al. 1987; but see Birkhead and Clarkson 1980). In the isopod *Asellus aquaticus*, large males sometimes take over the female from mate-guarding smaller males (Ridley and Thompson 1979), so large male size seems to be favored by male contest competition also in these crustacea.

Similar mechanisms partly explain the size-assortative mating of gammarids (Elwood at al. 1987), large females tending to mate with large males, and small females with small males. The importance of sexual selection in size-assortative mating in species where larger females lay more eggs was supported in Ridley's (1983) extensive review (also see Crespi 1989; Brown 1990a,b).

Large size need not always give an advantage in male contests among insects (e.g., Alcock and Pyle 1979; Litte 1979; McLachlan and Allen 1987; Steele and Partridge 1988). For species in which contests or mating take place in air or water, smaller, more maneuverable males may be favored (section 11.2). Females are often larger than males in such species (e.g., Ghiselin 1974; Ralls 1976; section 11.8 below). For example, in many ants, mating takes place in the air, and males spend much time in flight; they are much smaller than females in such species. In some *Pogonomyrmex* harvester ants, males instead wait for females in vegetation or on the ground (Hölldobler 1976; Davidson 1982; Hölldobler and Wilson 1990). In these latter species, male size relative to female size is much larger than in the aerial species. Davidson (1982) suggested the difference is explained by more economic flight energetics in small males among aerial species, and by a mating advantage of large males in the *Pogonomyrmex* ants that do not mate in the air. In *Chironomus* midges that mate in the air, mating success is higher for small than for large males, perhaps a consequence of higher maneuverability in smaller males (McLachlan 1986; Neems et al. 1990).

Dragonflies (Odonata) may be suitable for testing ideas on sexual size dimorphism in aerial species. The sexes do not differ much in size, but males are smaller than females in some species and larger in others, also within a single family (Libellulidae) (e.g., Walker and Corbet 1975). In three studies, small males were less likely than large males to hold a territory (Fincke 1984), or had territories of lower quality (Tsubaki and Ono 1987), suggesting that large size gives an advantage in male contests in territorial dragonflies (also see Marden and Waage 1990). In a comparative study, Anholt et al. (1991) found that adult female dragonflies are usually considerably heavier than males, but less so in territorial than in nonterritorial species. In *Libellula quadrimaculata*, however, satellite males with lower mating success were larger than active territorial males (Convey 1989). More detailed studies of sex dimorphism in relation to social behavior and life history in dragonflies are needed; Anholt et al. (1991), Gribbin and Thompson (1991), and Conrad and Pritchard (1992) discuss factors of potential importance.

Certain Hymenoptera also have larger males than females, among others the honey bee *Apis mellifera*. Darwin (1871) suggested that the male needs to be larger in order to carry the female during the nuptial flight. In dungflies *Scatophaga stercoraria*, larger males are better able to carry their mates in flight (Borgia 1981).

In a comparative study of sexual size dimorphism among twelve waterstriders (Gerrinae), Fairbairn (1990) concluded that much of the dimorphism pattern is explained by phylogenetic constraints and allometric relationships. In addition, selection for migration by flight tends to reduce dimorphism. Extended mating, during which the male rides on the back of the female, is associated with relatively larger female size. Male contest competition in territorial species apparently favors large male size (also see Vepsäläinen and Nummelin 1985). Female contest competition also seems to favor large female size in some species. The contested resource appears to be good foraging sites, not males (see Rubenstein 1984).

In some marine invertebrates with external fertilization, such as echinoderms, many mollusks, and some polychetes, males do not fight over mates, yet they are of similar size as females. One possible reason is that mating takes place during synchronized spawning, many males and females simultaneously releasing their gametes in the water (e.g., Thorson 1950; Williams 1975). This should result in severe sperm competition, favoring males producing the most sperm. There is no parental care, and the number of surviving offspring probably increases with the number of gametes in similar fashion for both sexes (Ghiselin 1974; Williams 1975; Warner 1984b). For example, in the purple sea urchin *Strongylocentrotus purpuratus*, the volume of the gonads is similar in males and females (Giese et al. 1959; also see Strathmann 1990). Reproductive success therefore may be similarly related to body size in males and females in marine organisms in which males and females simultaneously shed their gametes into the water.

In other marine animals where the sexes mate in pairs, sperm competition should be relaxed, especially in species with internal fertilization. Selection for large male size may then be reduced, resulting in sexual size dimorphism with larger females than males. Likely examples are found in zooplankton (Gilbert and Williamson 1983) and some *Ophryotrocha* polychetes (Åkesson 1972 and pers. comm.).

In some of the larger marine invertebrates, especially crustaceans, males fight over breeding burrows and females; males are larger and armed with greater claws (e.g., Crane 1975; Christy 1983). Males fight over females even in a gastropod, the fighting conch *Strombus pugilis* (Bradshaw-Hawkins and Sander 1981), but sexual dimorphism seems not to have been described in this species.

Additional examples of sexual size dimorphism and its explanations in invertebrates are discussed by Ghiselin (1974). The great variation and vast number of species, especially among insects, offer many opportunities for comparative analyses of sexual size dimorphism and its causes.

11.5 Fishes

Females are larger than males in most fishes. Darwin (1871) was not aware of any exception; he suggested that larger female size is related to a fecundity advantage. As he predicted, egg numbers increase with female body size in fishes.[2] Smaller male size may in part be a consequence of selection for early male maturation and reproductive effort in some (perhaps many) fishes, which reduces male growth compared to that of females (Ghiselin 1974; Warner and Harlan 1982; Endler 1983).

Since Darwin's time, it has become clear that a number of fishes have larger males than females (whether this holds also for individuals of the same age is often less clear, however). Species that change sex are of particular interest here (also see section 16.3). The existence of protogynous (female first) as well as protandrous (male first) species within some families, for example damselfish (Pomacentridae) and porgies (Sparidae), offers special opportunities for analysis of the factors that favor large size in the two sexes (Ghiselin 1974; Warner 1984b, 1988b). Like other organisms with sex change, these fishes have indeterminate growth. Change from female to male at increasing size seems most common. It occurs in at least fourteen fish families, and change from male to female is known in eight families (Policansky 1982; Warner 1984b). Protogyny seems to occur particularly in species where large males are able to monopolize matings by defending a spawning territory or a group of females (e.g., Robertson and Warner 1978; Thresher and Moyer 1983; reviewed by Warner 1984b). Large male size in such species seems to be favored by contest competition over females.

Many of the protandrous fishes on the other hand live in large schools, not closely associated with the substrate. Monopolization of females may be more difficult for males in such species, but their mating behavior is not well known (Warner 1984b). An exception is the anemone fishes (genus *Amphiprion*) that live in or near stinging sea anemones, apparently not triggering the stinging cells, instead receiving protection from them. Only two adults and a number of juveniles make up a social group in these species. The per capita production of offspring in such a pair should be greatest when the female is larger than the male, which may explain why these species are protandrous (e.g., Fricke and Fricke 1977; Warner 1978).

The existence of alternative male reproductive tactics in many cases is a consequence of the advantage of large size in contests over females or resources. In peacock wrasse *Symphodus tinca*, some of the largest males become pirates that fight and chase away smaller males from their nests and spawn there. By saving the time required for nest construction and attendance, the larger pirate males probably have higher reproductive success than other males (van den Berghe 1988). The only way for small males to get a chance to breed may be to circumvent the dominance of large males by other than aggressive behavior (see section 16.2 below).

Among schooling pelagic fishes, males and females do not differ markedly in sex roles; there is usually little contest competition among males,

[2] Ibid.

and little sex dimorphism (Noble 1938). In salmonids, for example, species that breed in schools show little sexual size dimorphism and lack the weapons found especially in males of territorial forms, with relatively larger head and jaws in males than females (Noble 1938).

In salmon, large male size is favored in contests over females (e.g., Gross 1985; Järvi 1990), and probably also in sperm competition. An advantage of large size in contest competition over breeding territories has also been found in female salmon. In coho salmon *Oncorhynchus kisutch*, large body size in females was favored by natural selection in at least three ways (van den Berghe and Gross 1989). Larger body size was associated with higher egg production, better territory, and more successful nest defense against other females. Female competition and defense of breeding sites and nests accounted for as much selection of female body size as did variation in egg production. This suggests that the fecundity advantage of large female size has sometimes been overemphasized compared to advantages in female contests over resources.

Mate choice can favor large male as well as female size. Male preferences for large, fecund females have been demonstrated in several fishes; females also prefer large mates in several species (e.g., table 6.A). In some cases, large males provide better parental care (chapter 8). In the redlip blenny *Ophioblennius atlanticus*, both males and females prefer large mates (Cote and Hunte 1989). Balancing factors that favor small size are less well known in this as in most other species.

To summarize, sexual contest competition favors large male size in territorial fishes. Fecundity advantages may explain larger female size. Where females defend territories, natural selection can favor large female size also through contests over suitable territories for offspring production. Mate choice favors large body size in some fishes.

11.6 Amphibia

Sexual size dimorphism in relation to male fighting in amphibia was reviewed by Shine (1979). Females are larger than males in 90% of anurans (belonging to nine families); clutch size increases with female size (section 11.2).[3] Males are larger than females in 13 of the 32 anurans in which fights between males had been recorded, but in only 9% of the other 557 species compared. Among urodeles (three families), males are larger than females in 13 of 15 species with male combat, but in only 18 of the other 64 species (Shine 1979).

Also at the family level, size dimorphism and male fighting are correlated in anurans (figure 11.6.1). Several studies of amphibia show that large

[3] Ibid.

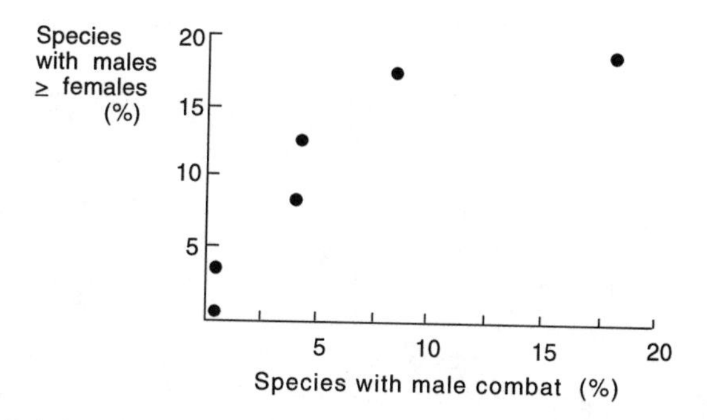

Figure 11.6.1 Relationship between the proportion of anuran species showing male combat, and the proportion in which males are larger than females. The points represent six families for which data on more than 20 species are available. (After Shine 1979)

body size confers an advantage in fights among males (e.g., nine cases in table 6.A). It seems that males are larger than females mainly in species where sexual contest competition is important (Trivers 1972; Wells 1978; Shine 1979). There are several species, however, such as the toad *Bufo bufo*, in which males have a clear advantage of large size in fights over females, yet are the smaller sex (Davies and Halliday 1979; Halliday and Verrell 1986; Arak 1988a).

From a model based on reproductive success in relation to body size in the two sexes, Arak (1988a) predicted that their difference in body size should be proportional to the difference in their reproductive selection gradient, for size. The results from nine anuran species agreed fairly well with the predictions (figure 11.6.2), suggesting that this approach may

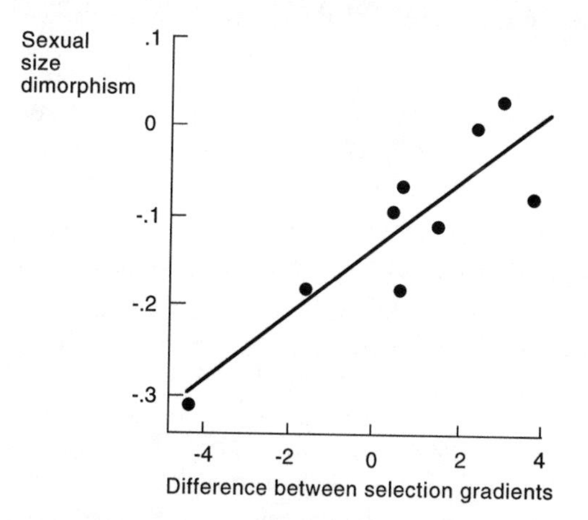

Figure 11.6.2 Regression of sexual dimorphism in body size of nine anurans on the difference between male and female reproductive selection gradients for body size. The dimorphism is measured as log(mean male length)–log(mean female length). (After Arak 1988)

contribute to quantitative understanding of the degree of sexual size dimorphism.

In some anurans, males have weapons such as tusks or spines and use them in contests (e.g., Balinsky and Balinsky 1954; Lutz 1960; McDiarmid and Adler 1974; Duellman and Savitzky 1976; Kluge 1981; section 4.2 above). Such weapons are often associated with large relative male size, as expected if large body size is favored by male contests. Like Wells (1978) in a comparison of eight species of *Rana* frogs, Shine (1979) and Howard (1981a) concluded that contests over mates among males is the main factor favoring larger male than female size in some amphibia (but see Halliday and Verrell 1986; Shine 1987). Other evidence that contests over mates favors large size is that females are markedly larger than males in two poison-dart frogs (Dendrobatidae) with paternal care, where females fight over males. In three species with maternal care, males and females have similar size (Summers 1989).

Two other suggested explanations are that larger male than female size enables him to cling to the female during amplexus (Pope 1931), and to fertilize more eggs because the cloacas are closer together than if the male is smaller than his mate (Licht 1976). If so, the type of amplexus (pelvic versus pectoral embrace) should influence the relative body sizes of the sexes. This is not the case: the type of amplexus follows taxonomic lines, but patterns of dimorphism do not (Shine 1979).

Females choose large males as mates in bullfrogs *Rana catesbeiana* and some other species (e.g., Emlen 1976; Howard 1978a,b; Ryan 1980b; Fairchild 1981; Forester and Czarnowsky 1985; Morris and Yoon 1989). Female choice may therefore in part explain why males are larger than females in some species.

Woolbright (1983, 1985) examined sexual size dimorphism in relation to the breeding system, with two extremes in anurans (Wells 1977b). In explosive breeders, most of the season's matings take place in dense groups during one or a few days. In prolonged breeders, found in more stable environments, the reproductive period often lasts a month or more. In explosive breeders, males scramble to first find receptive females; male territoriality is more common among prolonged breeders (Wells 1977b). Aspects of the breeding system may influence sex dimorphism: the shorter the breeding season, the lower the opportunity for many matings, each of which takes time (Emlen and Oring 1977). The chances of polygyny are therefore limited in explosive breeders, probably making large male size less important for mating success than in prolonged breeders. Woolbright (1983) also suggested that male anurans will be largest relative to females in species with medium-length breeding seasons, and that energetic stress may limit male body size in prolonged breeders. A comparison among eighteen species showed the predicted pattern (Woolbright 1983), but phylogenetic relatedness among several species makes the effective sample size small. Sullivan (1984) questioned several of the assumptions and pointed to contradictory evidence (also see Halliday and Verrell 1986; Partridge and Endler 1987). More detailed studies of an array of anurans are required for testing these ideas.

11.7 Reptiles

Lizards

Males are larger than females in many lizards (reviewed by Fitch 1981; Stamps 1983). In two families, Iguanidae and Agamidae, males defend territories that also contain the home range of one or more females. In three other families, Scincidae, Teeidae, and Lacertidae, males usually defend at most a minor part of the home range. These species tend to be less sex dimorphic than the territorial species. Dimorphism in territorial lizards increases with the degree of polygyny, estimated by the ratio of male to female home range size. In the few territorial species from mainly nonterritorial families, the ratio of male to female size is also larger than the average for these families (Stamps 1983). The results suggest that sexual selection is responsible for males being larger than females in many lizards (e.g., Trivers 1972, 1976; Schoener 1977). They refute the alternative that natural selection for reduced food competition between the sexes is responsible; this idea predicts that monogamous species are most dimorphic, contrary to the observed pattern (Stamps 1983).

Observations in a population of individually marked sand lizards *Lacerta agilis* showed that large males win contests over smaller males. Fight frequency and contest duration decreased with increasing size difference between the males (Olsson 1992b).

Among nine species of herbivorous Iguanine lizards, sexual dimorphism in head and body size was smaller in three species with little male aggression than in six polygynous species, where male contests are more common (Carothers 1984). Within these latter species, relative head size increased with body size in males, but it decreased in females. The jaws are used in male combat, and bite strength should increase with head size. Male relative head size also increased temporarily during the reproductive season (Vitt and Cooper 1985). The sex dimorphism in head size apparently is not explained by ecological resource partitioning, as males and females have a similar diet in these herbivorous lizards. Male contest competition is probably the reason why males have larger body and relative head size (Carothers 1984).

Several aspects of body size in *Uta palmeri* are favored by sexual selection via male contest competition over territories, crucial for attracting females. In addition, large head deapth is favored apparently because it enables the male to maintain a hold on the female during copulation (Hews 1990).

In the green iguana *I. iguana*, observations suggest that both female choice and male contests favor large males (Dugan 1982). Also among insectivorous lizards, sexual selection seems to explain why males are larger than females, but selection of foraging differences between males and females apparently has led to further size dimorphism in some cases (e.g., Schoener 1967, 1977; see Carothers 1984).

The reasons why females are larger than males in certain lizards are not known. Among the suggested explanations are size-dependent fecundity (but see Shine 1988) and female competition over nest sites (e.g., Wiewandt 1982).

In contrast with many other taxa, sexual size dimorphism among lizards appears not to increase with body size (Stamps 1983; see section 11.10 below).

Snakes

Among 224 species of snakes compared by Shine (1978), females were larger than males in two thirds, but in five of the eight families examined, some species had males larger than females. Male wrestling fights over females (e.g., Carpenter et al. 1976) are known from 15% of the species; there is a strong correlation between the proportion of species with male combat per family, and the proportion of species with males larger than females (Shine 1978).

Male combat does not always lead to larger male than female size: in the European adder *Vipera berus*, females are larger although fights between males are common (e.g., Andrén 1986; Madsen 1988; Forsman 1991; Madsen and Shine 1993). This is the case in 3% of the species reviewed by Shine (1979). As in many other taxa, female fecundity increases with body size in snakes (Fitch 1970; Madsen and Shine 1992).

Turtles

Sexual size dimorphism in turtles varies with the form of male competition over females. Reviewing 75 species from seven families, Berry and Shine (1980) distinguished four main types: terrestrial, semi-aquatic, bottom walkers, and swimmers. Among the swimmers, females are larger than males in 31 of 32 species (from four families). Male contests might favor small size because it increases agility in the three-dimensional habitat (e.g.,

Ralls 1976); males fight under water in at least some swimmers (e.g., Booth and Peters 1972).

Forced copulation is known from 18 of 19 semi-aquatic and bottom-walking species (from two families). The male bites the female in the head and legs until she withdraws into the shell; he then mates with her (Berry and Shine 1980). In some *Kinosternon* turtles, males have long, prehensile tails and specialized roughened scales on the rear legs, used for pulling the tail of the female aside during intromission. Males are larger than females in most of the species with forced copulation, a tactic that may favor large males. Male contests also occur in these species, however, and the relative roles of the two mechanisms in selection of large male size are not clear. Among terrestrial forms (mainly tortoises Testudinidae), males fight and are larger than females in more than four-fifths of the species.

Male mating tactics differ among habitats. Berry and Shine (1980) suggested that females can easily escape from males in swimming species, making forced copulation impossible. Female choice may therefore be important in these forms.

As in many other vertebrates (section 11.10), the degree of sexual size dimorphism among terrestrial and bottom-walking species with male combat increases with body size (Berry and Shine 1980).

11.8 Birds

Male birds are usually larger than females, in extreme cases being more than twice as heavy (e.g., capercaillie *Tetrao urogallus*, great bustard *Otis tarda*, the oropendola *Psarocolius decumanus*, and the Australian lark *Cinchloramphus cruralis*; Amadon 1959, 1977; Wiley 1974; Cramp et al. 1980; ffrench 1980). There are also conspicuous cases of "reversed" sexual size dimorphism, for example in jacanas and birds of prey; females are almost twice as heavy as males in some *Accipiter* hawks (e.g., Amadon 1959; Jenni and Collier 1972; Snyder and Wiley 1976; Newton 1986). In spite of all that has been written on sexual size dimorphism and its variation among birds, there is no consensus about its explanation. Some authors think that sexual selection explains most of the variation (e.g., Jehl and Murray 1986; Mueller 1990); others ascribe sexual selection a more limited role beside several other important factors (e.g., Selander 1966).

Payne (1984) reviewed body size, mating system, and sexual size dimorphism in sixteen families that contain both lekking and other species. Males are markedly larger than females in most lekking and other polygynous birds, but they are slightly larger also in most monogamous species (Verner and Willson 1969; Höglund 1989b; Trail 1990; Oakes 1992; and section 7.5 above). In grouse Tetraonidae, bustards Otididae (figure

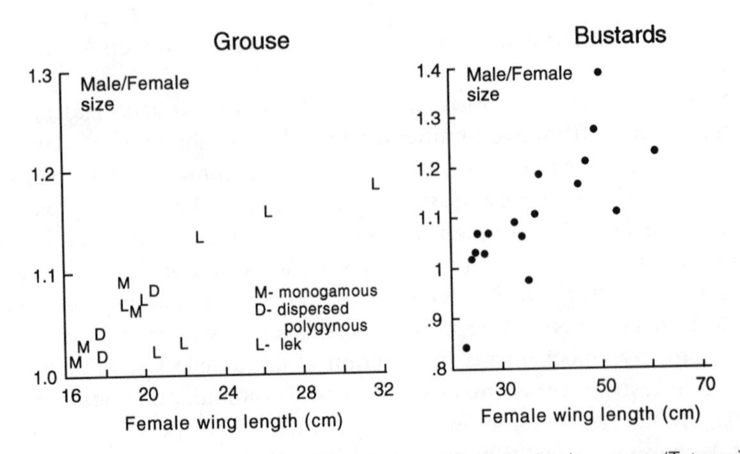

Figure 11.8.1 Sexual size dimorphism in relation to female size in grouse (Tetraonidae) and bustards (Otididae). In both groups, dimorphism increases with body size. Among grouse, several lekking species are larger and more dimorphic than species with other mating systems. (After Payne 1984; also see section 7.5)

11.8.1), and birds of paradise Paradisaeidae, dimorphism increases with the body size of the species. In the latter two families, females are larger than males in some of the smallest forms.

Reversed size dimorphism occurs among small species also in some other families, for instance hummingbirds (Trochilidae) and manakins (Pipridae). These cases include polygynous and lekking species in which males compete over females, so they are not explained by reversed sex roles. Some such species (two bustards, a snipe, a woodcock, small hummingbirds and manakins, and a few cotingids) have in common a display flight by the male. Payne (1984) therefore suggested that agility in aerial display, enhanced by small body size, has been selected in these species, not male fighting prowess. For geometrically similar animals, aerodynamic models indicate that agility and maneuverability increase with reduced size in five out of six aspects. A smaller bird has faster linear and angular roll acceleration, smaller turning radius, and faster climbing rate and horizontal flight; only terminal speed in dive increases with body size (Andersson and Norberg 1981).

Based on these results and on analysis of the variance in mating success among males, Payne (1984) concluded that the reason for larger size dimorphism in polygynous and lekking species is stronger sexual selection than in monogamous birds. Another likely contributing reason why monogamous forms are less sex dimorphic is that the sex roles, and hence the attributes that make for breeding success, are more similar in males and females of monogamous species (Clutton-Brock 1983, 1988a; Howard 1988b).

In a comparative analysis of sixty-five species of Fringillidae and Emberizidae, Björklund (1990) found that polygynous species had larger body size than others. There was also a tendency for sexual size dimorphism to increase with body size. When body size was standardized, polygynous species were not more size-dimorphic than monogamous species. Polygynous species did, however, show stronger sexual dimorphism in tail size than monogamous species (also see Björklund 1991a).

Webster (1992) reached different conclusions in a thorough analysis of new world blackbirds (Icteridae). His results refuted a number of alternative explanations and indicated that sexual size dimorphism is produced mainly by sexual selection in this group. Webster (1992) suggested that the differences in conclusions arose due to differences in methods of analysis; Björklund used the phylogenetic approach of Cheverud et al. (1985; see discussion of primates in section 11.9 below).

In some species, stronger sexual selection among males does not seem to be the only reason why they are larger than females. For example, sexual size dimorphism in ducks Anatinae is particularly large among hole-nesting forms. Only the female incubates, and small body size probably makes it easier to find a suitable nest cavity (Bergman 1965; Sigurjonsdottir 1981). Selander (1966) reviewed cases in which sexual size dimorphism is related to foraging differences between the sexes, for example in some *Centurus* and *Dendrocopus* woodpeckers, where sexual dimorphism may broaden the food resources of a pair (Selander 1966; Ligon 1968).

Strong support for the idea that sexual selection favors large body size in birds comes from some polyandrous species. There, sexual selection should be strongest in females. In accordance, females are much larger than males in, for example, strongly polyandrous jacanas (e.g., *Jacana spinosa*; Jenni and Collier 1972). Another likely example is coucals (Centropodidae). This subfamily of Cuculidae is often described as having similar sexes, but females are larger than males in most or all species (Andersson 1994). The sex roles are reversed and, as far as is known, mainly or only the male builds the nest, incubates, and raises the young. The most dimorphic species, the black coucal *Centropus grillii*, is sometimes polyandrous (Vernon 1971). This, and the reversed parental roles, suggests that sexual selection is the reason why females are larger than males in coucals.

Waders

Waders (Charadriiformes), especially sandpipers Scolopacidae and their allies, show more diversity in sexual size dimorphism than perhaps any other family of birds. In a comparative study, Jehl and Murray (1986) concluded that sexual selection is the reason for size dimorphism among waders as well as birds in general. Larger male than female size in birds has usually been explained by advantages in male contests over females, or over territories required for attracting females (e.g., Darwin 1871; Selander 1972; Searcy 1979b,c; Robinson 1986b; Eckert and Weatherhead 1987b).

Reversed dimorphism, which is common in waders, has often been explained by either of two hypotheses:

1. Contest competition among females, especially in polyandrous species (e.g., Jenni 1974; Erckman 1981; Oring 1986; Jönsson 1987). Severe female fights are common in the spotted sandpiper *Actitis macularia* and the American jacana *Jacana spinosa* (e.g., Jenni and Collier 1972; Maxson and Oring 1980; Emlen et al. 1989; section 7.5 above).

2. Male competition over females or territories in species with display or contests in the air (e.g., Schmidt-Bey 1913; Ralls 1976; Erckmann 1981; Payne 1984; Jehl and Murray 1986; Jönsson 1987; Höglund 1989b). Jehl and Murray (1986) showed that reversed sexual size dimorphism among nonpolyandrous waders is associated with aerial display, and suggested that sexual selection for male aerial agility is the reason. Erckman (1981) and Jönsson (1987) pointed out that the advanced display flight of waders such as dunlin *Calidris alpina* requires less energy in small than in large males, and that this might favor smaller male than female size in many waders. Female waders that choose a mate capable of performing much flight display may obtain mates in good nutritional condition, with high foraging and parental ability, of crucial importance after the female leaves the family (Erckman 1981; Jönsson and Alerstam 1990). Flight energetics might in part explain why males are smaller than females also in several other waders, and in many birds of prey (e.g., Reynolds 1972; Mosher and Matray 1974; Erckman 1981; Jönsson and Alerstam 1990).

Compared to the basal metabolic rate, the relative energy cost of active flight in general decreases with reduced body size, other things being equal (figure 11.8.2; Norberg 1990). From condor to colibri, the relative cost reduction may be as large as sixfold (Greenewalt 1975; McMahon and Bonner 1983). It might therefore be easier, in the sense of requiring a lower proportional increase in the foraging time, to cover the energy costs of flight display and other active flight in small than in large birds. A small bird may also have more energy headroom for power-intensive aerial display, which may require more extra power than a larger bird can raise. Such aspects of energetic efficiency might in part explain (1) why males tend to be smaller than females in species where males fly much more, in display or foraging; (2) why aerial display is often most developed in the smallest species of a taxon; and (3) why the smallest species in taxa such as bustards, hummingbirds, and manakins have flight display and reversed sexual size dimorphism. Flight energetics may also partly explain why males are smaller than females in most birds of prey, where males forage for the family and therefore fly much more than females (see below in this section).

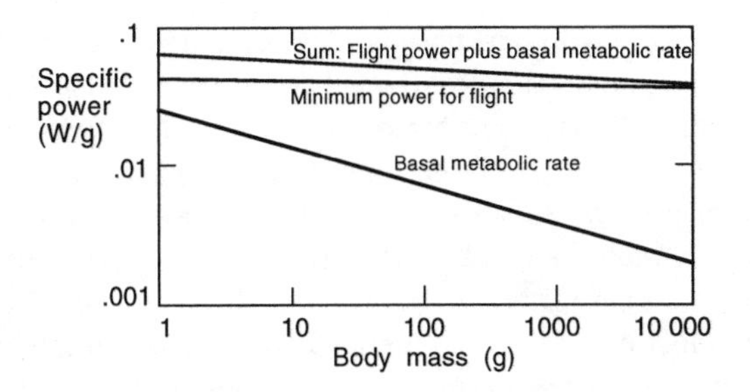

Figure 11.8.2 Specific metabolic power required for flight, and basal metabolic rate, in relation to body mass in birds. The relative energy cost of active flight compared to the basal metabolic rate decreases with reduced body size. (After Greenewalt 1975; McMahon and Bonner 1983)

Support for these ideas also comes from other groups of flying animals, such as midges, and the reproductive caste among certain ants where males are smaller than females (section 11.4 above). There are several puzzling exceptions, however. For example, the display flight of the sexually size-reversed curlew *Numenius arquata* and whimbrel *N. phaeopus* does not seem particularly acrobatic and agile in comparison with that of the lapwing *Vanellus vanellus*. Yet the lapwing shows normal size dimorphism, in contrast with *Numenius* species (Cramp et al. 1983). Three other puzzling cases are the jack snipe *Lymnocryptes minimus*, with flight display but normal size dimorphism; the great snipe *Gallinago media*, with ground display but slightly reversed dimorphism; and the subantarctic snipe *Coenocorypha aucklandica*, apparently the most terrestrial among all snipes, without aerial display, yet with reversed dimorphism (Jehl and Murray 1986).

The aerial display hypothesis of Jehl and Murray (1986) as well as the energetic efficiency hypothesis predicts that species with male song flight, such as some larks (Alaudidae) and pipits (*Anthus*), should have smaller relative male size than related species without song flight. This does not, however, seem to be the case (Cramp et al. 1988). The most puzzling exception is perhaps the most size-dimorphic of all birds. Males of the Australian brown song lark *Cinclorhamphus cruralis* are about 2.2 times heavier than females; yet males apparently range over a wide area in flight display (Amadon 1977). This bird offers some food for thought to those of us who think that flight display favors small male size; a detailed field study is needed.

Factors other than aerial display have probably also contributed to the reversed sexual size dimorphism among waders. Female body size is related to aspects of egg laying, such as double-clutching (e.g., Ross 1979; Oring 1982; Saether et al. 1986). In many waders, the mass of the clutch is large relative to that of the female. In some species she lays several clutches in quick succession, and the capacity for rapid egg production should be crucial. If egg laying is in part based on stored body reserves, large female size may be advantageous (e.g., Downhower 1976; Erckman 1981; Jönsson and Alerstam 1990). Competition over mates therefore need not be the only reason why females are larger than males in role-reversed phalaropes and some other waders. Advantages in egg production may contribute, as in many lower vertebrates and invertebrates, where fecundity rather than ability for parental care limits female reproductive success (section 11.2). Compared to most other birds, waders show reduced female parental care (e.g., Erckman 1981, 1983; Oring 1986; Jönsson and Alerstam 1990).

Another possible influence is differences in diet, but the causal relations are not obvious. Jehl and Murray (1986) found no clear association between sexual dimorphism and food in waders. Some of their results for wing, bill, and tarsus (their figures 2 and 9) suggest, however, that there may be such associations, and that foraging differences perhaps contribute to the sexual dimorphism. Selander (1966) and Ligon (1968) provide examples from other birds in which sex dimorphism is related to foraging differences.

There is a strong relationship between sexual size dimorphism and parental care patterns in Scolopacidae (figure 11.8.3). Focusing on differences in sex roles, Jönsson (1987) and Jönsson and Alerstam (1990) suggested that sex differences in bill size in dunlin *Calidris alpina* may have evolved because males remain with the chicks in their terrestrial habitat,

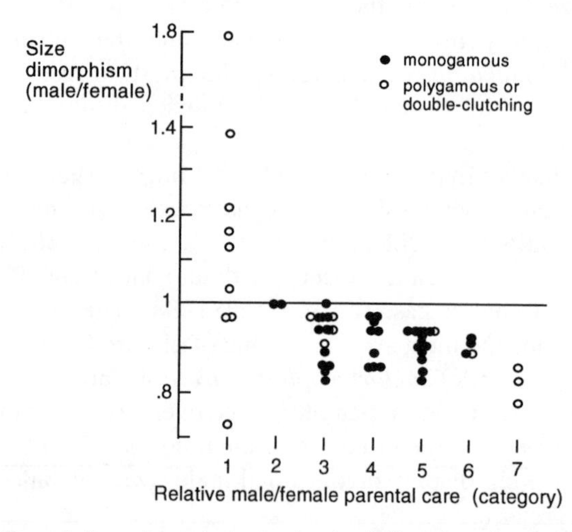

Figure 11.8.3 Sexual dimorphism in body size (male/female ratio of the cube of wing length) in relation to relative parental care in 57 species of Scolopacidae. The dimorphism is largest in species with little male parental care, and it is reversed in species with much male parental care. Categories range from incubation and brood rearing by the female alone (1), over increasing male and reduced female participation, to incubation and brood rearing by the male alone (7). (After Jönsson and Alerstam 1990)

feeding by surface-picking, for some time after females leave for marine littoral habitats. There, the longer bill of the female is probably more effective since most food is obtained by mud probing (also see Erckman 1981).

Jehl and Murray (1986) concluded that "sexual size dimorphism is the result of sexual selection"; "the initial morphological divergence between the sexes cannot arise from ecological selection but must be the result of sexual selection. . . If this view is correct, then the great diversity of combinations of sexual dimorphism and mating relationships in shorebirds (and other animals) must be explainable in terms of Darwinian sexual selection." Even if sexual selection in one sense may be ultimately responsible for all sexual dimorphism (Power 1980; see section 1.2 above), other factors such as foraging differences may, however, sometimes become more important in the evolution of the dimorphism. Even if sexual selection initiated the dimorphism, this need not imply that variation in sexual selection explains most of the variation in sex dimorphism that exists today among waders (or any other taxon). Experiments and quantitative multivariate analyses should help clarify the relative importance of factors such as relative egg size, clutch size, laying pattern, and sex differences in parental roles and foraging, in addition to sexual selection (e.g., Saether et al. 1986). The evolution of parental roles, mating system, and sexual size dimorphism in shorebirds still presents unsolved, fascinating problems.

Birds of Prey

The "reversed" sexual size dimorphism of predatory birds, with females in some cases almost twice as large as males, has long puzzled ecologists. The species in this category are mainly predators of vertebrates: raptors (Accipitriformes and Falconiformes), owls (Strigiformes), skuas (Stercorariidae), frigatebirds (Frigatidae), and boobies (Sulidae). Only some of the many ideas on this problem can be discussed here; a review by Mueller and Meyer (1985) listed twenty hypotheses![4] As body size is hard to manipulate experimentally, the ideas have been difficult to test.

I belong with those who think that the sex roles, especially the male's function as forager for the pair, and the female's egg production and parental care, partly determine direction as well as degree of dimorphism in birds of prey. The conflict between the needs for aerial agility and efficiency in energetically expensive hunting, and for accumulation of nutrients and energy for egg production and parental care, may partly explain why female

[4] For discussion and references, see, e.g., Earhart and Johnson 1970, Balgooyen 1976, Snyder and Wiley 1976, Newton 1979, 1986, Walter 1979, Sigurjonsdottir 1980, Andersson and Norberg 1981, von Schantz and Nilsson 1981, Cade 1982, S. M. Smith 1982, Widén 1984, Mueller and Meyer 1985, Mueller 1986, 1989a, 1990, Temeles 1985, Jehl and Murray 1986, Lundberg 1986, Olsen and Olsen 1987, Norberg 1987, Jönsson and Alerstam 1990.

birds of prey remain at the nest while the male hunts (e.g., Andersson and Norberg 1981; Newton 1986). This personal bias inevitably colors the following account. Critical different views are found, for example, in Mueller and Meyer (1985), Jehl and Murray (1986), and Mueller (1990).

SEX ROLES AND THE DIRECTION OF DIMORPHISM

Large size may enable the female to

1. Lay larger eggs or clutches (e.g., Reynolds 1972; Selander 1972)
2. Incubate more efficiently (Snyder and Wiley 1976; Cade 1982)
3. Better protect her developing follicles (Walter 1979)
4. Better defend her eggs and young against nest predators (Storer 1966; Reynolds 1972; Snyder and Wiley 1976; Andersson and Norberg 1981)
5. Put on relatively less mass, which impairs flight and hunting, prior to egg laying (Wheeler and Greenwood 1983)
6. Better endure food shortage on stored resources during the period when the male hunts for the family.

Some of the previous ideas seem to be refuted by recent work. Point 4, female nest defense, does not seem to be crucial. Field studies (Wiklund and Stigh 1983; Andersson and Wiklund 1987) and a comparative analysis (Mueller and Meyer 1985) suggest that males in some birds of prey are the main defenders of the nest. Also, since the male feeds the female when she puts on weight before egg laying (Newton 1986), in practice she has little disadvantage of reduced flight performance; this speaks against hypothesis 5.

FORAGING EFFICIENCY

Another set of ideas emphasizes foraging advantages arising from the dimorphism, such as (1) reduced competition for food between the sexes (Brüll 1937 and many later authors); (2) larger available prey biomass for smaller males (Storer 1966); (3) reduced energy costs for smaller males (e.g., Reynolds 1972; Mosher and Matray 1974; Jönsson and Alerstam 1990); (4) greater agility and skill in catching agile prey for smaller males (e.g., Reynolds 1972; Newton 1979; Andersson and Norberg 1981; von Schantz and Nilsson 1981).

Agility should increase with reduced size (Andersson and Norberg 1981; see waders above). Among raptors, attack success seems to be lowest for those that prey upon birds (Temeles 1985). Agility should therefore be particularly critical in predators of birds, which may in part explain why they are the most sex-dimorphic birds of prey (figure 11.8.4). The male role as food provider should impose stringent selection of males toward the most effective body size for foraging. The relative importance of male agility compared to other factors may also explain the trend toward larger size dimorphism with increasing prey agility. Species that feed on carrion show

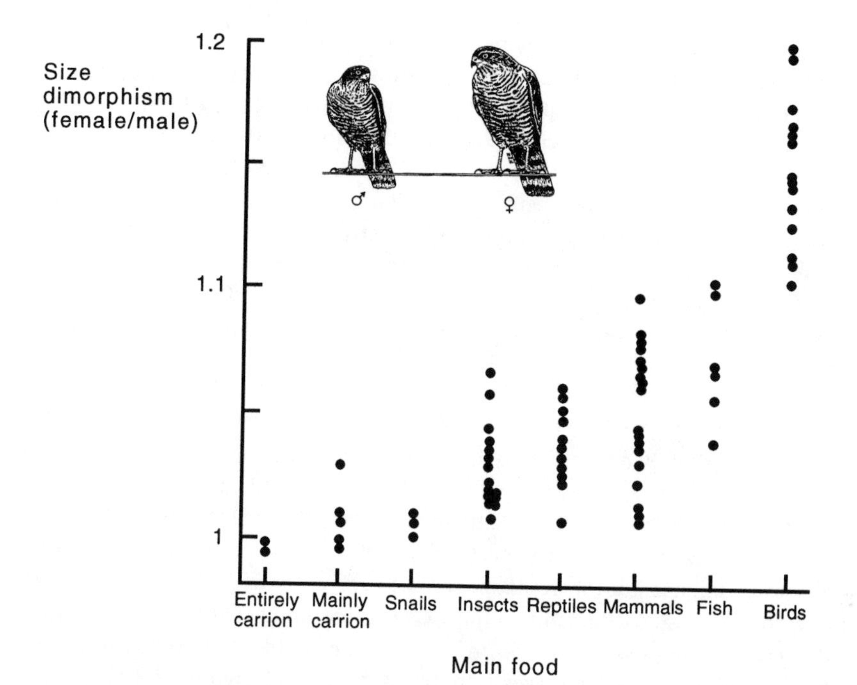

Figure 11.8.4 Sexual size dimorphism (female/male wing length) in raptors in relation to diet. Each point represents one species. The degree of dimorphism increases with the agility of the prey. (After Newton 1979; sparrowhawks by Å. Norberg in Andersson and Norberg 1981)

little size dimorphism, sometimes with males larger than females. The other extreme, with females twice as heavy as males, is found in some specialized bird hunters.

The aerial agility and other hypotheses related to foraging have been strongly criticized. Mueller and Meyer (1985) and Mueller (1986, 1990) argue that they are refuted mainly because trophic structures used in foraging, such as bill and claws, are no more dimorphic than other structures. This is also held up against the idea that competition for food between the sexes explains the degree of dimorphism (as was suggested, e.g., by Selander 1972; Newton 1979; Andersson and Norberg 1981). Among the nontrophic structures, Mueller and Meyer (1985) include body size. Aerodynamic theory and marked contrasts in food choice between the sexes in some size-dimorphic raptors suggest, however, that predator body size partly determines the sizes of prey taken by the sexes (e.g., Newton 1979, 1986; Andersson and Norberg 1981). Changes in body size therefore may suffice for adapting male and female to partly different diets.

Energy savings by smaller males was emphasized in particular by Reynolds (1972) and Mosher and Matray (1974), and by Jönsson and Alerstam (1990), who presented new evidence that energetic efficiency is important.

It seems doubtful, however, whether energy savings explains more of the trend in figure 11.8.4 than does the need for aerial agility. The most dimorphic raptors are small accipiters that pursue their agile prey in the air, with low capture success (Temeles 1985). Agility and maneuverability are then crucial for prey capture. One the other hand, raptors hunting rodents in open country, such as the kestrel *Falco tinnunculus*, fly much and often use energy-expensive hovering flight. Yet kestrels are much less dimorphic than specialized bird hunters, contrary to what is expected if energy savings is the main factor favoring small male size, but in line with the aerial agility hypothesis. More work is needed to estimate the relative importance of these factors.

SEXUAL SELECTION: MALE TERRITORY DEFENSE

Territory defense and display in birds of prey often takes place in the air (Schmidt-Bey 1913; Payne 1984; Widén 1984; Jehl and Murray 1986). Small males then have an advantage from greater aerial agility. This idea has several problems, however. Contrary to what it predicts, the share in territory defense by the female seems to increase with the degree of reversed sexual size dimorphism in raptors (Mueller and Meyer 1985). Most display and aggression at nesting sites of the highly dimorphic sparrowhawk *Accipiter nisus* involves females, not males (Newton 1986).

Moreover, as Widén (1984) and Mueller (1989a,b) note, this hypothesis fails for owls. They show similar patterns of dimorphism in relation to diet as do raptors, yet territorial behavior in most owls seems to be based mainly on nocturnal calling from perches, not on display and contests in the air (e.g., Mikkola 1983). As some evidence from raptors also runs counter to this idea, male display and territory defense in the air seems unlikely to be a major cause of reversed sexual size dimorphism in birds of prey. In addition, studies of dragonflies, another group with aerial contests, indicate that territoriality in this group favors large male size (section 11.4 above).

SEXUAL SELECTION: FEMALE DOMINANCE

As birds of prey have dangerous weapons, the male might perhaps injure the female unless she is larger and easily dominates him (Willoughby and Cade 1964; Amadon 1975; Cade 1982; S. M. Smith 1982). Mueller and Meyer (1985) and Mueller (1990) suggested that rapid, nonaggressive pair formation when the female is larger explains reversed sexual dimorphism in raptors, small males being favored by sexual selection through female choice (also see Jehl 1970).

A test between this and the foraging hypotheses (aerial agility and energetic efficiency) can be done by comparison with birds that do not hunt in or from the air, yet have dangerous weapons used in killing vertebrate prey.

The female dominance idea predicts that such species are reversed size dimorphic. Since they do not hunt in the air, the foraging hypotheses do not predict them to be so. Unfortunately, there are few suitable species for tests. Two species that sometimes kill relatively large vertebrate prey and have dangerous weapons are secretary bird *Sagittarius serpentinus* and marabou stork *Leptoptilus crumeniferus*. The male is larger than the female in both cases (Brown et al. 1982). These species contradict the female dominance hypothesis, but as they do not hunt in the air, they are compatible with the foraging hypotheses for smaller male size.

Other evidence against the female dominance hypothesis comes from experimental breeding of American kestrels *Falco sparverius* (Willoughby and Cade 1964). Artificially created pairs with larger males than females mated and reproduced as successfully as did pairs where the female was largest. Evidence from carnivorous mammals also speaks against the hypothesis. A male carnivore could clearly injure the female, and copulation sometimes involves aggression, for example in lions (e.g., Schaller 1972). Yet the male is larger than the female in most Carnivora (Ralls 1977). In stoat *Mustela erminea*, the much larger male dominates the female except during the breeding season, when females become dominant over males (Lockie 1966; Erlinge 1977), as in birds (e.g., S. M. Smith 1982). This dominance reversal seems no harder to achieve in stoat than in birds of prey, even though the male stoat is almost twice as heavy as the female and kills prey that may be his own size (e.g., Erlinge 1979; Ralls and Harvey 1985). In sum, it seems unlikely that enhancement of female dominance has been a main factor in the evolution of reversed sexual size dimorphism in birds of prey.

SEXUAL SELECTION: FEMALE CHOICE OF EFFICIENT FORAGERS

Females may prefer males that prove their foraging ability by frequent courtship feeding (e.g., R. Simmons 1988a,b). If small males are the most efficient foragers (see above), female choice may favor small males (Cade 1982; Safina 1984).

The evidence on this idea is mixed. Female sparrowhawks *Accipiter nisus* seem to remain only with males that provide them with food in the pre-egg stage (Newton 1986). In Tengmalm's owl *Aegolius funereus*, breeding males were smaller than nonbreeding males, and early-breeding males were smaller than late-breeding males. On the other hand, the surplus of nonbreeding males in this species is considerable, making it less likely that larger female size has been favored by competition over males (Korpimäki 1987a,b; Hakkarainen and Korpimäki 1991). In the barn owl *Tyto alba*, mated males were not smaller than unmated males (Marti 1990). Bowman (1987) also concluded that female American kestrels (*Falco sparverius*) did not favor the smallest males. In addition, the often polygynous harriers (*Circus*) are not more size dimorphic than other raptors with similar diet (e.g., Picozzi 1984; R. Simmons 1988a). Harriers have, however, more sex-dimorphic plumage, the male being paler and more conspicuous than the female (e.g., Cramp et al. 1980). This suggests that sexual selection has had marked effects in harriers (also see R. Simmons 1988a), but mainly on male plumage color, not body size (but see below).

For these reasons, there is as yet no strong evidence that female choice has been important for the evolution of reversed sexual size dimorphism in birds of prey.

SEXUAL SELECTION: FEMALE CONTEST COMPETITION OVER MALES

Another possibility is that competition over mates may be strongest among female birds of prey, particularly in species where great flight skill is needed in catching prey, and males may differ much in foraging success, depending, for instance, on age. This idea was suggested by Olsen and Olsen (1984, 1987) and by Newton (1986, 1988), who in addition stressed the need for the hunting male to be small in order to be agile, and for the female to be large in order to store body reserves for egg production.

Supportive evidence is that in the sparrowhawks *Accipiter nisus* studied by Newton (1986), higher male mortality produced a skewed sex ratio of 1.6 adult females per male. Females, not males, were involved in most observed aggression and display near the nest in spring. Moreover, among yearling females, those breeding were larger than nonbreeders, and offspring production increased with female size (Newton 1988). Female competition over mates might also in part explain the correlation between degree of size dimorphism and proportion of birds or other difficult prey in the diet (figure 11.8.4). Birds are particularly hard to catch (Temeles 1985). This might make variance in male foraging success larger, and female competition over skilled hunters stronger, as the proportion of birds in the diet increases, favoring large female size by sexual contest competition, particularly in raptors that specialize on birds (Olsen and Olsen 1987).

Experiments in the kestrel *Falco tinnunculus*, using teathered birds of both sexes near a nest, do not support the female contest hypothesis (Wiklund and Village 1992). In responses to birds of the same sex, males tended to be more aggressive than females. On the other hand, this species hunts mainly rodents and insects and shows little size dimorphism, so female competition over skilled hunters is not expected to be as strong as in bird-hunting raptors.

Another kind of evidence also seems to speak against stronger sexual selection in female raptors. The pattern of sexual color dimorphism is not as expected from strong female competition over males. If there is a scarcity of high quality mates for one of the sexes, traits that improve success in competition over mates should evolve in that sex. This applies to ornaments favored by mate choice, as well as to weapons, body size, badges, and other traits favored in contests over mates. I am not aware of any published evidence that one of the sexes has the largest ornaments, and the other sex the largest weapons or other traits favored by contests over mates. If a species shows sex dimorphism in both ornaments and weaponry, the largest ornaments as well as weapons occur in the same sex, usually males. Some examples are the conspicuous plumage, leg spurs, and larger body size among males in Phasianidae (section 12.3, and Delacour 1951), the larger canines, body size, and mane or facial hair among male primates (sections 4.3, 11.9, and 13.7), the larger claws, body size, and more conspicuous coloration in male fiddler crabs *Uca* (Crane 1975), and the brighter colors and larger body size of female jacanas (Jenni and Collier 1972).

In raptors with sexual color dimorphism, males usually have the most conspicuous colors, for example in harriers, some falcons, and hawks, including species with extreme reversed sexual size dimorphism, such as the sparrowhawk *Accipiter nisus* (e.g., Cramp et al. 1980). This pattern suggests that males are subject to strong sexual selection, and casts doubt on the idea that female contest competition over males is the primary cause of reversed sexual size dimorphism.

On the other hand, the previous argument is perhaps not correct. There is no logical necessity that all sexually selected traits in a dimorphic species are more pronounced in one and the same sex. If there is competition over mates in both sexes (for example because there is quality variation in both; see section 7.4), and if they have markedly contrasting sex roles, different kinds of traits might be favored by sexual selection in males and females. For instance, the presence of the female at the nest much of the time in raptors might impose strong selection for cryptic coloration. Conspicuous colors might then evolve only in males. The pattern of coloration discussed above is therefore not necessarily evidence against large female size being favored by sexual selection. It seems entirely possible that female contest competition might contribute to the dimorphism in some species, such as the sparrowhawk, with a strongly female-biased sex ratio (Newton 1986, 1988). An interesting possibility is therefore that sexual selection might favor different traits in males and females of some raptors: conspicuous colors in males, and large body size in females. Sex differences in the targets of sexual selection in principle might arise in any species with competition over mates in males as well as females, if combined with sufficiently different sex roles.

More field and comparative work is needed to test the previous hypotheses, most of which are not mutually exclusive, and to assess the relative importance of various selection pressures behind the reversed sexual size dimorphism of predatory birds. Some species do not seem to fit any hypothesis, such as the larger among the Australian *Ninox* owls. They hunt mammals and birds, yet the male is larger than the female (e.g., Schodde and Mason 1980). Clarifying their biology in detail will perhaps shed new light on the causes of reversed sexual size dimorphism. The largest species, the powerful owl *Ninox strenua*, hunts large, sluggish arboreal marsupials that are easy to catch, but hard to find and transport. Perhaps males therefore do not need to be agile, which might explain why they are larger than females in this and some other *Ninox* owls (Andrew Cockburn, pers. comm.).

Other groups of interest are swifts and swallows. They spend much of their life foraging on the wing, and it seems likely that competition over mates occurs in the air as well. Why do not these species show similar sex roles and reversed size dimorphism as birds of prey?

In sum, differences in sexual selection may be one important cause behind the variation in sexual size dimorphism among birds. Sexual selection alone seems not, however, to explain all the variation. Differences between species in foraging ecology, parental roles, and demands imposed by egg production also affect sexual size dimorphism. The relative importance of the different factors largely remains to be determined. In particular, the roles of sexual and other selection in monogamous birds with females larger than males are still obscure.

11.9 Mammals

Males are larger than females in most mammals, but forms with females larger than males are by no means rare; such species occur in 12 of 20 orders, and 30 of 122 families (Ralls 1976, 1977). Females are the largest sex in, for example, Vespertilionid bats, rabbits and hares, three families of baleen whales, Lobodontine seals, and two tribes of small antelopes (Cephalophini and Neotragini).

Ralls (1976, 1977) rejected the idea that sexual selection on the whole explains which sex is largest. Sexual selection is probably the reason why males are largest in most mammals (see below), but it does not seem to explain why females are larger in some species. Females in most of these forms are not polyandrous, more aggressive, or armed with better weapons than males. The males seem to be more strongly sexually selected, whereas females show greater parental investment. Natural selection may therefore be the cause of larger female than male size in certain mammals. In some cases, larger female size may be related to strong competition over resources other than males. For example, in the flying squirrel *Glaucomys volans*, females defend territories with food and nest cavities. Males are smaller and nonterritorial (Madden 1974). Ralls (1976) also suggested that competition over females in aquatic or aerial species may sometimes favor smaller, not larger males (section 11.8). In mammals where large size clearly is important for male mating success, for instance elephant seals and red deer (sections 5.5–5.6), males grow for a longer time and start breeding later in life than females (also see section 10.5, and Jarman 1983).

Ralls's (1976) main explanation for larger females than males is the Big Mother hypothesis: "A larger mother may produce a larger baby with greater chances of survival, she may enable it to grow more rapidly by providing more or better milk, and she may be better at such aspects of maternal care as carrying or defending her baby." Many of the species in which females are the largest sex bear offspring of large relative size, which may select for large female size.

In Vespertilionid bats, where females are usually the same size as males or larger (Ralls 1976), there is no evidence of strong female competition over mates. Among species, the size of the female relative to the male increases with litter size (Myers 1978; figure 11.9.1). Such a trend is also found within one species, *Eptesicus fuscus*, where litter size varies geographically. Females in several bats have larger wings than males, also after sex differences in body size are corrected for. Myers (1978) concluded that their need to fly with and nourish large fetuses is the main reason why females are larger than males in many species. Jehl and Murrary (1986) objected, suggesting that males are smaller in bats owing to sexual selec-

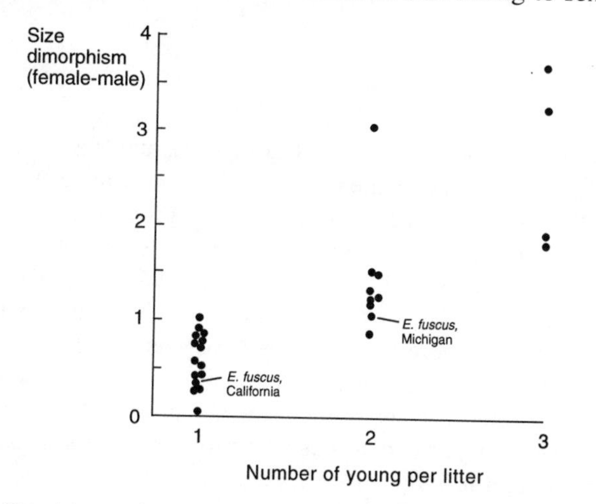

Figure 11.9.1 The degree of sexual size dimorphism (female/male forearm length) in Vespertilionid bats increases with litter size (see text). (After Myers 1978)

tion for aerial agility. But even if sexual selection contributes, it does not explain the strong correlation between dimorphism and litter size (Myers 1978).

With decreasing female size, the relative size of the offspring tends to increase, and that of the male to decrease (Ralls 1976). Most mammals with larger females than males are smaller than the average for the taxonomic group; a possible explanation is suggested in section 11.10. The reasons for females being the larger sex in some mammals are complex and still poorly known, as in birds (Ralls 1976).

Does sexual selection and size dimorphism depend on the relative parental investment of the two sexes, as suggested by Trivers (1972; section 7.2 above)? As in many other taxa, the strongest size dimorphism tends to occur among mammals with large body size. Ralls's (1977) qualitative comparisons among twenty mammalian orders suggested that parental investment explains only part of the variation in sexual selection.

As expected, the most polygynous species are strongly dimorphic (Clutton-Brock and Harvey 1977; Alexander et al. 1979; Clutton-Brock et al. 1982; Jarman 1983; Gittleman 1989). Whereas large parental investment by males is correlated with monogamy and little sex dimorphism, small paternal investment is, however, a poor predictor of extreme polygyny and dimorphism. Other aspects are clearly important. Following Jarman (1974) and Stirling (1975), Ralls (1977) suggested that the distribution of food and other resources affects the degree of clumping of females, and hence in part controls the mating system and the strength of sexual selection in males (also see Bradbury and Vehrencamp 1977; Emlen and Oring 1977; Jarman 1983).

Large Herbivores

Bovids (antelopes, sheep, goats), Cervids (deer), and Macropodids (kangaroos and wallabies) make up the majority of the large terrestrial herbivores. Jarman (1983) compared their mating systems, growth patterns, and sexual dimorphism in body size, weapons, and other aspects. In all three taxa, sexual size dimorphism is slight among the smallest species, females being larger than males in a few of them. In some species with medium to large body size, males are more than twice as heavy as females. Among bovids and cervids, male weapons tend to be largest relative to body size in the most size dimorphic species, which are not always the biggest in their group. Similarly, kangaroo males use their forelimb in fights, and its length relative to the hind foot increases in parallel with sexual size dimorphism among species (Jarman 1983). Large weapons do not seem to be only allometric side effects of large body size; large male weapons and body size have probably both been favored by selection.

For antelopes, Jarman (1974) suggested that the male mating tactic reflects the dispersion, grouping, and predictability of females, which in turn depends on habitat, food, and risk of predation. Sexual monomorphism is associated with monogamy and dimorphism with polygyny, in ways that depend on these ecological aspects (Clutton-Brock et al. 1982; Jarman 1983).

Although there are clear trends in sex dimorphism among large herbivores, with dimorphism increasing with the degree of polygyny, there are exceptions. For example, horses and zebras largely lack sex dimorphism in size and weaponry (but males have canines) (e.g., Feh 1990). Males in these species fight by biting and kicking, and agility may be more important than large size and strength. Perhaps this explains why equids do not show sexual size dimorphism, in contrast with most other ungulates, where males fight by pushing (Feh 1990).

Marine Mammals

Seals, sea lions, and other pinnipeds show larger contrasts in sexual size dimorphism than any other higher vertebrates. Males are 4–6 times as heavy as females in some species, whereas females are larger than males in others (e.g., Alexander et al. 1979). There is a strong correlation between degree of polygyny and dimorphism (figure 11.9.2). The mating system in turn depends on habitat and ecology (Bartholomew 1970; Stirling 1975).

Extreme polygyny in elephant seals *Mirounga* and some fur seals (Otariidae) is possible because females gather in dense groups on predator-free islands to give birth and mate. Owing to scarcity of suitable sites, a successful male can defend a territory with many females (e.g., Bartholomew 1952, 1970; Ghiselin 1974; Stirling 1975; see section 5.5 above). Large male size in these species, favored by fights over females, in part reflects

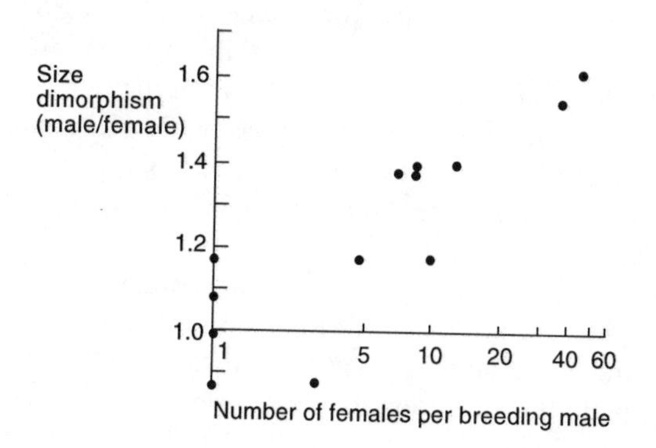

Figure 11.9.2 Sexual size dimorphism (male/female body length) in seals, sea lions, and other pinnipeds increases with the degree of polygyny. (After Alexander et al. 1979)

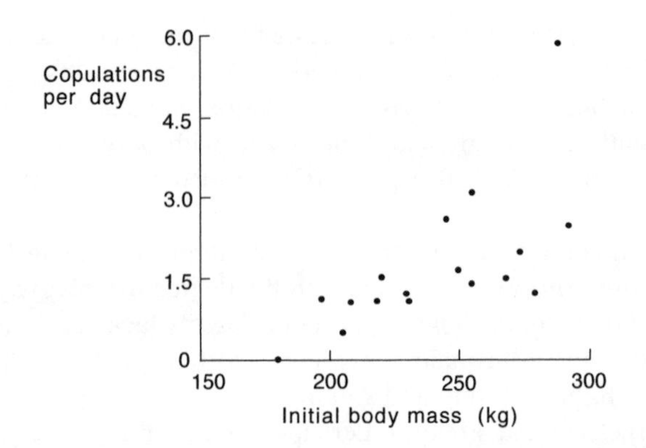

Figure 11.9.3 Mating success in male gray seals *Halichoerus grypus* increases with the body size of the male at the start of his reproductive season. (After Anderson and Fedak 1985)

stored energy in form of blubber, which enables a territory owner to fast and remain longer on land, mating with many females during the long breeding season (e.g., Bartholomew 1952, 1970; Le Boeuf 1974; McCann 1981; Deutsch et al. 1990). Also in gray seal *Halichoerus grypus*, male mating success increases strongly with body mass (figure 11.9.3). Large males are less likely to lose contests, have greater energy reserves, and can be sexually more active than small males during the long breeding season (Anderson and Fedak 1985).

In seals breeding on pack ice, females are dispersed and the mating season is short, so the number of mates that a male can acquire is low (Stirling 1975). Sexual size dimorphism in these species accordingly is slight (Le Boeuf 1986). The Weddell seal *Leptonychotes weddelli* shows slightly reversed size dimorphism in spite of being polygynous. Females give birth on ice, and several females use the same diving hole (e.g., Kaufman et al. 1975). Males compete underwater for females; this may favor male agility and hence small size. There are other possible reasons why females are larger, for instance that they are selected for extreme fatness, or that males on average are younger, as females survive better (Kaufman et al. 1975; Testa and Siniff 1987). The likewise Antarctic leopard seal *Hydrurga leptonyx* also shows reversed size dimorphism (Alexander et al. 1979), but its breeding biology is poorly known.

Walrus *Odobenus rosmarus* males seem to compete for females in lek-like groups, in water near ice floes on which females gather and rest (Fay et al. 1984). Males are larger in body and tusk size. It does not yet seem possible to explain the variation about the trend in figure 11.9.2. Only the general correlation between size dimorphism and polygyny, and the importance of large male size for mating success in contests and endurance rivalry in species that mate out of the water, seems clear. Further study of species such as the harbor seal *Phoca vitulina*, which shows large variation in breeding group size, might be revealing (Stirling 1975; Le Boeuf 1986). Copulation in many of the arctic and antarctic seals takes place under ice, in water of about −1°C, so quantification of mating success in these species is likely to progress at modest pace.

As in birds, there are problems with the agility hypothesis for smaller males than females in some marine mammals. Whales show several unexplained patterns of sexual size dimorphism. In some tooth whales (Odontoceti), for instance sperm whale *Physeter macrocephalus* and killer whale *Orcinus orca*, adult males are much larger than females, but females are larger than males in baleen whales (Mysticeti) (Lockyer 1976; Ralls 1976). Tooth whales tend to be more social and have larger group size than baleen whales (Gaskin 1982). Several tooth whales seem to be polygynous, with females, calves, and a few adult males in some groups, young males in other groups. Perhaps more females are at stake in contests between male tooth whales; this might select for large male size in these species, some of which use the teeth as weapons in fights among bulls (Best 1979). Male sperm whales seem to spend much time searching for groups with females in estrus (Whitehead and Arnbom 1987). There are also indications that males do not reproduce every year, instead foraging and building up reserves in higher latitudes (Best 1979). Large size may then be favored by endurance rivalry, permitting long periods of fasting when the male attempts to reproduce (section 7.2). Endurance rivalry might also favor large size in bull elephants *Loxodonta africana* which, like male sperm whales, search extensively for estrus females rather than defend a permanent group (Barnes 1982). But male sperm whales, like bull elephants, apparently also fight and wound each other in competition over females (Best 1979; Poole 1989b); large size is favored by male contests over females in elephants (Poole 1989a,b).

Perhaps one reason why larger male than female size has not evolved in baleen whales is that they lack teeth or other weapons for fighting, and compete for females in more peaceful ways. Unless combined with suitable weapons, large size might not be advantageous in contests. In addition, in baleen whales, for example in the fin whale *Balaenoptera physalus*, females migrate north to warm breeding waters and seem to raise their young on energy reserves previously stored in rich Antarctic waters (Brodie 1975). As storage capacity should increase with body size (Calder 1984, and section 7.2 above), this can perhaps explain why females are larger than males in baleen whales. The patterns of sexual size dimorphism in whales and other marine mammals offer challenging problems for quantitative comparative analyses.

Primates

Some of the first quantitative analyses of relationships between sexual size dimorphism, mating system, and other factors were made in primates (section 4.3) (e.g., Clutton-Brock and Harvey 1977, Clutton-Brock et al. 1977, Leutenegger and Kelly 1978). In a comparison of twenty-two genera of primates, relative size dimorphism (male:female body mass) increased with the socionomic sex ratio (number of adult females per adult male in breeding groups) (Clutton-Brock and Harvey 1977) (figure 11.9.4). Further analyses showed that the main difference is found between monogamous and polygynous forms. In a more detailed analysis of forty-two species, Clutton-Brock et al. (1977) showed that terrestrial forms are more dimorphic than tree-living forms, perhaps because the best food in trees is often located on fine end twigs, hard to reach for a heavy animal (also see section 4.3).

As in many other taxa, sexual size dimorphism in primates increases with body size. In a partial regression analysis, the degree of dimorphism was more strongly related to body weight ($b = 0.41$) than to socionomic sex ratio ($b = 0.28$); 56% of the variance in size dimorphism was accounted for by these two variables. Almost half the variation hence remained unexplained. One reason may be that the socionomic sex ratio underestimates sexual competition in multimale troups, for example in baboons and macaques, where breeding is not shared equally among the males (Clutton-Brock et al. 1977). The crudeness of the measure of sexual selection may reduce the amount of variation explained by it (see Gaulin and Sailer 1984). Also, reduced major axis analysis is more appropriate than regression analysis of this kind of data (Harvey and Pagel 1991).

In two comparative analyses, Leutenegger and Cheverud (1982) and Cheverud et al. (1985) concluded that body mass accounts for much of the variance in sexual size dimorphism among primates, and that sexual selection is relatively unimportant. They suggested that males and females may differ in genetic variance for body size, and that this suffices for directional selection to create sexual size dimorphism, even if selection is identical in the two sexes (Leutenegger and Cheverud 1982). Gaulin and Sailer (1984) pointed to problematic aspects of this study, such as omission of all species where size dimorphism is absent or reversed, and use of absolute sex differences (male mass – female mass) for the measure of dimorphism. Ecological and sex role differences are, however, more likely related to relative than to absolute size differences between male and female. Relative dimorphism (male:female mass), which has been used in most other studies of size dimorphism, for this and other reasons seems more relevant in comparative analyses of selection and other factors behind the evolution of sex dimorphism. The conclusion that body size explains most of the variation

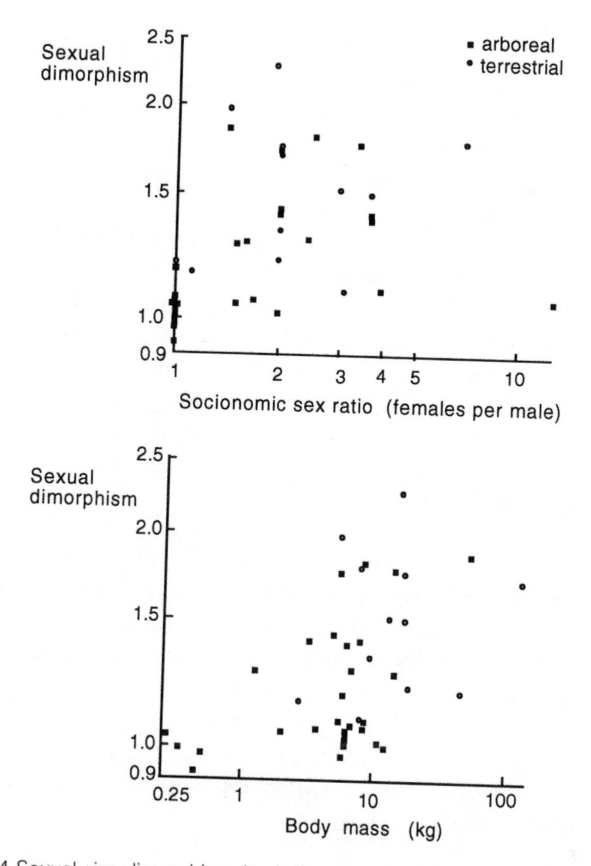

Figure 11.9.4 Sexual size dimorphism (male/female mass) in arboreal and terrestrial primates in relation to socionomic sex ratio (*top*) and body mass (*bottom*). (After Clutton-Brock et al. 1977)

in sexual size dimorphism (Leutenegger and Cheverud 1982) seems to be mainly a consequence of the use of absolute size differences between males and females, and of some other misleading procedures (Gaulin and Sailer 1984; Rogers and Mukherjee 1992; Webster 1992).

Cheverud et al. (1985) attempted to solve the problem of species not being independent sample units by estimating the effects of phylogeny. Under the assumptions of the model, phylogenetic constraints explained roughly half of the variation in sexual size dimorphism among the forty-four species examined. Body size explained another third, whereas habitat, mating system, and diet explained minor amounts. The model in total accounted for 92% of the variation. "Thus, in attempting to explain why a particular modern primate species is very dimorphic compared to other primates, we would say first because its ancestor was more dimorphic than average, second because it is a relatively large species, and third because it is terrestrial, polygynous, and folivorous" (Cheverud et al. 1985).

These conclusions differ markedly from most other studies, which emphasize the importance of sex differences in selection related to habitat, food, predation, competition over mates, and other factors (e.g., Crook 1972; Clutton-Brock and Harvey 1977; Clutton-Brock et al. 1977; Ralls 1977; Leutenegger 1978; Alexander et al. 1979; Gaulin and Sailer 1984, 1985; Harvey and Pagel 1991). One reason for the discrepancy is probably that Cheverud et al. (1985) attempted to quantify the role of phylogeny (see section 4.3, and Gittleman and Kot 1990). Unfortunately, the approach used seems to assume that any similarity between related species is due to phylogenetic constraints, and not at all to responses to common selection pressures. This need not be so: closely related species may also experience similar selection regimes (see Webster 1992; Williams 1992). Another reason for the differences in conclusions is probably that Cheverud et al. (1985) used absolute size differences between the sexes. Most other studies of sexual size dimorphism use relative differences (see above).

In another multivariate analysis of sixty primates, Gaulin and Sailer (1984) used relative (as well as absolute) size dimorphism, and also included species with reversed dimorphism. They did not, however, estimate the effect of phylogeny. Gaulin and Sayler (1984) concluded that sexual selection is the main cause behind sexual size dimorphism, and suggested that also the correlation between dimorphism and body size reflects sexual selection (see also Rogers and Mukherjee 1992; Webster 1992). The tendency for sex dimorphism to be expressed in size may increase with body size, for reasons discussed below.

11.10 Why Does Sexual Size Dimorphism Increase with Body Size?

Sexual size dimorphism increases with body size in many of the vertebrates with males larger than females reviewed above. This trend was examined by Rensch (1950, 1959), who explained it with allometric growth; there are many other hypotheses (table 11.10.1; e.g., Ghiselin 1974; Wiley 1974; Clutton-Brock et al. 1977; Webster 1992). Some taxa for unknown reasons show a reverse trend, for example lizards (Stamps 1983) and mustelids (Moors 1980; Ralls and Harvey 1985).

In species with strong sexual selection for large male size, genetic correlations between the sexes should lead to larger female body size as well (hypothesis 1; Maynard Smith 1978b). (Similarly, strong selection for small male size should lead to smaller females as well, which may partly explain why birds with aerial display and reversed size dimorphism are

Table 11.10.1
Suggested Causes of the Increase in Relative Size Dimorphism (Male : Female) with Body Size

1. Genetic correlations between the sexes leads to larger size also in the smaller sex when selection increases the size of the larger sex.
2. Lower risk of predation in larger species permits more extreme expression of sexually selected traits.
3. Fewer competing species in animals with large body size permits more extreme expression of sexually selected traits.
4. Higher energetic efficiency in larger species permits more extreme expression of sexually selected traits.
5. Weapon efficiency increases with body size.
6. Selection for small female size is strong in large species.
7. Selection for large female size is strong in small species.

often among the smallest species in their groups; section 11.8.) As the correlation between the sexes is not perfect, the size increase in females will be relatively smaller than in males. For this reason, larger than average sex dimorphism then becomes associated with larger than average body size. Lande (1980) and Rogers and Mukherjee (1992) discuss this possibility in detail. Empirical support is that artificial selection for larger size in one sex makes the other sex larger as well in, for example, *Drosophila*, house mouse *Mus musculus*, and turkey *Meleagris gallopavo* (Shaklee et al. 1952; Frankham 1966; Eisen and Hahnrahan 1972). Field studies of birds point to strong genetic correlations between the sexes (e.g., van Noordwijk et al. 1980; Price 1984a), and so does analysis of human size data (Rogers and Mukherjee 1992).

Other evidence that sexual selection explains the correlation between body size and sexual size dimorphism is that both aspects are correlated with the mating system in blackbirds (Icterinae). This is predicted by the sexual selection hypotheses, but not by the alternatives (Webster 1992).

Larger species tend to be more sex dimorphic not only in body size, but also in weapons (e.g., horns; section 12.2). Shine (1979) suggested that these trends are related to a lower risk of predation as body size increases. Sexual selection can then have stronger effects, as it conflicts less with survival in large than in small species (hypothesis 2). In line with this idea, male fights in anurans seem to occur mainly in two categories of species with better than average defense against predators. One category is species with large body size, for example the unrelated forms called bullfrogs in four continents. The other category is species with toxic skin secretions. In snakes, male fights seem to occur mainly in large and venomous species (Shine 1979).

Another possibility is that with increasing body size there are fewer competing species; this might relax constraints on body size (hypothesis 3; Clutton-Brock et al. 1977).

Size-related trends in energetic efficiency, and the ratio of body volume to surface, might make the costs of increased body size lower and the benefits higher in large species (hypothesis 4). Gaulin and Sailer (1984) also suggested that the energy delivered by a blow increases directly with body mass, and hence with length cubed, and that the area of protective dermal shields increases only with length squared. If so, the advantage of a relative increase in size over an opponent might increase with the size of the species, perhaps in part explaining the correlation between body size and degree of dimorphism (hypothesis 5).

Small females in fluctuating environments may be able to breed earlier than large females; this advantage may be most important in large species that start reproducing relatively late in life (hypothesis 6; Downhower 1976; Clutton-Brock et al. 1977). An advantage of early reproduction for small females is evident in some polygynous mammals with large sexual size dimorphism, for instance small mustelids (e.g., Erlinge 1979; King and Moors 1979). Such species often show sexual bimaturism (section 10.5), with females reproducing at a younger age than the larger males (Wiley 1974; Ralls 1976).

Finally, reversed dimorphism, with females larger than males, tends to occur in relatively small species within some taxa; this suggests that selection for large female size is stronger in small than in large forms (hypothesis 7; see, for mammals: Ralls 1976; turtles: Shine 1980; birds: Payne 1984). One possible reason is that the relative size of eggs and newborn young tends to be largest in small species (birds: Lack 1968; mammals: Ralls 1976; Peters 1983). Selection for increased offspring size should also favor larger size of the mother, other things being equal.

These hypotheses are not exclusive, but might apply in combinations; most of them remain to be tested.

11.11 Summary

Many comparative studies suggest that differences in the nature and strength of sexual selection is one of the main reasons for the enormous variation in sexual size dimorphism among animals, but many other factors are also involved. Females are larger than males in most animals, probably often because of fecundity advantages for large females, especially in species where female reproduction is based on resources accumulated over a long time. Energy storage capacity should increase with size faster than maintenance costs. In the minority of species with reversed sex roles, sexual selection may favor larger females, for example in certain birds and fishes. In many animals where females are larger than males, including fishes and invertebrates, sexual selection may often favor large males, but fecundity advantages for large females may be stronger. Contest competition can also favor large body size in females that defend breeding territories or other resources, even if mainly males compete over mates, for example in salmon.

Where males are larger than females, the reason is usually sexual selection by male contest competition over females, territories, or other resources needed to gain mates. There is abundant evidence from terrestrial species that large size gives an advantage in male contests over females or territories. This is probably the reason why males are larger than females in some beetles and amphibia, many reptiles, and most birds and mammals. In some species, large body mass in males is also favored by endurance rivalry, if greater stored energy reserves permit males to spend a longer time competing for females.

In some aerial and aquatic species where contests do not take place on the ground but in three-dimensional space, sexual selection may favor smaller, agile males. Small males can also be favored by more rapid maturation and efficient search for females, for example in some arthropods and deep-sea anglerfish.

There have been few attempts to quantify the relative importance of the different selection pressures and other factors influencing the relative sizes of the sexes. Primates are one exception; apparently both phylogenetic constraints and sexual and other selection pressures contribute importantly to the pattern.

More detailed work on particular species as well as comparative analyses are needed for tests. Taxa that have been relatively little studied in these respects, yet offer promising opportunities, are fishes and arthropods. The recent advances in comparative methods should be of great help for clarifying the causes behind the bewildering variety of sexual size dimorphism among animals. A life history perspective based on the main fitness components that depend on body size in males and females will probably be essential.

In sum, advantages in contests over mates seem to explain many cases of larger male than female size. The reasons why females are larger in most animals are less clear. Fecundity advantages may often be involved; sexual selection is probably responsible only in exceptional cases.

Figure 12.1.1 The fossil skeleton of an Irish elk, *Megaloceros giganteus*. (From Millais 1906)

12 Weapons

12.1 Introduction

Horns, antlers, tusks, spurs and other weapons are some of the striking male traits that led Darwin to put forth his theory of sexual selection (figures 12.1.1, 12.1.2). Horns occur in beetles, reptiles, and ungulates, tusks in anurans and various mammals, and spurs in arthropods and birds. Field work and comparative studies have helped clarify their function and variation in several taxa (reviewed, e.g., by Geist 1966, 1978; Clutton-Brock 1982; Goss 1983; Davison 1985; Lincoln 1992). The evolution of weapons by sexual selection was modeled by Parker (1983a) and Maynard Smith and Brown (1986) (see box 11.1.1).

At least four alternative explanations have been suggested for horns and hornlike organs in ungulates (Clutton-Brock 1982). Following Geist (1966), such organs are here taken to include "all hardened excrescences projecting from the head," also antlers and enlarged teeth (tusks). They may function as

1. Weapons against predators,
2. Weapons against other males,
3. Indicators of male strength and fighting ability in display toward other males,
4. Indicators of sexual vigor and quality, assessed by females in their choice of mate.

The same hypotheses are relevant for sex-dimorphic weapons in other taxa, such as claws, horns, and mandibles in various arthropods, and spurs in birds. Horns and antlers have received most study and provide some of the most extreme examples of sex dimorphism; they are the main subject of this chapter.

12.2 Mammals

Besides the four ideas just mentioned, several other explanations for horns have also been raised, for instance that they serve as metabolic valves for storing excess minerals (Beninde 1937), or as temperature regulators

Figure 12.1.2 Horns and hornlike structures have evolved in a variety of organisms, exemplified here with a reptile *Chameleon Owenii*, an antelope *Tragelaphus strepsiceros*, and a pig *Babyrousa babyrussa*. (From Darwin 1871)

(Stonehouse 1968). These suggestions have been criticized on several grounds (e.g., Geist 1966, 1971b; Ghiselin 1974; Clutton-Brock 1982; Goss 1983). For instance, sheep and deer tend to have larger horns and antlers in temperate regions than in lower, hotter latitudes, contrary to what is expected if temperature regulation were the main function. Several species of deer in addition grow their antlers during the cold winter and spring, and juveniles and females in most deer lack antlers. The four hypotheses listed above therefore remain the major alternatives.

Weapons for Defense against Predators

Males as well as females in several ungulates sometimes use their horns in defense against predators (Packer 1983). This need not imply that horns evolved or are currently favored mainly by an antipredator function, the importance of which is debated (e.g., Packer 1983; Kiltie 1985). In some species with large antlers, such as moose *Alces alces*, they are not used against predators, which are instead held back by kicks (Mech 1966). Many cervids shed their antlers in winter when the risk of predation may be highest. These patterns reject the antipredator hypothesis at least for some species (Geist 1966; Clutton-Brock 1982).

On the other hand, a comparative analysis of African antelopes suggests that horns in females are important defense weapons against predators (Packer 1983). Females have horns in only half of the genera, mainly among the larger forms. The possibility that female horns result from correlated responses to selection in males does not seem to explain this pattern. Small antelopes rely on crypsis or flight, whereas large species often stay and defend themselves and their calves (Jarman 1974). In spite of being as long as in males of the same size, female horns are straighter and thinner, which should make them more effective stabbing weapons. Male horns are much thicker at the base, which should reduce the risk of breakage in clashes and wrestles with other males.

Alternatively, female horns might be thinner because of the need to shunt minerals into the developing embryo. If this were the case, the thinner horns of females should break more often than those of males; horn breakage is more common in species with thin than with thick horns. In practice, however, female horns break less often. They therefore seem to be used differently than male horns; field studies show that males usually fight more than females. Horn shape also differs between the sexes in a number of genera. Like Geist (1966) and Clutton-Brock (1982), Packer (1983) for these reasons concluded that horn shape in males is adapted for contests with rivals, whereas female horns are more specialized as stabbing weapons used against predators.

In a comparative analysis of eighty-two genera of ungulates, Kiltie (1985) found no unifying adaptive reason for horns in females. They occur mainly among large, herd-forming ungulates, perhaps because female contests are more common in such species. On the other hand, there are many observations of females using their horns in defense of calves against predators (Packer 1983). Jarman (1983) suggested that female horns are useful in defense against undesired mating attempts by young males. There is a need for studies of populations containing both horned and hornless females, and for comparisons of related species among which females are horned in some, and lack horns in others (Kiltie 1985).

The only cervid with female antlers is reindeer *Rangifer tarandus*. By retaining their antlers longer, females become dominant over males at feeding sites during the severe, critical winter period; this may improve foraging for the female and her calf (Espmark 1964). Winter herds are large and contain many bulls, perpaps making social competition unusually strong in reindeer. In addition, females do not hide their calves but defend them against predators (Clutton-Brock et al. 1982).

In sum, there is little evidence that use against predators is the major factor selecting for horns in male ungulates, nor can such a function explain why male horns are more massive and differently shaped than female horns. It seems likely, however, that defense against predators is an important function of female horns in some antelopes.

Weapons Used against Rivals

This and the next two hypotheses assume that male horns and antlers are used in sexual competition over females. The yearly growth of antlers ends just before the rutting season, as expected if they have their main function at that time.

A widespread idea is that horns and antlers are used as weapons against rival males (e.g., Darwin 1871; Geist 1966). Critics have argued that fights among males are rare (e.g., Darling 1937; Gould 1974). Where adequate fieldwork has been carried out, however, it is usually clear that males often do use their horns or antlers in contests (e.g., Walther 1958; Geist 1971b; Bützler 1974; Clutton-Brock et al. 1979). Such fights during the rut are common and dangerous enough to make them a serious source of permanent injury and even death in males of several cervids (reviewed by Geist 1971b; Clutton-Brock 1982).

Removal of antlers in male reindeer and red deer reduces their fighting ability and dominance status. This indicates that antlers are important in competition among males, even if other factors such as age, body size, and condition also play a role (e.g., Espmark 1964; Lincoln 1972; Prowse et al.

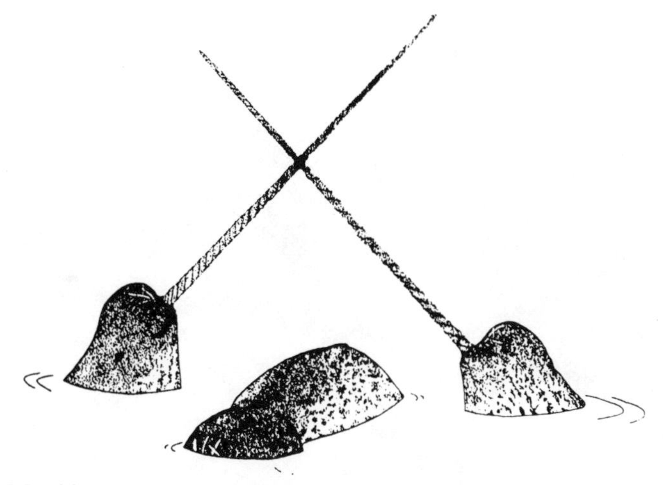

Figure 12.2.1 Male narwhals use their tusks in contests with other males. (From Silverman and Dunbar 1980)

1980; Clutton-Brock et al. 1982). Horns and antlers are used in fights not only in species with large, well-developed weapons, but also in forms with small, simple spikes (e.g., Geist 1964; Barrette 1977). Field studies of African elephants *Loxodonta africana* also show that the tusks are used in (sometimes lethal) fights between bulls (e.g., Poole 1989b).

One of the most remarkable of all hornlike organs is the 2.5 m long tusk of the male narwhal *Monodon monoceros* (figure 12.2.1). Some of its suggested functions are to stir up benthic prey, pierce prey, pierce ice, transmit sound, thermoregulate, and ward off predators. The function with best empirical support is, however, use in display or fights with other males (Darwin 1871; Silverman and Dunbar 1980). Narwhals often cross tusks, and many males are full of scars in the forehead; such scars are rare in females and subadult males. Adult males have been found with a broken tip of the tusk from another male jammed and embedded in the jaw, suggesting that the tusk is used as a weapon in fights among males (Silverman and Dunbar 1980).

The presence and supposed success of antlerless "hummel" stags in some deer has been raised as evidence against the importance of antlers as weapons (e.g., Darling 1937). No study has, however, compared the relative success of hummels versus antlered males, and the reasons why some stags lack antlers are not known (Clutton-Brock et al. 1982). Possibly, hummels represent an alternative reproductive tactic (Gadgil 1972; section 16.1 below), but lack of horns might simply be a nonadaptive developmental response to poor growth conditions (Lincoln and Fletcher 1977).

Figure 12.2.2 Antlers reach extreme size and complexity in some deer, such as the highly branched antlers of the extinct bush-antlered deer *Eucladoceros dicranios*. (From Rütimeyer 1881; a broken tine is indicated in the drawing)

A third objection against a primary importance of horns and antlers as weapons is their size and shape: had hornlike organs been weapons for stabbing an opponent, a single or paired sharp spike might be expected, not horns and antlers with complex, diverse shapes (figs. 12.1.1, 12.2.2; e.g., Colbert 1955). Field studies have shown, however, that horns and antlers are used not only as weapons for stabbing, ramming, or wrestling an opponent, but also as protective shields for catching blows and gripping the rival's weapons. Complex antlers and horns therefore have important defense as well as attack functions; adaptation for defense may partly explain their complexity (e.g., Walther 1958; Geist 1966, 1978; Clutton-Brock 1982).

Additional evidence that horns and antlers are serious weapons is that areas with particularly thick skin in adult males cover parts of the body that receive blows in male fights. In herbivores where males strike with teeth or feet instead of horns, such as kangaroos, vulnerable areas of the body are also covered with thick skin (Jarman 1989; also see Jarman 1983).

To sum up, field work in many species shows that horns, antlers, and other similar weapons are often used in fights between males during the mating season. This applies also for large, complex antlers such as those of red deer. Male horns and antlers clearly often have a main function as weapons for attack and defense against rivals; they probably evolved in large part owing to this function (see below in this section).

Display of strength and vigor to other Males

If large horns and antlers are efficient weapons in fights between males, they may also be used in display for intimidating opponents (e.g., Beninde 1937; Geist 1966; Lincoln 1972; Gould 1974). The agonistic broadside display of ungulates with small horns is replaced in forms with large horns or antlers by display emphasizing the head and its weapons (Geist 1966). Horn and antler size is correlated with body size and nutritional condition, and hence probably also with strength (e.g., Huxley 1932; Suttie 1980; Suttie and Kay 1983; Clutton-Brock et al. 1982; Ullrey 1983). Males might therefore assess each other in part by the size of horns and antlers, avoiding fights with superior competitors (e.g., Geist 1966, 1971b; Clutton-Brock and Albon 1979; Clutton-Brock 1982).

Among the evidence put forth to show that large antlers function as display organs is the following:

1. Antlers are used in aggressive display,
2. Loss of antler parts leads to more frequent challenges and drop in dominance,
3. Antler size correlates with dominance or fighting ability.

All three kinds of evidence have been questioned (Clutton-Brock 1982). For example, it is not clear that antler size affects fighting success or dominance in the wild, because other factors such as age and body mass have not been fully controlled. Body mass is perhaps the decisive factor, with which antler size is correlated (Clutton-Brock et al. 1979; Suttie 1980; Clutton-Brock 1982). Dominance rank is closely related to body mass in captive reindeer (Espmark 1964) and red deer. When Suttie (1980) removed the antlers of all stags, dominance ranks did not change. In addition, antlers may not provide the best basis for assessment of an opponent. In red deer, roaring ability in contests among stags more reliably mirrors male condition, which declines over the rut as stags loose mass. The parallel walk display of rival stags may also permit assessment of body condition (Clutton-Brock and Albon 1979; Gibson and Guinness 1980b; Clutton-Brock et al. 1982).

A field study of a population of caribou *Rangifer tarandus caribou* suggests, however, that antlers may play a role in male assessment of competitors. Barrette and Vandal (1990) distinguished between fighting and sparring. Fighting is very violent, occurs between matched males, and can cause injury and death. It was rare (6 cases) compared to sparring matches (1308 cases), which usually start slowly and are gentle, the animals carefully adjusting their antlers before pushing and twisting. In 713 sparring matches between bulls with unequal antlers, the animals with smallest antlers initiated about 50% of the matches, but withdrew in almost 90% of them. Barrette and Vandal (1990) suggest that frequent sparring is a form of tactile assessment of fighting ability, which might later permit visual assessment of other males based on their antler size.

Female Choice in Relation to Male Weapons

Female choice of mate in ungulates might be based in part on horn size (e.g., Darwin 1871; Geist 1971b). Choosing a male with large weapons might lead to immediate benefits, such as a reduced risk of being courted by young, inexperienced males with inept mating behavior that may harm the female (Clutton-Brock et al. 1982). Hinds seem to avoid young stags (Gibson and Guinness 1980a,b). Choosing a male with large antlers that reflect good condition, a female might also make it likely that her offspring will be of high quality, to the extent that it is heritable (section 3.5). There is, however, no strong evidence in mammals that females choose a mate in relation to the size of his weapons (Clutton-Brock 1982).

Changes in group membership by red deer hinds did not suggest that antler size was important for female choice (Clutton-Brock et al. 1982). Gibson and Guinness (1980a,b) found no evidence that hinds actively select breeding partners among the mature stags. There was a positive correlation between the number of antler points and male success in holding hinds, but the number of points is closely correlated with body mass and age, so the real cause may be differences in male fighting ability rather than female choice (Clutton-Brock et al. 1982). The idea that females choose a mate in relation to the size of his horns or antlers remains to be substantiated.

The russian palaeontologist L. Davitashvili (1961, cited in Gould 1974), in a book on the theory of sexual selection suggested that the antlers of Irish elk *Megaloceros giganteus* (figure 12.1.1) were primarily display organs used for attracting females or deterring rivals. He made the same suggestion for some of the puzzling structures of extinct vertebrates, such as the sail of pelycosaurian reptiles (e.g., *Dimetrodon*) and the frill of ceratopsian dinosaurs (e.g., *Triceratops*). These organs have often been suggested to function in heat regulation and defense against predators, respectively (e.g., Romer 1966). The idea that they were favored by sexual selection was also expressed by Bakker (1986; also see Ostrom 1986). This prediction—or, really, postdiction—might be tested to some extent if a sufficient number of fossils can be sexed. If the puzzling structures are much smaller or absent in females, sexual selection is a likely explanation. This does not necessarily imply that the traits were directly favored by male contests or female choice, however. If they were sex dimorphic, they may have been so owing to sex differences in, for example, predation risk or foraging.

Unfortunately, samples are rarely large enough to permit identification of the two sexes of fossil reptiles and amphibia, but in a few cases there is suggestive evidence of sex-dimorphic weapons in fossile reptiles several hundred million years old (Olson 1969).

The Evolution of Horns in Ungulates

How did horns and antlers get their diverse shapes and sizes? They include such extremes as the coiled horns that form over 10% of the total body mass in certain mountain sheep (figure 12.2.3), and the complex, enormous antlers of some pleistocene extinct cervids, for instance the bush-antlered deer *Eucladoceros dicranios* (figure 12.2.2) and the Irish elk *Megaloceros giganteus* (fig. 12.1.1; Geist 1971a; Gould 1974).

Geist (1966, 1971a, 1978) suggested the following picture of the evolution of horns and antlers. In small mammals, fights often involve wrestling and rolling on the ground; the teeth are important weapons. With larger body mass, the risk of injury from a fall increases, so it becomes important to remain on the feet. But larger mass also makes use of the head more effective for striking an opponent (also see Gaulin and Sailer 1984a). Alternatively, the tail can become a formidable weapon, as in extinct glyptodontid edentates and certain dinosaurs (e.g., Bakker 1986).

The first stages of mammalian horns may have been small knobs on the skull, which helped concentrate the energy of a blow onto a small area and make it more painful to the rival. As the weapons grew larger and more efficient, better defense became important. One possibility is to develop thick, protective fur or dermal shields; another is to prevent or catch the rival's blow with a counterblow. This requires a thick skull and suitable horn shape. Early stages of this kind of defense are found in *Bos* and *Bison* (figure 12.2.3, Geist 1966). The final stage is represented by the (partly convergent) evolution of large horns and antlers in sheep, antelopes, and deer. Following Walther (1958), Geist (1966) suggested that the complex curvature and ridges of antelope horns, and the branched structure of antlers, permit catching the weapons of the opponent and preventing attacks from the side. The contest becomes a less dangerous wrestling match where strength is decisive. Another path, toward ramming contests, has been followed in sheep, the heavy horns of which give further impact to the blow. Geist (1966) suggested that the weapons in all three lines have also assumed a display function (figure 12.2.3).

Also canines and other teeth are mammalian weapons that often seem to be sexually selected. In many primates, males have much larger canines than females, probably at least in part owing to sexual selection (Harvey et al. 1978; section 4.3 above). Tusks, which are smaller or absent in females, have evolved in many families (e.g., Odobenidae, Suidae, Cervidae,

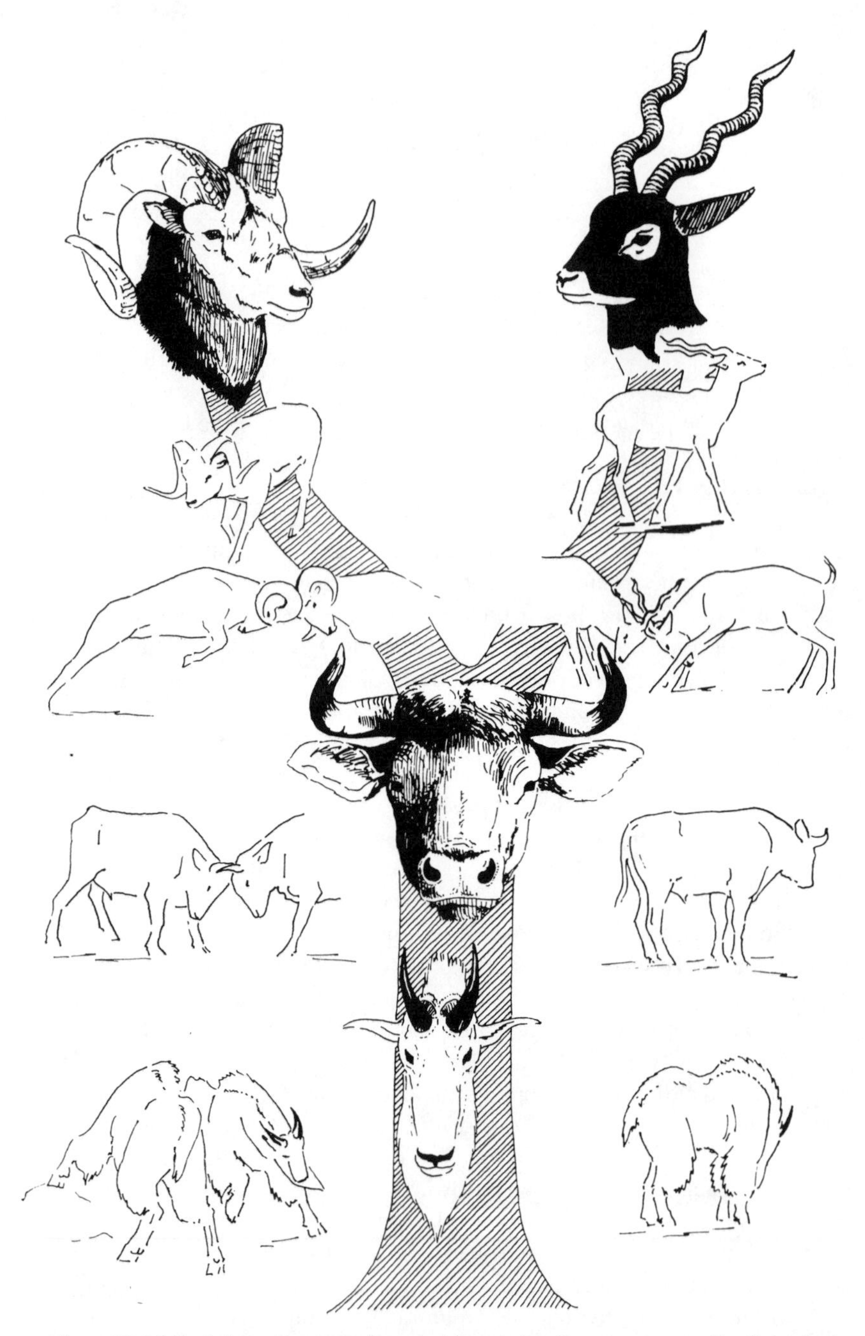

Figure 12.2.3 Evolution of horns and combat form in bovids, as suggested by Geist. In the early stages, exemplified by mountain goat *Oreamnos* (*bottom*), the horns are short and sharp, the skull thin and fragile. Contest engagement is lateral, and broadside display is prominent. Next, head to head combat evolves as a defensive mechanism. The skull is heavy, and horns curve out to hold the opponent (*Bos*). The most advanced development is represented by two horn and combat forms. Wrestling is illustrated by *Antelope*, where the strength of opponents is tested in frontal pushing matches. The other combat tactic, ramming, is found in sheep *Ovis*. (From Geist 1966)

Tragulidae, Elephantidae). In pigs, cervids, and elephants, the tusks are known to be used in contests between males (e.g., Geist 1966; Poole 1989b). Also the bizarre tusks of male *Babyrousa babyrussa*, a pig endemic to some of the Indonesian islands, seem to be used in fights among boars (figure 12.1.2, MacKinnon 1981).

Large Male Weapons in Pleistocene Mammals

Members of the Pleistocene northern mammal fauna carried some of the most spectacular horns, antlers, and tusks ever evolved, such as the 3 m long straight tusks of the mastodon *Anancus arvernensis*, which were as long as the rest of the body; the 2 m frontal horn of the giant rhinoceros *Elasmotherium sibiricum*; the branched antlers of the bush-antlered deer *Eucladoceros dicranios*; the huge antlers of some moose *Alces*; and the celebrated 3 m antler span of the Irish elk *Megaloceros giganteus* (figures 12.1.1, 12.2.2). In spite of their enormous weapons, some of these species were no larger than related forms today. The Irish elk was about the size of a moose, and the mastodon *Anancus* was smaller than present-day elephants.

Steppe mammoths *Mammuthus trogontherii* had curved tusks that sometimes exceeded 5 m in length (Osborn 1942). Males in many proboscids, including the woolly mammoth *M. primigenius*, have relatively larger tusks than females (Osborn 1942). This sex dimorphism suggests that tusks in male proboscids were at least in part sexually selected; field studies confirm that tusks are important weapons used in fights among elephant bulls (e.g., Poole 1989a,b).

What were the reasons for the evolution of huge male weapons during the Pleistocene? Geist (1971a,b) suggested that large weapons were favored in species colonizing recently glaciated land. Some time after glacial withdrawal, the new ground offers abundant high-quality forage, permitting body growth that approaches limits set by physiology rather than nutrition. Several other aspects may also have been important. Ungulates in the temperate north experience highly seasonal food abundance. Food is superabundant during the summer, the main growth period, so the development of large horns and antlers may be less constrained by energetics in temperate areas than in the tropics (Geist 1977a). In addition, the breeding season is short in the north. A dominant male that defends a group of females for some weeks during the rut may be able to fertilize many or even most of the females, as each copulation takes little time (Gibson and Guinness 1980b). The sexual-selection advantage of male weapons may therefore be stronger in temperate than in low-latitude areas, in species in which a male can control access to several females.

Ungulates with large male weapons and body size seem to be geologically less long-lived than forms with smaller weapons and body size (Simpson 1953). Geist (1974) suggested the reason is that evolution of large weapons is irreversible. Males with the largest weapons may have a mating advantage even if the population declines. The same argument may apply for large body size (section 11.1). Owing to genetic correlations between the sexes, selection for large size in males may also increase the size of females over their ecological optimum (see section 11.10). There is evidence from male ungulates that large body size leads to higher mortality during energetic stress (Clutton-Brock et al. 1982, 1985). This might partly explain why forms with large sex-dimorphic weapons and body size seem more prone to go extinct than smaller forms (see also Simpson 1953; Lande 1980; Maynard Smith and Brown 1986).

Allometric Relationships

Some of the pleistocene mammals were larger than extant related species. If weapons show allometric growth ($y = bx^a$; Huxley 1932; Reiss 1989; Harvey and Pagel 1991) with allometry constant $a > 1$, these species are expected to have larger relative weapon size than living related forms. The question then is why weapons should grow in such fashion.

Gould (1974) showed that there is strong positive allometry (with $a \approx 1.7$) of antlers with body size among deer (figure 12.2.4). The Irish elk falls on the extrapolated curve based on other cervids. It might be argued that the large relative antler size in Irish elk need not have been selected for as such; it could simply arise from selection for large body size which, through preexisting growth mechanisms, inevitably leads to larger relative antler size. As allometric relationships can be changed by selection, this would, however, imply that the large antlers were not selected against. Gould (1974) pointed out that developing 40 kg of antlers each year is likely an energetic burden, which should favor smaller antlers. He suggested that large antlers were selected in other ways, for example as display organs in dominance contests among stags (also see Wallace 1987; Harvey and Pagel 1991). Geist (1986) proposed that they were favored by female choice. Based on observations of many serious fights in red deer, fights that had previously been described as rare and nondamaging, Clutton-Brock (1982) on the other hand suggested that Irish elk might also have used its antlers as effective weapons in fights.

An allometric relationship is no complete explanation, and does not preclude the possibility that the pattern is an outcome of selection (Eberhard and Gutierrez 1991; Harvey and Pagel 1991). For a given body size, antler size increases with the size of the breeding group, perhaps as a consequence of stronger competition over females in larger groups (figure

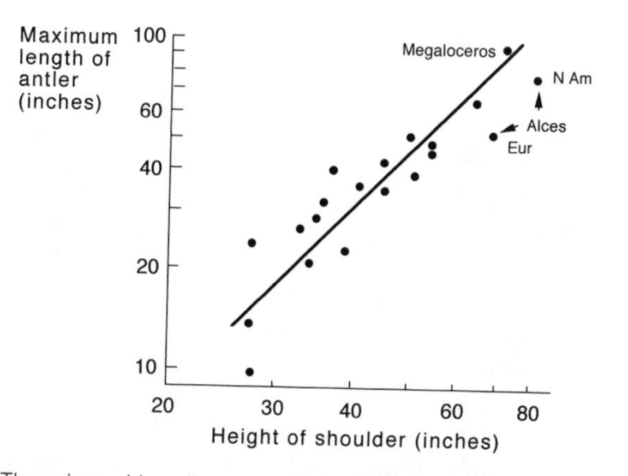

Figure 12.2.4 There is positive allometry of antlers with body size in cervine deer. The regression line is based on reduced major axis analysis of the species values. The Irish elk (*Megaloceros*) is close to the predicted antlers for its body size, whereas European and American moose (*Alces*) have smaller antlers. (After Gould 1974)

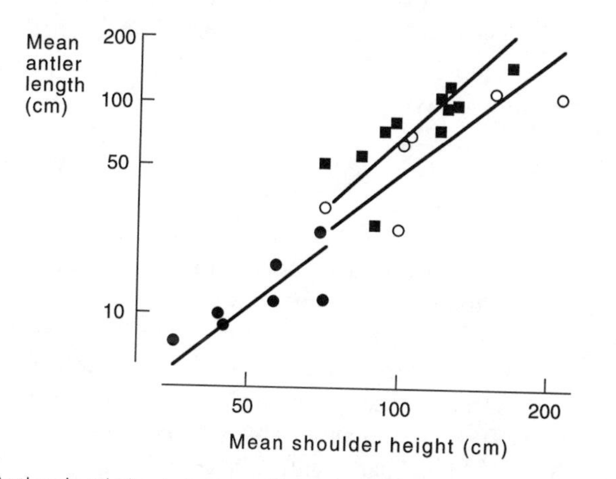

Figure 12.2.5 Antlers in relation to body and breeding group size in deer (Cervidae). Antler size increases not only with body size, but also with the typical size of breeding groups of the species. Breeding group sizes: dots, < 2; open circles, 3–5; squares, > 5. Lines based on major axis analysis. (After Clutton-Brock et al. 1980)

12.2.5). There does remain a positive allometry when group size is taken account of, but the allometry relation is far from perfect. The largest relative horn and antler sizes are found in species of intermediate body size, not in the largest species (Jarman 1983). These and other results suggest that the variation in antler size among deer is in part related to sexual and other selection factors. For discussion of intraspecific allometric growth of secondary sex traits, see, for example, Reiss (1989), Harvey and Pagel (1991), Green (1992), Petrie (1992a).

12.3 Avian Spurs

Horns, antlers, and tusks have their counterparts in birds in the leg or wing spurs found in some families (reviewed by Davison 1985). Leg spurs occur in 113 species in three subfamilies of Phasianidae among Galliformes (Phasianinae, Numidinae, Meleagridinae), and wing spurs in a number of aquatic birds of five families (Anhimidae, Anatidae, Jacanidae, Charadriidae, and Chionidae). More than two thirds of the species with leg spurs are monogamous. Davison (1985) proposed that spurs originated and were favored initially among monogamous species through competition over resources other than females. However, 46 out of 125 monogamous species lack spurs in the three subfamilies of Phasianidae mentioned above, whereas only 2 out of 36 polygynous species do so. Polygynous species have the longest and sharpest spurs. In addition, in 97 out of the 113 species with spurs, they occur only in males, but never only in females. These patterns suggest that competition over females, or resources needed to attract them, has been crucial in the evolution of spurs in Phasianidae. A proper test should, however, be based not on species numbers, but on a more adequate comparative method (see chapter 4.3, and Sullivan and Hillgarth 1993).

Wing spurs are slightly larger in males than females, except in polyandrous jacanas, where they are slightly larger in females (Davison 1985).

Sexual selection of spurs has been studied in detail in the pheasant *Phasianus colchichus*. In an introduced population in Sweden, females favored males with long spurs, but surprisingly, long spurs did not seem to be selected by male contests (von Schantz et al. 1989; Göransson et al. 1990; Wittzell 1991a,b). In an experiment where the spurs of seven males were shortened, those of another seven were elongated, and further eight males were used as controls, dominance rank was not affected. Nor did male dominance in the field correlate with spur length, attraction of females, or reproductive success. Spur length in the experiment as well as in a larger sample of observational data was, however, correlated with the attraction of females, and with reproductive success. Contrary to the prediction from Fisherian models, male survival increased with the size of the preferred trait. Female reproductive success was also correlated with the spur length of her mate. The mechanism producing this effect is unknown, as is the genetics of spur size and female preference.

Hillgarth (1990b) found no female preference for male spur length in British pheasants. One possible reason is the difference in criteria: she measured female copulation solicitation displays in mate choice pens, whereas von Schantz et al. (1989) took female presence less than 100 m from the male in the wild as a measure of female preference (see Hillgarth 1990b and references therein for discussion).

12.4 Beetle Horns and Other Arthropod Weapons

No less remarkable than the antlers and horns of ungulates are those of some Lamellicornid and other beetles: in some species the horns are longer than the rest of the body (figure 12.4.1, Hercules beetle). Darwin (1871) remarked that "if we could imagine a male *Chalcosoma* . . . with its polished, bronzed coat of mail, and vast complex horns, magnified to the size of a horse or even a dog, it would be one of the most imposing animals in the world."

Various parts of the body have become the site of horns or hornlike structures in beetles: head, prothorax, thorax, and even elytra (wing covers). In many species, for example stag beetles *Lucanus* sp., the "horns" are enlarged mandibles (figure12.4.2). Females usually have smaller or rudi-

Figure 12.4.1 Two male Hercules beetles *Dynastes hercules*, a large member of Scarabeidae, fighting over a female on the floor of a South American rain forest. Fights may lead to injuries, as each male tries to lift the opponent and slam him to the ground. (Drawing by Sarah Landry in Wilson 1975. Copyright © 1975 by the President and Fellows of Harvard College)

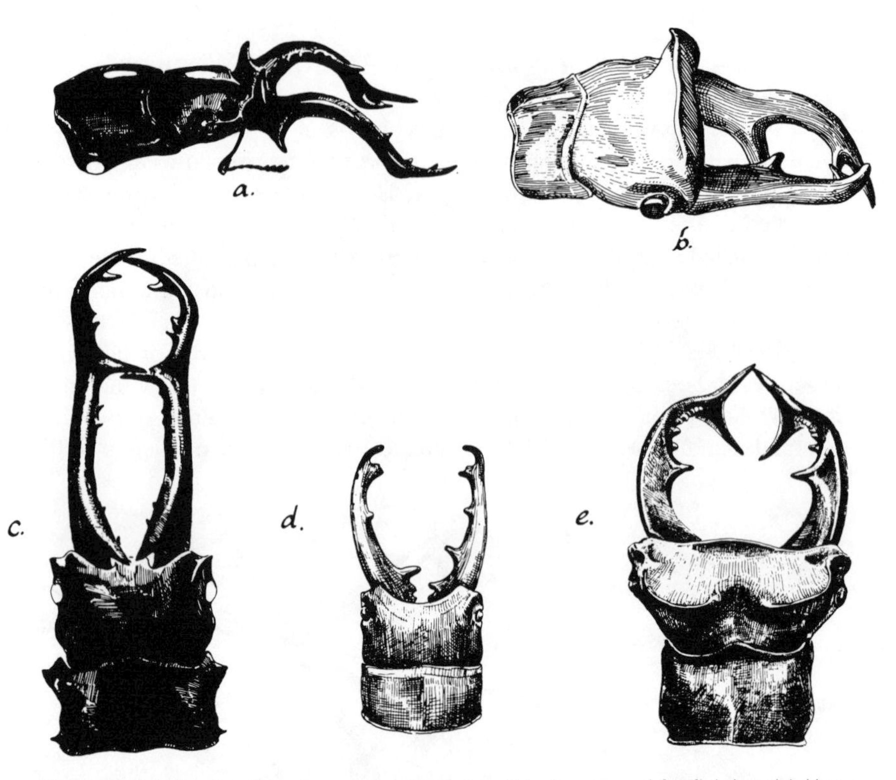

Figure 12.4.2 Stag beetles show a variety of mandible types used for fighting. (a) *Hexarthrius fosteri*, (b) *Cyclommatus kaupi*, (c) *Cladognathus giraffa*, (d) *Leptinopterus tibialis*, (e) *Lucanus elephas*. (From Otte and Stayman 1979)

mentary horns, but in some cases there is no sexual dimorphism. Lack of sex dimorphism might indicate that the horns are not favored by sexual selection, that sexual selection is equally strong in both sexes, or that it is stronger in one sex but the genetic correlation between the sexes has not been broken up. In, for example, the burrowing beetle *Coprophanaeus ensifer*, both sexes are horned and use the weapons in fights (Otronen 1988b).

Beetle horns have provoked much speculation, reviewed by Eberhard (1979, 1980b) and Otte and Stayman (1979). Referring to a lack of fights between males, and to large variation in horn size within and among species, Darwin (1871) proposed that the horns are ornaments favored by female choice. Other authors have suggested that the horns are weapons for defense against predators or other males, or tools for digging, or for perforating plants to feed on sap (see Eberhard 1979). In his monograph on horned beetles, Arrow (1951) suggested that horns lack function and are neutral outcomes of allometric growth, coupled with selection for larger

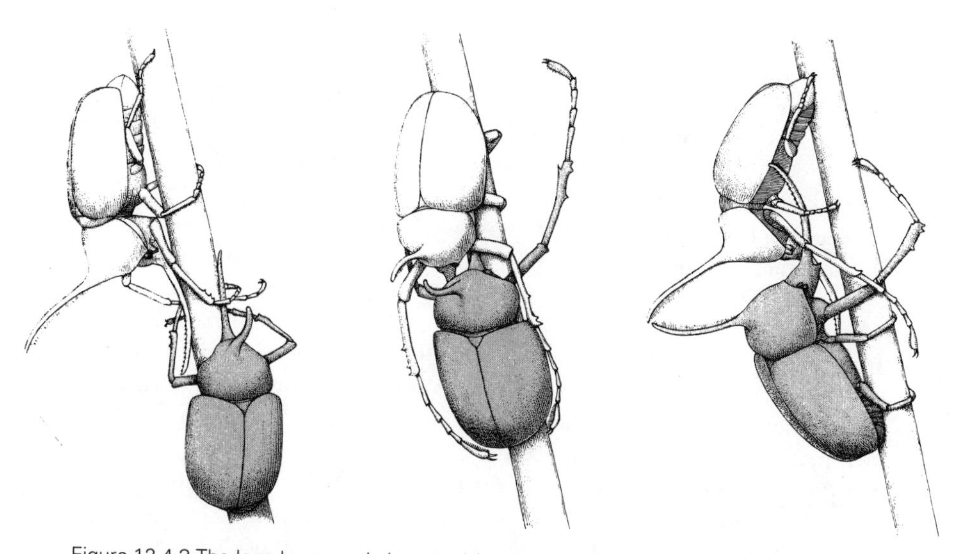

Figure 12.4.3 The long horns and elongated front legs of the scarab beetle *Golofa porteri* serve as male weapons in contests over shoots of the plants on which this species feeds. When he meets a rival (*left*), the male inserts his head horn under the body of the rival and wraps the long front legs around him (*middle*). The attacker (*right*, shaded lower male) jerks his head up to throw the opponent off the stem. (From Eberhard 1980b. Drawing by T. Prentiss; Copyright © 1980 by Scientific American Inc.)

size in some species. This explanation is highly unlikely (see Eberhard and Gutierrez 1991).

In Darwin's time there were few detailed observations of beetle behavior in the wild. It has since become clear that males in most of the horned species studied do fight (e.g., Beebe 1944, 1947; Mathieu 1969; Palmer 1978; Eberhard 1979, 1980; Hamilton 1979; Conner 1989b). Observations of the horns being used for foraging or defense against predators are still rare or absent, so these two suggested functions are refuted as general explanations for beetle horns.

Eberhard (1979, 1980) clarified how the horns are used in fights in seventeen species of beetles. Most horns fall in two broad functional categories: they are used either to pry up and push a rival off his site (figure 12.4.3), or to grasp, lift and drop him to the ground (figure 12.4.1). In spite of their superficial similarity to the horns or antlers that ungulates use for stabbing, wrestling, or ramming the rival, beetle horns therefore are used rather differently. Eberhard (1980) suggested this is because the hard armorlike cuticle of a beetle makes stabbing or ripping less effective than in soft-bodied ungulates. Ramming or stabbing also requires swift attacks, but beetles are relatively slow and clumsy. To this explanation can be added the huge differences in body size: ungulates are 10^3–10^7 times as heavy as most horned beetles. Body mass increases with body length cubed, whereas the cross-sectional area of muscles and tendons, and hence their tensile strength, increases only with length squared. As a consequence, relative athletic ability tends to increase with reduced body size, and beetles are strong enough to use the head to pry up and lift a rival, a feat that is out of the question for most ungulates as they lack the necessary strength.

The function and selection of horns has been studied in detail in the North American forked fungus beetle *Bolithotherus cornutus*. This species spends its entire life cycle on or near fungi growing on tree bark in deciduous forest (Brown and Rockwood 1986; Conner 1988, 1989a,b). Like many other beetles, it varies much in body size and even more so in horn length (section 16.2). The variation according to Brown and Rockwood (1986) is phenotypic, not genetic. The paired horns are used in fights when two males meet on a fungus, ending when one of them retreats or falls to the ground. Males also use the horns to pry a copulating rival off the female. To examine the importance of relative horn size, Brown and Bartalon (1986) placed pairs of males of equal body but unequal horn size together with a female and food in a terrarium. The male with longest horns tended to court the female a higher proportion of the time. This and other evidence suggested that interactions between males favored the one with larger horns; there was no suggestion that females choose males in relation to horn size (Brown et al. 1985). Sample sizes were small, however, and Brown and Bartalon (1986) emphasized that further experiments are desirable.

Conner (1988, 1989b) reached similar results in a field study of three different populations followed for three years. Selection gradient analysis (section 4.5) showed that larger horns were favored directly by sexual selection. Six estimates all showed a positive correlation between male horn length and fighting success. Higher access to females by long-horned males for unknown reasons was more pronounced at low population density. Sexual selection of horn size therefore was density dependent. Zeh (1987b) described a similar case, but with opposite effect of population density.

Why have horns evolved repeatedly among beetles but not among other insects? Arrow (1951) pointed out that adult beetles in some species eat no solid food. Selection of mandibles for feeding efficiency is then reduced: it is mainly among such species that they have evolved into large weapons. In addition, many beetles are adapted for digging, and for a life in relatively enclosed spaces. The associated traits, such as a thick cuticle and short legs, make kicking and biting ineffective. Instead, ability of prying or pushing competitors should be important (Eberhard 1979; also see Otte and Stayman 1979).

In addition to the enlarged horns and mandibles in male beetles of many species, sex-dimorphic weapons occur in other insects such as fig wasps, bees, flies, pentatomid bugs, and earwigs (reviewed by Hamilton 1979; Thornhill and Alcock 1983; Wirtz et al. 1988; Eberhard and Gutierrez 1991; also see section 16.2). In the thrips *Holothrips pedicularis* and *Elaphrothrips tuberculatus*, males fight by grasping and stabbing with their forelegs, which are enlarged compared to those of females. Large males tend to win fights and become dominant at oviposition sites, and obtain most of the matings (Crespi 1986a,b).

Fiddler crabs (e.g., *Uca*) are well known for their males having one of the two claws greatly enlarged, using it as a signal and weapon in contests over mates and breeding burrows, and also in attracting females (Crane 1975; Christy and Salmon 1991).

An experimental study of pseudoscorpions by Zeh (1987a,b) also demonstrated an advantage of large male weapons in contests over females. In *Dinocheirus arizonensis*, males fight over females by grasping and pulling each other's chelae; male fighting success increased with chela size, and so did male reproductive success (Zeh 1987b). The advantage was present only at high population density, however. Similar density effects have been found in dung flies *Scatophaga stercoraria* (Borgia 1979) and wood frogs *Rana sylvatica* (Howard and Kluge 1985; also see Ghiselin 1974). In less dense populations, female pseudoscorpions tend to be larger than males, perhaps because small male size is favored by shorter maturation time when scrambles to find females are more important than contests (Zeh 1987a,b).

A puzzling character that seems to be favored by male contest competition is long eye stalks, which have evolved at least nine different times among Diptera (e.g., McAlpine 1979). These flies have their eyes on the ends of stalklike projections that may be almost as long as the rest of the body. Rival males face each other with spread wings, sparring or grabbing the opponent with the forelegs (figure 12.4.4). During these contests, males can compare their eye span, which is closely related to overall body size and, hence, probably also to strength. McAlpine (1979) therefore proposed that long eye stalks may help impress rivals. Borgia (1979) and Spieth (1981) suggested that large head size has been favored for similar reasons in certain *Scatophaga* and *Drosophila* flies. Behavioral observations in the Asian stalk-eyed fly *Cyrtodiopsis whitei* (figure 12.4.4) show that males, which have longer stalks than females, display their eye stalks and forelegs in contests with rivals over females at mating sites. Long eye stalks seem to increase success in such contests, but female choice also favors long stalks (Burkhardt and de la Motte 1987, 1988).

Some spiders also have stalk-eyed males. An extreme case is *Walckenaera acuminata*, where the eyes of the male sit on a slender spire at the front of the cephalothorax (Bristow 1929). Its function appears not to have been studied (e.g., Foelix 1982).

Figure 12.4.4 Male stalk-eyed flies *Cyrtodiopsis whitei* during contests face the rival eye to eye, sparring and grappling with the forelegs. (From de la Motte and Burkhardt 1988)

12.5 Summary

Among the explanations for sexually dimorphic horns, antlers, tusks, and spurs, the empirical support is strongest for the idea that they have evolved and are favored in males as weapons in contests over females. Horns may also play a role in display of strength and condition to rivals and perhaps also in female mate choice, but this remains to be shown. Sex-dimorphic weapons in some species may have additional functions, for example in defense against predators. Also in beetles, the use of horns in fights over females is the function for which there is best evidence. Sex-dimorphic weapons in a variety of other arthropods and vertebrates play similar roles.

13 Coloration and Other Visual Signals

13.1 Introduction

Conspicuous colors, feather plumes, enlarged fins, and other visual signals have played a prominent role in the theory of sexual selection since Darwin (1871), Wallace (1889), Poulton (1890) and others began debating their functions (1.5). Darwin claimed that female choice has generally been responsible for the evolution of bright male colors and other sex-dimorphic signals. The debate persists today; this chapter reviews evidence on the various mechanisms that favor visual secondary sex signals.

Some of the most convincing studies of sexual selection have shown by combinations of field observations and experiments that such traits improve mating success. But comparative evidence and patterns of usage of the signal can also be revealing, as made clear by Darwin (1871; also see Ghiselin 1969b). Although more than a century old, the simple kinds of evidence he put forth are still relevant, in the era of sophisticated experiments and computerized multivariate analyses. Their usefulness in clarifying selection pressures has been far from exhausted, especially if combined with modern comparative methods (e.g., Harvey and Pagel 1991).

Darwin (1871) pointed to patterns suggestive of sexual selection in the expression of visual display in relation to sex, age, season, and situation. For example:

1. The trait is not aquired before sexual maturity;
2. Females develop the trait less than males, or not at all;
3. Males develop the trait only during the mating season;
4. The trait is displayed mainly to potential mates or sexual rivals.

Taken together, such patterns suggest sexual selection. Point 4 is particularly revealing; the usage of a trait can often show whether it is favored by mate choice, contests over mates, or other mechanisms (e.g., Darwin 1871; Butcher and Rohwer 1989). Quantification of the context in which a trait is used is a powerful means of clarifying its function, a means that seems not to have been used to full potential in the study of sexual selection.

The hypotheses reviewed in chapter 1 (e.g., table 1.4.1) are relevant for most traits discussed here. Broader reviews of the communicative functions of conspicuous colors and other visual signals are found in, for example, Hailman (1977), Smith (1977), Burtt (1979), and Butcher and Rohwer (1989). A particularly detailed study of the selection of coloration has been carried out in the guppy by Endler (e.g., 1978, 1983, 1991, 1992b; also see section 5.3 above). He analyzed coloration in relation to environmental aspects such as ambient light and water transmission spectra, courtship and predator attack distances, as well as the visual characteristics of predators and prey, also considering which colors are conspicuous under what circumstances, an often neglected problem (also see Marchetti 1993). The degree of conspicuousness should preferably be measured at the relevant time in the natural environment, or the conclusions may be misleading (Endler 1990, 1991). The previously common approach of scoring visual conspicuousness from paintings in field guides can often be misleading (see section 13.8).

13.2 Invertebrates

The sexes are similar in many brightly colored ascidians, echinoderms, planarians, nemerteans, and molluscs. With exception mainly for arthropods and cephalopods, which use conspicuous visual signals in courtship and threat (e.g., Moynihan and Rodaniche 1977), many invertebrates at most have simple light-sensitive organs and probably lack color vision. Intraspecific signaling therefore seems unlikely to explain their coloration. Darwin (1871) proposed that it is an incidental by-product of the chemical composition or fine structure of their tissue. Aposematism is a more likely reason in some cases (e.g., Edmunds 1974; also see Guilford 1990).

Among crustaceans, there are many examples of sex-dimorphic colors that may be sexually selected, for example in fiddler crabs (e.g., Crane 1975). In the snapping shrimp *Alpheus armatus*, the spines of the uropod are conspicuously black in adult males (Knowlton 1980).

Many spiders are sexually color dimorphic, the male usually being most conspicuous. In one of the earliest studies supporting Darwin's ideas on sexual selection by female choice, Peckham and Peckham (1889) pointed out that the male displays his colors in courtship (figure 13.2.1).

The colors of insects and other animals were reviewed a century ago by Poulton (1890), strongly supporting Darwin's (1871) ideas. Modern reviews of insects are found in Thornhill and Alcock (1983) and West-Eberhard (1984). Sex-dimorphic colors occur in many dragonflies and damselflies (Odonata). In some genera, for example *Libellula*, emerging males are drab like females, but later aquire more conspicuous colors (e.g., Jacobs 1955; Gibbons 1986). West-Eberhard (1984) suggested that drab colors in young males reduce the risks of predation and attacks from older males (also see section 13.6).

In an early field experiment on sexual selection, Jacobs (1955) found

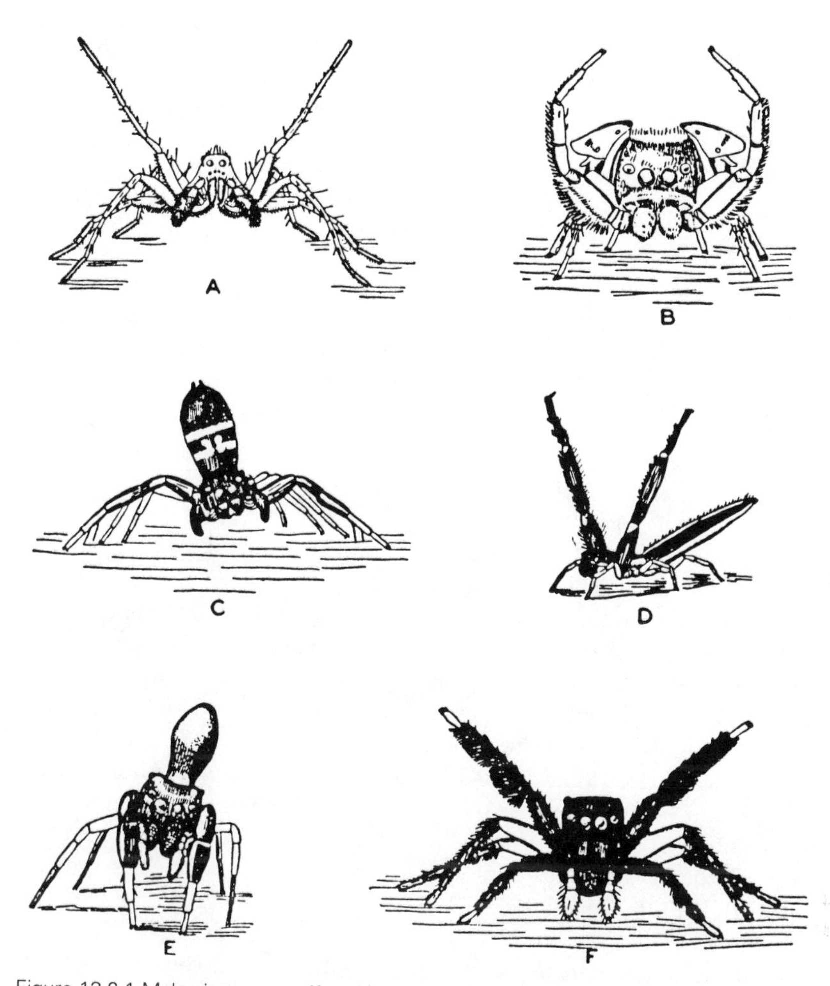

Figure 13.2.1 Males in many wolf and jumping spiders display their color patterns to the female during courtship display. (A) *Pardosa milvina*, (B) *Pellenes viridipes*, (C) *Peckhamia noxiosa*, (D) *Marpissa pikei*, (E) *Peckhamia picata*, (F) *Euophrys monadnock*. (From Gertsch 1979)

that males of the dragonfly *Plathemis lydia* became less able to hold territories if the white dorsal part of the abdomen was painted dark to resemble that of females. Such males had reduced mating success compared to white-painted control males. The white color is apparently favored by sexual contest competition, not female choice: it was exposed in display to other males, not females (Jacobs 1955). Also in several other dragonflies, the male displays his conspicuously colored abdomen in aggressive encounters (e.g., Campanella 1975).

In some genera, for example *Aeschna*, *Lestes*, and *Coenagrion*, females have several color morphs. In some species, one female morph resembles the male. In the damselfly *Argia vivida*, the female morphs do not differ in attractiveness to males, and the factors that maintain the color polymorphism in this and other similar cases are poorly known (Conrad and Pritchard 1989). In *Ischnura*, some of the female morphs seem to be more colorful than males (see, e.g., Gibbons 1986; Sandhall 1987). Odonates show parallels with certain coral reef fishes in that both sexes may be conspicuous yet differently colored; the reasons are not clear in either case (section 13.3).

From a selection analysis based on the approach suggested by Schluter (1988; section 4.5 above), Moore (1990a) concluded that the brown and white wing patches in male *Libellula luctosa* are favored by female choice. The white patch is also favored by male contest competition over territories; so is large body size.

Carrying some of the most conspicuous colors among animals, butterflies (Lepidoptera) are of great interest as regards sexual selection (e.g., Darwin 1871; Poulton 1890; Rutowski 1984; Silberglied 1984; D.A.S. Smith 1984). In butterflies with sex-dimorphic colors, males are usually most colorful, and differ more among closely related species than do females. This pattern speaks against Wallace's (1889) view that males represent a more ancestral state from which females have evolved cryptic colors.

Darwin (1871) was aware that male butterflies sometimes fight and injure each other, but he thought that their colors are favored by female preferences. In mimetic butterflies, both sexes, or only females, mimic the unpalatable model. Male-limited mimicry is rare or nonexistent (e.g., Vane-Wright 1975). One possible reason is that sexual selection prevents the evolution of mimetic morphs in males, either because female preferences are conservative, or because male contests select for species-specific colors (Belt 1874; Turner 1978; Rutowski 1984; Silberglied 1984; Krebs and West 1988). In some other contexts, however, for example speciation by sexual selection, female preferences are not assumed to be conservative, but rather directional or even diversifying. It would therefore be of interest to know whether female mate choice or male contest competition is responsible for the more conservative colors of male butterflies.

Also, in many other insects, some spiders, fiddler crabs, and lizards, color variability within species is greater in females than males (figure 13.2.2). Reviewing the pattern in relation to four hypotheses, Stamps and Gon (1983) concluded that it is best explained by sexual selection. Greater female color variability is most prevalent in species with visual communication. For this and other reasons, several other explanations (higher, frequency-dependent predation risk for females, higher environmental heterogeneity for females, and female heterogamy) seem unlikely to account for much of the pattern. Stamps and Gon (1983) outline additional desirable tests of these hypotheses.

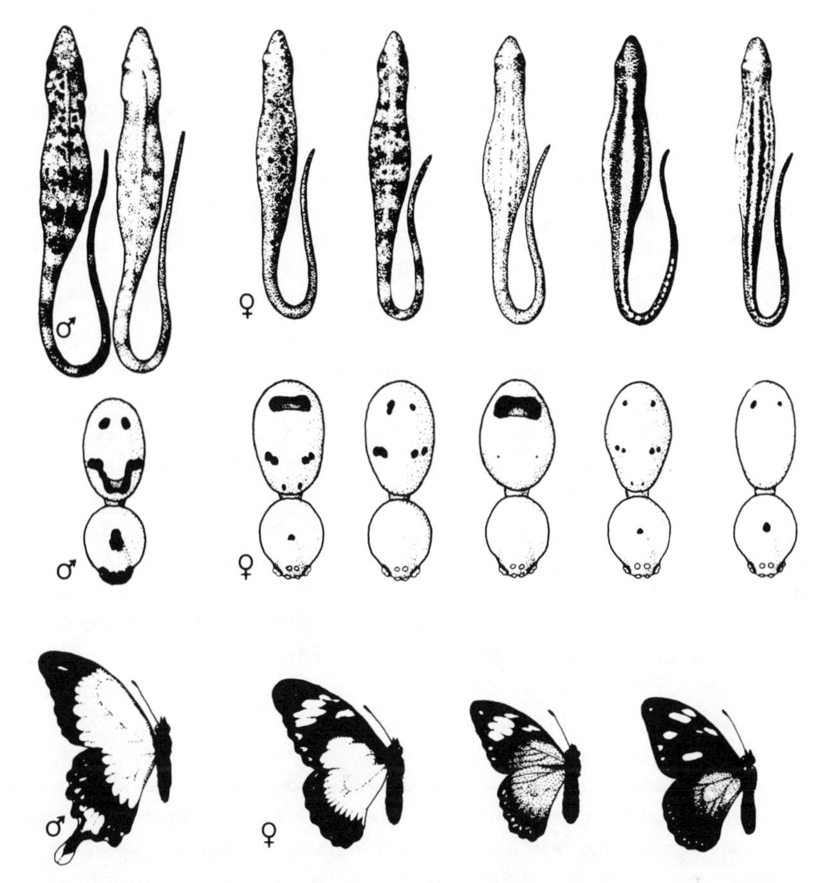

Figure 13.2.2 Females vary more in color than males in some insects and other animals, shown here in a lizard *Anolis aeneus*, a spider *Theridion grallator*, and a butterfly *Papilio dardanus*. (From Stamps and Gon 1983. Copyright © 1983 by Annual Reviews Inc.)

Butterflies have excellent color vision, ranging from ultraviolet to red in the color spectrum (Silberglied 1984). Female butterflies therefore should have the sensory capacity to discriminate among males in relation to color patterns. A painting experiment in *Papilio glaucus* indicated that females respond to male differences in wing colors (Krebs and West 1988). In *Pieris occidentalis*, females choose among males in relation to wing melanin patterns, discriminating against males with reduced melanin areas. This reduces the risk of hybrid matings with the closely related *P. protodice* (Wiernasz 1989; Wiernasz and Kingsolver 1992; section 9.3 above). These results support Darwin's (1871) suggestion that female butterflies choose mates in relation to their coloration (also see D.A.S. Smith 1984), and Belt's (1874) ideas on sex-limited mimicry in butterflies.

Colias butterflies respond to UV-absorbing or reflecting wing patterns in potential mates (Silberglied and Taylor 1978; Rutowski 1982b). Painting of wings in different colors had, however, no effect on male mating success (Silberglied and Taylor 1978). Following Hingston (1933), Silberglied (1984) suggested that conspicuous coloration in male butterflies is selected by contests over mates. There is observational evidence for this view, but strong experimental support is still lacking. The contrasting results call for more tests of the role of male and female colors in contests as well as mate choice; the roles might differ among butterfly species.

13.3 Fishes

A variety of hypotheses have been raised to explain why males, and in many species also females, have conspicuous colors in certain fishes (e.g., Brockmann 1973; Endler 1983). Besides his main explanation (sexual selection by female choice), Darwin also noted that "with some fishes, as with many of the lowest animals, splendid colors may be the direct result of the nature of their tissues and of the surrounding conditions, without any aid of selection." In addition, warning to predators (aposematism) might favor conspicuous colors (Darwin 1871). There is supporting evidence from a few poisonous fishes; this may be a common reason, but more studies are needed (e.g., Edmunds 1974; Ehrlich et al. 1977).

Wallace (e.g., 1889) proposed that bright colors in fishes are cryptic: "It may well happen that many of the colors of tropical fishes, which seem to us so strange and conspicuous, are really protective, owing to the number of equally strange and brilliant forms of corals, sea-anemones, sponges and seaweeds among which they live." Darwin (1871) argued against this idea, for "in the fresh-waters of the Tropics there are no brilliantly colored corals or other organisms for the fishes to resemble; yet many species in the Amazon are beautifully colored." (Darwin's point is only partially correct; some tropical freshwater fishes with bright colors, such as guppies, are found in clear waters with bright background, whereas other species occur in blackwaters; J. Endler, pers. comm.). Even if true in some cases, Wallace's explanation therefore does not hold for all colorful tropical fishes. Nor was Darwin impressed by M'Clelland's (1839, cited in Darwin 1871) suggestion that conspicuous colors in fishes serve as a "better mark for kingfishers, terns and other birds which are destined to keep the numbers of these fishes in check."

Reviewing sexual selection of conspicuous coloration in fish, Noble (1938) concluded that female choice favors the most strongly colored males in several species. He also pointed out that male nuptial colors may have several other functions. In particular, they are used for threatening rivals in many territorial fishes. In such species, females also tend to be conspicuously colored if they defend nest or young. In most school breeders without territories, there is usually little contest competition over females, less pronounced or no change to nuptial colors, and no sexual color dimorphism.

Experimental work on female choice in relation to male secondary sex traits in fish was pioneered by Pelkwijk and Tinbergen (1937), and by Noble and Curtis (1939), who found that females in aquarium tests preferred males with experimentally emphasized nuptial colors over duller males. Possible effects of the manipulation on male behavior were, however, not controlled for.

Lorenz (1962) suggested that conspicuous "poster colors" in coral reef fishes enhance species recognition and reduce territorial strife between different species. He argued that such coloration evolves in these fishes because the diversity of species specialized for different kinds of food makes territorial exclusion of other species unnecessary. Zumpe (1965), Fricke (1966), and Reese (1975) gave supportive evidence.

A religious fundamentalist might like to view the Beaugregory *Eupomacentrus leucostictus* as a practical joke by the Lord on Darwin's theory of sexual selection. Juveniles, females, and nonbreeding adult males are bright yellow and blue; juveniles are particularly bright (Brockman 1973; for other similar cases, see Barlow 1974). Breeding males on the other hand develop dull, gray-blue to dark brown colors. Brockmann (1973) suggested the reasons may be that (1) juvenile as well as adult Beaugregories defend food territories, favoring conspicuous colors; (2) only the adult male defends the nest and offspring, perhaps making him (and then the offspring) particularly vulnerable to predation. It is not clear, however, why defense of territory and offspring should prevent conspicuous colors in male Beaugregory but not in, for example, male stickleback *Gasterosteus aculeatus* (e.g., Rowland 1982b) or pupfish *Cyprinodon pecosensis* (Kodric-Brown 1983). The puzzle remains.

Some of Brockman's (1973) results contradict Lorenz's poster coloration idea. For example, territorial chases were directed against other species as often as against Beaugregories. Comparing a number of butterfly fishes (Chaetodontidae), Ehrlich et al. (1977) found no correlation between conspicuousness and territoriality, and for this and other reasons rejected the poster coloration hypothesis (also see Peterman 1971; Brown et al. 1973). Reviewing evidence from butterfly fishes, Neudecker (1989) concluded that their conspicuous colors are not explained by Lorenz's (1962) hypothesis, as most chaetodontids are not territorial. Nor do the colors seem to be explained by species recognition or crypsis (disruptive coloration). On the other hand, there is much evidence that coloration in butterfly fishes is aposematic, warning predators that these fishes make a low-quality meal. They are bony, deep-bodied, and difficult to swallow. The

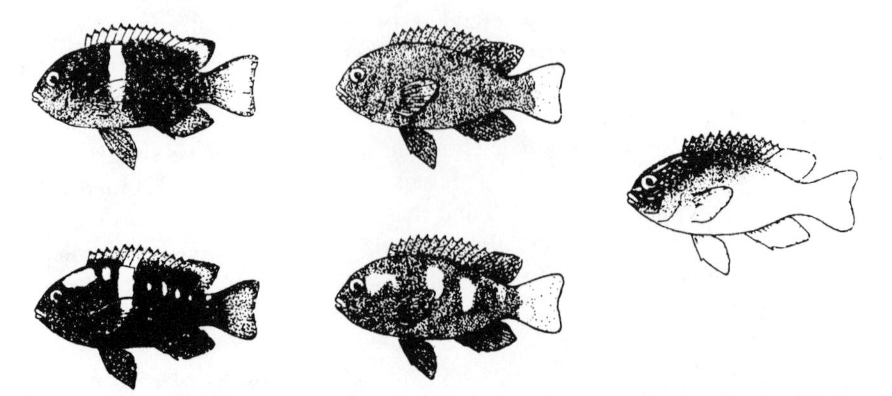

Figure 13.3.1 Normal and courtship colors in males of *Glyphidodontops biocellatus* (left), *G. cyaneus*, and *G. rollandi* (right). In the last species, male color is unchanged during courtship and spawning. (From Thresher and Moyer 1983)

reasons for the conspicuous colors in many other coral reef fishes remain poorly known.

Comparing dimorphism and mating system among tropical labroids, Robertson and Hoffman (1977) suggested that female choice is responsible for the usually brighter colors of males. In species with permanent harems, the sexes have the same coloration, whereas males are more conspicuous in many species where males attract roving females from a distance. In some species the male can change colors rapidly, and shifts to conspicuous colors during interactions with females, not males.

In a detailed study of three damselfishes (Pomacentridae), Thresher and Moyer (1983; figure 13.3.1) concluded that permanent color dimorphism in one species is favored by contest competition among males, and temporary dimorphism in another species by female choice (also see Schmale 1981). In *Chrysiptera cyanea*, the orange color of the caudal fin seems to be favored by female choice (Gronell 1989). More species need to be studied in detail before general conclusions can be drawn.

Bright coloration in some fish may be favored in contests over other resources than mates. For example, the conspicuous "gold" morph of the midas cichlid *Cichlasoma citrinellum* dominates the cryptic morph in contests over food, and grows more rapidly (Barlow 1973).

Experiments in killifish *Notobranchius guentheri* (section 10.1), guppies (section 5.2), and sticklebacks provide strong evidence for female choice of secondary sex ornaments such as enlarged fins and conspicuous colors. Next to the guppy, the three-spined stickleback *Gasterosteus aculeatus* is perhaps the most thoroughly studied fish as regards sexual selection. Several experiments have shown that reproductive females prefer males with the brightest red breeding coloration (e.g., Pelkwijk and Tinbergen 1937; McPhail 1969; Semler 1971), also when male behavioral differences are controlled for (Milinski and Bakker 1990). The sensitivity of the female visual system increases in the red part of the spectrum during the mating season (Cronley-Dillon and Sharma 1968; section 2.7 above). The red coloration of the male is perhaps also favored in contests among males (e.g., Pelkwijk and Tinbergen 1937; Rowland 1982b; Bakker and Sevenster 1983; but see Rowland 1989b). Females prefer large males; even supernormal male dummies 25% larger than normal-sized dummies are preferred. Males also prefer large females (Rowland 1989a,b).

In several stickleback populations from western North America, males lack red coloration, being instead black. Increased risk of detection in areas where predators are common has been suggested to explain the pattern (e.g., Semler 1971; Moodie 1972). This idea now seems unlikely, however. In some cases, convergence in male threat display with another, competing species is perhaps involved in the shift from red to black (Hagen et al. 1980). An extensive study of sixty-six populations showed that loss of red coloration is correlated with heavily stained waters, reduced visibility, and perhaps carotenoid deficiency in the diet. There was no association between color morph and presence or absence of vertebrate predators (Reimchen 1989); further analysis is desirable.

Milinski and Bakker (1990) found that the intensity of red breeding coloration, which varies markedly among males (McLennan and McPhail 1989), reflects his general physical condition. The preference for bright red males should therefore help a female obtain a mate in good nutritional condition and health (also see sections 3.6 and 3.7). Such males will be most likely to provide good paternal care for the offspring, which are tended solely by the father (also see Knapp and Kovach 1991). Factors that reduce male condition should make him duller and less attractive as a mate. Experimental infection with a ciliate that causes "white spot disease" led to reduced color intensity. Even after the males had become free from infection, females responded to the differences in coloration in controlled tests of their preferences (figure 13.3.2). In addition to direct benefits in form of better paternal care for their offspring, the preference for the brightest red males might also lead to favorable genetic constitution of offspring, to the extent that there are heritable differences in parasite resistance (Milinski and Bakker 1990). Whether this is the case in the stickleback is not known. The genetic correlation between male coloration and female preference demonstrated by Bakker (1993; section 2.7 above) is compatible with Fisherian and other alternatives, in addition to indicator models.

In contrast with the species just discussed, female blue-headed wrasse *Thalassoma bifasciatum* do not choose males but rather the spawning sites

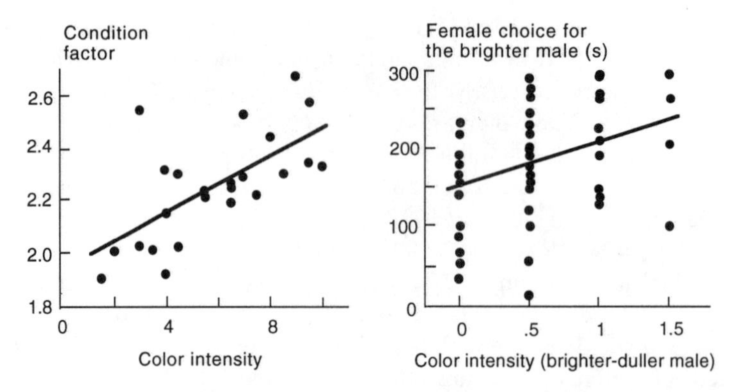

Figure 13.3.2 *Left*: Correlation between the intensity of red breeding coloration and body condition of 24 reproductive male sticklebacks. *Right*: Correlation between the difference in color intensity in pairs of males (brighter–duller male) and female choice of the brighter male (time spent near the brighter male during 300s choice sessions). (After Milinski and Bakker 1990)

they defend. Warner (1987) therefore suggested that the bright coloration and display of the male may demonstrate that the site is relatively safe from predators.

The female is more colorful than the male in some pipefishes (Syngnathidae) with reversed sex roles (e.g., Berglund et al. 1986b). The male can only accommodate roughly the number of eggs produced by one female; females may therefore compete over males (Berglund et al. 1989). In *Nerophis ophidion*, males prefer females with the most extensive blue nuptial colors (Berglund et al. 1986b; section 7.6 above). For sexually selected coloration in fish, see also section 5.2 (guppies).

In sum, there is good evidence that conspicuous colors are favored by mate choice in several fishes. Male contest competition may also be involved; the relative importance of these two and other mechanisms needs to be assessed in a variety of species in the field. There are many puzzling cases, and the bright coloration of many coral reef fishes remains to be explained.

For coloration in relation to parasites, see section 13.8.

13.4 Amphibia

In many newts and salamanders (Urodela), males develop conspicuous colors during the mating season. In *Triturus* newts, they also develop a dorsal crest and a deep tail (Halliday 1977a; figure 13.4.1). In the crested newt *T. cristatus*, females favor males with large crests as mates (Green 1991; also see Malacarne and Cortassa 1983; Hedlund 1990). There is no aggres-

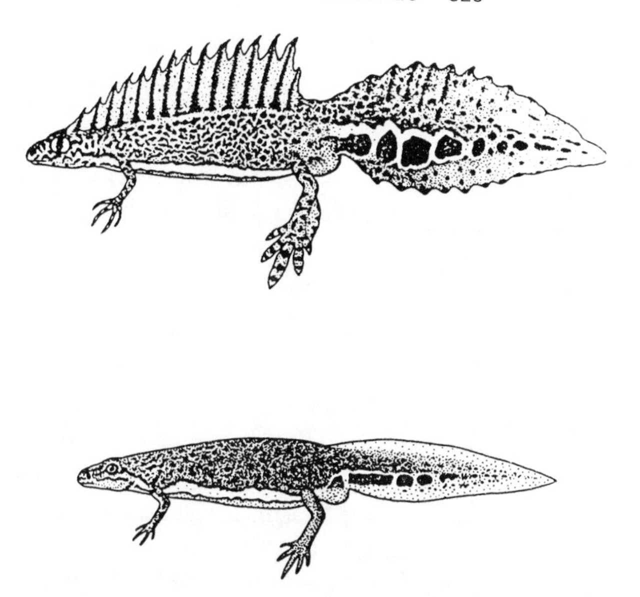

Figure 13.4.1 During the breeding season, male newts develop conspicuous nuptial colors, a dorsal crest, and a deep tail. The figure shows male (*top*) and female *Triturus vittatus*. (Drawing by Tim Halliday, in Raxworthy 1989)

sion between males, so the crest and deep tail are probably favored by female choice in this species. A visual function in display seems likely, but the larger tail may also enhance the transfer of male pheromones to the female during courtship. The height of the crest and tail is correlated with male body condition, and might therefore be used as an indicator by females in their choice of mate (Green 1991; Baker 1992). Other selective processes could also be responsible; for example, the crest and tail should enhance cutaneous breathing in water. Higher male than female activity levels and oxygen consumption might therefore favor a large male crest (Green 1989, 1991; Baker 1992). The banded newt *T. vittatus* shows the most extreme sexual dimorphism among urodeles (figure 13.4.1). The crest and tail are displayed in courtship, but also in threat between males; this is one of the few newts in which males fight (Raxworthy 1989).

In frogs and toads (Anura), bright colors are usually aposematic, being associated with toxicity or distastefulness. Dart-poison frogs (Dendrobatidae) have some of the strongest toxins known from the animal kingdom (e.g., Myers and Daly 1983). Some salamanders are also poisonous and conspicuously colored (Duellman and Trueb 1986). In such species, both sexual selection and predation might favor conspicuous colors (Baker and Parker 1979; Götmark 1992). There are a few Anurans with more conspicuous colors in males than females, for example the Central American *Bufo periglenes* (Savage 1966). Males of this species perch conspicuously on the forest floor and are visited by females for mating during daytime, suggesting that the bright male colors are sexually selected. Bright coloration appears to have replaced male calls in this species, which also lacks auditory apparatus.

In the normally light brown dendrobatid *Colostethus trinitatis*, males turn black when calling, and become aggressive to other black males, suggesting that contest competition may favor the color change (Wells 1980).

13.5 Reptiles

Conspicuous Male Colors

Males are more colorful than females in some turtles (Chelonia); in a few species the color dimorphism is seasonal. In *Callagur borneoensis*, male color changes are associated with breeding, and are induced by testosterone. Moll et al. (1981) suggested that brighter male colors are sexually selected by female choice. Berry and Shine (1980) also suggested that females choose mates in some turtles that show elaborate premating display behavior.

A few snakes have more conspicuous males than females, for example the European adder *Vipera berus* (e.g., Andrén 1986; figure 13.5.1). Male colors become particularly distinctive after molt, at the onset of the mating season, so sexual selection is perhaps involved, but the mechanism is not known. Male contests are common in this species (section 11.7). In most snakes, however, the sexes are similarly colored. Snakes differ from lizards in this respect, for unknown reasons. Perhaps it has to do with less well developed vision in snakes, which seem to depend more on other senses, such as olfaction.

Many lizards show strong sexual color dimorphism (e.g. Cooper 1988). In some forms it is permanent; in others the male becomes more conspicuous during the mating season. In some species the colors can be changed within minutes (e.g., Harris 1964; Stamps 1977; Madsen and Loman 1987). Based on laboratory studies, Noble and Bradley (1933) suggested that "the bright colors of many male lizards . . . function not in stimulating females to breed, but in frightening away rival males." Noble (1934) drew similar conclusions from pioneering field experiments in fence lizards *Sceloporus undulatus*, painting males and females to resemble the other sex.

Detailed later studies have given more complex results. The brightly colored "dewlap" of male *Anolis* lizards is probably the most thoroughly studied secondary sex trait among reptiles (figure 13.5.1; e.g., Fitch and Mills 1984). Observations suggest that it functions in threat display (e.g., Carpenter 1967; Fleishman 1992), but no experiment has yet shown that it

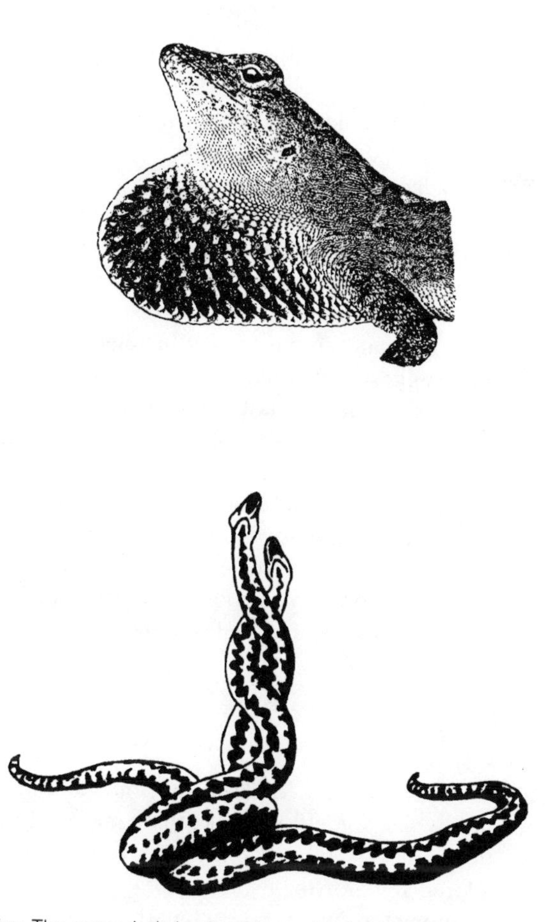

Figure 13.5.1 *Top*: The extended dewlap of a male *Anolis* lizard. *Bottom*: Male adders *Vipera berus* during the reproductive season shed skin and become conspicuously patterned in black and white, whereas the female is a more cryptic gray and brown. The figure shows two fighting males. (Drawing by A. Falck-Wahlström in Andrén 1986)

enhances male contest success. In some cases it clearly influences female responses to males. For example, males prevented from extending the dewlap could not induce ovarian activity in females, in contrast with normal control males (Crews 1975). They were also less likely to be accepted for mating, and experimental change of dewlap color reduced male mating success (Crews 1975; Sigmund 1983; also see Greenberg and Noble 1944). The brightly colored dewlap therefore is favored at least in part by female choice.

An experiment in the tree lizard *Urosaurus ornatus* showed that the size of the central blue area of the dewlap reflects male aggressiveness, and is favored in contests (Thompson and Moore 1991). In laboratory tests, males with a large blue central area dominated males without blue. This was the case both for pairs of males with natural differences in the amount of blue, and for pairs in which size- and color-matched males were painted differently.

Olsson (1992a) reached similar results in an experimental study of the sand lizard *Lacerta agilis*. Males in this species develop green flank and ventral coloration in the mating season. In paired contests, male success increased with the size of the green area. Males with larger painted green area also tended to dominate other, age- and size-matched males. Females did not seem to discriminate among males in relation to the amount of green. These results suggest that conspicuous male colors in some lizards are favored by contest competition, not female choice.

Dewlap color plays a role in species recognition in some lizards; in others, bright male colors function in sex recognition (e.g., Webster and Burns 1973; Ferguson 1977; Williams and Rand 1977; Cooper and Burns 1987; Cooper and Vitt 1988). For example, male *Tropidurus delanonis* painted to resemble females apparently were treated as such by other males, which often intruded upon and sometimes took over their territories (Werner 1978). Females painted to resemble males were attacked and injured by larger males.

Conspicuous Female Colors

In some iguanids, reproductive females develop ventral bright yellow, orange, or red hues, which sometimes make them more colorful than males (e.g., Cooper 1984; Olsson 1994). Contrary to most other animals with more brightly colored females than males, these lizards do not show reversed sex roles. Among the suggested explanations, the rejection hypothesis seems most consistent with the data (Cooper 1984). When a gravid female aggressively rejects a male, she displays her bright, mainly ventral colors, which are otherwise less visible. This is the case also in the Australian agamid *Ctenophorus maculosus* (Mitchell 1973; Olsson 1994). Trying to avoid mating attempts by males, unreceptive females of this species turn on their backs, displaying the orange ventral side. Predation risk appears to be very high in the extremely open habitat: dried-up salt lakes. This may make it particularly important for females to avoid unnecessary, conspicuous matings (Olsson 1994).

The bright coloration of the unreceptive females may reduce molestation by males. Bright female colors and aggression toward males are induced by the same hormones, progesterone and testosterone, in *Holbrookia propingua* (Cooper 1984; Cooper and Crews 1987). Some details do not seem to fit simple forms of the rejection hypothesis, however. For example, females develop maximal brightness around ovulation and mating, and males court bright females in a number of species (Ferguson 1976; Cooper 1984). Cooper and Crews (1987) suggested that the initial brightening may function as an advertisement of breeding condition. Further work is needed to clarify the reasons for bright colors of reproductive females in some lizards.

13.6 Birds

Birds are the most thoroughly studied group as regards conspicuous colors and other visual signals. Focus here is on their sex-dimorphic colors, where sexual selection seems most likely to be involved. Butcher and Rohwer (1989) gave a broader review of the functions of conspicuous colors in birds; physical and physiological aspects were treated by Burtt (1986).

Based on patterns of occurrence of sex-dimorphic colors in birds, Darwin (1871) concluded that they are favored by female choice. He provided a list of empirical rules of age and sex differences in coloration, indicating that adult males have been more modified than females and juveniles from the ancestral state. Wallace (1889) proposed that conspicuous colors function in species recognition, or, in impalatable species, as warnings to predators. He concluded that sexual plumage dimorphism in birds has evolved under natural selection for female crypsis in open-nesting species. These insights were important; Darwin (1871) tried to explain duller female colors as a neutral consequence of an initial sex limitation to males of bright colors. Wallace claimed both that selection can change the inheritance of a character, making it sex limited if maladaptive in one of the sexes, and that adaptive changes in the mode of inheritance often help explain dimorphism, which Darwin (1871) doubted (see Kottler 1980, 1985). Genetic models now suggest that a change from equal to sex-dimorphic or sex-limited expression of a trait is a feasible and probably common, if slow, process (e.g., Fisher 1930; Lande 1980; Lande and Arnold 1985; section 2.5 above). In this respect, Wallace read a more correct message in the comparative data than Darwin, whose mistaken views of inheritance (e.g., Provine 1971; Hull 1973) may have been a hindrance.

Wallace (1889) clarified much of the variation in female coloration. For example, he pointed out that females have dull colors in most species where they are exposed to predators while incubating. Many birds in which both sexes are brightly colored nest in tree holes or other protected cavities (see Baker and Parker 1979; Johnson 1991). On the other hand, many species nesting in tree holes live in the canopy, against which their bright colors may be cryptic; a possible example is some parrots. Association with the canopy rather than with hole nesting, or in addition to it, might therefore in part explain the bright colors of some forest birds (J. Endler, pers. comm.).

Wallace (1889) did not, however, satisfactorily explain why males have conspicuous colors. He regarded large male ornaments as an outcome of "a surplus of vital energy, leading to abnormal growths" (p. 293). If "ornament is the natural product and direct outcome of superabundant health and vigour, then no other mode of selection is needed to account for the presence of such ornament." He therefore rejected Darwin's ideas on female choice; Wallace's strong adaptationist views left little room for selection that reduces survival (e.g., Ghiselin 1969b; Kottler 1980, 1985; Burkhardt 1985).

Sex and Species Recognition

Sex-dimorphic colors sometimes play a role in sex recognition (Rowland 1979). For example, female budgerigars *Melopsittacus undulatus* with their normally brown cere painted blue, as in males, were attacked by males (Cinat-Thomson 1926). In the woodpecker *Colaptes auratus*, a female with an experimentally added black mustache streak typical of the male was attacked by her mate; she was accepted when the black feathers were removed (Noble 1936). Bright signal colors occur in both sexes of many woodpeckers, but usually there is at least one clear difference between males and females, which may function in sex recognition (Short 1982; W. S. Moore 1987).

There is no clear demonstration that selection for species recognition has led to more conspicuous male than female colors in birds. The pattern is suggestive in, for example, ducks and parulid warblers (section 9.2), but more comparative and experimental work is needed to test this possibility.

Mate Choice

Darwin's (1871) suggestion that female choice underlies conspicuous male plumage colors is now supported by several experimental studies. For instance, in a laboratory colony of the village weaverbird *Ploceus cucullatus*, males with the normally yellow wing lining dyed dark, reducing its contrast with the rest of the wing, attracted fewer mates than did control males (Collias et al. 1979).

One of the most thorough studies of female preferences for bright male colors, also demonstrating a selective advantage for the preference, has been done in the house finch *Carpodacus mexicanus* (Hill 1990–1993). In this monogamous, sex-dimorphic species, males develop orange or red crown, ventral, and rump feathers. The extent and brightness of the colored areas vary among males in the wild. Feeding experiments showed that the intake of carotenoid pigments with the diet strongly affects coloration; the evidence suggests that it is condition-dependent and reflects foraging success (Hill 1992). In controlled laboratory experiments, wild-caught females preferred the most colorful among four males. In the field, mated males were more colorful than unpaired males. A field experiment where the coloration of forty males was brightened, that of another forty lightened, and twenty were left intact as controls showed that brightened males were more likely to get a mate than the two other categories. Control males were also more likely to mate than lightened males (proportions mated: 100%, 60%, and 27%, respectively).

Male house finches feed their incubating mates as well as the young. Male feeding rate was positively correlated with his coloration, suggesting that it is a reliable indicator of his performance as a provider for the family (figure 13.6.1). Females should therefore have a direct material advantage in choosing a colorful male. Some females abandoned their nest before the eggs hatched, perhaps in part because of being less well fed than others. The females that abandoned had less colorful males than other females. In addition, males returning after the winter had been more colorful the previous year than males not returning, suggesting that colorful males had higher survival. There was also a correlation between the coloration of fathers and sons. Whether these results reflect genetic or purely phenotypic quality differences is not known. In sum, Hill's (1990–1993) work shows that female house finches prefer colorful males, and that they obtain direct material benefits as a consequence. A similar direct benefit may favor female great tits *Parus major* choosing males with a large black breast stripe (Norris 1990a,b, 1993; section 8.8 above).

Other, surprising evidence on the role of colors in mate choice comes from color-ringed populations. Working with captive zebra finches *Poephila guttata*, Nancy Burley found that the colors of the leg rings influence the attractiveness of the bird as a mate, its reproductive success, longev-

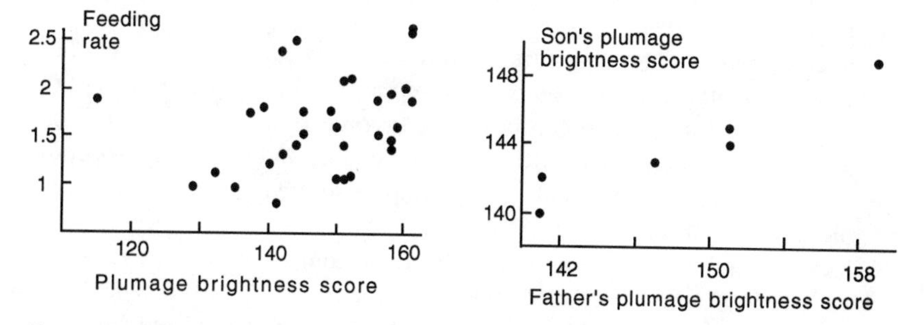

Figure 13.6.1 *Left*: The rate of chick feeding by male house finches *Carpodacus mexicanus* increases with male plumage brightness. *Right*: The plumage brightness of one-year-old males is correlated with the brightness of the father. (After Hill 1991)

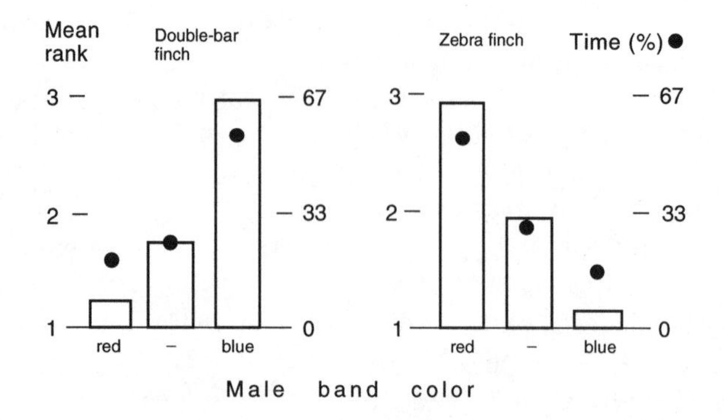

Figure 13.6.2 Female preferences for blue and red leg bands in male double-bar finches *Poephila bichenovii* (*left*) and zebra finches *P. guttata* (*right*). Female double-bar finches prefer blue over red, whereas female zebra finches show the opposite preference. Bars represent within-trial preference rank, dots represent time allocation by females. The central bar and dot in each figure represent unbanded control males. Sample sizes are 9 and 15 females, respectively. (After Burley 1986)

ity, and even the sex ratio of its offspring (e.g., Burley 1981b, 1986b,c, 1988a,b; Burley et al. 1982; but see Ratcliffe and Boag 1987). Both males and females in this monogamous species with biparental care benefit from having rings of a preferred color. Red rings are most preferred, and green or blue rings are avoided, in domesticated as well as wild-caught birds. Burley (1985) suggested that the preference is related to the coloration typical of the species: zebra finches have orange to red bills. Females prefer red bill colors in males of this species (Burley and Coopersmith 1987). Burley (1986a) tested color preferences of the zebra finch and the closely related, partly sympatric double-bar finch *P. bichenovii*, which has a blue bill. This species preferred blue, but avoided red color rings, the opposite response of zebra finches (figure 13.6.2). The color preferences may reflect selection for species recognition in this group, where changes in color patterns may have been crucial in speciation (Burley 1985, 1986a; also see Clayton 1990a,b).

Such effects of color rings are not limited to *Poephila* finches. In a field study of rock ptarmigan *Lagopus mutus*, cocks marked with leg rings of the same color as the red eye comb, a trait favored by sexual selection, had higher mating success. The effects of mate choice and male contests could not be distinguished; the comb, and perhaps also the color rings, may play a role in both (Brodsky 1988).

Rings with the same colors as a male signal trait sometimes have deleterious effects. In red-cockaded woodpeckers *Picoides borealis* and red-winged blackbirds *Agelaius phoeniceus*, males with several red rings had reduced breeding success. In the woodpecker, such males fledged fewer young, and in the blackbird, they more often lost their territories. A possible reason is that addition of red color rings leads to continuous display of the same color as in the coverable male badge of these two species, rendering this signal unreliable (Hagen and Reed 1988; Metz and Weatherhead 1991, 1992). Another possibility is that males with many red bands appear abnormal enough to be ignored (W. Searcy, pers. comm.). These results show that great care is sometimes needed in the choice of color marks for studies of animals with color vision. Normal color-banding practice may, however, not usually have such pronounced effects (e.g., Beletsky and Orians 1989; Weatherhead et al. 1991b).

In several studies of pied flycatchers *Ficedula hypoleuca*, females chose mates mainly in relation to nest site quality (Alatalo et al. 1984, 1986; Askenmo 1984; Slagsvold 1986). Black males in some populations seem to have an advantage over browner males in attracting mates, but they also tend to be older and arrive earlier on the breeding ground, so the effect of plumage color in mate choice is uncertain. There is no clear evidence that coloration in itself, rather than age and experience, affects male mating success (e.g., Järvi et al. 1987; Lifjeld and Slagsvold 1988; Slagsvold and Lifjeld 1988; Alatalo et al. 1990c; Lundberg and Alatalo 1992). In a field experiment where brown males were dyed black, their mating success did not improve (Alatalo et al. 1990c). An experiment using stuffed birds suggests that black coloration in male pied flycatchers leads to fewer predator attacks (Götmark 1992; see below in this section).

Black males seem to be favored by mate choice in several Darwin finches (Price 1984b; Grant and Grant 1989; section 4.5 above). In the house sparrow *Passer domesticus*, the black throat patch of the male is apparently favored by female choice as well as by male contest competition, during winter as well as the breeding season (Møller 1987, 1988b). Females treated with estradiol (see section 14.3) responded with copulation solicitation display to a male taxidermic mount in proportion to the size of his black badge. Male mating success in the field also increased with badge size, and males with large badges had better territories than others (Møller 1988b).

In the lekking great snipe *Gallinago media*, female choice seems to favor males with much white in their tail (Höglund et al. 1990b; section 7.5 above). There was no evidence that males during contests paid attention to the amount of white, which increases with age during the first years. A remaining question is why not all males develop a white tail, if it leads to higher mating success.

A comparative study of waterfowl by Scott and Clutton-Brock (1990) suggests that sexual selection has favored conspicuous male colors. Earlier studies found no correlation between male plumage and degree of polygyny, contrary to what is expected from sexual selection theory. A likely reason is that many species had been lumped together as monogamous;

refined analysis showed that plumage dimorphism is related to details of the mating system. The least dimorphic species have lifelong monogamy. Those with a change of mates between years, and hence with yearly competition over mates, are more dimorphic. The most sex-dimorphic species have serial monogamy, with mating competition several times each season, some males gaining more than one mate in a year, others none (Scott and Clutton-Brock 1990).

This pattern indicates that sexual selection favors conspicuous male plumage, but it is not known whether mate choice or contests are responsible. Other studies show that female ducks base their choice of mate in part on his conspicuous plumage (e.g., Klint 1980; Williams 1983; Holmberg et al. 1989; Weidmann 1990). As the risk of hybridization is considerable in ducks, species recognition may be an important aspect of such responses (section 9.2). It is also possible that females gain direct benefits from choice of male plumage (Williams 1983). For example, forced copulation is common in ducks, and females are sometimes injured or drowned (McKinney et al. 1983). Defense of the female by her mate might help protect her. The males most successful in fights tend to have the least damaged plumage (Titman and Lowther 1975); females might therefore benefit directly from choosing such males as mates. There is evidence that females favor males with undamaged plumage in mallard *Anas plathyrhynchos* (e.g., Holmberg et al. 1989; Weidman 1990) and black grouse *Tetrao tetrix* (Alatalo et al. 1991; section 7.5 above).

Some studies have failed to find an effect of male colors on the attraction of mates. Manipulations that reduced color contrasts in male yellowthroats *Geotyphlis trichas* and Bullock's orioles *Icterus galbula bullockii*, respectively, did not reduce mate attraction compared to controls (Lewis 1972; Butcher 1991). Similar results were obtained in white-crowned sparrows *Zonotrichia leucophrys* by Götmark (1993b). Nor did complete blackening of the yellow head and breast reduce mate attraction in male yellow-headed blackbirds *X. xanthocephalus* (Rohwer and Røskaft 1989; see below).

Long Tails

A visual ornament in form of a long tail in males has evolved independently in many birds from different families (reviewed by Balmford et al. 1993). Its function has been studied experimentally in four species (table 13.6.1). In addition, observational evidence from malachite sunbirds *Nectarinia johnstoni* and kestrels *Falco tinnunculus* suggest that female choice favors males with a long tail (Evans 1991, Evans and Hatchwell 1992b; Palokangas et al. 1992). Jones (1992) discussed approaches and possible pitfalls of such studies.

In the four species of table 13.6.1, male attractiveness to females increases with tail length (for details, see sections 5.4, 7.3, 7.5). There was no

Table 13.6.1
Birds in Which Female Choice Has Been Shown Experimentally to Favor Long Tail in Males

Species	*Mating System*	*Reference*
Jackson's widowbird, *Euplectes jacksoni*	Lek polygyny	Andersson 1989, 1991
Long-tailed widowbird, *E. progne*	Resource defense polygyny	Andersson 1982a
Swallow, *Hirundo rustica*	Monogamy	Møller 1988, H. G. Smith and Montgomerie 1991
Shaft-tailed whydah, *Vidua regia*	Polygyny	Barnard 1990

indication that the long tail is also favored in male contests, but this possibility cannot be ruled out entirely. In the whydah, higher display rate in long-tailed males may also contribute to their attractiveness (Barnard 1990). The evidence suggests that female choice is usually, but not always, the main mechanism selecting for long tails in birds. In the pheasant *Phasianus colchicus*, however, the long tail is displayed in dominance interactions among males, and quantitative analysis gave no evidence that it is favored by female choice (von Schantz et al. 1989; section 12.3 above).

The classic epitome of male tail ornamentation, the peacock *Pavo cristatus*, was studied by Petrie et al. (1991, 1992) at a lek of 10 males in a feral population at Whipsnade Park, England. Male mating success increased with the number of ocelli ("eye-spots") in his tail train (figure 13.6.3). Females in captivity also laid more eggs if mated to males with large trains (Petrie and Williams 1993). Females usually visited a number of males on the lek before returning to one of them for copulation. In 10 out of 11 visitation sequences observed in detail, the female returned for mating with the male with the highest number of ocelli. In the remaining case, the favored male had only one eye-spot less than the visited male with most spots (about 150 spots).

Female choice therefore favors trains with many ocelli, as suggested by Darwin (1871). The number of ocelli increases with train length. The ocelli are perhaps favored by female choice not directly, but through correlation with some other aspect, such as train length (Petrie et al. 1991). Male contest competition may also contribute, for peacocks display their trains also in threat. Males without lek display site were lighter and had shorter trains than lek males. Fox predation seemed to fall disproportionately upon males with low mating success (Petrie 1992b).

Females may respond in part to the symmetry of male ornaments (Møller, e.g., 1990b, 1992a). The peacocks with largest numbers of ocelli also show highest symmetry of ocellus numbers in the two sides of the train

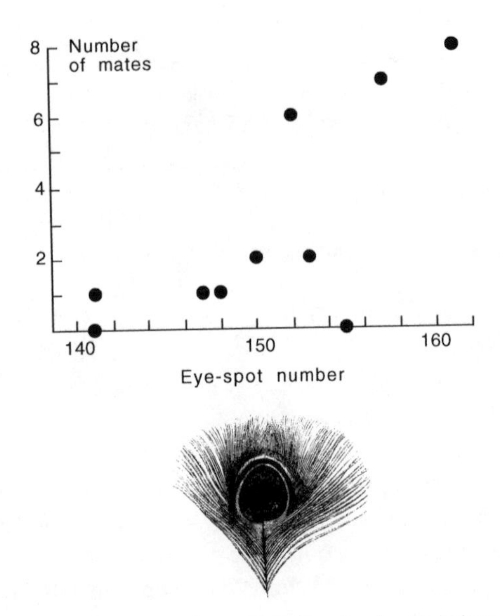

Figure 13.6.3 The mating success of peacocks in a feral population at Whipsnade Park, England, increased with the number of eye spots in the male tail train. (After Petrie et al. 1991; eye spot from Darwin 1871)

(Manning and Hartley 1991). The degree of symmetry may be related to the capacity for developmental homeostasis of the male (section 3.6). As in other lekking species, there are several hypothetical advantages for peahens mating with the most ornamented males (Petrie et al. 1991).

The size of tail ornaments often reflects male age (e.g., L. H. Smith 1965, 1982; Manning 1987, 1989) and condition (Møller 1989a), making a long tail a possible candidate for an indicator function (Andersson 1982a,b; section 3.6 above). Long tail ornaments have also evolved in both sexes in some monogamous birds, for example skuas (Stercorariinae; e.g., Cramp et al. 1983). In the long-tailed skua *Stercorarius longicaudus*, the two central tail streamers may be as long as the rest of the body. They are used in courtship display by both sexes, and tail length appears to increase with age also in adult birds (Andersson 1971, 1976, and unpublished observations). It may reflect breeding experience and parental ability in this and some other species with biparental care. (Whether the long tail in skuas is important for mate choice is not known.) This idea predicts that long tail ornaments in both sexes should occur mainly in species with biparental care, and that the degree of similarity of the tail (or other indicator traits) in males and females should be correlated with the similarity of their parental roles, predictions that should be testable with comparative methods.

In the organismal world it seems possible to find exceptions to every rule. Inevitably, there is an Australian bird in which males shift to a *shorter* tail before the mating season. Cisticolas in general molt to a breeding plumage with shorter tail. In *Cisticola exilis*, tail-shortening is exaggerated in males, which use the tail in conspicuous, tail-fanning courtship flight. Cockburn (1991, and pers. comm.; also see Darwin 1874, p. 494) pointed to circumstantial evidence that the short tail is favored by female choice. As improved body condition seems unlikely to result in growth of shorter feathers, Cockburn suggested that the exaggerated tail shortening in males has probably not arisen by an indicator mechanism based on condition dependence. Perhaps there is a general mating preference for short tail in cisticolas, for example because it reflects reproductive condition. If so, male *C. exilis* may have taken advantage of the preference by further tail reduction, perhaps through Fisherian coevolution with the female preference. This possibility might be tested by measuring female responsiveness to different male tail lengths in *C. exilis*, and in closely related species without sex-dimorphic tail shortening.

Analyzing variation in the length of tail ornaments in birds, Alatalo et al. (1988b) found that tail length varies much more within populations than do other body size characters. This suggests that long tails are likely candidates for indicator functions (section 3.6–3.7). Tail length did not show greater geographical variation than other aspects of body size. Craig (1989) obtained similar results as regards geographical, but not within-population, variation in tail length in the long-tailed widowbird *Euplectes progne* (also see Cherry 1990; Barnard 1991).

Balmford et al. (1993) reviewed the evolution of long tails in birds. Comparative analysis combined with aerodynamic models suggested that long tails of shallow fork type are aerodynamically optimal, show little sex dimorphism, and may have evolved largely by natural selection. Other types of long tail impair flight, show greater sex dimorphism, and may have evolved by sexual selection through mate choice (also see Evans and Thomas 1992; Thomas 1993).

Contests

Bright colors might increase the efficiency of threat signals used in contests over mates or other resources (e.g., Fisher 1930; Hingston 1933; Noble 1936; section 1.7 above). There are at least two possible mechanisms (Butcher and Rohwer 1989). (1) The *priority hypothesis* emphasizes that the owner of a territory or other resource usually wins conflicts over it, for example because superior fighters tend to become owners. The resource may also have greater value to the owner, making it more likely to escalate a conflict to a dangerous level. Conspicuous colors and other signals can help show at a distance that the resource is occupied, or that the owner is prepared to fight for it (e.g., Huxley 1938a; Borgia 1979; Andersson 1982b; Slagsvold and Lifjeld 1988; Butcher and Rohwer 1989).

According to (2) the *fighting-ability hypothesis*, bright colors are favored if they make good fighters recognizable (Rohwer 1982; Butcher and Rohwer 1989; Rohwer and Røskaft 1989). The coloration may be correlated with the phenotypic condition and fighting ability of the possessor, permitting strongly ornamented individuals to win contests through display, without escalated fights (Baker and Parker 1979; Andersson 1982b; Parker 1982; Maynard Smith and Harper 1988).

Experiments show that conspicuous male colors in several species are favored by male contests over mates, or resources that attract mates. Manipulations of the red shoulder patch in male red-winged blackbirds *Agelaius phoeniceus* provide such strong evidence (section 5.4). The coverable patch, which is important in competition over breeding territories (e.g., Hansen and Rohwer 1986; Røskaft and Rohwer 1987; see section 5.4 above), is displayed to conspecific males and females, but not to predators, refuting predation-based hypotheses (Butcher and Rohwer 1989). There remains a possibility that it is also favored by female choice. To the extent that male dominance ability is heritable, a female should obtain fitter offspring by preferring male traits that reflect dominance.

In a study of the malachite sunbird *Nectarinia johnstoni* on Mount Kenya, Evans and Hatchwell (1992a) tested the function of the scarlet pectoral tuft of the male (figure 13.6.4). Like the red badge of the red-winged blackbird, the coverable tuft is displayed in contests. Males with reduced

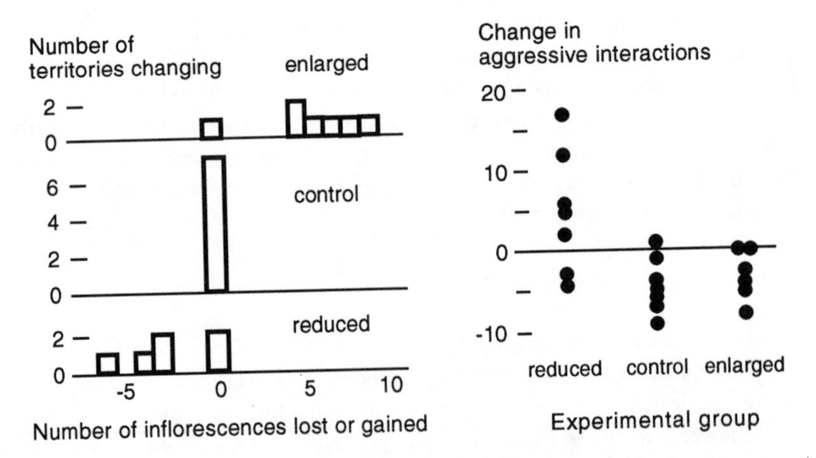

Figure 13.6.4 Male malachite sunbirds *Nectarinia johnstoni* display their scarlet pectoral tuft in contests over territories. Experimental reduction of the size of the tuft in males led to reductions in the size of their territories with defended *Lobelia* inflorescences (*left figure, bottom*), whereas males with enlarged tufts gained territory area (*top*). The changes were associated with increasing numbers of contests with intruders for males with reduced tufts, and fewer contests for males with enlarged tufts (*right*). (From Evans and Hatchwell 1992a)

tuft size lost territory area; males with enlarged tuft increased the size of their territories. The size of the tuft therefore seems important in male-male assessment during territorial contests. Similar experimental results were obtained by manipulation of the wing-bar in territorial male warblers (*Phylloscopus inornatus*) by Marchetti (1993).

Coverable wing badges occur in several other birds, for example black grouse *Tetrao tetrix*, whinchat *Saxicola rubetra*, and chaffinch *Fringilla coelebs*; they have apparently evolved many times independently. Their function needs to be tested in more species to clarify whether the role in male contests is general.

Surprisingly, the mating success of male yellow-headed blackbirds *X. xanthocephalus* dyed all black was similar to that of control males (Rohwer and Røskaft 1989). Remarkably, five of twelve blackened males took over better territories from other males, whereas none of eleven control males did so. There was no indication that blackening of the yellow head and breast reduced male success in competition over territories or mates. Rohwer and Røskaft (1989) concluded that the result is best explained by the "arbitrary identity badge hypothesis" (fighting-ability hypothesis mentioned above), but other interpretations are not entirely ruled out. The unexpected result suggests that some important aspect of the selection of conspicuous secondary sex colors in birds may not be understood, in spite of much experimental progress (see also Eckert and Weatherhead 1987a,b).

In the pied flycatcher *Ficedula hypoleuca*, the males with darkest plumage are most successful in contests over nest sites. As plumage darkness is correlated with age, arrival date, wing length, and condition, this need not imply that black color in itself is important (Järvi et al. 1987; Slagsvold and Lifjeld 1988, 1992). The relative darkness of males compared to females varies over the species' range. Males are brown and female-like in areas of overlap with the closely related collared flycatcher *F. albicollis* (Røskaft et al. 1986; Alatalo et al. 1990a; Lundberg and Alatalo 1992; Røskaft and Järvi 1992). Males of this species dominate male pied flycatchers in contests over nest sites. In an experiment comparing responses to black versus brown pied flycatcher dummies, male collared flycatchers were most aggressive toward black dummies. Female-like pied flycatcher males may therefore benefit from lower aggression in areas of sympatry with the collared flycatcher (Kral et al. 1988; also see Røskaft et al. 1986). Another possible advantage is that brown males evoke less aggression and can nest closer to darker, older males (Slagsvold and Saetre 1991). In sum, it seems that competition with collared flycatchers at least partly explains the geographic variation in the nuptial plumage of the pied flycatcher, but the reasons for the dark plumage of the males is debated (e.g., Lundberg and Alatalo 1992; Slagsvold and Lifjeld 1992; also see unprofitable prey, below).

Evidence for a function of conspicuous colors in contests has also been found in, for example, *Colaptes* woodpeckers (Noble 1936; Short 1972, 1982; Moore 1987; section 7.4 above), chaffinch *Fringilla coelebs*, great tit *Parus major*, yellow warbler *Dendroica petechia*, and ruff *Philomachus pugnax* (Marler 1955; Järvi and Bakken 1984; Studd and Robertson 1985a,b; Höglund and Lundberg 1989; Norris 1991a,b, 1993). Johnston (1966) suggested that the steel-blue coloration of adult males in the purple martin *Progne subis* is favored by male contest competition, not female choice (also see Stutchbury 1991). Contest competition may be an important selection factor behind many conspicuous color signals in birds, a possibility that needs more empirical testing. This seems especially likely in some species where juveniles are more conspicuously colored than adults of one or both sexes, for instance certain woodpeckers, hummingbirds, and also some coral reef fishes (e.g., Short 1982; Bleiweiss 1992; Barlow 1974). In these species, bright juvenile coloration, also in females, seems to function in contest competition over food (reviewed by Bleiweiss 1992; also see section 13.3 above).

Some form of indicator mechanism may have been involved in the origin and early evolution of wing epaulets and other similar badges (see section 1.4). For example, brighter coloration might indicate better condition and fighting ability (Baker and Parker 1979; Andersson 1982b; Rohwer 1982). This is related to the question how conspicuous colors, such as melanins and red and yellow carotenoids, became secondary sex traits in so many birds, fishes, and other animals. Carotenoids are obtained from the food, and hence may reflect foraging ability and nutritional condition (e.g., Brush 1978; Endler 1980; Kodric-Brown 1989; Hill 1992). The first stages of such coloration may have arisen as a neutral consequence of high carotenoid intake by individuals in particularly good nutritional status. Initially, new badges are perhaps even favored by reduced predation (Götmark 1994b). Responding to coloration might then be advantageous if it helps a male avoid contests with rivals of superior strength and condition. This possibility might be tested if a population can be found where carotenoid colors occur in some but not all adult males. In addition, choice of mates with carotenoid (or other) colors that reflect high foraging success might be selected for several reasons (see above in this section, and section 5.2).

Unprofitable Prey

Conspicuous coloration in many animals is aposematic, informing predators that the potential prey is poisonous, weaponed, alert, or unsuitable to attack for some other reason (e.g., Cott 1957; Edmunds 1974). Conspicuous, possibly aposematic colors occur in many unpalatable birds (Cott and Benson 1970; Götmark 1994a). From a multivariate analysis of Western Palaearctic species, Baker and Parker (1979) drew the radical conclusion that "bird coloration has evolved almost entirely in relation to predation-based selection pressures . . . they have not evolved in relation to sexual selection."

This conclusion can, however, be turned around. The theory predicts that sexually selected conspicuous colors often evolve to a stage where they are disfavored by predation (Darwin 1871; Fisher 1930; for evidence, see section 10.1 above). In species that are relatively invulnerable, for instance owing to habitat choice or large body size, predation is a less severe check on the evolution of conspicuous colors. They may therefore occur mainly in species with low risks of predation, even if not used for signaling to predators (Krebs 1979; Promislow et al. 1992). For this reason, sexual selection theory does not necessarily predict that predation on males is higher in species where they are conspicuous than where they are not. Correlations among species between plumage conspicuousness and risk of predation (Baker and Hounsom 1983; Baker 1985; Baker and Bibby 1987) for this and other reasons are not unequivocal support for the unprofitable prey hypothesis; such correlations may arise in other ways (Lyon and Montgomery 1985; Promislow et al. 1992). For instance, body size rather than coloration, which can be correlated with body size, may explain much of the variation in predation risk (Reid 1984; Butcher and Rohwer 1989). A test in New World orioles (*Icterus*) refuted two predictions and suggested that the unprofitable prey hypothesis does not explain conspicuous coloration in these birds (Flood 1989). Other evidence against the hypothesis, and support for the importance of sexual selection, was presented by Sigurjonsdottir (1981).

According to Baker and Parker (1979), "In species that are sexually dimorphic, the male is not brightly colored as a result of female choice or male-male competition but because he represents a less profitable prey to predators than females and juveniles." For example, he might have better chances of escaping an attacking predator. Among the birds analyzed by Baker and Parker (1979), sexual color dimorphism was associated particularly with (1) small body size; (2) guarding of the incubating female by the male; (3) uniparental care; and (4) polygyny. Trends 2–4 are probably associated with reduced male parental care, slacking selection for male crypsis and hence making evolution of conspicuous male colors more likely (Butcher and Rohwer 1989). There is, however, no evidence that sex-dimorphic conspicuous colors in male birds are displayed mainly to predators. As already Darwin (1871) pointed out, such colors are usually emphasized in communication within the species, often in the context of mating or in competition over mates (see descriptions of display behavior in, e.g., Armstrong 1965; Cooper and Forshaw 1977; Snow 1982; Johnsgard 1983). Such evidence on the use of conspicuous traits in display can give strong indications about their function and selection.

Development of conspicuous plumage before the spring mating season also suggests that sexual selection is involved. Bright seasonal colors in males do not develop first in the summer, when easier alternative prey in form of vulnerable young become available. This is contrary to what the unprofitable prey hypothesis predicts. Seasonal plumage changes in birds with unusual pairing times are particularly revealing (Andersson 1983). For example, most ducks (Anatinae) begin to form pairs in autumn or winter, long before the nesting season. In these species, as expected if sexual selection is responsible, males shift from inconspicuous "eclipse" plumage, put on during the summer, to conspicuous plumage already in autumn. In the aberrant ruddy duck *Oxyura jamaicensis*, however, which forms pairs late in spring, the drake molts to conspicuous nuptial plumage first at this time.

Such patterns fit well with sexual selection, but run counter to the unprofitable prey hypothesis (Andersson 1983). For these and other reasons, it seems unlikely to be a general explanation for conspicuous coloration in male birds. It, or similar mechanisms such as aposematism, may explain certain cases, as suggested by Götmark's (1992, 1993a; see below) study of the pied flycatcher. The unprofitable prey mechanism is not likely to be important in cases where the colors are emphasized in display to individuals of the same species, but not to predators. It seems more likely to apply for some of the species with conspicuous colors that are permanently visible. A role for the unprofitable prey mechanism in the evolution of sex-dimorphic traits remains to be demonstrated. It might work in concert with sexual selection in some cases (Baker and Parker 1979; Götmark 1992). Perhaps a novelty effect may reduce predation on males with a new conspicuous trait (Götmark 1994b).

The possibility that conspicuous plumage may be aposematic in some birds has recently been substantiated in the New Guinean passerine genus *Pitohui*. These birds are black and contrastingly orange-brown, have a strong odor, and contain a batrachotoxin, the class of toxins found in South American dart-poison frogs (Dendrobatidae). The toxins have apparently evolved independently as defense substances in frogs and pitohuis (Dumbacher et al. 1992; also see Myers and Daly 1983; Diamond 1992). Other surprising discoveries will perhaps be made as the possibility of aposematism is studied in detail in more birds.

Experimental results suggest that the dark coloration of male pied flycatchers *Ficedula hypoleuca* may be favored by reduced predation. Previous work suggested that dark color may not make a male more successful in contests over territories, or in attracting mates (e.g., Slagsvold and Lifjeld 1988; Alatalo et al. 1990c; Lundberg and Alatalo 1992; see above in this section). Following a different tack, Götmark (1992) tested the effect of coloration on predation risk, exposing 15 pairs of stuffed male and female pied flycatchers in the field to autumn-migrating sparrowhawks *Accipiter nisus*, an important predator of this species. In contrast with expec-

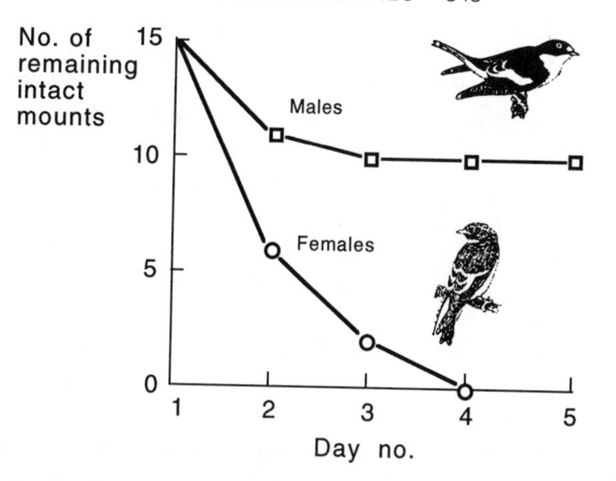

Figure 13.6.5 Distribution of predator attacks on mounted male and female pied flycatchers *Ficedula hypoleuca* during five days from the start of the experiment. The duller brown females were attacked more often than the conspicuous black-and-white males. (From Götmark 1992)

tations, there was a clear tendency for sparrowhawks to attack females rather than the darker and more conspicuous males (figure 13.6.5). This outcome suggests that the conspicuous coloration of male pied flycatchers does not increase the risk of predation, contrary to the assumption of sexual selection theory. A second experiment at another site in spring gave similar results (Götmark 1993a). On the other hand, sparrowhawks in parallel tests with chaffinches *Fringilla coelebs* at the same site attacked the more conspicuous red males more often than the gray-green females. The conspicuous male plumage in the chaffinch therefore seems to increase the risk of predation, as predicted by sexual selection theory. Why the pied flycatcher shows the opposite pattern is not yet clear (Götmark 1992, 1993a).

Delayed Plumage Maturation

In most passerine birds with sex-dimorphic plumage, males develop adult colors in their first year. In some species, one-year-old males retain a plumage similar to that of the female, and get bright colors first the next year. Yet one-year-old males are sexually mature in most passerines and can breed if a chance arises. The timing of development of adult plumage may in part be sexually selected. Of the proposed explanations for this delayed plumage maturation, as it is often called, most are based on adaptations in the breeding season (reviewed, e.g., by Rohwer et al. 1980; Lyon and Montgomery 1986; Rohwer and Butcher 1988; Björklund 1991b):

1. The *crypsis hypothesis* (Selander 1965) suggests that although a young male would have higher chances of mating if he wore adult plumage, the risk to survival would rise too much for there to be any fitness gain.

2. According to the *female mimicry hypothesis* (Rohwer et al. 1980; Slagsvold and Saetre 1991), a female-like plumage makes subadult males harder to distinguish from females, and therefore less likely to be attacked by older males. This helps young males establish territories and perhaps breed, or achieve copulations with mated females.

3. The *status-signaling hypothesis* (Rohwer 1975; Lyon and Montgomery 1986) suggests that the female-like plumage is a reliable signal of subordinance, making subadult males more likely to be tolerated in the breeding area by adult males.

4. The *winter adaptation hypothesis* (e.g., Rohwer and Butcher 1988) suggests that female-like plumage is favored in subadult males outside the breeding season. It is kept in the second summer because a special spring molt to adult plumage costs too much energy. Female-like plumage in subadult males is not advantageous as such in the breeding season.

It is not yet possible to distinguish among these alternatives (see, e.g., Lyon and Montgomerie 1986; Rohwer and Butcher 1988; Hill 1989; Enström 1992). The functions and consequences of subadult male plumage differ among species (e.g., Grant 1990; Enström 1992). The comparative evidence is perhaps strongest for the winter adaptation hypothesis. For example, among 105 sexually dichromatic North American passerines, 42 species have an extensive spring molt. In all these species, subadult males after the molt are more similar or identical to adult males. This is not predicted if female-like plumage in young males is an adaptation to the breeding season (Rohwer and Butcher 1988; also see Stutchhbury 1991; Enström 1992).

Björklund (1991b) suggested that the term "delayed plumage maturation" is a misnomer. He presented evidence from a comparative analysis of North American Fringillidae (nine-primaried oscines) that female-like plumage in one-year-old males is often an ancestral rather than derived state. Sexual plumage dimorphism in this group may initially have evolved from an ancestral state of conspicuous plumage in both sexes, by selection for female drabness (Wallace, 1889, presented similar views). Björklund's (1991b) study, which may not be the last word in the debate, underscores the importance of a phylogenetic perspective in analyzing the evolution of a trait (section 4.3).

13.7 Mammals

The most conspicuous sex differences in coloration among mammals are found in primates, which have more advanced color vision (Jacobs 1981) and probably poorer nose than most other mammals, in which odor may play an important role (section 15.3 below). Two kinds of sexual coloration are especially conspicuous in primates (Wickler 1967). In some species, particularly those where females live in groups with several adult males, females develop pink or red anogenital skin swellings, usually during estrus. The coloration is exposed in sexual invitation display. Possibly, it has been favored because it enhances sperm competition among males, and makes the female more likely to bear sons that are successful in sperm competition (Harvey and Bennett 1985).

In some primates, males also have similarly colored skin areas, using them in appeasement display to group members higher in rank. The trait is taxonomically more widespread in females, and is often formed from nonhomologous structures in the two sexes, suggesting that it arose in males by convergent evolution toward the female pattern (Wickler 1967).

Colorful skin markings also occur in the male genitalia of Old World monkeys. A red penis in many species contrasts with a blue scrotum. The male mandrill *Mandrillus leucophaeus* carries the same color in perhaps the most striking face of any mammal. Darwin (1876) suggested that the coloration evolved by female choice. Like the blue and red colors of penis and scrotum of other male primates, however, the face colors of the mandrill are displayed in ways that rather suggest a role in contests among males (Wickler 1967).

Many aspects of male hair growth in primates may also be related to threat (Guthrie 1970). In large mammals, conspicuous and often contrastingly colored hair or other structures that increase apparent size often occur in parts of the body that are emphasized in social display. In ungulates with lateral threat, their dewlaps, manes, or hair crests along the neck and back enlarge the apparent size viewed from the side. In mammals with frontal threat, such as many primates, the face is often enlarged by conspicuous ruffs, whiskers, or other hair (Guthrie 1970; figure 13.7.1). Their emphasis in male threat display and conflicts over females, or resources that attract females, suggests that sexual selection by contest competition is responsible for the evolution of such adornments. Mate choice might also play a role; these possibilities need to be tested in suitable species by experimental manipulation of such structures. Guthrie (1970) suggested that also the male beard in many human populations has been favored by competition over dominance rank and resources, in turn influencing mating success and reproduction.

In spite of their visual signals, which often involve conspicuous colors and structures, few primates approach the strong dichromatism found in many birds. Perhaps one reason is that males and females in most primates live in social groups, with more similar selection of coloration in males and females than in some of the strongly sex-dimorphic birds, with markedly different sex roles, for instance incubation by the female alone.

Figure 13.7.1 Some types of facial hair and color patterns among primates. The species are, columnwise from the left: *Cercocebus aterrimus, Pongo pygmaeus, Homo sapiens, Gorilla gorilla, Saguinus mystax, Alouatta seniculus, Cercopithecus aethiops, C. hamlyni, Cacajao rubicundus, Colobus guereza.* (From Guthrie 1970)

13.8 Parasites, Mate Choice, and Conspicuous Colors: Comparative Evidence

The idea of Hamilton and Zuk (1982) that bright colors, birdsong, and some other secondary sex traits permit choice of mates with heritable parasite resistance has been tested by comparative analyses, in particular of bird coloration (tests within species were reviewed in section 3.5). One of the predictions is that good health is most important to advertise in species often subject to debilitating parasites. Among species, male plumage conspicuousness should then correlate with parasite burden. Hamilton and Zuk (1982) reported such a correlation with blood parasites among 109 species of North American passerines. The conclusion was also supported in studies by Zuk (1991b) and by Read (1987, 1988), who also included European passerines. He reduced the problem of statistical dependence among sample species by examining trends within genera; 22 out of 32 genera showed the predicted trend.

Renewed analysis based on a new scoring of conspicuousness by six independent ornithologists did not, however, confirm the previous conclusions (Read and Harvey 1989a,b). The conspicuousness scorings were not concordant, illustrating the problem of using subjective human judgment in such cases. In analyses that controlled for phylogenetic effects, the results were no longer statistically significant for the North American sample. For the European passerines, there was a significant association between male brightness and blood parasites, but for unknown reasons only among species in which few individuals had been examined for parasites. The reassessment did not support the prediction of the Hamilton-Zuk (1982) hypothesis (Read and Harvey 1989a,b; Read 1990, 1991; but see Zuk 1989, 1991b, 1992). Another independent analysis of North American passerines (Johnson 1991) likewise did not support the model. A detailed comparative analysis of wood warblers (Parulinae) found no significant correlation between parasitism and plumage brightness (Weatherhead et al. 1991a). Pruett-Jones et al. (1991) studied parasitism in relation to sexual selection in ten species of birds of paradise. As predicted, parasite infection was correlated with the degree of sexual dimorphism, being most common in promiscuous species with strong dimorphism. There was also, however, a possible correlation between population density and parasite burden, and the authors concluded that the cause of the observed correlation between plumage showiness and parasite burden is not yet resolved.

Comparative analyses of other traits have also produced conflicting results (see Read 1990; Clayton 1991). Hamilton and Zuk (1982) reported an association beween male song complexity and blood parasites. A test that controlled for phylogeny effects found, however, no association between song elaboration and parasites (Read and Weary 1990). Another study compared European freshwater fishes, where the number of parasite genera reported per species increased with sexual color dimorphism (Ward 1988b, 1989). The analyses were, however, affected by similar problems as the bird studies. A larger analysis of North American fishes suggested that the correlation found in European species was, again, caused by phylogenetic effects (Chandler and Cabana 1991). A comparative study of male conspicuousness versus level of parasite infection in iguanine lizards found an inverse correlation between brightness and parasite prevalence (Lefcort and Blaustein 1991; also see Ressel and Schall 1989).

In sum, there are several negative as well as positive results, but no comparative work that adequately controls phylogenetic effects has yet demonstrated a correlation of the kind suggested by Hamilton and Zuk (1982) (Read 1990; Harvey and Pagel 1991). A problem in many of these studies is that they were done across several families or orders, which may confound variables, and conspicuousness was scored subjectively by human subjects from field handbooks. More objective measurements of conspicuousness against the natural background are needed (Endler and Lyles 1989; Endler 1990; Götmark and Unger 1994).

More critical tests can be done through detailed work on suitable species. Such studies, several of which concern bird coloration, were reviewed in section 3.5. Many of them confirmed the assumptions and predictions tested. As with other versions of genetic indicator mechanisms, however, no study has yet demonstrated that the assumed genetic mechanism is at work; alternative explanations are possible (section 1.7). This need not mean that genetic indicator mechanisms are unimportant, but they are difficult to test in ways that exclude alternatives.

13.9 Summary

Conspicuous colors in animals are selected by a variety of mechanisms. Aposematism or mimicry explains some cases where both sexes are brightly colored, but in most such species, including many insects, fishes, and birds, the selective mechanisms have not been studied in detail. Observations, experiments, and comparative analyses suggest that sexual selection is the main factor favoring conspicuous sex-dimorphic colors and other visual ornaments, which are usually most pronounced in males.

Field studies have shown that female choice of mate is involved in some of these cases in butterflies, fishes, salamanders, lizards, and birds. Empirical work therefore supports Darwin's ideas on female choice favoring male ornaments in many cases. The evidence also shows, however, that male contest competition over mates, dominance rank, or resources needed to attract mates often favors male visual ornaments. There is experimental support from dragonflies, lizards, and birds, and other supportive evidence from fishes, mammals, and other groups. Detailed studies of a wide range of species are required before the relative importance of female choice versus male contests in the evolution of conspicuous colors and other visual ornaments can be reliably assessed. Empirical knowledge of the mechanisms by which female preferences evolve for male ornamental traits is scant, in contrast with the profusion of models and hypotheses.

There are several exceptions to the general pattern of more conspicuous colors in males. Most of these species remain to be studied, but in a role-reversed fish, male choice of mate favors conspicuous colors in females. In certain lizards, bright colors in reproductive females seem to be favored by mechanisms other than sexual selection.

Natural selection for species and sex recognition, and contest competition over resources other than mates, may also be involved in the evolution of sex-dimorphic signals. So may perhaps signaling to predators that the male is unprofitable as prey; but the evidence suggests that sexual selection is the main mechanism favoring sexually dimorphic visual signals.

14 Acoustic Signals

14.1 Introduction

Song and other acoustic signals are the traits most often shown to be sexually selected (tables 6.A, 6.2.1). The present chapter deals mainly with the evidence for sexual selection of song; species recognition was treated in chapter 9. Song is here defined broadly as long-range acoustic signals produced mainly during the breeding season. It occurs especially in three groups: insects, frogs, and birds.[1] Song is also found in some mammals (see below) and fishes (e.g., Myrberg et al. 1986). For example, the mormyrid fish *Pollimyrus isidori* uses sound as well as electric signals in courtship (Crawford et al. 1986; Bratton and Kramer 1989; Kramer 1990). Female choice in relation to male song in fish has not yet been statistically tested, but such choice seems likely in certain species where song attracts females (Myrberg et al. 1986).

Song is often conspicuous and frequent; a red-eyed vireo *Vireo olivaceus* has been recorded singing over 22,000 times in one day (de Kiriline 1954), and red deer stags may roar over 1000 times each day during the rut (Clutton-Brock and Albon 1979). Speculations about the functions of song date back at least to classical times (Thorpe 1961). In his *Natural History of Selborne*, Gilbert White (1789) suggested that bird song is used in contests between males. Darwin (1871) claimed that it is also used for attracting females, and that the two functions may occur together; he held similar views on the song of insects and frogs. Both functions are related to sexual selection, as the territory is important for mating success (e.g., Howard 1974; Davies 1978).

In spite of early studies pointing to dual functions of song (Tinbergen 1939; Marler 1956), its role in territory defense came to be emphasized almost exclusively until the 1970s. Since then, many studies have provided support for effects on female choice as well (reviewed by Searcy and Andersson 1986; Catchpole 1987; Kroodsma and Byers 1991). Evidence for sexual selection of song comes from patterns of song use, effects on recipients of the signal, and correlations with the mating system. Loudspeaker

[1] Reviews of song from a sexual selection perspective are found in, e.g., Catchpole 1982, 1987, Thornhill and Alcock 1983, Searcy and Andersson 1986, Gerhardt 1988, Ryan 1988a, Fritzsch et al. 1988, Kroodsma and Byers 1991, Searcy 1992.

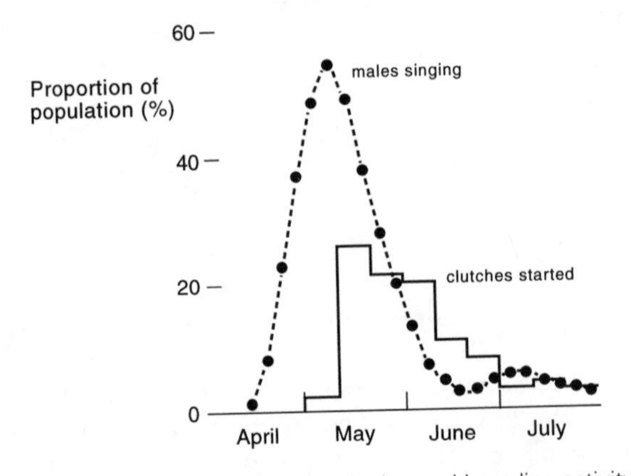

Figure 14.1.1 Relationship between seasonal singing and breeding activity in the sedge warbler *Acrocephalus schoenobaenus*. Male singing reaches a peak just before egg laying begins. (After Catchpole 1973)

experiments using recorded or synthesized calls have been particularly clarifying, as has the technique of treating females with estradiol to enhance and test their responses to song (Searcy and Marler 1981). Male song can have several different effects on a female: attracting her from a distance, priming her into physiological condition for breeding, and stimulating her for copulation (e.g., Kroodsma and Byers 1991).

Temporal patterns of song use suggest a relation to reproduction, as song often reaches a peak just before mating and egg laying (e.g., Duellman and Trueb 1986). Analysis on a finer time scale provides more detailed clues to song functions. For example, male sedge warblers cease singing after getting a mate (figure 14.1.1), suggesting that they sing to attract females, a conclusion also supported by experiments (section 14.3). Males in this and some other species also resume singing if they lose the mate (Catchpole 1973). This effect was also found in several species where females were experimentally removed from pairs (e.g., Wasserman 1977; Cuthill and Hindmarsh 1985). There are similar observational results in a number of species (see Björklund et al. 1989).

Patterns of use have revealed that in some birds there are several kinds of song with different functions. For example, two distinct song types in some Parulid warblers appear to be directed mainly at rivals and females, respectively (e.g., Morse 1970; Lemon et al. 1987; Kroodsma et al. 1989). A similar dichotomy is found in several other birds (e.g., Baptista 1978; Payne 1979; Järvi et al. 1980; Catchpole 1980, 1983; Catchpole et al. 1986; Temrin 1986; Lemon et al. 1987; Bensch and Hasselquist 1992). In some species, advertisement song is longer and more complex than aggressive song (see Catchpole 1987; Capp and Searcy 1991a,b). In the willow warbler *Phylloscopus trochilus*, a special song syllable becomes more common just before attack (Järvi et al. 1980). The aquatic warbler *Acrocepha-*

lus paludicola even has three main types of song, one used in short-range threat, another as a long-range territorial signal, and the third possibly for mate attraction (Catchpole and Leisler 1989).

Many frogs also have more than one type of song; there is evidence for territorial versus mate-attracting functions of two different types of call in about a dozen anurans (reviewed by Duellman and Trueb 1986; Wells 1988). The situation is similar in some crickets and other orthopterans (e.g., Alexander 1961, 1962; Burk 1983). In field crickets Gryllidae, males in addition to a long song have a shorter call used in contests, comparable to the short song of birds. Short chirps evoke subordination behavior in males that have just lost a fight, and aggression in males that have won. There is also a special male courtship song that elicits copulation behavior in females (Alexander 1961).

A possible explanation for the differences in form between mate-attracting versus aggressive song was suggested by Slater (1981) and Ince and Slater (1985). In birds that sing to attract a mate, males need not listen for a reply, as females simply respond by silent approach. The more continuous the song, the more likely it is to be heard by a female moving through the area. But when song is used in contests, males need to listen for responses by competitors; the song is then shorter and simpler. This idea needs to be tested by detailed analyses in a variety of species. Some birds use the same song types for both mate attraction and repulsion of rivals (e.g., Searcy 1988); it is not clear why species differ in this respect.

A statistical relationship between female choice and some aspect of male song has been documented for 45 species in table 6.A: 11 insects, 13 frogs, one deer, and 20 birds. If other display behavior with strong acoustic components is also included, five more birds are added to this list. A relationship between a male's contest success and some aspect of his song has been found in ten species in table 6.A (one insect, four anurans, four birds, one mammal).

Whereas many studies have examined present-day selection of song, quantitative analyses of its evolution are rare. A start in such a direction, using cladistic comparative analysis of calls in frogs, was made by Ryan (1986, 1988b). A genetic model of sexual selection of imitative song was presented by Aoki (1989).

14.2 Male Contest Competition and Song

Insects

In some Orthopterans, song has been shown to influence the spacing of males (e.g., Bailey and Thiele 1983). Increasing the loudness of song playback beyond a certain level silences a singing male *Teleogryllus oceanicus*

(Cade 1981b). In a laboratory study of the gregarious cricket *Amphiacusta maya*, experimentally silenced males were often interrupted in their courtship. They had to fight more and had lower mating success than calling males. Song in this species therefore deters other males. There was no evidence of female choice of song (Boake and Capranica 1982; Boake 1984).

Anurans

Calling leads to spacing out in many anurans as well (e.g., Whitney and Krebs 1975b; Fellers 1979; Robertson 1984; Sullivan 1982a; Telford 1985; Arak 1988b). Males in some species are less likely to sing if they hear rivals, or adjust their song in other ways to the presence of competitors (e.g., Whitney and Krebs 1975a; Perrill et al 1982; Telford 1985; reviewed by Wells 1988; Narins 1992).

The pitch (dominant frequency) of the call in many anurans is inversely related to male size (e.g., Ryan 1981, 1985; Gerhardt 1982; Robertson 1986b). A copulating male in the common toad *Bufo bufo* calls when attacked by rivals that try to take over the female. Davies and Halliday (1978) prevented males from calling by fitting a rubber band through their mouths like a horse's bit. Playback of calls then showed that attacking males respond to call pitch, being most likely to withdraw from the low-pitched calls of large males (figure 14.2.7). Male mating success in the wild increased with body size and fighting ability (Davies and Halliday 1979).

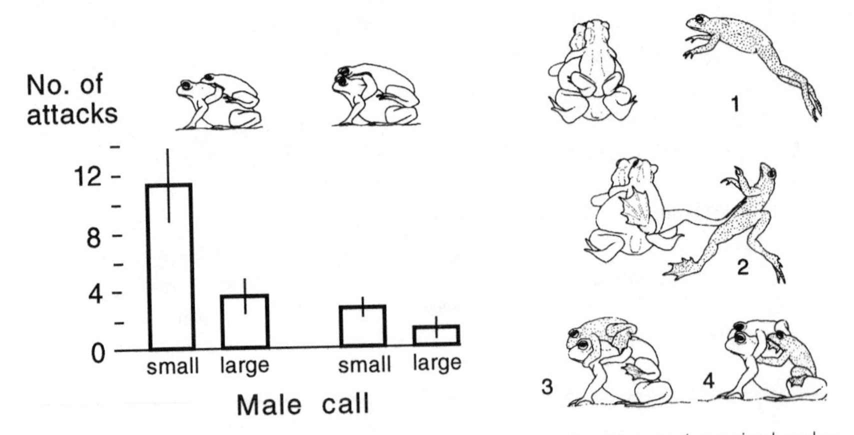

Figure 14.2.1 *Right*: Male toads *Bufo bufo* compete over females, and unpaired males often attack pairs in amplexus (1). The paired male tries to kick the rival away (2), but if he gets a hold, he tries to take over the female by pushing off the paired male backwards (3) or by squeezing between the pair from behind (4). *Left*: Males use calls in contests over females. Medium-sized males are more likely to attack a silenced defending male if he is small (left two bars) than if he is large (right two bars), and if there is playback of a call from a small male (left bar in each pair) instead of a large male (right bar). (After Davies and Halliday 1978, 1979; toads by Tim Halliday)

In natterjack toads *B. calamita*, large males produce louder calls with lower pitch than small males. In playback experiments, males attacked high-pitched calls and retreated from low-pitched or loud calls. Larger males had wider exclusive zones and higher fighting and mating success than smaller males in natural breeding ponds, indicating that sexual contest competition favors loud, low-pitched calls (Arak 1983a, 1988b; also see female choice below). Also, in the Australian frog *Uperoleia rugosa*, males are more likely to retreat from low-pitched calls (Robertson 1986b; section 7.5 above).

In *Acris crepitans*, males sometimes lower the pitch of their own call when confronted with the call from a larger opponent (Wagner 1989).

Birds

Loudspeaker playback experiments on territories of five species of birds suggest that song helps keep out other males (reviewed by Falls 1988). Such speaker occupation tests in, for example, the great tit *Parus major* and the red-winged blackbird *Agelaius phoeniceus* have shown that song effectiveness increases with repertoire size, that is, with the number of types of songs or syllables sung by a male (Krebs 1977; Krebs et al. 1978; Yasukawa 1981a; Yasukawa and Searcy 1985). Muted male red-winged blackbirds in several experiments were less likely to gain and more likely to lose territories than control males (Peek 1972; D. G. Smith 1976, 1979).

Temporarily muted males of Scott's seaside sparrow *Ammodramus maritimus peninsulae* obtained territories later than controls, and attracted females only after voice recovery (M. V. McDonald 1989). In a second, midseason test, all muted males lost their mate, and their territories shrunk or were lost. The muted males were involved in more close-range aggression than controls, probably because intruders ignored the song attempts of muted males, which instead turned to other defense behavior. After song recovery, the males re-expanded their original territories or established new ones (table 14.2.1), and some males attracted new mates. As the muted

Table 14.2.1

Effects of Losing (and Regaining) Singing Ability in Seaside Sparrows *Ammodramus maritimus**

	Change upon Losing (Regaining) Song					
Treatment	*Sample Size*	*Lost (Won) Territory*	*Smaller Territory*	*Larger Territory*	*No Change in Territory*	*Mean Size Change (%)*
Muted	21	6(3)	15(0)	0(14)	0(4)	−79(+76)
Sham-operated	13	0(0)	1(8)	10(0)	2(5)	+44(−13)
Undisturbed	30	0(0)	0(12)	17(0)	13(18)	+31(−15)

* Effects of regaining song are given in parentheses. After M. V. McDonald 1989.

males could produce other calls normally, McDonald's (1989) experiments demonstrate that song is necessary for holding a territory in this species. Her results also suggest that song is required for mate attraction and retention, but the reduced territory size of muted males may also have contributed to their lower mating success.

Possession of a good territory is crucial for male mating success in many birds (section 8.4), so the previous results imply that song is sexually selected by male contest competition. The great reed warbler *Acrocephalus arundinaceus* provides another example (Catchpole 1983, 1988; Catchpole et al. 1985, 1986).

Mammals

Playback experiments in the gibbon *Hylobates agilis* suggest that song plays a role in male contests over territories (Mitani 1988). The howling of the howler monkey *Alouatta seniculus* and the loud call of the male orangutan *Pongo pygmaeus* may be used similarly (Sekulic 1982; Mitani 1985). In the spotted hyena *Hyaena crocuta*, adult males are most active in territorial calling or direct competition over females (East and Hofer 1991). In red deer *Cervus elaphus*, male roaring rate is correlated with fighting ability and reproductive success through contest competition (Clutton-Brock and Albon 1979; Clutton-Brock et al. 1979), and also with female choice (see sections 5.6 and 14.3).

14.3 Song and Mate Choice

Insects

Crickets and other Orthopterans dominate the examples of female choice of song among insects (table 6.A). Their song intensity (loudness) has often been shown to affect female responses (e.g., Bailey et al. 1990). For example, in a controlled field experiment, male mole crickets *Scapteriscus acletus* with loud calls attracted more females than did other males. Tests with calling males as well as loudspeakers showed that the attraction of flying females increased with call loudness more strongly than predicted by simple sound propagation laws. Possibly, females compare several males before making a choice, favoring males with the loudest calls (Forrest 1980, 1983; also see Ulagarej and Walker 1975).

Female field crickets *Gryllus integer* in laboratory tests preferred males giving long, uninterrupted songs over males with shorter songs (Hedrick 1986, 1988). Father-son as well as sib correlations pointed to a strong genetic component of song length, with heritability estimates of about 0.7.

Even if this figure is higher than in natural, more variable situations, the result suggests a substantial heritability of song length, even though females strongly prefer long song (Hedrick 1988; also see section 3.4).

Song (produced by wing vibration) is favored by female choice in two flies, *Anastrepha suspensa* and *Drosophila melanogaster* (Webb et al. 1983; Sivinski et al. 1984; Bennet-Clark and Ewing 1969). While these and other studies suggest that females in some insects choose a mate in relation to aspects of his song, the evidence in many cases is based on female phonotaxis in the laboratory; more work is needed on male mating success in relation to call properties in the wild.

In homopteran plant hoppers Delphacidae, females as well as males produce acoustic vibrational courtship signals (e.g., Claridge 1985, 1990; DeWinter and Rollenhagen 1990). Contrary to expectations, males appear to be more discriminating than females in choosing between mates of their own and other species, although other aspects of sex roles do not seem to be reversed in plant hoppers (DeWinter and Rollhagen 1990). Further study of sexual signaling, mate choice, and sex roles in these and other insects with male discrimination in relation to female signals are needed.

Acoustic and vibrational signals are also used in courtship by many spiders (reviewed in Witt and Rovner 1982).

Anurans

Most studies of anuran calls have examined female responses to calls played through loudspeakers. Mating preferences have been examined in relation to a number of song features, such as call length, call rate, loudness, and frequency. Gerhardt (1991) distinguished between static call properties, such as dominant frequency, with coefficient of variation typically less than 4% within males, and dynamic call properties such as call rate, with c.v. of 12% or more. Playback experiments in treefrogs (Hylidae) showed that female preferences will result in stabilizing or weakly directional selection on static call properties, and highly directional selection on dynamic properties. The attractiveness of a signal is affected by variation in both static and dynamic properties (Gerhardt 1991).

Call rate

A song property often subject to female choice is call rate (seven anuran species in table 6.A). This is the case, for example, in Woodhouse's toad *Bufo woodhousei*, where male copulation success in the wild increases with call rate (Sullivan 1982a,b, 1983b, 1987). The possibility that call rate reflects male vigor, as suggested by indicator models (chapter 3), was tested but not supported (Sullivan and Walsberg 1985).

CALLING TIME

In many frogs, male mating success is correlated with chorus attendance and time spent calling (e.g., Forester et al. 1989; Morris 1989; and references in table 6.A). Endurance rivalry therefore seems to be important in anurans.

LEADING CALLS

In playback tests with four loudspeakers, female Pacific tree frogs *Hyla regilla* approached chorus leaders (Whitney and Krebs 1975a). The three females observed choosing in the field paired with leaders, which not only start and end calling bouts but also call louder and faster than other males. There was also a preference for leading calls in laboratory tests with *Hyperolius marmoratus* (Dyson and Passmore 1988a,b).

CALL LOUDNESS

Call loudness is often favored by female choice (in seven anurans in table 6.A). In the natterjack toad *B. calamita*, females placed halfway between two loudspeakers approached the loudest calls (Arak 1983a). When females were placed where the perceived loudness was the same from the two speakers, they showed no preference. Therefore, they do not seem to actively select the loudest callers (which could have been done by moving toward the sources to assess sound pressure gradients). Females seem to approach the calls of highest perceived intensity, tending to favor the loudest callers. This should lead a female to the nearest male and minimize the time during which she is vulnerable to predators. Many dead, partly eaten females at a breeding pond had probably been killed by predatory birds, so there should be a premium on rapid mating and spawning before daybreak (Arak 1983, 1988c).

CALL PITCH

Female choice in relation to male call pitch (dominant frequency) has been subject of much study and debate. Females discriminate with respect to pitch in loudspeaker tests in several frogs, for example tungara frog *Physalaemus pustulosus* (Ryan 1983, 1985; section 5.3 above), *Uperoleia laevigata* (Robertson, 1986b, 1990; section 7.5 above), and *Hyla chrysoscelis* (Morris and Yoon 1989). Within the normal range of frequencies, female tungara frogs preferred low-pitched calls characteristic of large males, which also had the highest mating success. Such a preference may be adaptive for females, as many eggs are left unfertilized if the male is too small; this is the case also in *U. laevigata* if the male deviates too much from the preferred size of 70% of the female's body mass (Ryan 1985; Robertson 1990). In *Hyla crucifer*, females in laboratory tests preferred low-pitched calls typical of large males, but Forester and Czarnowsky (1985) found no mating advantage for large males in the field.

Female cricket frogs *Acris crepitans* discriminate among males in relation to their call pitch, and the frequency sensitivity of the auditory system is correlated with the dominant frequency of the call, which differs among local populations (Nevo and Capranica 1985; Ryan and Wilczynski 1988).

The previous results may not be typical of most anurans. Field studies looking for a relationship between call pitch and male mating success in several other species gave negative results (Doherty and Gerhardt 1984; Gerhardt et al. 1987; Jacobson 1985, Arak 1988c; Höglund and Robertson 1988). So did a number of studies of male size and mating success (reviewed by Gerhardt et al. 1987). Frequency resolution precise enough for auditory discrimination between males of different sizes seems to be present only in some species. The physiology of the auditory system, especially the amphibian papilla, differs considerably among anurans, in ways that may constrain female response abilities (see Ryan 1988b; Zakon and Wilczynski 1988).

PULSE REPETITION RATE

Some of the most detailed experimental tests of female choice of calls have been done in Hylid frogs. Their calls are simple enough to be synthesized for tests of responses to various call properties. Females in several Hylids and other frogs are sensitive to pulse repetition rate, which often differs between closely related species (e.g., Gerhardt 1982; Klump and Gerhardt 1987; reviewed by Gerhardt 1988, 1991; also see section 9.2).

In *Hyla versicolor*, variation in pulse repetition rate is much more important than call loudness. For example, in tests between synthetic calls with pulse repetition rates of 20 and 40 pulses per second, females responded exclusively to the call with lower pulse rate, even if the faster call was up to eight times (18 dB) louder. Females preferred a call with pulse rate typical of a conspecific male over alternatives that differed by 25%. These responses are probably related to the risk of mismating with another sympatric species (Gerhardt 1982; Klump and Gerhardt 1987; Gerhardt and Doherty 1988; see section 9.2 above).

CALL DURATION

Loudspeaker tests show that females in several Hylids are not very selective with respect to call duration (reviewed by Gerhardt 1988). In the gray tree frog *H. versicolor*, however, females strongly prefer calls that are 1.5–4 times longer than typical for the species (Klump and Gerhardt 1987). The response in part depends on the higher duty cycle ratio (the average proportion of time calling) of the longer calls, but this is not the only factor. When long calls (18 pulses) were played with a duty cycle 2% longer than that of

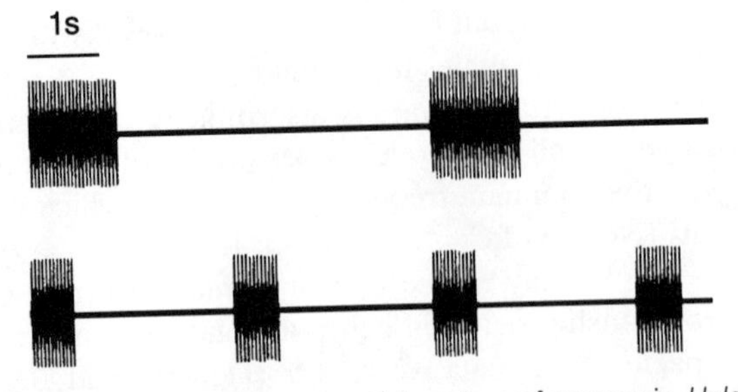

Figure 14.3.1 Synthetic calls used in tests of female preferences in *Hyla versicolor.* Females strongly preferred the longer calls (24 pulses per call, *top*) over the shorter calls (12 pulses). (From Klump and Gerhardt 1987)

short calls (12 pulses), females preferred long calls. A difference of only 1% was enough to make females favor 24-pulse over 12-pulse calls (figure 14.3.1). When the duty cycle was larger for short calls, even by as much as 10%, there was no preference. Females clearly were more responsive to long calls.

CALLING is energetically costly. Variation in call rate and duration together account for over 80% of the variance in metabolism during calling in gray tree frogs *H. versicolor*. Energy consumption may be twenty times higher at calling than at rest (Taigen and Wells 1985; Wells and Taigen 1989; Prestwich et al. 1989; also see section 10.4 above). Female gray tree frogs prefer energetically costly calls, which should favor males in good physical condition, but the consequences in terms of female fitness are not known (Klump and Gerhardt 1987).

Male mating success in several species is strongly related to the proportion of time he spends singing and is present at the song site, lek, etc. (table 6.A). The time he is able to spend at the song site is likely to be limited by energy requirements. Halliday (1987) and Ryan (1988a) reviewed the cost of call production in anurans, and pointed out that females in several species prefer calls with high energy content. Additional studies of such species might allow tests of sexual selection hypotheses. If male song production, for example call rate, is correlated with his general physiological fitness, an indicator process could be at work. If male call rate, on the other hand, is correlated only with his capacity of sustained calling, but not with more general aspects of his physiological fitness, a Fisherian process seems more likely (Ryan 1988a).

BIRDS

There is quantitative evidence that song or other acoustic display influences female responses to males in twenty birds in table 6.A. Some of these studies have identified song properties of importance for female choice. The reasons why females respond the way they do are still poorly known, however.

SONG RATE

There is a variety of evidence for female choice based on male song in the pied flycatcher *Ficedula hypoleuca*. Von Haartman (1956) found that males rarely used song in disputes, but sharply reduced their singing after gaining a mate. In a field test of this species and *F. albicollis*, with a male dummy at each of fourteen nest boxes with playback of male song, and at fourteen boxes without song, nine of ten females were caught in boxes with song, demonstrating its importance for mate attraction (Eriksson and Wallin 1986; also see Eriksson 1991).

Examining song phrase length and repertoire size, Alatalo et al. (1986) found no correlation with male mating success, which was, however, correlated with aspects of the territory, particularly nest box quality (also see Askenmo 1984; Slagsvold 1986). But at least one aspect of song, its rate of delivery, may influence female choice in the pied flycatcher. Male mating success was correlated with song rate in one year of Gottlander's (1987) study. In a feeding experiment a second year, the rate of singing increased if males were given extra food, especially at low temperatures, when few insects are active. Male song rate therefore likely reflects food abundance and territory quality, but the possibility was not excluded that females respond directly to the food situation, male song rate and mating success being indirectly related via food abundance. This possibility was controlled for in another experiment, where males provided with extra food during cold days sang at twice the rate of control males, which spent more time searching for food (Alatalo et al. 1990b). The experimental males attracted females more quickly. It is not entirely clear whether females compare several males in making their choice, but this seems likely; at least some females visit several nest boxes defended by different males before settling (Dale et al. 1990).

Willow warblers *Phylloscopus trochilus* use song in contests as well as for mate attraction. In a removal experiment, the song rate of replacement males was correlated with that of removed previous owners, suggesting that territory quality partly determines male song rate (Radesäter and Jakobsson 1989). In theory, a female can get a reasonable estimate of a male's singing rate in 10–15 minutes which might help her judge the quality of his territory. Male mating and reproductive success was also correlated with song versatility, which in turn was correlated with male age and arrival date (Järvi 1983). Females probably choose mates based on song rate or some other male trait, but a direct effect of territory quality also seems likely (Järvi et al. 1980; Järvi 1983; Radesäter et al. 1987; Radesäter

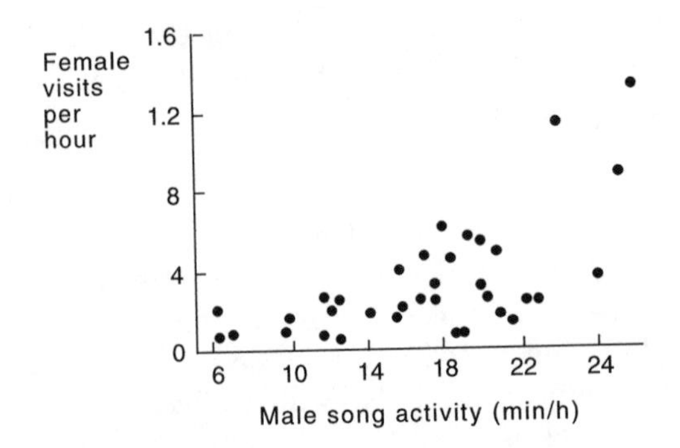

Figure 14.3.2 Success in attracting females increases with singing rate in male village indigobirds (*Vidua chalybeata*). (After Payne 1983)

and Jakobsson 1988, 1989; Arvidsson and Neergaard 1991; also see section 8.8).

Song rate correlates with the food supply also in, for example, dunnock *Prunella modularis* and Carolina wren *Thryothorus ludovicianus* (Davies and Lundberg 1984; Strain and Mumme 1988). Dunnock males with extra food sang more and had higher mating success than control males. In these and other species, male song rate should reflect territory quality and male condition, which perhaps influence female reproductive success (e.g., Yasukawa 1981b,c; East 1982; Greig-Smith 1982; Gottlander 1987; Radesäter et al. 1987; Radesäter and Jakobsson 1988; Arvidsson and Neergaard 1991). If feasible, experiments that manipulate male song rate and food abundance independently might clarify the relative importance of the two factors for female settlement.

In the village indigobird *Vidua chalybeata*, male mating success is closely correlated with his song rate and time spent singing (figure 14.3.2; Payne 1983). In this species, it seems unlikely that the correlation arises indirectly by female choice of territory quality. The indigobird is a nest parasite of an estrildid finch, and a female is not restricted to the resources of a single male (Payne and Payne 1977; Payne 1983).

SONG REPERTOIRE SIZE

One of the first studies showing that song repertoire size affects female responses was done in female canaries *Serinus canarius* (Kroodsma 1976). Females exposed to playback of large song repertoires in the laboratory nested earlier and laid more eggs than females hearing small repertoires. The development of female reproductive behavior is enhanced by male courtship calls or song also in budgerigar *Melopsittacus undulatus* and ring dove *Streptopelia risoria* (e.g., Brockway 1965; Lehrman and Friedman 1969; Nottebohm and Nottebohm 1971).

Song can also have a short-term effect, stimulating the female to copulate. For example, female cowbirds *Molothrus ater* in the laboratory show most copulation solicitation displays to the song of dominant males, which also obtain most copulations (West et al. 1981). Searcy and Marler (1981) and Searcy (1984) showed that estradiol-treated female song sparrows *Melospiza melodia* solicit more for playback of conspecific than for alien song, and more for repertoires of several song types than for repetition of one type. This suggests that females favor males with large repertoires. Such males did not pair up earlier than others in the field, however. Song repertoire size was not correlated with male pairing date, male quality (age, size, weight, dominance), or territory quality (Searcy 1984).

Laboratory tests of responses to song repertoires have shown that females usually prefer large repertoires. Among ten passerine species reviewed by Searcy (1992), eight gave positive results (also see Searcy and Marler 1984). Field tests have produced more mixed results. There was no preference for large repertoires in four out of six species when territory quality was controlled for. Searcy (1992) concluded that repertoire size usually has at most a weak effect on female settlement patterns in the field; the effect may be more important in extra-pair copulations. Male mating success was, however, correlated with song repertoire size in the field in the sedge warbler *Acrocephalus schoenobaenus*; laboratory tests also support this conclusion (Catchpole 1980; Catchpole et al. 1984). There were similar results also in the great reed warbler *A. arundinaceus* (Catchpole 1986).

In the great tit *Parus major*, large song repertoires give an advantage in repelling other males (see above), but several field studies found no effect of repertoire size on male mating time or mate quality (Krebs et al. 1978; McGregor et al. 1981; Lambrechts and Dhondt 1986). Yet estradiol-treated females tested with repertoires ranging from one to five song types responded with solicitation displays in proportion to repertoire size (Baker et al. 1986). The reason for the different results in laboratory and field are not known. In two field studies of great tit, male lifetime reproductive success increased with song repertoire size (McGregor et al. 1981; Lambrechts and Dhondt 1986). The reason appears to be correlations between repertoire size and male dominance, survivorship, and territory quality.

In the red-winged blackbird *Agelaius phoeniceus*, song repertoire size appears to be more important for attracting or stimulating females than for territory defense. Males use larger repertoires in the presence of a stuffed red-winged blackbird if the dummy is a female than if it is a male (Searcy and Yasukawa 1990). Repertoires of four song types are more effective than a single type in triggering solicitation displays in estradiol-treated females (Searcy 1988). So far, however, there is no direct evidence that song repertoire size in red-winged blackbirds influences female choice of mate in the field.

Many kinds of advantages of large song repertoires have been suggested (see Kroodsma 1990; Harper 1991; Kroodsma and Byers 1991; Searcy 1992 for reviews). For example, large repertoires may

1. Stimulate females in mate choice or other phases of the reproduction (e.g., Catchpole 1980);
2. Intimidate or deter rivals in contests (section 14.2);
3. Permit song matching of more neighbors (which might lead to advantages in territory defense (Falls et al. 1982; McGregor and Krebs 1989);
4. Allow specialization of song for sending different messages (section 14.1);
5. Lead to less exhaustion in the singer than does repetition of the same song (Lambrechts and Dhondt 1988a; but see Weary and Lemon 1988).

There is some support for each of these hypotheses, which are not mutually exclusive, but more work is required to assess their relative importance within and among species (see Harper 1991; Kroodsma and Byers 1991; Searcy 1992). For example, comparative analyses may shed light on the causes of variation in song repertoire size among species. Four of the hypotheses in table 1.4.1 seem plausible in this context (B1-2 and B4-5); their role is likely to vary among species, in ways that remain to be clarified (also see section 14.4).

FEMALE SONG

In species where females sing, this is usually not a correlate of reversed sex roles. Most such birds are duetters, the mates singing together. Duetting occurs mainly in monogamous species with year-round territoriality, and it seems to function in territory defense and communication between the mates (Farabaugh 1982).

In species with reversed sex roles, females are vocally more active than males (Thorpe 1961). Females sing also in some nonreversed species, for example song sparrow *Melospiza melodia* and yellow warbler *Dendroica petechia*, apparently in defense of the territory against other females (Arcese et al. 1988; Hobson and Sealy 1990). Female red-winged blackbirds *Agelaius phoeniceus* have two types of song, one apparently used against other females, the other in communication with the mate (Beletsky 1983).

In some species, for instance the European robin *Erithacus rubecula*, both males and females sing and defend individual territories outside the breeding season. This is the case in the wintering area of some migratory species, in which winter territories provide resources for survival, but not for reproduction (e.g., Schwabl 1992). Song in these species therefore clearly is favored also by natural selection, not only by sexual selection.

SONG AND ACOUSTIC DISPLAY IN RELATION TO MATING SYSTEM

Certain song properties are related to the mating system in birds. Among North American wrens Troglodytidae, song complexity increases with the degree of polygyny (Kroodsma 1977). In European *Acrocephalus* warblers and *Emberiza* buntings, on the other hand, the most complex songs are sung by monogamous species (Catchpole 1980; Catchpole and MacGregor 1985). The wrens are a much larger group, ranging in habitat from marsh to desert; their songs may have evolved in communities with very different species diversities, which might influence song complexity (Kroodsma 1977). *Acrocephalus* warblers are more homogeneous, mainly inhabiting eutrophic lake shores and marshes. Catchpole (1980) suggested that female choice favors long, complex song in monogamous species of this genus, perhaps because song complexity reflects male capacity for providing food to the offspring (also see Bensch and Hasselquist 1992). Further comparative studies of other groups are needed to test these ideas.

In a comparative study of nine families of birds, Loffredo and Borgia (1986b) examined acoustic display in relation to the mating system. Lekking and other polygynous species in which males provide no parental care have more harsh, broadband, "noisy" sounds that are less tonal and melodic than those of their monogamous relatives. This pattern also holds within families, and so has arisen by convergent evolution, probably representing adaptations to the mating system. Monogamous species often have large territories, where highly tonal sounds should be effective long-distance signals. At short range, on the other hand, for example on a lek with many males, harsher broadband sounds may be more locatable; harsh sounds are characteristic of aggressive behavior (e.g., Morton 1977, 1982; Wiley and Richards 1982). Also in bowerbirds, however, where male display sites are scattered, the calls have similar properties as in lekking species (e.g., Borgia et al. 1985; Loffredo and Borgia 1986a,b). Loffredo and Borgia (1986b) suggested that both male contests and female choice have shaped these displays, and that the acoustic similarities among the polygynous species have arisen in part through female choice of the most dominant males.

These similarities of calls in polygynous birds are not predicted by Fisherian models, which suggest that male display evolves in rather arbitrary directions by female choice. The similarities seem to be more in line with assessment hypotheses, suggesting that choice of a dominant male increases the chances for the female to gain some direct phenotypic benefit of her mate choice (chapter 8), or to have her offspring fathered by a high-quality sire (e.g., Borgia 1979; Loffredo and Borgia 1986b).

PARASITE RESISTANCE

Hamilton and Zuk (1982; also see sections 1.5, 3.7, 13.8 above) reported a correlation in North American passerines between hematozoa prevalence and song complexity (as well as plumage brightness). The song relationship was reanalyzed in a more detailed study of European and North American passerines by Read and Weary (1990), who found it to be confounded by phylogenetic associations. Within taxa there were no associations between any of seven song complexity measures and hematozoa, contrary to the prediction.

MAMMALS

In red deer *Cervus elaphus*, roaring is important not only in contests among males (Clutton-Brock and Albon 1979); it also influences female responses in important ways. In two-choice loudspeaker tests with farmed red deer, females were attracted to high roaring rates and to leading bouts of roars. In another experiment, females exposed to playback of roaring ovulated earlier than control females (McComb 1988, 1991). This suggests that roaring may enhance male success both by attracting hinds and making them more likely to ovulate and mate before being taken over by another stag. In the wild, roaring rate reflects male condition and is positively correlated with fighting ability and reproductive success (Clutton-Brock and Albon 1979; Clutton-Brock et al. 1982).

FEMALE INCITATION OF MALE COMPETITION?

In some mammals and birds, sexually active females attract the attention of males by calls. Female elephant seals *Mirounga angustirostris* usually protest loudly against being mounted, especially if the male is young. The behavior of the female alerts and attracts other competing males and makes her more likely to mate with an older, dominant male (Trivers 1972; Cox and LeBoeuf 1977).

Female African elephants *Loxodonta africana* show similar responses to young males (Poole 1989a). Estrus females solicit guarding from large bulls in musth and accept mating with them (section 3.6), but avoid younger, nonmusth males. Chased females give a loud pulsated roar that attracts nearby higher-ranking bulls, which prevent the younger male from mating. Poole suggested that some of the very low-frequency calls of elephants (14–35 Hz; e.g., Poole et al. 1988) may be used by estrus females to attract bulls from long distances and incite competition among them. This should make females more likely to mate with large, dominant males in musth, which seems to reflect good condition.

Female chipmunks *Eutamias obscurus* and *E. merriami* before estrus spend long periods each day calling loudly from exposed perches, attracting the attention of males. Callahan (1981) suggested that this long-range display fills a similar function of inciting male competition. So did Montgomerie and Thornhill (1989) for certain calls given by females of some birds just after laying an egg, when copulation may be most likely to fertilize the next egg (also see Arvidsson 1992). Their review suggested that such calls occur mainly in other than strictly monogamous species, as predicted.

14.4 Selection of Responses to Song

There is now evidence from many species that song is sexually selected by male contest competition, female choice, or both. It is less clear why song has come to play these roles in competition over mates and territories, and why listeners respond to song the way they do (Searcy and Andersson 1986).

Some possible reasons why males use or respond to song in competition over breeding territories are that it provides information about the local density of males, the singer's species identity, and his aggressive intentions and defense ability. In many frogs, the dominant frequency reflects the size of the caller (section 14.2). In some insects, frogs, and birds, males use different song types for contests and mate attraction, and the likelihood of attack in some species is reflected by male song properties (section 14.1).

Empirical evidence for Fisherian sexual selection of song is weak, although it seems a plausible possibility in some species, such as the sedge warbler *Acrocephalus schoenobaenus*, where females favor males with large repertoires (Catchpole 1980). There are, however, several alternative explanations for the advantage of large song repertoires in mate choice, such as reduction of habituation in listeners (e.g., Nottebohm 1972; Krebs 1976). In combination with stimulus specificity, this might give rise to a female response bias favoring large song repertoires in some species (Searcy 1992; also see section 10.7).

The comparative study of Loffredo and Borgia (1986b) found patterns of song properties in relation to mating system that are not predicted by Fisherian models (section 14.3). Other aspects have clearly been important, but this does not refute the possibility that a Fisherian mechanism has also played a role. As ever, it is difficult to exclude a particular mechanism of sexual selection, even if others can be shown to be at work.

The situation is similar for indicator mechanisms. In Woodhouse's toad *Bufo woodhousei*, however, a genetic indicator mechanism seems unlikely to be involved in the maintenance of a female preference for high call rates.

There was no correlation between male call rate and his maximum aerobic capacity (Sullivan and Walsberg 1985). Other negative evidence comes from the great tit *Parus major*, in which the possibility for an indicator process based on repertoire size seems to be at hand, but females did not choose mates in the predicted way in the field (section 14.3). Song properties in many other species are correlated with male size, dominance status, or other qualities, but their relations to mate choice and fitness are usually unknown (Searcy and Andersson 1986).

Female attraction to the strongest perceived signal has been demonstrated in the natterjack toad *Bufo calamita* (Arak 1983, 1988c), and may occur also in other species. In general, such a response should minimize the time during which a female is exposed to predators while searching for a male (section 1.4). Such a simple mechanism does not, however, explain female choice in, for example, *Hyla versicolor* and some other frogs, where mating preferences seem to have been shaped in part by selection for avoidance of mismatings with closely related species (e.g., Gerhardt 1982; Klump and Gerhardt 1987; Gerhardt and Doherty 1988; also see section 9.2).

Material benefits may favor female preferences for song properties that reflect male quality, such as condition or experience, especially in species where males defend breeding territories, or provide parental care or other resources (section 8.8). There is another direct benefit of female choice in relation to male song in some frogs: choosing a male of suitable size (reflected by call pitch), a female can improve the fertilization of her eggs as compared to random choice of mate (Ryan 1985; Robertson 1990; sections 5.3 and 7.5 above).

14.5 Some Aspects of Method

Much of the understanding of sexual selection of song has come from playback experiments in laboratories, combined with field studies. For example, in the sedge warbler *Acrocephalus schoenobaenus*, playback tests with estradiol-treated females agree with field results on male pairing date, strongly suggesting that females prefer large song repertoires (Catchpole 1980; Catchpole et al. 1984). In several studies of song repertoires and female choice, however, mating advantages of large repertoires in the laboratory are not supported by field results, for example on pairing date (section 14.3). This does not necessarily mean that there is no such advantage. The most attractive males need perhaps not be the first to pair, if they can afford to be more critical than others in their choice of mate (Baker et al. 1986; also see Burley 1977). There may also exist as yet undiscovered advantages of large repertoires for mate choice.

There is, however, a possibility that some laboratory playback tests have little relevance to the field situation, where females will simultaneously hear many males at different sound pressure levels (see, e.g., Gerhard 1982, 1988, 1991; Searcy and Andersson 1986, for discussion). Uncontrolled influences from intensity and temporal relationships of the test calls can also complicate interpretation (Dyson and Passmore 1988a,b). A host of experimental details may influence the outcome of playback experiments (McGregor et al. 1992).

Laboratory tests using copulation solicitation displays by females measure the close-range, copulation-stimulating effect of song. In the field, long-range attraction of females by song is also likely to be important, but copulation solicitation tests do not seem well suited for exploring this aspect. There is a need for more studies that distinguish between long-range attraction and close-range stimulating functions of song. In addition, long-term priming of reproductive behavior needs to be distinguished from short-term stimulation of copulation.

Statistical and experimental problems in playback experiments, which also apply to some analyses of visual and other signals, were reviewed by Kroodsma (1989, 1990) and McGregor et al. (1992). For example, whereas the sample of responding individuals is usually adequate, often only one stimulus of each category studied is used in playback tests. In some cases it is more appropriate to use a sample of stimuli of each category (see Kroodsma 1989a,b, 1990; Catchpole 1989; Searcy 1989, for discussion). Pseudoreplication, with statistical testing based on formally larger samples than the actual number of independent units measured, is a problem in certain playback experiments, as in some other areas of animal behavior and ecology (Hurlbert 1984; see section 4.2 above). Kroodsma (1989a,b) and McGregor et al. (1992) discuss approaches for avoiding such problems, stressing the need for clear specification of the hypothesis to be tested.

14.6 Summary

Sexual selection of male song has been demonstrated in many insects, frogs, and birds, by female choice as well as male contest competition. Song form is related to function. In some species, short, simple song is used in male contests, and longer song for mate attraction. Monogamous territorial birds often use tonal sounds effective in long-distance communication, whereas polygynous, mainly lekking birds use harsher sounds, characteristic of aggressive behavior.

Experimental song playback and other evidence show that song helps keep out other males from the territory, attract a female to it, and stimulate her to copulation. Song has been shown to deter other males in crickets, anurans, and birds. Calls with low fundamental frequency, reflecting large male size, are particularly effective in male contests and mate stimulation in some anurans; so are large song repertoires in some birds.

Playback tests have also demonstrated that song attracts mates in crickets, fish, anurans, birds, and deer. Two of the important call variables are song loudness (crickets, anurans) and rate (anurans, birds, deer), both of which may reflect male condition.

The selective reasons for male and female responses to song are poorly known. There is some support for several hypotheses of female choice, mainly direct attraction to the strongest signal, species recognition, and material benefits. Studies that examine the relative roles of the various mechanisms are needed to clarify the evolution of song, and the responses to it.

15 Chemical Signals

Scent is the third major channel of animal communication. Being harder to detect and measure, it has been less studied in a sexual selection context, but results from a variety of animals now point to a role of scent in competition over mates; nine examples are listed in table 6.A.

It has long been known that scent is used for mate attraction in insects, and for territorial marking in mammals (reviewed, e.g., in Jacobson 1972; Birch 1974; Gosling 1982; Brown and Macdonald 1985; Duvall et al. 1986; Macdonald et al. 1990; Alberts 1992). Scent is apparently involved in sexual selection by mate choice as well as by contests over mates, or resources needed to attract mates. Also in flowering plants, sexual selection may influence the evolution of pollinator-attracting scent (section 17.6).

15.1 Insects

There is strong evidence of sexual selection of male scent signals among insects. In contrast with visual and acoustic signals, however, long-range pheromones that attract mates are usually emitted by females, not males (e.g., Jacobson 1972; Thornhill 1979; Greenfield 1981; Hölldobler and Wilson 1990). (There are exceptions, such as the cabbage looper moth *Trichoplusia ni*, in which both sexes produce long-range attraction pheromones; Landolt and Heath 1990).

At first sight, this seems at odds with the rule that males show the greatest mating effort (section 7.2). It is possible, however, that males do so also when females produce attraction pheromones. (Williams 1992 even questioned whether the emitted female substances are adapted for signal function.) Compared to most visual and acoustic signals, pheromones require more specific receptors, and few predators may be able to locate the female. The risk incurred by signaling for this and other reasons should be lower with pheromones than visual or acoustic signals. Searching for the pheromone source, which is harder to locate than visual or acoustic stimuli, on the other hand will make an insect more vulnerable to predators such as birds, bats, spiders, and other hazards. This risk is taken by males (chapter 10). In those cases where predators have broken into a pheromonal system of mate attraction, the victims are also mostly males. The main cost of acquiring a mate may therefore fall upon males also when females emit attraction pheromones (Alexander and Borgia 1979; Thornhill 1979; Greenfield 1981; Thornhill and Alcock 1983). Predation risk and energetic as well as other fitness costs need to be quantified to test whether this is the case.

Lloyd (1979) and Greenfield (1981) suggested that a form of female choice is possible also when females release attraction pheromones. By emitting pheromone at a low rate, a female should make it likely that she obtains a male highly efficient at detecting the signal. To the extent that receptor sensitivity and search efficiency is heritable, sons of the female will then also have high mating success. Energetic constraints and avoidance of specialized predators should also favor restraint in signaling, however, so this idea is difficult to test. In addition, the rate of pheromone release is probably also influenced by many other selective factors, such as advantages of rapid mating. Reduced pheromone production will also reduce the distance from which the female is perceived by males, hence reducing her pool of potential mates. Modeling might help clarify what rate of pheromone production is optimal under given conditions.

There has been exciting progress in the understanding of genetic variation in moth pheromone systems (reviewed by Baker 1989, Löfstedt 1990). For example, the European cornborer *Ostrinia nubilalis* has two pheromone strains. Hybridization tests and genetic analyses show that the strain differences depend on pairs of alleles on at least three loci. One autosomal locus influences the female pheromone, another the male receptor, and a third, sex-linked locus influences male search flight behavior. Probably all three loci reside on different chromosomes; there is no evidence of linkage (Klun and Maini 1979; Hansson et al. 1987; Roelofs et al. 1987; Löfstedt et al. 1989).

These and other results suggest that reciprocal selection of genetically separate sender and receiver mechanisms explain the match between mate-attracting pheromones in one sex and preferences in the other. There is no evidence of genes with pleiotropic effects on both sender and receiver (Löfstedt 1990; also see Doherty and Hoy 1985; Boake 1991; and section 2.7 above).

Competition over mates is probably stronger in males than females in these insects as well. With female emission of long-range pheromones, the receptor (male) side is therefore more likely than the sender (female) side to be subject to sexual selection. A wide male "response window" permitting detection of females with different pheromone compositions may reflect sexual selection of male responses in such long-range systems (Löfstedt 1990). Corresponding genetic and other analyses of cases in which females respond to male close-range pheromones (see below in this section) should also be of interest.

Figure 15.1.1 During a stage of the courtship sequence in the queen butterfly *Danaus gilippus* the male flies with extruding pheromone glands close above the female. (From Brower et al. 1965)

Courtship pheromones used after the sexes have met in most cases are emitted by males (reviewed, e.g., by Birch 1974; Arnold and Houck 1982; Boppré 1984). Early experiments with grayling butterflies *Hipparchia semele* by Tinbergen et al. (1942) indicated that chemical stimulation of the female is necessary for male mating success. In the queen butterfly *Danaus gilippus*, the male, after overtaking a flying female, induces her to alight by extruding his hair-pencil glands, releasing pheromones near her antennae (figure 15.1.1). The hovering male emits pheromones until he achieves copulatory position. Experimental prevention of pheromone delivery reduces male mating success, suggesting that the pheromone is favored by sexual selection through female choice (Brower et al. 1965; Myers and Brower 1969; Pliske and Eisner 1969; reviewed by Boppré 1984).

A similar courtship system is found in the arctiid moth *Utetheisa ornatrix*, which shows several parallels with some Danainae butterflies (Conner et al. 1981, 1990). In both cases, male mating success depends on emission of close-range pheromones by evertible glands. The pheromones are derived from food plant alkaloids (figure 15.1.2). The alkaloids seem to provide protection against some herbivores for the plants, and against insectivores for the butterflies. Conner et al. (1981) suggested that females might benefit from preferring males with high defense vigor (reflected by pheromone emission), either through offspring quality, or through protection during copulation. It also seems possible that males have evolved these courtship pheromones by sensory exploitation of a preexisting female preference for defense alkaloids and related substances.

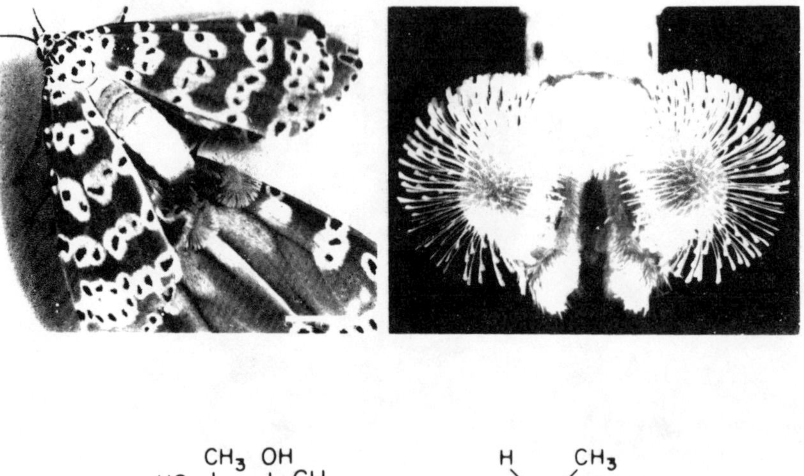

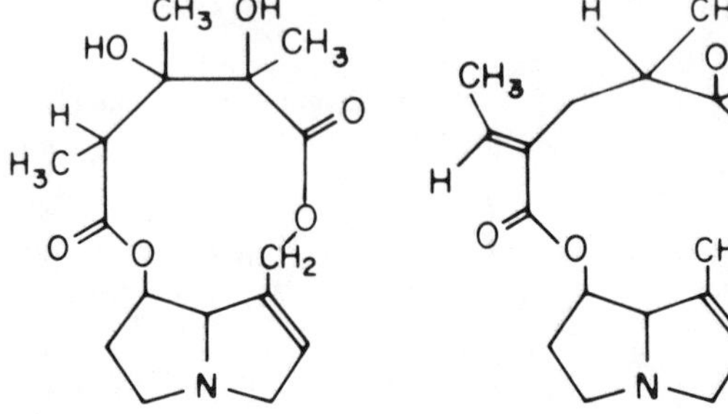

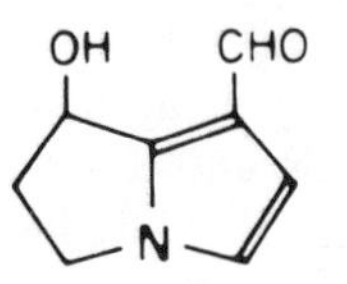

Figure 15.1.2 The male courtship pheromone (hydroxydanaidal, *bottom*) in the arctiid moth *Uthetheisa ornatrix* is emitted from paired abdominal glands (*top*, right). The pheromone is derived from food plant alkaloids, two of which are also shown (monocrotaline and usaramine). (From Conner et al. 1981)

In the oriental fruit moth *Grapholita molesta*, females choose their mate based in part on his quality or quantity of courtship pheromones (Baker and Cardé 1979; Löfstedt et al. 1989). As in other moths, the female attracts the male by a long-range pheromone. He courts her at close range by extruding scent-emitting hair pencils and vibrating the wings, propelling pheromones to her (e.g., Baker and Cardé 1979; figure 15.1.3). Females discriminate

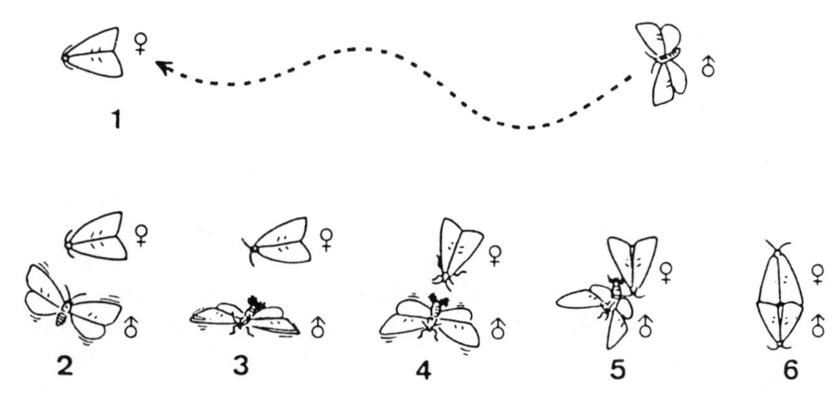

Figure 15.1.3 Courtship sequence of *Grapholita molesta*. Upon locating the female through her long-range pheromone and landing beside her (1–2), the male turns away and creates an airstream by his vibrating wings, propelling pheromones from his abdominal hairpencils to the female (3). The female walks toward the male and touches the tip of his abdomen (4), evoking a copulatory attempt by the male (5, 6). (From Nishida et al. 1985)

among males in relation to experimentally manipulated composition of the scent (Löfstedt et al. 1989). Presence in the male scent of ethyl-transcinnamate, normally ingested with the fruit diet, improves the attraction of females. Also here, it seems possible that males in evolving this courtship pheromone have taken advantage of an already present, food-related female preference for certain scents. Perhaps sensory exploitation has been common in the evolution of Lepidopteran courtship pheromones.

Baker and Cardé (1979) suggested that the male scent-emitting organ and its display have evolved by female choice of mate; so did Müller already in 1877. In *G. molesta*, the organ may derive from an extrudable copulatory clasper that incidentally emitted an odor providing the basis for signal evolution. Comparative evidence suggests that selection against hybrid matings has played a role in the evolution of male scent organs in moths (Phelan and Baker 1986; section 9.2 above).

In the butterfly *Colias eurytheme*, female preferences for male pheromones lead to disruptive sexual selection. There are two different female morphs, white and orange, which prefer males at opposite ends of the spectrum of courtship pheromones (Sappington and Taylor 1990).

In some cases, Lepidopteran male pheromones perhaps also function in contest competition over females. Such a function seems plausible because in contrast with females, who cannot detect their own scent, males have receptors for the scent of their own sex (Boppré 1984).

Odors play a role in sexual selection also in the cockroach *Nauphoeta cinerea*. Laboratory experiments showed that females discriminate among males based on odor, which is related to male dominance status (Breed et al. 1980; Moore and Breed 1986; Moore 1988). In the sweat bee *Lasioglossum zephyrum*, males discriminate among females based on pheromonal cues; avoidance of inbreeding may favor such discrimination (Smith and Ayasse 1987).

15.2 Fishes, Amphibians, and Reptiles

Pheromones play an important attracting or priming role in the reproduction of many fishes (reviewed by Stacey et al. 1986). Male pheromones have been shown to attract females in a variety of species, and to stimulate ovarian growth and ovulation in a Belontiid and a cichlid. Stacey et al. (1986) suggested that male territoriality and parental behavior in turbid waters have favored chemical rather than visual male signals for attracting females in, for example, Belontiids.

There is suggestive evidence that glandulocaudine fishes of the family Characidae produce courtship pheromones in glands at the base of the tail. Extreme radiation of the glands and other secondary sex traits in these fishes may have occurred by sexual selection (see Nelson 1964).

In a number of teleosts, for example Poeciliids, chemical stimuli from females attract males and increase their sexual activity, apparently at the time most appropriate for successful fertilization (Stacey et al. 1986).

Courtship pheromones with short-term stimulating effects are common in salamanders (reviewed by Arnold and Houck 1982; Houck 1986). Observations in the salamander *Notophtalmus viridescens* suggests that the male stimulates the female chemically during courtship (figure 15.2.1). Reproductive males of this species develop "hedonic glands" on the cheeks, secretions from which are rubbed over the nares of the female during courtship (Rogoff 1927; Verrell 1982). As Arnold has remarked, we shall never know whether the glands really do promote salamander pleasure, but recent experimental tests at least indicate that secretions from these glands promote sexual receptivity (P. Watson; cited in Houck 1986).

In some *Desmognathus* salamanders, the male has sexually dimorphic teeth that are used to pierce the skin of the female, in an area where male glandular secretions are applied, probably facilitating "injection" of courtship pheromones into the female's circulatory system (Arnold and Houck 1982).

In *Triturus* newts, experiments suggest that chemical signals from the cloacal glands of the male, transmitted by the water current created by his tail-fanning display, have a cumulative stimulatory effect on the female (Halliday 1975, 1977a,b). In the great crested newt *T. cristatus*, the current has been shown to carry pheromones from the male abdominal gland to the

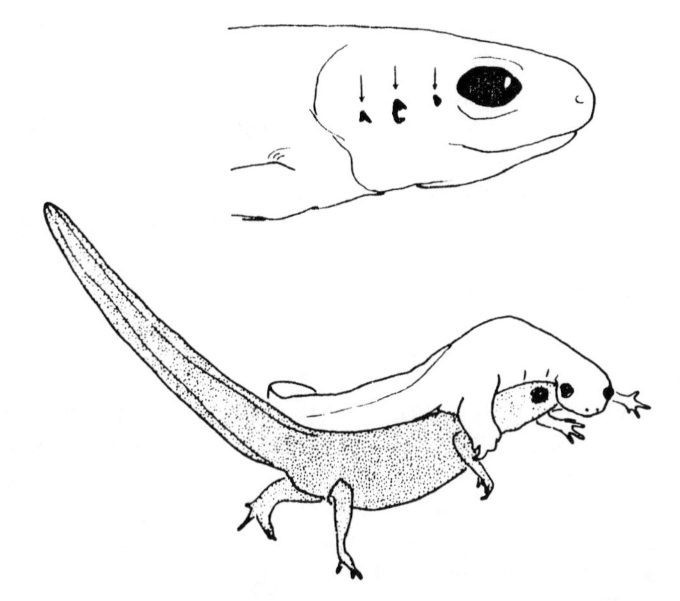

Figure 15.2.1 The male red-spotted newt *Notophtalmus viridescens* produces a courtship pheromone in his cheek glands (*top*), applying it to the female by rubbing the glands across her nares during courtship (*bottom*). (From Arnold 1977)

female's nostrils (Malacarne et al. 1984; Malacarne and Vellano 1987). Reviewing a variety of evidence, Houck (1986) concluded that it best supports the idea that male courtship pheromones in salamanders affect female receptivity.

Terrestrial *Plethodon* salamander males (and perhaps females) scent-mark their territories with fecal pellets, which probably function in territory defense, and perhaps also in attraction of females (e.g., Jaeger 1986; Wall et al. 1989; Mathis 1990).

In some snakes and lizards, female scent is important for male courtship behavior (e.g., Garstka and Crews 1986; Ford 1986). Scent also seems to influence specific mate recognition in lizards (e.g., Duvall et al. 1980; Cooper and Vitt 1986).

15.3 Mammals

Scent plays a major role in the social life of mammals (e.g., Doty 1976; Brown and Macdonald 1985; Alberts 1992). There is usually a marked sex dimorphism in mammalian odor production. Males in many mammals produce a larger amount and a wider repertoire of odors, and scent-mark more than females do (e.g., Stoddart 1974; Thiessen and Rice 1976; Bell 1983; Brown and Macdonald 1985; Jannett 1986). Blaustein (1981) suggested that scent in small mammals may take the role of other sexually selected traits in larger species, such as body size, weapons, or visual ornaments. This prediction was corroborated by Jannett (1986) in a comparison of twenty-six species of *Microtus* voles. Scent glands were larger in males than females; they were most developed at sexual maturity and during the reproductive season; the glands were larger in polygynous than in monogamous or polyandrous species.

Male Contest Competition

Many male mammals scent-mark their home range or defended territory (reviewed in Brown and Macdonald 1985). An array of hypotheses has been raised to explain scent marking, whose functions are still debated. Among the hypotheses are that scent marks (1) inform territory intruders about the status of the owner, and deter or intimidate them, (2) help orientate the owner within his territory, (3) attract or stimulate mates, and (4) assist in pair-bond formation (reviewed by Gosling 1982, 1986, 1990). There are problems with several of these explanations. Deterrence seems an unlikely general explanation (but see, e.g., Erlinge et al. 1982), as animals without defended territories often scent-mark (Ralls 1971). In addition, the intimidation and deterrence hypotheses do not explain why intruders should respond in this fashion (Gosling 1982).

Territorial scent marking might play a role in attraction or physiological priming of mates in some species but this remains to be demonstrated. Several aspects suggest that male scent marking is related to contest competition over females, or resources needed to attract females. Males mark more than females, dominant males and territory owners mark more than others, and males often mark conspicuously during contests with others (reviewed, e.g., by Ralls 1971; Gosling 1990). Scent marks may function in competitor assessment (Erlinge et al. 1982; Gosling 1982; 1990). For example, if an intruder compares the memorized scent from marks with the scent of an opponent, the intruder may be able to find out if the opponent is the holder of the territory or females at stake. If he is, he may be likey to escalate a contest to a dangerous level, and his status suggests good "resource holding power" (Gosling 1982, 1990).

In line with this idea, males in many mammals mark their own body with the same substances used to mark the territory. In some species, males also directly scent-mark females (figure 15.3.1; reviewed by Gosling 1990). As predicted, male house mice were less likely to fight another male if his scent matched that of the substrate than if it did not match (Gosling and McKay 1990).

Figure 15.3.1 Male gerenuks *Litocranius walleri* mark females with secretion from the antorbital gland during mating interactions. (From Gosling 1985)

Attraction and Priming of Mates

There is evidence from lagomorphs (e.g., Reece-Engel 1988) and rodents that females discriminate among males in relation to their odors. For example, in the brown lemming *Lemmus trimucronatus*, estrus females tested in Y-maze olfactometers preferred the odor of a dominant male over that of a subordinate. Androgens may be involved in the odor difference between males (Huck et al. 1981). Females with access to tethered dominant and subordinate males spent more time with dominants. Free-moving dominant males had higher copulatory success than subordinates, an effect of male agonistic behavior as well as female choice (Huck and Banks 1982).

As with other kinds of signals (section 1.5, chapters 13, 14), there is also a long-term priming effect of female reproductive functions by male scent signals. Male odors can induce (or suppress) estrus or ovulation in rodents (reviewed by Brown 1985). In the house mouse *Mus musculus*, urine from dominant males is more effective in stimulating female maturation than urine from subordinates (Lombardi and Vandenberg 1977; Labove 1981). What benefits, if any, females gain from mating with dominant males is not known.

The physiological priming effects of mammalian odors have been clarified particularly in rodents (see Brown 1985 for review). Such priming effects may play important roles in reproductive competition and sexual selection (section 1.5). Odors from adult males can accelerate puberty in juvenile females, induce estrus and ovulation and facilitate pregnancy in adult females, delay puberty in juvenile males, and perhaps reduce testosterone levels in other adult males. Female odors may also stimulate the male sex hormone system. For the male, such effects on females should often increase his mating and fertilization success. Why females have evolved responses to male odors is less clear. Among the suggested reasons are that these responses ensure that ovulation does not occur in the absence of potential mates, and that they trigger female reproductive condition at a suitable time of the year (see Brown 1985).

The Bruce (1959) effect, where the pregnancy of a female mouse is blocked if her mate is replaced by a strange male, making her receptive again, depends on odor differences between males (reviewed by Schwagmeyer 1979; Labove 1981). It leads to an obvious advantage for a new male that manages to replace the former male, but some adaptive female tactic is probably also involved. For example, pregnanacy blocking might reduce female investment in offspring that risk being killed by the new male. It also shortens the time until production of a litter together with the new male—offspring that he will likely accept and perhaps help rear (e.g., Schwagmeyer 1979; Labove 1981).

15.4 Summary

There is much evidence that sexual selection favors male scent signals. Scent is used for mate attraction and stimulation in insects, salamanders, mammals, and some other taxa. In moths, females attract males by long-range pheromones; genetic variation has been found at separate loci that affect the female signal and male response. In some moths and butterflies, males court females by emission of short-range pheromones, the composition of which influences male mating success.

Females have been shown to prefer the odor of dominant males in some species. Males of many mammals scent-mark their territories or ranges. An assessment function in male contests over females, or resources needed to attract females, is likely, but attraction or stimulation of females may also be involved.

16 Alternative Mating Tactics

16.1 Individual Differences and Mating Tactics

In many populations there are large, discontinuous differences in mating behavior and morphology among males. Such differences seem to be strongly influenced by sexual selection in the form of male contest competition. For example, small males in some species may have such low success in contests over mates that they do better by using entirely different tactics, such as female mimicry, "sneaky" matings, or searching for females away from larger males.

With exception for obvious sex and age differences, the members of a population have traditionally been regarded as similar in ecology and ethology. Such typological thinking prevailed until a few decades ago, when the emergence of evolutionary and behavioral ecology brought a radical change in outlook, emphasizing individual selection. Variation in morphology and behavior within populations came to be viewed as interesting phenomena, with important ecological and evolutionary consequences. A similar shift had previously occurred in the species concept and speciation theory (e.g., Mayr 1963). The analysis of individual differences gained further momentum during the 1970s from evolutionary game theory and related ideas (e.g., Hamilton 1967; Parker 1970a, 1984b; Gadgil 1972; Maynard Smith and Price 1973; Maynard Smith 1982; Charnov 1982). Game theory has been used especially to understand animal behavior in conflict situations, when the success of one alternative depends on the tactics used by other members of the population. Early examples of implicit game-theoretic reasoning are found in Darwin's (1859) and Fisher's (1930) ideas on sex ratio evolution, and there are now many examples in sexual selection.

Alternative mating tactics can be separated into cases where the alternatives yield equal success, and those where they do not but follow from unavoidable constraints. Other classifications focus upon whether the alternatives are mainly genetically or environmentally determined, and on whether they are reversible or not. In many cases, such classifications may, however, enforce dichotomies upon more or less continuous variation (see, e.g., Austad 1984, Caro and Bateson 1986 for discussion).

When the alternative tactics yield similar success, they may represent a mixed evolutionarily stable strategy (ESS). An ESS is a strategy such that when it is established in the population, no other strategy can invade, given the present constraints. If the success of the strategies are negatively frequency-dependent, a mixed ESS can arise in several ways (see Maynard Smith 1982, 1988), two theoretical extremes being:

1. The polymorphism is purely genetic, each individual following a particular tactic, and the population consists of that mixture of genotypes that makes for an overall evolutionarily stable state.
2. Individuals are phenotypically flexible, respond to what others do, and adopt the presently best tactic. No individual is expected to adopt the mixed ESS, but the population as a whole approaches an evolutionarily stable state. Which of the tactics is best depends on their relative frequencies, on the environment, and on the age, size, condition, strength, and other abilities of the individual itself, and of the other members of the population (e.g., Parker 1984b). Choice of tactic in such situations, which are probably common, can depend on many factors. In practice, genetic and phenotypic influences often seem likely to interact in the choice.

Age-related tactics that follow a similar ontogenetic course in most individuals are found in many species (Caro and Bateson 1986). Tactics may also vary with individual condition or size rather than age (e.g., West-Eberhard 1979; Dawkins 1980; Austad 1984; Dominey 1984b; Waltz and Wolf 1984; Caro and Bateson 1986). Rather than being equal from a fitness point of view, these alternatives may represent the best courses of action under constraints that the individual can do little about, such as food shortage that severely limits body size, preventing any chance of success in contests over mates. Alternative tactics, such as "sneaky matings," can then be viewed as doing the best of the situation, which is sometimes a poor one, leading to low fitness. Inevitably, animals will also make maladaptive mistakes that contribute to the variation.

This chapter reviews some examples of alternative mating tactics of particular interest in relation to sexual selection theory. Competition over mates may be involved in many alternative reproductive tactics; the present review only touches upon a few aspects of this vast and rapidly growing field.[1]

Age- and Size-dependent Differences

The best course of action in competition over mates will often depend on the relative condition, strength, and experience of the individual (e.g., Dawkins 1980; Eberhard 1982; Austad 1984). This may be the reason why

[1] For reviews of various aspects of alternative reproductive tactics, see, e.g., Hamilton 1979, Dawkins 1980, Charnov 1982, Maynard Smith 1982, 1988, Thornhill and Alcock 1983,

sexual display and contest behavior change with age, particularly among sexually mature males in species where growth continues after maturation. For example, young, small males in some anurans do not call or defend a territory, but are silent satellites on the territories of larger males (e.g., Emlen 1976; Wells 1977; Howard 1978, 1984; Fellers 1979; Arak 1983a, 1984; Sullivan 1982a). In some species, males switch between periods of calling and energetically cheaper noncalling tactics (e.g., Fellers 1979; Sullivan 1982a; Robertson 1986b; Arak 1988b; Krupa 1989). Satellite behavior usually seems to yield lower mating success. Although large calling males may be subject to higher predation (e.g., Howard 1984; Ryan 1985), calling versus satellite tactics is probably no mixed ESS. Small silent males seem to have lower success than large calling males. A similar situation is found in elephant seals *Mirounga angustirostris* (Le Boeuf 1974) and red deer *Cervus elaphus* (Clutton-Brock et al. 1979): young, small males do not defend harems but stay in their vicinity, occasionally getting a chance to mate.

Alternative tactics are common also when males do not grow as adults, yet differ in size. Genetic or nutritional differences may lead to size differences among adults. In the dungfly *Scatophaga stercoraria*, large males fight over females on attractive cow pats, whereas small males try to find mates in the surrounding grass. The large males have higher mating success (e.g., Parker 1970c; Borgia 1979, 1980). In the anthophorid bee *Centris pallida*, some males weigh three times as much as others (Alcock et al. 1977; Alcock 1979a). Large males, which win over small males in fights, patrol the ground where virgin females emerge. When a male discovers an emerging female, he digs her up and mates with her, provided that he can keep competitors at bay. Small males instead hover over or near emergence areas and try to mate with flying females not discovered by the patrolling, larger males. The smaller males have lower mating success (figure 16.1.1). The size of a male depends on how much food he was fed by his mother. Possibly, a female can therefore have the same fitness whether she produces many small or few large sons, but the reason why females produce sons of widely different sizes remains unknown (Alcock 1984a).

Males of *Panorpa* scorpionflies may use three different mating tactics (Thornhill 1979, 1980b, 1981). In order of decreasing success, they are (1) defending a dead insect to attract the female; (2) producing a nutritious salivary gift for the same purpose; and (3) catching a female and forcing copulation upon her. Large males are more able than small males to use the successful tactic of food presentation, and also seem to run the lowest risk

---

Arak 1984, Austad 1984, Dominey 1984b, Parker 1984b, Waltz and Wolf 1984, Caro and Bateson 1986, Krebs and Davies 1987, Eberhard and Gutierrez 1991. The symposium edited by Austad and Howard 1984 deals entirely with alternative reproductive tactics.

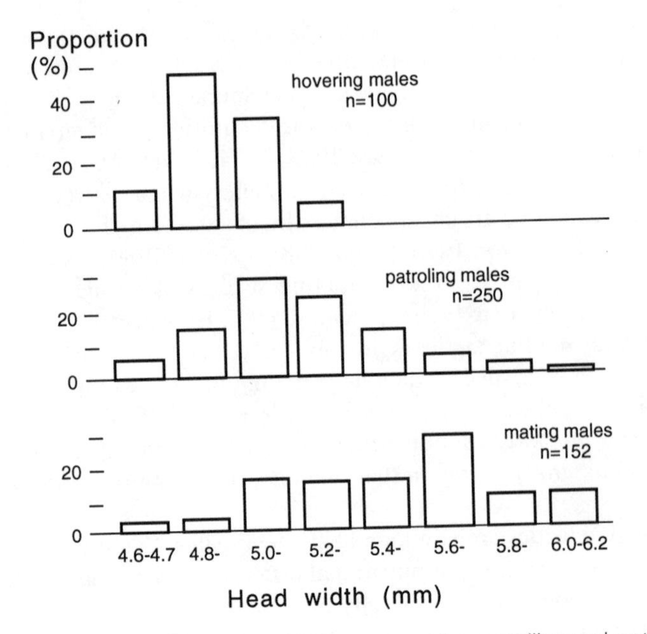

Figure 16.1.1 Size distribution (head width) among hovering, patrolling, and mating males of the bee *Centris pallida*. Hovering males tend to be smaller than patrollers, which in turn tend to be smaller than males captured while mating. (After Alcock 1979)

of predation when foraging for a nuptial gift. The different tactics are conditional and can be used by males of all size classes, but the competitive advantage of large males usually restricts the most successful tactic to their use (Thornhill 1981, Thornhill and Alcock 1983).

Responses to Changes in the Environment

The use of alternative behaviors often depends on the environment. For example, male speckled wood butterflies *Pararge aegeria* patrol the forest for females when there is no bright sunlight, but on clear days they are stationary and defend sunny patches preferred by females (Davies 1978b).

Behavioral changes may also be related to the social environment. Some male natterjack toads *Bufo calamita* switch from stationary calling to searching actively for females as the number of males in the chorus increases (Arak 1983a). In addition, large males are more likely to call, so this is probably no mixed ESS; small males apparently are at a disadvantage, which they may partly compensate by flexible behavior. Also in green tree frogs *Hyla cinerea*, males adjust their behavior to what others are doing (Perrill et al. 1978, 1982). When the calling male was removed from 19 caller-satellite pairs, the silent male in 11 of the pairs began calling within five minutes. When mating calls on the other hand were broadcast from a loudspeaker, 11 of 14 calling males became silent (no caller became silent when a control sound was broadcast). Fellers (1979) and Sullivan (1982a) made similar observations.

Alternative male mating tactics also occur in socially monogamous species. For example, in rock ptarmigan *Lagopus mutus*, the male remains on his territory and guards his mate during her receptive period. When the female begins to incubate, the male often leaves the territory and seeks copulations with other receptive females. Similar mixed tactics have been found in many other species (e.g., Brodsky 1985, and references therein).

16.2 Dimorphism in Male Morphology and Behavior

Male Weapons

At about the same time as evolutionary game theory was invented, Gadgil (1972) suggested that several puzzling cases of dimorphism in male weapons might represent alternative genetic strategies with equal fitness. The morph with large weapons may have higher competitive mating success, the morph with small weapons better survival. Such dimorphism might be stabilized by equal fitness of the alternatives, particularly if there is negatively frequency-dependent selection in at least one of the natural or sexual selection components of fitness.

The extreme male dimorphism in fig wasps of the genus *Idarnes* was analyzed by Hamilton (1979) (figure 16.2.1) (also see Murray 1990). Some males lack wings but have strong mandibles and perhaps also poisonous glands; these males stay inside the figs in which they were born and fight, sometimes to death, over the females that also hatch there. Other males have wings and small mandibles like females; they fly away from their birth site and mate with emerged females. The success of each morph is likely to be negatively frequency-dependent: when most males disperse, competition is low among those few remaining in figs; when most males stay, competition is low among those dispersing. Among species, the fraction of winged males as predicted is correlated with the fraction of females expected to leave the birth figs before mating. This suggests that the success may be similar for the two morphs, but the extent to which this extreme dimorphism is genetic or environmental is not known.

In addition to fig wasps, many beetles and other insects show large intraspecific size variation in weapons such as horns or mandibles, sometimes of more or less distinct dimorphic character (e.g., Darwin 1871; Huxley 1932; Gadgil 1972; Hamilton 1979; Otte and Stayman 1979; Eberhard 1979, 1980b, 1982; Eberhard and Gutierrez 1991). There is marked dimorphism in, for example, the sugarcane beetle *Podischnus agenor* (Eberhard 1980b, 1982; figure 16.2.2), and in several species studied by Eberhard and

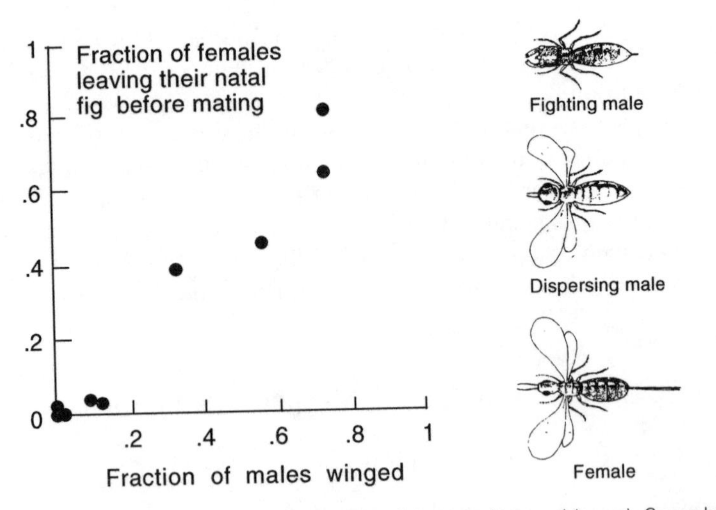

Figure 16.2.1 Males are extremely dimorphic in fig wasps (genus *Idarnes*). Some have a small head and can fly; others are flightless and have large mandibles used in fights with other males in the natal fig. Among ten species compared by Hamilton (1979), there is a positive relationship between the proportion of males that are winged and the proportion of females that leave their natal fig before mating; these females will mate with winged males. The results suggest that the two male morphs have equal fitness. (After Hamilton 1979; wasps from Krebs and Davies 1987)

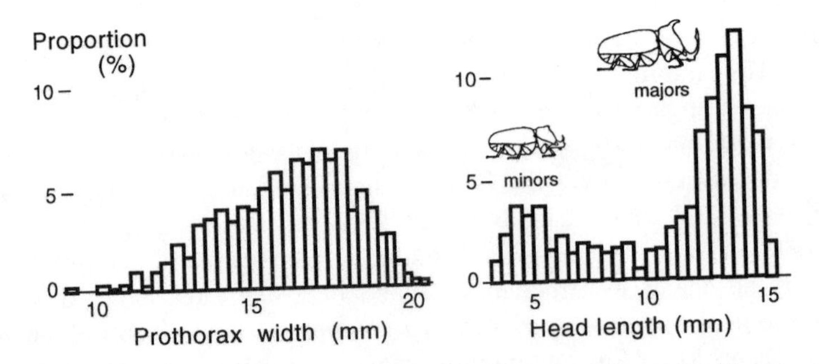

Figure 16.2.2 In some horned beetles, small "minor" males with minimal horns coexist with larger "major" males that have large horns. In *Podischnus agenor* shown here, the minor and major males seem to represent two partly distinct head morphologies rather than extremes of a continuum. Whereas body size (prothorax width) shows a unimodal distribution in the population, head (and horn) size shows two peaks, one for small, another for large values. (After Eberhard 1980b. Drawing of beetles by T. Prentiss. Copyright © 1980 by Scientific American Inc.)

Gutierrez (1991). Eberhard (1980b) suggested that the best competition tactic varies with male size. Large males can get high mating success by using their horns in fights over mates, but small males cannot compete effectively by fighting. As most of the competitors will be larger and stronger, developing large horns is not advantageous for a small male. By not doing so, he can save time and energy for better purposes, such as earlier emergence (Eberhard 1982). This situation might lead to the evolution of relatively smaller weapons in small males. Dimorphism in these beetles does not seem to be an ESS with equal fitness for small and large males. Eberhard (1982) suggested that the dimorphism arises from differences in larval food, which limits adult size (see, e.g., Brown and Rockwood 1986). Small males are probably at a disadvantage; their behavior and smaller relative horn size can then be regarded as making the best of a bad lot.

The extent to which horned beetles show dimorphism rather than continuous variation in horn size is debated, and varies among species (Otte and Stayman 1979; Eberhard and Gutierrez 1991). Several patterns of size variation in beetle horns are poorly understood—for example, why, among related species, the largest forms tend to have largest relative horn size (e.g., Otte and Stayman 1979).

Genetic Variation in Cricket Mating Tactics

Genetic variation in male mating behavior is known from a few species. Some males of the cricket *Gryllus integer* attract females by singing; other males are silent satellites that try to intercept females approaching a caller (Cade 1979). By artificial selection for singing or silent behavior, Cade (1981a) could change the amount of singing markedly. This suggests that singing behavior is genetically variable in the population; the estimated heritability was about 50%. It seems unlikely, however, that male crickets fall into two discrete categories; environmental influence and continuous variation appears more likely (Dominey 1984b). Cade (1979) also found that callers attract parasitic flies in addition to females, which may in part explain why not all males call (section 10.1 above). Whether the success of calling versus silent behavior is negatively frequency-dependent remains to be tested.

Male Trimorphism in an Isopod

The marine isopod *Paracerceis sculpta* is genetically trimorphic in male morphology and mating tactics (Shuster 1989, 1992; Shuster and Wade 1991a,b). The large, ornamented α males defend groups of females within intertidal sponges. The smaller β males resemble mature females, and can

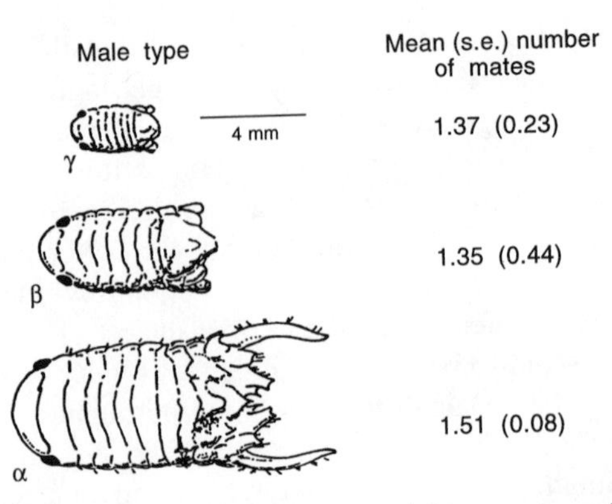

Figure 16.2.3 The three male morphs of the isopod *Paracerceis sculpta* (heads at left) have roughly equal mating success, and seem to represent a genetic mixed evolutionarily stable strategy. (Data from Shuster and Wade 1991a; animals from Shuster 1989)

enter the spongocoel as they apparently are taken for females by the α male. The tiny γ males seem to enter by stealth. Genetic experiments suggest that the male morphology is influenced mainly by a single autosomal locus, with allelic dominance relations $\beta > \gamma > \alpha$.

Fertilization success of the three male morphs in various combinations of males and females in sponges was studied by experiments using genetic markers that influence external pigmentation. The large α males guard females and sired most of the young when there was only one female in the sponge. When more females were present, β males sired about 60% of the young. The reproductive success of γ males increased roughly linearly with the number of females. From these results, and from monthly samples of female group size and male morphs in a natural population over two years, Shuster and Wade (1991a) concluded that the average reproductive success was roughly equal among the three male morphs (figure 16.2.3). The estimated total opportunity for sexual selection occurred almost entirely within morphs, and to less than 0.1% among morphs. This and other evidence suggested that there is no marked total selection difference among the morphs. This species therefore is a possible example of a genetic mixed ESS in male morphology and mating tactics.

Alternative Mating Tactics in Salmon and Sunfish

Several species of fish provide remarkable examples of alternative behavioral and morphological tactics. In some salmonids and North American sunfishes (*Lepomis*), large males compete aggressively over females, whereas small males try to mate by "sneaking" and mimicking females. These alternatives possibly reflect different genetic or phenotypic morph

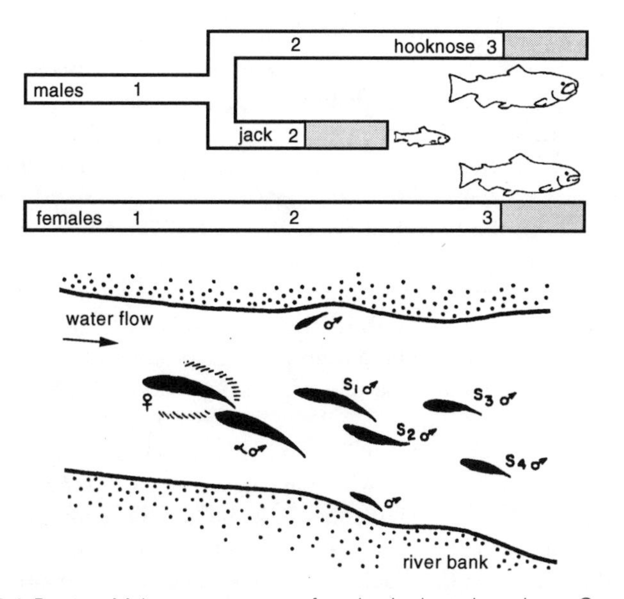

Figure 16.2.4 *Bottom*: Males compete over females in the coho salmon *Oncorhynchus kisutch*, which dig spawning nests in the gravel bottom of the stream. The large "hooknose" males form a dominance hierarchy, with the alpha male closest to the female and the subordinates (S_1, S_2. . .) farther away. At the shores of the stream, small "jack" males also await the spawning. Like the other males, they will then rush into the nest and release sperm. Males tend to enter the nest in their order of proximity to the female. *Top*: There are two alternative male life histories. One year is spent in the birth stream before migration to the sea, where jack males spend 5–8 months before returning to the stream. Males that follow the hooknose life history stay for 17–20 months in the sea and grow much larger before returning for spawning; so do females. The shaded parts of the life histories indicate sexual maturity. (From Gross 1984, 1985)

tactics with equal fitness (Ghiselin 1974; Gross 1979–1991a,b; Gross and Charnov 1980; Dominey 1980; but see Dominey 1984b).

Males of the Pacific coho salmon *Oncorhynchus kisutch* follow either of two different life histories (Gross 1984, 1985, 1991b; figure 16.2.4). Spawning takes place in rivers where the fry remain for one year (in some cases up to three years) until migrating to the sea. Surviving females usually return to spawn on the gravel bottom of their home stream after another one and a half years. So do the large hooknose males, with conspicuous red colors, hooked jaws, and enlarged teeth used in fights over females (the evolution of dangerous weapons in this semelparous species parallels that in the gladiator frog; see sections 4.2, 10.5; all coho salmons die after spawning). Another segment of the male population, known as jacks, returns already after half a year in the sea. These much smaller males develop no secondary sex characters. They attempt to mate by sneaking from a sheltered position behind boulders or other structures near females (figure

16.2.4). When the female spawns in her simple nest in the gravel, the closest male (usually the largest hooknose) enters and sheds his sperm. Male order of entering, and hence probably fertilization success, decreases with increasing initial distance from the nest. For this reason, also the jack males in the best position may fertilize eggs. Since sheltered hiding sites close to the female are few, the fitness of jack males will be negatively frequency dependent, as will the fitness of hooknose males. Conditions for a mixed ESS therefore seem to be at hand. Selection of morphology and behavior is probably disruptive as the traits that make for success as a sneaker (small size, inconspicuousness) are unsuitable for the fighting tactic (Gross 1984).

Gross (1985) calculated male fitness as the product of survivorship to maturity, breeding lifespan, and relative mating success. The estimated relative fitness of jack versus hooknose males was 0.95. Gross (1985) concluded that jack and hooknose males probably represent a mixed ESS, perhaps being genetic alternatives with equal fitness. Some of the assumptions have not been tested, for example that jack and hooknose males for a given distance from the female have equal fertilization success. Hooknose males are several times larger than jacks and may produce more sperm per ejaculate, which should give a sperm competition advantage. Fertilization success in relation to male type, size, and position might be tested for example by genetic markers.

In the seatrout *Salmo trutta*, some males remain in their birth stream throughout their lives, spawning as small, sneaky satellites in competition with the larger sea-going males. Although they are smaller as adults, the satellites tend to be the males that grew best as juveniles (Bohlin et al. 1986; Dellefors and Faremo 1988). The satellite tactic therefore is probably not a poor alternative used by inferior males, but the primary tactic of juveniles that grow rapidly. This lends indirect support to the ESS idea of equal fitness for the two conditional tactics (Bohlin et al. 1990).

Also, in the bluegill sunfish *Lepomis macrochirus*, there are at least two different male reproductive tactics (Gross 1979, 1982, 1991a; Dominey 1980; Gross and Charnov 1980). Parental males do not begin to reproduce until about seven years old, when they build and defend nests to which females come for mating and spawning. Other, satellite males early in life use a sneaking tactic. Later on, they mimic females and release sperm in pairings with a parental male and a female. Satellite males usually die before the age when parental males begin to reproduce, and the two tactics appear to be exclusive alternatives, not simply age-dependent differences (Gross and Charnov 1980; Gross 1982). The evidence suggests that the two tactics have negatively frequency-dependent success (Gross 1991a). Preliminary crossing results from the F_1 generation suggest that the existence of alternative male tactics in sunfish is not due to genetic differences (Dominey 1984b).

Plumage and Behavior Polymorphism in the Ruff

A wader bird, the ruff *Philomachus pugnax*, has received much attention owing to its conspicuous lek behavior and male plumage ornaments. In addition, males are polymorphic in coloration and behavior (figure 16.2.5) (e.g., Hogan-Warburg 1966; van Rhijn 1973, 1991; Lank and Smith 1987, 1992; Höglund and Lundberg 1989; W. L. Hill 1991). The neck collar and head tufts range in color from black over brown and gray to white. Most dark males defend lek territories. White-collared males are often subordinate satellites on the lek territories of dark males; intermediate males use either tactic. The differences in plumage and behavior are possibly genetic, but this has not been studied in detail. Nor has the relative reproductive success been measured in sufficient detail to show whether the morphs have equal fitness. Lank and Smith (1987) found that male tactics to some extent are conditional, varying over time. Lekking territorial and satellite males may switch to following females as they move about over their home range, and intercepting them at resource-rich sites. These tactics seem to be used more by satellite than by territorial males.

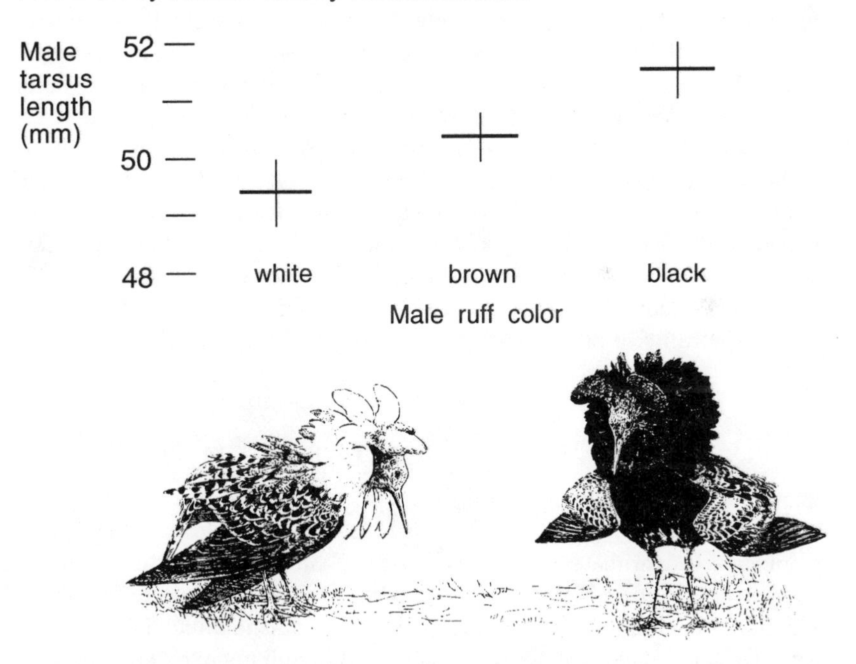

Figure 16.2.5 Male ruffs *Philomachus pugnax* are polymorphic in their nuptial coloration, ranging from mainly white over brown to black head tuft and neck collar. Coloration seems to be related to male size, black males being largest and white males smallest. There are also behavioral differences between the morphs (see text). (After Höglund and Lundberg 1989; birds from Cramp et al. 1983)

Differences between the morphs may also depend on age, size, or condition. Black males tend to be largest, brown intermediate, and white smallest (figure 16.2.5). It therefore seems that the largest males, which may be in best condition, develop dark collar and aggressive defense of a lek territory. Dark pigmentation and level of aggressiveness in part may be under similar hormonal control. Aggressive behavior in birds is partly controlled by testosterone level, which in turn depends on nutritional condition (Wingfield 1980; Wingfield et al. 1987). Castration in early winter prevents the development of male nuptial plumage in ruff (van Oordt and Junge 1936), and testosterone injections lead to growth of nuptial plumage in phalaropes (Johns 1967). The nuptial plumage of waders therefore appears to be under androgenic control. Whether this applies also to the darkness of ornamental feathers in the ruff might be tested with hormonal implants and other experiments. Androgens control the deposition of dark melanic pigments in some but not all birds that have been studied in this respect (Witschi 1961; Ralph 1969).

Höglund and Lundberg (1989) suggested that the different morphs may have unequal fitness. Males with white collars may be doing the best of a bad situation if they cannot compete aggressively with the larger, darker males, instead developing a white collar that is met with less aggression by territorial males. Perhaps the behavioral differences among the color morphs also depend on learning from experiences on leks (Caro and Bateson 1986). These possibilities might be tested by long-term studies of body condition, coloration, and behavior in a population of individually marked males. Studies that control parentage are needed to examine the degree of genetic influence on the color polymorphism. Experimental manipulation of the collar and head tufts might show whether they are favored by female choice, male contests, or both. In spite of the many results already obtained, the ruff still presents fascinating unsolved problems.

6.3 Sex Allocation, Sex Change, and Sexual Selection

Theory

Competition over mates is important in the evolution of sex allocation: the ways in which resources are partitioned to male and female reproductive function. For example, sexual selection often seems to affect the pattern of sex change in sequential hermaphrodites. Although many other factors are also important, such as female fecundity advantages of large body size, it does not seem possible to understand all the variation in patterns of sex change without considering the nature and strength of sexual selection (e.g., Ghiselin 1969a, 1974; Charnov 1979a, 1982; Lloyd and Bawa 1984;

Goldman and Willson 1986; Bull and Charnov 1988; Schlessman 1988; Warner 1988c).

Three of the main problems in sex allocation theory concern sex ratios, sex change, and relative investment in male and female function in hermaphrodites. Competition over mates can play a role in all three areas, influencing the evolutionarily stable strategy. Sex change is discussed here, and sexual investment in hermaphrodites in section 17.9. Sex ratio theory is dealt with in, for example, Hamilton (1967), Charnov (1982), Maynard Smith (1982), Bull and Charnov (1988), Clutton-Brock and Godfray (1991).

The evolutionarily stable strategy in sex allocation theory is calculated from the genetic contributions through male and female function. Charnov (1982) showed that surprisingly many sex allocation problems can be analyzed with the Shaw-Mohler equation or related mathematical models. The basic rule is that selection favors a gene for a mutant strategy if it leads to a relative gain in fitness through one sex function that exceeds the loss through the other sex function. In symbols,

$$\frac{\Delta m}{m} + \frac{\Delta f}{f} > 0,$$

where f is the fitness through female function for the common strategy, Δf is the change in fitness through female function for the new, rare strategy compared to the original strategy, and m and Δm are the corresponding fitnesses through male function.

The ESS allocation to male versus female function is often the one that maximizes the product of the fitness gains through eggs and sperm, mf. Charnov (1982) suggested that selection tends to maximize this product in the choice of sex state (for example, hermaphroditism versus separate sexes) as well as in the allocation to male and female function in hermaphrodites. The models for sex allocation in hermaphrodites are formally identical to those for sex ratio evolution in organisms with separate sexes, and Queller (1984a) suggested that similar principles apply in both situations. Sex allocation theory offers a unified perspective on several problems that previously seemed unrelated.

In sequential hermaphrodites, individuals may change sex in relation to age, size, or the environment. Examples are found in many aquatic invertebrates and fish (e.g., Ghiselin 1969a; 1974; Policansky 1982) and in some plants (section 17.9). The main explanation for sex change is the "size advantage hypothesis" of Ghiselin (1969a, 1974), further developed in mathematical models (e.g., Warner 1975; Warner et al. 1975; Leigh et al. 1976; Charnov 1979b). The idea is that reproductive success increases with body size in different ways in the two sexes (also see sections 11.1–11.2). It may then be best to begin life as one sex, and change when the fitness/

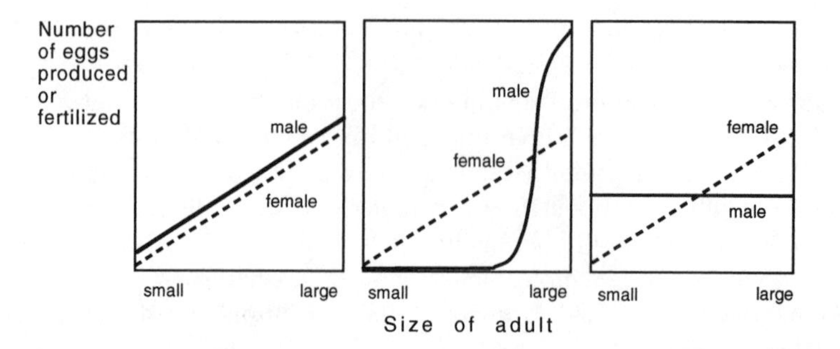

Figure 16.3.1 *Left*: When reproductive success increases similarly with size in both sexes, there is no selection for sex change. *Center*: If females show a steady increase in fecundity with size, the largest males monopolize females, and smaller males have little or no mating success, the tactic that yields highest lifetime success may be to start life as a female, and change to male sex at large size. *Right*: If female fecundity increases more strongly with body size than does male success, for example if mating occurs in monogamous pairs with little competition over females, it may be best to start life as a male, and change to female sex at larger size. (After Warner 1984b)

body size curves for the two sexes cross (figure 16.3.1). More exactly, the individual should change sex when the reproductive value (Fisher 1930) of the other sex exceeds that of its present sex. Whether sex change is favored over permanent sexes should depend on costs of changing sex, and on other life history trade-offs (Charnov 1982).

Sex change may be favored if female fecundity increases faster with body size than does the number of eggs fertilized by a male. It would then be best to start life as a male (protandry), and later switch to female sex, as done by some crustaceans, mollusks, and plants. For example, in the protandrous polychaete *Ophryotrocha puerilis*, large males in laboratory tests have a reproductive disadvantage compared to females (Berglund 1986). The reverse pattern is found in some fishes where males defend spawning sites or groups of females, large males being most successful in competition over matings. The best tactic is then to start life as a female (protogyny) and change sex at a body size that permits successful competition as a male (figure 16.3.1) (see, e.g., Ghiselin 1969a, 1974; Warner 1975, 1988c; Charnov 1982). The nature and strength of sexual selection therefore affects sex change and its life history pattern.

Sex Change in Fishes

The theory has been tested for example in coral reef fishes (reviewed by Warner 1984b, 1988b). Sex change in some damselfishes (Pomacentridae) follow the predicted pattern. *Amphiprion* anemone fishes live in or near large sea anemones without being injured, enjoying protection by the anemone and its stinging cells. A social group of anemone fishes contains one adult monogamous pair plus juveniles. As usual in fish, larger females produce more eggs (section 11.2), which can be fertilized even if the male is smaller. The production of offspring by the pair is therefore higher when the female is largest, so protandry is advantageous to both adults (Warner 1978). As expected, anemone fishes are protandrous hermaphrodites, the smallest member of an adult pair being male (Fricke and Fricke 1977). In the closely related *Dascyllus* damselfishes, social groups contain several adult males and females. Sexual competition should therefore be stronger, favoring large males. Accordingly, these species are protogynous, adults changing from female to male sex as they grow larger (e.g., Fricke and Holzberg 1974; Coates 1982).

Effects of mating competition and social control of sex change have also been demonstrated in group-living fishes of other families (reviewed by Charnov 1982; Policansky 1982; Warner 1984b; 1988b; Shapiro 1987). In some species where groups consist of a single male and several females, the male is the largest member. For example, in the cleaning wrasse *Labroides dimidiatus*, a species that feeds by cleaning the skin and mouth of other fishes and lives in groups of one male and a handful of females, removal of the male makes the largest female change sex. After only a few hours she shows male behavior such as spawning attempts with females, and within ten days he produces sperm (Robertson 1972). In this species, sexual selection therefore seems to prevent all group members except the largest one from turning male. A second, smaller male would probably have low mating success; this might be possible to test by adding second, larger and smaller, males to established groups.

Whereas many observed cases of sex change fit theoretical expectations, there are other cases where sex change might be expected, yet does not occur. More detailed studies of mating systems, life histories, and constraints are needed to better understand the pattern. (For discussion, see Shapiro 1987, 1989; Warner 1988b,c). Families that contain protandrous as well as protogynous species should be especially suitable for tests (Warner 1984b).

Sexual selection and protogyny have been studied in detail in the bluehead wrasse *Thalassoma bifasciatum* by Robert Warner and his colleagues (e.g., Warner 1975, 1984a; Warner et al. 1975; Warner and Hoffman 1980). There are two alternative life histories in this coral reef fish (figure 16.3.2). A proportion P of each cohort matures as *initial phase males*, and the rest, 1-P, as females. Initial phase males resemble females and reproduce in spawning groups. Later in life, in their *terminal phase*, males become brightly colored and defend spawning territories. Females, when they grow larger, change color and sex to terminal phase males. Theory predicts that females should change sex at the size where male and female success is equal. Observations are in reasonable agreement: males that have just entered the terminal phase spawn about 1.5 times per day, compared to once

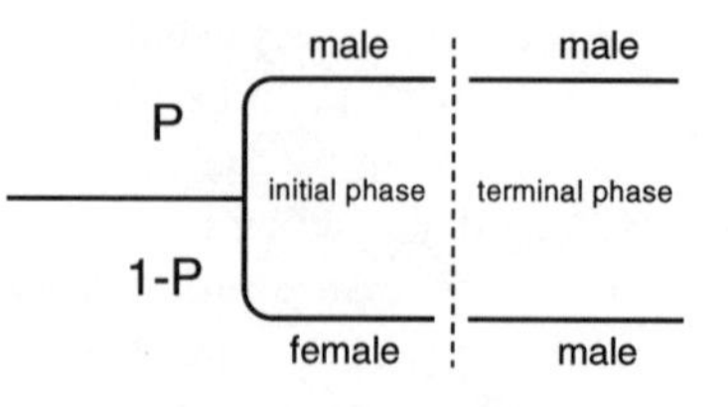

Figure 16.3.2 There are two alternative life histories in the bluehead wrasse *Thalassoma bifasciatum*. A proportion P of a cohort matures as males, whereas the rest $(1 - P)$ mature as females. Later in life, during the terminal phase, males change color, take up territories, and spawn in pairs with females. Initial-phase males often spawn in groups. Females later in life change both sex and color and become terminal-phase males. (After Charnov 1982)

per day for females. The largest males spawn over 40 times per day (Warner et al. 1975).

For the outcome to be evolutionarily stable, initial-phase males should occur in a proportion P such that the two life histories have equal fitness. It can be shown that this rule applied to the biology of the bluehead wrasse yields $P \approx (1 - TS)$, where T is the proportion of the population in terminal phase with a territory, and S is the average number of females with which a territorial male spawns per day (see Warner and Hoffman 1980; Charnov 1982). The empirical estimates of P agree fairly well with this prediction, suggesting that the two life histories may represent a mixed ESS (Warner and Hoffman 1990; Charnov 1982). Accurate, unbiased estimates of the lifetime fitness of the alternative tactics are, however, difficult to obtain in this as in other cases (section 16.4).

Males in some mammals, such as elephant seals and certain ungulates and primates, for example gorilla, also defend groups of females. Young males have low mating success in these species. Why are they not protandrous hermaphrodites? Warner (1978) suggested the reason is that the two sexes are much more differentiated in mammals than in fish. Sex change would require major restructuring for a mammal, but it is relatively simple in fishes with external fertilization and no specializations for embryo growth within the mother. It is harder to understand the absence of sex change in some fishes with a male size advantage, especially as closely related species show sex change; there is scope for refinement of the theory (Warner 1988b).

16.4 Some Empirical Problems

The analysis of alternative mating tactics is often difficult, especially if the aim is to measure the relative lifetime fitness of the alternatives. This task requires huge sample sizes for the confidence limits of the estimates to be reasonably narrow, and for the risk of type I as well as type II error in statistical tests to be low. These and other problems of testing hypotheses on alternative reproductive tactics are discussed for example, by Austad (1984), Dominey (1984b), and Caro and Bateson (1986).

In the light of these problems, there is a risk that variation in male morphology and behavior within species is sometimes overinterpreted as mixed evolutionarily stable strategies. The variation in some cases may turn out to be partly nonadaptive consequences of environmental or other variation. Yet, without hypotheses of adaptation, it seems unlikely that many of the fascinating discoveries of alternative mating tactics would have been done. Probably few researchers would be prepared to do long-term, heavy field work on intraspecific variation in behavior and morphology, if there were not even a theoretical possibility of interesting adaptive aspects about the variation. If combined with alternative hypotheses and considerations of phylogenetic and other constraints, hypotheses of adaptation should continue to play a constructive role and lead to further insights.

16.5 Summary

More or less discontinuous variation in male structures and behavior occurs in many species; the cause often appears to be sexual selection in form of male contest competition over mates. Some examples in fishes and arthropods are aggressive behavior, large body size, weapons, or other secondary sex traits, versus small size and sex traits, and inconspicuous mating behavior. Other examples are calling versus silent satellite behavior in anurans and acoustic insects.

In some cases, the alternatives may have similar fitness, and represent a mixed evolutionarily stable strategy. Whether they really do lead to equal fitness is, however, difficult to test critically. The alternatives can differ genetically, but usually, alternative mating tactics are related to variation in age, size, condition, or other phenotypic differences. They may not be equal from a fitness point of view, but arise from unequal constraints or environmental influences upon individuals.

Other examples of alternative tactics favored by sexual selection are found in sequential hermaphrodites, for instance certain crustacea, mollusks, fishes, and plants. Such situations are dealt with in sex allocation theory. Whether a species is protandrous (male first) or protogynous in part depends on the nature and strength of competition over mates. Strong contest competition over females usually seems to favor protogyny, with change to male sex as the individual grows larger, for example in some fishes. When contest competition and the advantage of large size is weaker among males, fecundity advantages of large size in females may dominate, favoring protandry, with change to female sex at large size.

17 Sexual Selection in Plants

17.1 Introduction

One of the most vigorous research areas in sexual selection today concerns plants. This is a new development. Except for a footnote on protandry, Darwin (1871) did not apply his ideas on sexual selection to plants, an exciting opportunity left for the present generation of botanists.

Why did sexual selection theory take so long to find its way into botany? One of the main reasons is perhaps that male reproductive success is harder to quantify in plants than animals. Traditionally, reproductive success in plants has therefore been studied in terms of seed production (Bertin 1988; Willson 1991). But in higher plants as in animals, male gametes are smaller and more numerous than female gametes, which means that sexual selection is likely to be strongest in males (Bateman 1948). In addition, sexual selection may often be more difficult to distinguish from fecundity selection in plants than in animals. If taken too literally, some of the important concepts from sexual selection in animals, such as mate choice and mating preferences, do not seem applicable to plants. If reasonably interpreted, however (section 1.1), they can also be used in botany (see, e.g., Bertin 1988; Lyons et al. 1989; Willson 1991; Waser 1992).

This chapter briefly reviews sexual selection and related phenomena in higher plants. After sections on the forms of sexual selection, sex roles, and secondary sex differences in plants, the text follows the path of pollen on its journey from the anther via pollinators to the stigma and ovules. During these stages, pollen scrambles are likely to be important, favoring traits that improve pollination and fertilization success (figure 17.1.1). The last two sections compare sexual selection in plants and animals, and discuss resource allocation in hermaphrodites and how it might be influenced by sexual selection.

Already Haldane (1932) and Huxley (1942) suggested that pollen competition has important consequences, and Bateman (1948), in his classical paper, pointed to sexual selection also in plants. There was, however, little interest in this area until the 1980s. Some early exceptions were Willson and Rathke's (1974) suggestion that male pollen export may benefit from larger inflorescences (flowering shoots) than required for full fruit production, Gilbert's (1975) suggestion that sexual selection favors copious pol-

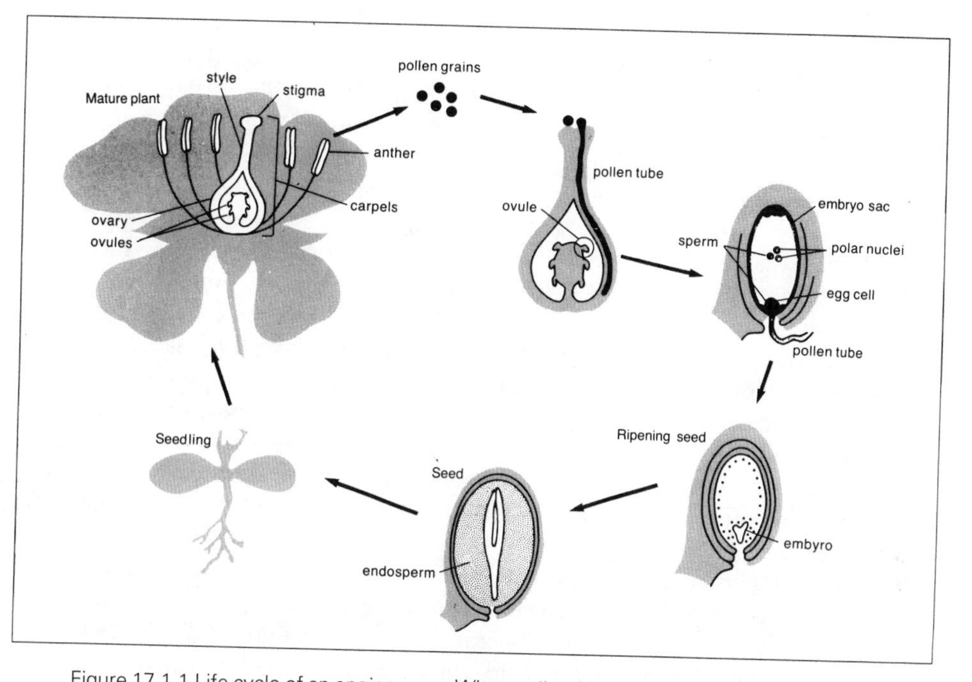

Figure 17.1.1 Life cycle of an angiosperm. When pollen from the anther of a mature plant reaches and germinates on a stigma, it produces a fast-growing pollen tube that penetrates down the style toward the ovary. As the pollen tube enters an ovule, one of the two sperm cells fuses with the egg cell, forming a diploid zygote. The other sperm cell fuses with the polar nuclei to produce the endosperm, nutrition for the embryo. The pollen, pollen tube, sperm, egg, polar nuclei, and embryo sac are haploid. The stages from zygote to mature plant form the diploid (sporophyte) part of the life cycle. The stages that seem most likely to be influenced by sexual selection are pollen production and dispersal, pollen tube growth, and fertilization. (From Mulcahy and Mulcahy 1987)

len and nectar production in male flowers, and Janzen's (1977) emphasis on "mating preferences" in plants. Ideas presented at a crucial time by Charnov (1979b) and Willson (1979) helped shape and spread the new perspective. Since then, a rapidly growing number of botanists are exploring sexual selection in plants.[1]

Some important differences from most animals are that plants are sessile, and their male gametes are brought to females by wind, water, or pollinating animals subject to other kinds of selection than the plants they help fertilize. This might leave less control in plants over which female becomes

[1] Reviews are found in, e.g., Charnov 1982, Stephenson and Bertin 1983, Willson 1983, Willson and Burley 1983, Charlesworth et al. 1987, Queller 1987, Lyons et al. 1989, Marshall and Folsom 1991, Willson 1990, 1991, Walsh and Charlesworth 1992, and in the volume edited by Lovett Doust and Lovett Doust 1988, which aims at integrating plant and animal reproductive ecology.

fertilized by which male, but it does not eliminate the potential for competition over mates. There is little scope for male contest competition: in spite of the relevance of behavioral ecology to botany (e.g., Willson 1983; Charnov 1984), no male plants have yet been found wrestling over females. This need not imply that male plants never compete repressively in some way over females (*Catasetum* orchids may provide an example; see section 17.4 below). The most obvious form of mating competition is, however, scrambles among pollen to first reach stigmas and fertilize ovules. Such scrambles apparently have favored, among other things, production of large amounts of pollen, attraction of pollinators by conspicuous flowers, scent, and nectar, and rapid growth of pollen tubes. The traditional explanation for most such characters is that they are favored by natural selection through fecundity or other advantages. In addition, they are likely to be favored by sexual selection, that is, competition over mates (see sections 1.1 and 17.2). A major problem in plants as in animals is to measure the relative importance of the various forms of natural and sexual selection for such traits.

Other important differences are found in the mechanisms of fertilization, some aspects of which have no counterpart in animals. After reaching the stigma at pollination, the angiosperm pollen rapidly grows a tube that penetrates the style toward the ovary (figure 17.1.1). The tube carries two sperm cells, one of which fertilizes the ovule. The other sperm fuses with the polar nuclei and forms the endosperm, which provides nutrition for the seed. Possible consequences of the remarkable genetic relationships among the different tissues in a developing seed were reviewed by Queller (1984b) and Haig and Westoby (1988). Using comparative methods, Donoghue (1989) traced the evolution of this system and some other elements of sex expression, such as dioecy, in plant reproductive biology.

An interesting possibility is that the female influences which among the competing pollen grains will fertilize her ovules and give rise to mature seeds (e.g., Janzen 1977; Charnov 1979b; Willson 1979; Stephenson and Bertin 1983; Willson and Burley 1983; Lyons et al. 1989; Marshall and Folsom 1991). Except for the special form of mate choice involved in self-incompatibility, which is better understood, such female choice is one of the most debated aspects of sexual selection in plants. Usually only a proportion of the zygotes mature into ripe seeds. Mechanisms of female choice as regards which embryos to bring to maturation are presently the subject of much research (see Charlesworth et al. 1987; Queller 1987; Waser et al. 1987; Lyons et al. 1989; Marshall and Folsom 1991).

Compared to animals, higher plants show much more selfing. The scope for sexual selection decreases with increased selfing, which makes ovules less available to pollen from other plants (Willson 1979; Stephenson and Bertin 1983). In accordance, pollen production declines with increasing self-fertilization (section 17.9).

Table 17.1.1

Forms of Sexuality in Angiosperms, Based on Flowering Morphology

A *unisexual* plant or flower produces either pollen or ovules. Unisexual plants, with separate male and female individuals, are *dioecious*.

A *cosexual* plant or flower produces both pollen and ovules. Among cosexual plants, those with unisexual (separate male and female) flowers are *monoecious*, those with cosexual flowers are *hermaphroditic* (*perfect*).

The approximate percentage of species of the various forms are:

dioecious	4	
cosexual	96	
perfect		72
monoecious		5
other forms of cosexuality		19

Plants in the last category have mixtures of cosexual and male or female flowers (Yampolsky and Yampolsky 1922; Charnov 1984). The proportion of dioecious species may be underestimated (Bawa 1980b).

Another difference from arthropods and vertebrates is that plants are usually cosexual and show more variable sexuality (reviewed, e.g., by Bawa and Beach 1981; Willson 1983). The rich flora of botanical sex terms is exotic to an outsider; I have plucked a small bouquet for use here (table 17.1.1; see, e.g., D. G. Lloyd 1980; Stephenson and Bertin 1983; Lovett Doust and Lovett Doust 1988).

The variety of forms of sexuality in higher plants offers fascinating evolutionary problems. As in animals, the form and strength of sexual selection, particularly male (pollen) competition over mates, apparently is one of the main variables that explain the diversity of sex habits among higher plants (e.g., Charnov 1979b, 1982; Willson 1979, 1983; Bawa 1980b; Bawa and Beach 1981; Lloyd 1982; Wyatt 1983; Emms 1993).

17.2 What Is Sexual Selection in Plants?

Sexual selection has been defined in perhaps even more ways for plants than animals. To facilitate comparison between the two groups, the term will be used here in Darwin's (1871) sense of competition over mates (see section 1.1 above). This definition applies to scrambles and contests as well as mate choice, which leads to competition to be chosen. Competition here has similar meaning as in ecology, occurring whenever an organism, by using a resource, makes it less available to others; competition over mates therefore is the unifying aspect of all forms of sexual selection (section 1.1). To demonstrate sexual selection of a plant trait, it is important to show that variation in the trait leads to variation in mating success (more precisely, in pollination and fertilization success) owing to competition among rivals. This has rarely been convincingly done for scrambles or mate choice in plants. It is also desirable to show that the variation carries over to offspring production, which may require study of genetic markers.

In an important review, Stephenson and Bertin (1983; also see Bertin and Stephenson 1983) used a broader concept of sexual selection than the Darwinian one. They defined it as differential reproductive success among the surviving members of a sex that are physically capable of reproduction. But reproductive success can be influenced by many traits not involved in competition over mates, for example by nutrients that the parent bestows upon the seed (section 1.1). If a female parent with high nutrient reserves is more able than others to invest resources in offspring, her probably higher reproductive success need not have anything to do with Darwinian sexual selection. In order to distinguish sexually selected traits from reproductive adaptations in general, it seems preferable to use competition over mates as the defining aspect of sexual selection in plants as well as animals; Stephenson and Bertin (1983) largely did so in practice.

Charlesworth et al. (1987) pointed out that Darwin's (1871) concept of sexual selection in some respects is difficult to apply in flowering plants, as most of them are cosexual and have flowers that contain both primary sex organs and structures for attracting pollinators. This makes some issues more complex, but it does not render Darwin's definition useless for plants. It is still often meaningful to ask and measure to what extent such traits as floral display influence male and female mating success, and to what extent these traits are selected by competition over matings (also see Lloyd 1988; Stanton and Preston 1988).

Competition over mates is likely to occur if the reproductive output of one sex is limited by the availability of the other sex. The limitation is usually stronger in males, but it can occur in females as well (section 7.6). It is therefore desirable to define sexual selection such that the concept is applicable to both sexes, not only males. Seed set in some female plants seems to be limited by the availability of pollen (reviewed by Johnston 1991b; section 17.6 below). Such cases may involve female competition over pollen and fertilization: Darwinian sexual selection of females (Bertin and Stephenson 1983; Campbell 1989b).

This possibility can be tested, for example by removing some of the female flowers in experimental plots to see if those remaining on average have more fertilized and maturing ovules than females in control plots (see Stephenson and Bertin 1983; Lyons et al. 1989). This is no necessary outcome; removal of neighbors under some circumstances might reduce the mating success of a plant, for example if pollinators preferentially visit stands of many individuals (e.g., Rathke 1983; Feinsinger et al. 1991; Kunin 1993).

A complementary approach is to increase or reduce a floral trait in some individuals to see if (1) their mating success changes in the predicted way, and (2) if there are also corresponding inverse changes for neighbors. This might show if the trait is favored by competition over mates. A control could consist, for example, of measuring the mating success of the same plants before manipulation. The female function of flowers can also be tested by changing the size of floral display, and estimating the relationship between seed set and floral traits. With suitable choice of species, for example orchids or milkweeds (asclepiads) with countable pollinia, the removal of pollen in relation to floral display can be quantified (section 17.6; also see Mitchell 1992; Mitchell and Waser 1992). It may then be possible to estimate the degree to which floral display is sexually selected by male or female function.

The strength of sexual selection might vary with population density of plant as well as pollinator. In widely dispersed plants, pollinators may develop "trap-lining" patterns of flower visits (Janzen 1977). The presence of neighbors in such species need not necessarily decrease the pollination success of an individual, but might even increase it if pollinators require some minimum population density of a plant species to visit it (e.g., Feinsinger et al. 1991; Kunin 1993). On the other hand, strongly male-biased sex ratios of flowers in some such species suggest that they are subject to male competition for pollinator services (e.g., Gilbert 1975; Bawa 1980a). Studies of communities of plants that share pollinators have also shown that there may be strong competition over pollinators among species (e.g., Feinsinger and Tiebout 1991). The relationships between population density, degree of clustering, and strength of sexual selection need to be further clarified in plants as well as animals.

17.3 Sex Roles

In Bateman's (1948) explanation of sexual selection in animals, males compete over females because the reproduction of a male is limited by the number of mates he can fertilize. Female success on the other hand is limited by the number of eggs or young she can produce (section 7.2). This situation occurs also in plants, and Bateman suggested that "the production of microspores in far excess of the minimum required to produce effective fertilization is explicable in this way." Male and female sex roles may therefore differ in similar ways in plants and animals. For example, in the dioecious lily *Chamaelirium luteum*, males vary more in their number of mates than do females (Meagher 1986).

One contrast is that in plants, male reproductive investment is almost entirely mating effort (production and dispersal of pollen), whereas males in some animals make large parental contributions (section 7.2). The relative magnitude of female compared to male investment in mating among plants is debated, as is the limitation of female reproduction by scarcity of pollen (section 17.6). In dioecious plants, females obviously make almost all the parental investment, supplying resources for the developing seeds. Mature fruits have from one to five orders of magnitude higher dry weight and protein content than the ovary (Stephenson and Bertin 1983; also see Lloyd 1988). The energy cost of reproduction therefore is typically larger in females than males (e.g., Lloyd 1973; Wallace and Rundel 1979; Allen and Antos 1988; Popp and Reinartz 1988; Carr 1991). The scope for parental investment by males in outcrossing plants is limited by the lack of contact between the male and the embryo, and by the minute size of pollen, probably favored by dispersal efficiency and advantages in production of many pollen grains. Possible exceptions might occur among species with large pollinia, such as orchids and asclepiads which donate a package of many pollen grains to the female (e.g., Dressler 1981; also see Willson 1979; Willson and Burley 1983).

This does not mean that the only important variation in pollen quality is genetic. Young and Stanton (1990a) showed that pollen produced under good nutritional conditions had higher competitive success in fertilizing ovules than pollen produced under low nutrient conditions. The difference seemed to concern only male competitive mating success, however, and did apparently not influence the total seed set of the fertilized female.

Some animals show reversed sex roles, females competing strongly over males, which take care of the offspring (section 7.6). Parental role reversal seems not to be known among higher plants. One likely reason is that plant sessility and dependence on wind, water, or pollinating animals for outcrossing prevents females from having their ovules, much larger than pollen grains, brought to males. This is probably not the only reason, however, as pollen packages that may be larger than an ovule are transferred by pollinators in, for example, some orchids and asclepiads. A long evolutionary history of morphological adaptation for transfer of pollen to ovules, fixed and encased in the carpel, probably raises strong developmental and other constraints against a reversal of plant sex roles in gamete dispersal, even if it would sometimes be advantageous in some respect.

In sum, reproductive effort through male function in plants is largely mating effort, whereas females expend parental effort in form of nutrition for embryos, seeds, and fruits. The degree to which females also compete and invest in mating is debated and less well known (section 17.6).

17.4 Secondary Sex Differences

There is now much evidence that sexual as well as natural selection has contributed to the evolution of conspicuous flowers and several other traits in higher plants. In this respect, they are similar to animals. Sexual selec-

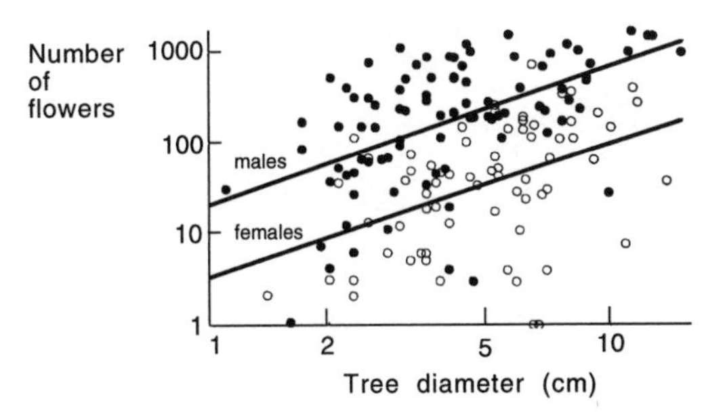

Figure 17.4.1 As in many other dioecious plants, males (dots) produce more flowers than females (open circles) in the American holly *Ilex opaca*. The lines show regressions of flower numbers on tree diameter for males and females. (After Carr 1991)

tion has usually not, however, led to nearly as conspicuous sex differences in plants as in animals. In the words of Lloyd and Webb (1977), "the marked sexual differentiation present in many animals which allows the sexes . . . to be recognized by their secondary sex characters has no parallel in the sporophyte generation of plants. Secondary sex differences in plants are less conspicuous" (also see Darwin 1877).

Lloyd and Webb (1977) defined secondary sex traits in plants as differences between the sexes in structures other than stamens and carpels. One common secondary sex difference in plants is production of more flowers and inflorescences in males than females (see Stephenson and Bertin 1983). For example, in the dioecious American holly *Ilex opaca*, males produce on average more than seven times as many flowers as females (figure 17.4.1; Carr 1991). A common explanation is that male pollination success benefits from large numbers of flowers (section 17.6). In several dioecious plants, individual males spread their flowering over a larger part of the season than do females. This should permit a male to fertilize more individual females, which vary in the timing of their flowering peaks (Allen 1986; Carr 1991).

Males also tend to flower more often and earlier in life and in the season than females. In addition, flowers are usually larger in males; comparative evidence suggests that they often invest more than females in conspicuous flowers. For instance, among 79 species measured by Knuth (1906), male flowers were larger than females in 74. They were almost twice as heavy as female flowers in seven insect-pollinated plants studied by Bell (1985), but the sexes had flowers of similar mass in three wind-pollinated species. Among New Zealand umbellifers, male plants bear more inflorescences per plant than females, and more flowers per inflorescence (Lloyd and Webb 1977). The ultimate reasons for these sex differences are doubtful, and there are exceptions. For example, in 14 out of 20 Costa Rican forest trees, female flowers were larger than males (Bawa and Opler 1975).

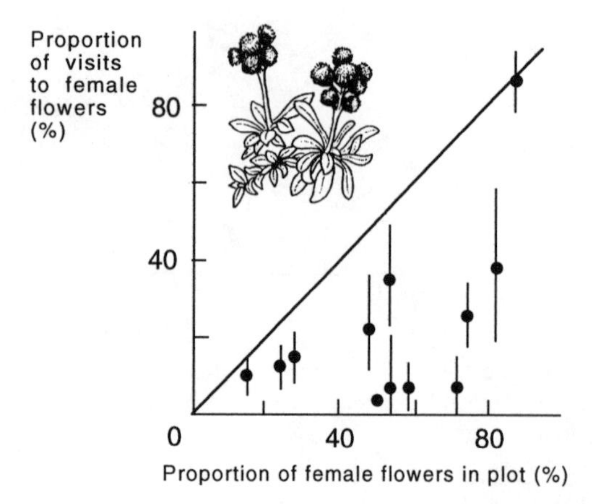

Figure 17.4.2 Pollinators prefer male flowers in *Antennaria parviflora*; the proportion of visits to female flowers is lower than their frequency in a plot. (From Cockburn 1991, after Bierzychudek 1987)

In some species, male flowers are more attractive than female flowers that offer no reward to pollinators, for example, in *Antennaria parviflora* (Bierzychudek 1987; figure 17.4.2). Male flowers may be more likely than females to suffer herbivore attacks (Bawa and Opler 1978; Willson 1991; section 17.6 below).

A few plants show conspicuous secondary sex differences. The flowers of male and female Neotropical *Catasetum* orchids are so different in shape and color that the sexes were once assigned to different genera. As in many sex-dimorphic animals, male flowers differ markedly between closely related species, whereas female flowers are similar. The sex differences may have evolved owing to competition among males over pollination (Romero and Nelson 1986). The pollinators, *Euglossum* bees, avoid male flowers after the experience of having the heavy pollinium rather brutally attached to their body. This should reduce the risk that a pollinium from another male flower will end up at the same pollinator and interfere or compete with the first pollinium. If female flowers resembled males, they would probably also be avoided by the pollinator. Romero and Nelson (1986) suggested that this harsh pollination tactic by male flowers is the reason why these orchids have evolved strongly sex-dimorphic flowers (figure 17.4.3).

In some plants, the sexes differ in other respects, such as the size of leaves. Bond and Midgley (1988) suggested that sexual selection of floral display may be responsible, as leaf size and other traits may be genetically correlated with the size of floral display.

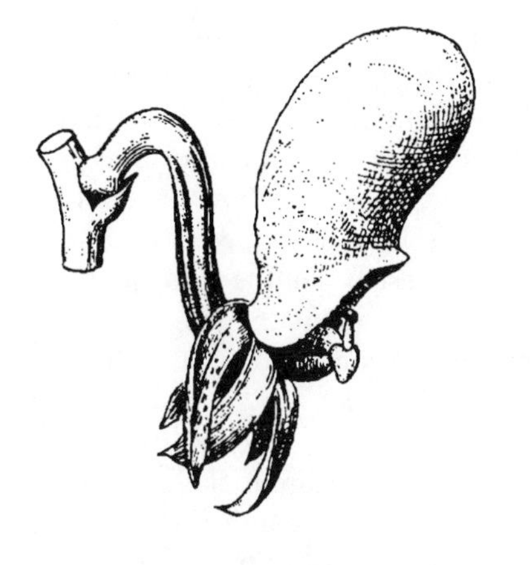

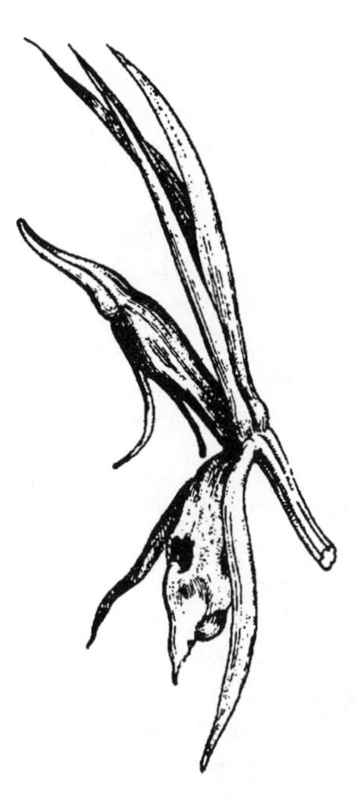

Figure 17.4.3 *Catasetum* orchids are among the few plants in which male and female flowers differ strongly, for reasons that may have to do with pollination efficiency and competition over pollinators (see text). The figure shows male (*right*) and female (*left*) flowers of *C. saccatum*. (From Wettstein 1935)

17.5 Pollen Production

Competition over pollination should favor production of large amounts of pollen, especially as it is also a nutrient reward for many pollinators. There are, however, few data on relationships between pollen quantity and paternal success (Bertin 1988). One of the most suggestive results comes from wind-pollinated white spruce *Picea glauca*. Studying male fertilization success by electrophoretic analysis of genotypes, Schoen and Stewart (1986) estimated the proportion of seeds sired by a clone in relation to its production of male cones. The results indicated that male fitness increased with the amount of pollen produced, at least up to a limit. In wild radish *Raphanus sativus*, the proportion of pollen removed by pollinators increased with pollen production (Young and Stanton 1990b).

Size of the pollen grain is probably under strong selection; by making it smaller, a plant can produce more pollen. In addition, the female might use nutrition supplied by a pollen grain even if its sperm does not fertilize an ovule, which should also favor low male investment in each grain (Willson and Burley 1983). Bertin (1988) reviewed variation among pollen in size and other respects. The lower size limit seems to depend on material requirements for tube growth; variation in pollen size among species is related to the tube length needed for reaching the ovule from the stigma.

Also the shape and other characteristics of pollen may be subject to strong selection for efficiency in reaching the target (reviewed by Campbell and Waser 1987). An example is the unusual convergent form of pollen and spores in aquatic plants, which in many cases are extremely elongated or large compared to terrestrial forms. Using search theory models, Cox (1983, 1988a; Cox and Knox 1988) suggested that the elongate form should increase the chances for aquatic pollen to find its target.

The total pollen dispersal of a plant depends on, among other things, how it distributes pollen among visiting pollinators. By "packaging," a plant divides its pollen production into separate units that are offered sequentially to pollinators; by "dispensing," the plant limits the amount of pollen that a pollinator removes from a package during a visit (Lloyd and Yates 1982). Of the pollen that a bumblebee removed from a flower of the lily *Erythronium grandiflorum*, it deposited on average only 0.6% on stigmas of the same species. Based on results from this species, Harder and Thomson (1989) developed a model of pollen transport, predicting the most effective means of packaging and dispensing pollen, and how it depends on patterns of pollinator visits and pollen removal. In another model of pollen dispersal, Campbell and Waser (1987) pointed out that dispersing pollen in small rather than large packages may be favorable especially when ovule number is low and seed production is not pollen-limited.

17.6 Pollinator Attraction

The attraction of pollinating animals is a stage in angiosperm reproduction where sexual selection is particularly likely. This was realized by Bateman (1948), who suggested that sexual selection may explain why male catkins in insect-pollinated sallow *Salix caprea* are bright yellow, in contrast with the gray female catkins (both produce nectar).

Competition over pollinators can affect male as well as female mating success. There is considerable evidence in plants for sexual selection of males, and some indications also for females. A whole spectrum of traits is involved in pollinator attraction (reviewed, e.g., by Proctor and Yeo 1973; Faegri and van der Pijl 1979; Waser 1983a,b). For example, Galen (1989) showed by a selection experiment that pollinating bumblebees can bring about the altitudinal clines in flower phenotypes observed in the alpine sky pilot *Polemonium viscosum*.

The focus here is mainly on visual stimuli provided by flowers, which have been the subject of most research. The likely role of plant-pollinator interactions in the profuse speciations of angiosperms was discussed in section 9.3 above.

Sexual Selection of Flowers

It has sometimes been suggested that the conspicuous sexually selected signals of animals have no counterpart or are poorly developed in plants (e.g., Stephenson and Bertin 1983; Moore 1984; also see Charlesworth et al. 1987; Queller 1987). Others have suggested that flowers are the botanical counterparts of sexually selected attractive traits in animals (e.g., Gilbert 1975; Bawa 1980 a,b; Lloyd and Yates 1982; Queller 1983). These contrasting views may in part reflect different definitions of sexual selection.

The classical, natural-selection explanation of conspicuous flowers is that they facilitate gamete transfer and outcrossing by attracting pollinators (see, e.g., Waser 1983b; Wyatt 1983). Cross-fertilization may lead to higher reproductive success by reducing inbreeding depression. In addition, there is likely competition among males over pollination, and perhaps among females over pollen (Bawa 1980a; Stephenson and Bertin 1983; Bertin 1988). Therefore, besides having a naturally selected role in cross-fertilization, attractive flowers are probably often sexually selected in the Darwinian sense of competition over mates. Such competition should lead to more extreme floral traits than favored by the outcrossing function alone. A challenging task is to quantify the relative roles of natural versus sexual selection in the evolution of pollinator attractants and rewards, such as flower and nectar.

Like conspicuous ornaments of animals, flowers may be costly to produce and drain resources from growth and maintenance (reviewed by Lloyd 1988). For example, a substantial part of a plant's available energy may be shunted into nectar production (Pleasants and Chaplin 1983; Southwick 1984). Experimentally increasing the rate of nectar production by removing nectar from flowers of Christmas bells *Blandfordia nobilis*, Pyke (1991) showed that higher nectar production reduces the plant's ability to produce seeds. (Hand-pollination ensured that differences in pollination success were not responsible for the pattern.) This suggests that there is a trade-off in nectar production between pollination advantages (see below in this section) and energetic disadvantages that limit seed production (Pyke 1991).

In some plants, herbivores selectively eat flower buds or flowers. Pollinators may eat quantities of pollen; some reproduce at the plant and their offspring eat seeds or other plant parts (reviewed by Willson 1983). Male flowers that offer pollen or other rewards, or that are more conspicuous than females, often seem to be subject to higher risk of attack. This was the case in ten species with sex differences in risk, reviewed by Willson (1991). There does not seem to be a similar sex difference in risk as regards vegetative parts.

In two herbs, *Ipomopsis aggregata* and *Bartsia alpina*, pollinators as well as seed predators seem to be attracted by the conspicuous inflorescences: pollination success and seed predation both increased with inflorescence size (Hainsworth et al. 1984; Molau et al. 1989). The evolution of conspicuous flowers and pollinator rewards for these reasons is likely to come to a halt at some point where mating advantages are balanced by losses in terms of energy, nutrients, predation, or other factors, as with sexually selected traits in animals. The main difference is that flowers attract mates indirectly via pollinators, which can have important consequences for the genetic evolution of the system (section 17.8).

MALE SUCCESS

The male contribution to fitness in cosexual plants has only recently become subject of quantitative study. It is difficult to measure in the field, as the parent of a pollen grain that fertilizes an ovule is hard to identify (e.g., Bertin 1988). Electrophoretic studies of natural populations of wild radish *Raphanus sativus* (figure 17.6.1) have shown, however, that individual maternal plants often have 6–8 mates, and that more than 50% of the fruits of a plant have multiple fathers (mean 2.1; Ellstrand 1984; Ellstrand and Marshall 1986). As in animals, DNA fingerprinting (section 4.1) may open new avenues to analysis of parentage in plants. So far, indirect measures of male success have dominated, measures that may be only weakly related to male fitness, such as degree of pollen removal from the anther by pollinators (Bertin 1988). More direct estimates of male fitness are needed.

An extensive study of the role of flowers in male and female function was done by Bell (1985). Many of his observations and experiments showed that the presence and size of conspicuous flowers are important for pollinator attraction. For example, insects visited flowers of strawberry *Fragaria virginiana* in proportion to their size (figure 17.6.2). The largest flowers received about ten times as many visits as the smallest (also see Willson and Price 1977; Schaffer and Schaffer 1977; Andersson 1988, Young and Stanton 1990b, Campbell et al. 1991). Removal of parts of the corolla in jewelwood *Impatiens capensis* and other species reduced insect visits. When all petals were removed, visits dropped to about one tenth of the normal value (Bell 1985).

Also in studies of several other angiosperms, the number of pollinators attracted increased with inflorescence size (e.g., Schaffer and Schaffer

Figure 17.6.1 Flowers of some of the plants discussed in this chapter. Top: *Asclepias exultata*. Center, left: *Delphinium nelsonii*. Bottom: *Raphanus sativus*. (From Gleason 1952 [top, bottom], Martin and Hutchins 1980 [center])

1977; Stephenson 1979; Schemske 1980b; Davis 1981; Schmid-Hempel and Speiser 1988; Larson and Larson 1990). In the orchid *Brassavola nodosa*, the number of pollinia removed per flower and inflorescence increased with inflorescence size (Schemske 1980a; table 17.6.1).

Pollen export is harder to quantify, but it has been measured in species with pollinia: coherent masses of pollen grains that are dispersed as count-

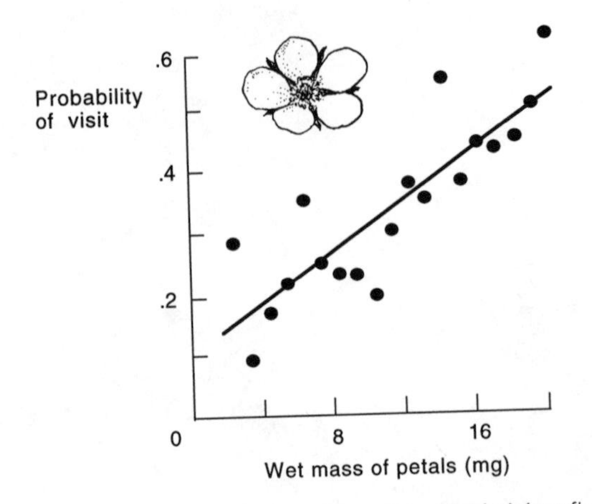

Figure 17.6.2 The attraction of insects to strawberry *Fragaria virginiana* flowers increases with their size. (Probability of visit was measured in 2-minute periods for plants within 1 mg intervals). (After Bell 1985; flower from Hitchcock et al. 1961)

Table 17.6.1

Relative Reproductive Success of *Brassavola nodosa* in Relation to Inflorescence Size*

	Flowers per Inflorescence (Sample Size)				
Relative Success	*1 (170)*	*2 (106)*	*3 (36)*	*4 (3)*	*5 (1)*
Pollinaria removed per inflorescence	1.0	2.7	5.5	11.1	9.5
Fruit set per inflorescence	1.0	2.5	7.6	8.4	25.0

* Relative success = absolute success, divided by that for one-flowered inflorescences. After Schemske 1980a.

able units (e.g., Nilsson et al. 1992). Milkweeds *Asclepias* have five pairs of pollinia per flower. In an experimental test of pollen removal, Queller (1983, 1985) cut off half the number of flowers from inflorescences of *A. exultata* (figure 17.6.1). Compared to controls, this led to half as many pollinia being removed by pollinators, probably reducing male success of the plant. Milkweeds mature fruits in only a minority of their flowers, and the number of fruits per plant was not reduced by the removal of flowers. Queller (1983) therefore concluded that the high number of flowers is selected mainly by male function. Other aspects of his study suggested that this is the case also for inflorescence longevity and seasonal distribution of flowers.

Bell (1985) made a similar experiment in *A. syriaca*, where he cut off various proportions of inflorescences in pairwise tests. The number of pollinia removed and received increased with inflorescence size, per flower as well as per inflorescence, again suggesting that conspicuous flower display increases male success. Several other studies of *Asclepias* indicate that male function benefits from larger inflorescences than needed for maximum fruit production, which may be limited by other ecological factors (Willson and Rathke 1974; Willson and Price 1977).

Comparative evidence also suggests that male function in hermaphrodites benefits from presence of more flowers than can result in fruits (Sutherland and Delph 1984; Bawa and Webb 1984). The hypothesis predicts that, compared to monoecious or dioecious species, hermaphrodites will have more flowers that yield no fruit. The reason is that with flowers of separate sexes, the plant can develop the number of male flowers favored by pollination success, without also raising the number of female flowers. This prediction was upheld in a comparison of 316 hermaphroditic and 129 monoecious or dioecious species, supporting the idea that male function favors excess flowers (Sutherland and Delph 1984). Phylogenetic constraints were not taken into account, however, as species were treated as independent sample units (see section 4.3).

An experimental study of wild radish *Raphanus sativus* (figure 17.6.1) showed that its floral display is favored mainly by male function. Using equal numbers of two Mendelian color morphs with white and yellow petals, Stanton et al. (1986, 1989) found that pollinators, mostly cabbage butterflies *Pieris rapae*, favored the yellow morph. (The mechanism that maintains the white morph at high frequency in the wild is unknown.) It received 74% of all pollinator visits. Male mating success was in close agreement with pollinator visits: 72% of 621 progeny were fathered by yellow plants. (The two aspects do not always agree; e.g., Schoen and Clegg 1985; Stanton et al. 1991). In contrast, maternal success was equal for the two morphs, yellow plants yielding 863, and whites 876 seeds (figure 17.6.3). The differences in attraction of pollinators thus had strong effects for male but not female fitness in this hermaphroditic plant (also see Stanton and Preston 1988; Young and Stanton 1990b).

Notwithstanding this and several other admirable studies of sexual selection of flowers, competition over pollinator attraction needs additional testing in several respects. For example, the results of Stanton et al. (1986, 1989) suggest that the white morph had reduced male success owing to competition with the yellow morph. This is not fully confirmed, however. The yellow morph might attract more pollinators to the area. If so, absence of the yellow morph need not necessarily increase the success of the white morph: fewer pollinators might turn up, leading to reduced success for the white morph. Such an effect was found by Waser and Price (1981, 1983), comparing pollinator attraction and seed set in two color morphs of Nelson's larkspur *Delphinium nelsonii* (figure 17.6.1) in the Rocky Mountains. The normal blue flower has better nectar guides and is visited by pollinators more than a rare white morph, which yields lower rates of en-

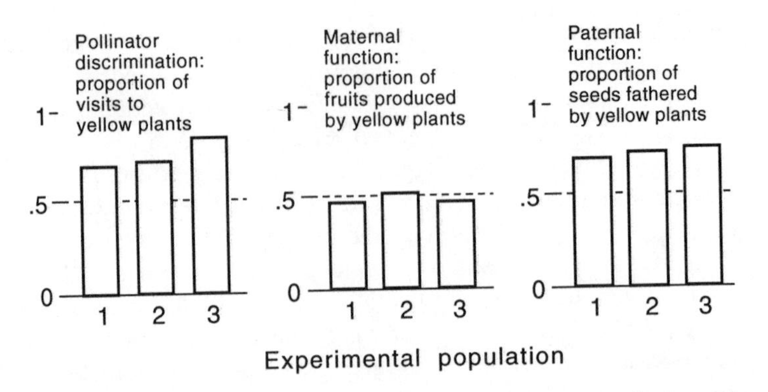

Figure 17.6.3 Floral display in wild radish *Raphanus sativus* appears to be favored mainly by male function. The figures show relative performances of yellow- and white-flowered homozygotes in three experimental populations, (1–3), each with equal numbers of yellow- and white-flowered plants. Yellows received most pollinator visits (*left*) and fathered most seeds (*right*), but yellows and whites produced similar numbers of fruits (*center*). (After Stanton et al. 1986. Copyright © by the AAAS)

ergy intake for hummingbirds and bumblebees. In experimental 1×1 meter plots with nine plants in each, pollinator attraction and seed set was higher for blue than white flowers. Yet the success of white flowers did not increase when there were fewer blue flowers, because fewer pollinators were then attracted to the plot. Perhaps pollinators instead visited other blue morph flowers elsewhere. If so, the white morph suffers in mating competition with the blue morph, but this is only one of several possibilities. The situation can be complex, especially if several plant species compete for the same pollinators (e.g., Rathcke 1983; Waser 1983a; Feinsinger et al. 1991; Kunin 1993).

Competition over pollinators is strongly suggested by some of Bell's (1985) results. In two species tested experimentally, the presence of neighboring plants of the same species tended to reduce the number of insect visits per flower and plant. Most pollinator visits per flower in strawberry *Fragaria virginiana* and jewelwood *Impatiens capensis* occurred at single flowers without immediate neighbors. Flowers in dense groups had fewer individual visits. Similar tests in other species are needed for assessing the importance of sexual selection by competition over pollinators. It is then desirable to follow pollen success all the way through to production of breeding offspring, which probably requires genetic analysis (e.g., Meagher 1986, 1991), such as DNA fingerprinting.

FEMALE SUCCESS

Many of the previous studies showed that variation in floral display has a stronger effect on male than female reproductive success. The display is, however, likely to play a role also for females, which must be visited by pollinators to set seeds in most outcrossing species. For example, Schemske (1980a) concluded that not only male but also female reproductive success per flower increases with inflorescence size in the orchid *Brassavola nodosa* (table 17.6.1; also see Schmid-Hempel and Speiser 1988; Campbell 1989a). In a sex allocation model by Lloyd (1988), the evolutionarily stable level of resource use for pollinator attraction turns out to be proportional to the additive marginal benefits for female as well as male function (section 17.9). Some of the investment in pollinator attraction in hermaphrodites is then accounted for not only by male, but also by female function.

Females invest many resources in seeds and fruits (Lloyd 1988). Even if floral display is crucial for the success of females, they may therefore have fewer resources than males to use on such traits. Lower investment in display by females need not imply that it means little for their fitness. There is evidence that floral display may be crucial for female function, as insufficient pollination limits seed or fruit set in several species (see below in this section). Also, receipt of more pollen than needed to fertilize all ovules might permit the female a "choice" of high-quality pollen or zygotes for maturation into seeds (section 17.7).

In some of Bell's (1985) experiments, removal of flowers from an inflorescence reduced the fruit set of its remaining flowers. In *Viburnum alnifolium*, each inflorescence in addition to the fertile flowers also bears a marginal ring of large, sterile flowers. Removal of these sterile flowers reduced the fruit set by 20% compared to intact controls. Female fertility in some cosexual plants therefore does vary with the floral display, which is then favored by female as well as male function. The effects of various levels of pollinator attraction on both male and female success need to be assessed experimentally in a range of species to clarify the situation.

Campbell (1989b; also see Campbell 1991; Mitchell 1993) used fluorescent powdered dye to track pollen transport by hummingbirds in *Ipomopsis aggregata*. The variances in male and female success (measured as relative donation and receipt of dye) were of similar magnitudes. Multivariate selection analysis showed that male function favored a wide corolla, late flowering, and short pistillate phase, whereas female function favored early blooming.

The limitation of seed production in plants is complex. Reviews by Stephenson (1981); Willson and Burley (1983); and Zimmerman and Pyke (1988) showed that seed set is often limited by environmental resources such as minerals, light, or water. The evidence is of several kinds. For example, the seed crop tends to increase with addition of mineral fertilizer and with removal of competing neighbors. After a rich fruiting year that probably drains stored resources, reproduction and vegetative growth are reduced. But even if seed production is resource limited, the quantity or quality of pollen received might be limiting as well. The extent to which pollen limits seed set is, however, debated and difficult to determine.[2]

Johnston (1991b) reviewed the evidence for pollen limitation of seed set, focusing on studies that tested this possibility by supplemental hand pollination of the whole plant. Studies that only supplement a subset of flowers are harder to interpret, as increased seed set by these flowers might occur at the expense of other flowers on the same plant. In addition, such experiments in perennials should ideally be carried out and followed up over several seasons, as increased seed set at one time may lead to reduced seed set later on (Janzen et al. 1980; Zimmerman 1988; Zimmerman and Pyke 1988; Ackerman and Montalvo 1990; Calvo and Horwitz 1990; Primack and Hall 1990). Among the studies reviewed by Johnston (1991b), seven demonstrated pollen limitation of seed set, whereas six found no such evidence.

In many hermaphroditic plants, only a fraction of the flowers produce seeds and fruits. A number of proximate and ultimate explanations for low fruit:flower ratios were reviewed by Ayre and Whelan (1989; also see, e.g., D. G. Lloyd 1980; Stephenson 1981; Bawa and Webb 1984; Sutherland 1986; Lee 1988; D. Charlesworth 1989). Four of the proximate hypotheses are

1. Insufficient amounts of pollen
2. Low pollen quality
3. Ecological resource limitation
4. Predation of flowers or ovules.

At an ultimate, evolutionary level, the question is what factors select for more flowers than usually result in fruit. Among the suggested explanations are that excess flowers benefit the plant in

1. Attracting pollinators that disperse its pollen
2. Producing large amounts of pollen
3. Permitting a "choice" of the highest quality zygotes for maturation
4. Permitting many fruits to be produced in an unusually good year.

The main question of interest here is whether seed set is often limited by pollen. If so, female function should favor traits that attract pollinators, and females might compete over pollen, leading to female sexual selection of traits that increase pollen import. The evidence reviewed by Johnston (1991b, see above) suggests that such limitation may often occur.

Comparative analyses suggest several patterns in seed:ovule and fruit:flower ratios. They seem to be higher in annuals than perennials, and

[2] See, e.g., Bierzychudek 1981, Bertin 1982, Stevenson and Bertin 1983, Willson and Burley 1983, Snow 1986, Charlesworth et al. 1987, Lee 1984, 1988, Haig and Westoby 1988, Horwitz and Schemske 1988, Zimmerman 1988, Zimmerman and Pyke 1988, Johnston 1991a.

in self-compatible than outbreeding species. Pollen limitation is probably not responsible for lower ratios in perennials, but it may contribute to higher values in self-compatible plants, which seem less likely to remain unpollinated than do outbreeders (Sutherland and Delph 1984; Wiens 1984; Sutherland 1986). Another factor contributing to reduced seed set might be high mutational load in outcrossing species (Wiens et al. 1987; also see D. Charlesworth 1989).

One example of pollen limitation of seed set is the dioecious, perennial herb jack-in-the-pulpit *Arisaema triphyllum*. Pollinating a number of plants by hand, Bierzychudek (1981) found that they produced more than ten times as many seeds as control plants. Seed set depended on plant size only in hand-pollinated females, suggesting that only they came close to ecological resource limits. There was no hint that higher seed set in hand-pollinated plants occurred at the expense of next year's reproduction. Evidence for such an effect is a higher likelihood for plants that produce many seeds to change to male sex the following year (Bierzychudek 1984).

Limitation of female success by insufficient pollination seems particularly likely in orchids (Orchidacae), perhaps the largest of all angiosperm families. Compared to other plants, orchids have tiny, numerous seeds, in extreme cases more than one million per flower (Darwin 1877a; Benzing 1981; Dressler 1981; Nilsson 1992). There is evidence that insufficient pollination often limits seed set in orchids.[3]

Darwin (1877a, p. 288) suggested that the high number of seeds explains both why most orchids disperse their pollen in two pollinia, each containing a large number of pollen grains, and why orchids have evolved highly specialized pollinator relationships, increasing the chances that the few pollinia will reach a stigma of the correct species. The dependence on special pollinators, with sensory abilities to discriminate among variants, perhaps in part explains why orchids have speciated so profusely (see, e.g., Benzing 1981; Dressler 1981; Kiester et al. 1984; and section 9.3 above). Many aspects of orchid speciation and reproductive biology still remain mysterious (Gill 1989; Nilsson 1992).

To summarize, observations and experiments show that conspicuous flowers increase male pollination success. Some studies suggest that also female fertilization and seed production depend on floral display. The relative roles of selection of male and female function in pollinator attraction need to be examined in a variety of species.

FEMALE FLOWERS MIMICKING MALES?

A Central American tree, *Jacaratia dolichaula*, provides a likely case of female mimicry of male flowers (Bawa 1980a). The stigmatic lobes of the

[3] E.g., Janzen et al. 1980, Nilsson 1980, Gill 1989, Snow and Whigham 1989, Calvo and Horwitz 1990, Ackerman and Montalvo 1990, Primack and Hall 1990.

female flower closely resemble the corolla lobes of the male. Females produce no nectar but have a similar sweet scent as males; sphingid moths are the likely pollinators. Male flowers were 4–16 times as common as females, perhaps owing to male competition. (The number of pollinators visiting a plant increases with its flower numbers in several species; see Male Success above.) Other evidence for likely scramble competition among males is that male *Jacaratia* start flowering earlier than most females. The high proportion of male flowers permits females to attract pollinators by mimicking male appearance and scent without producing nectar: only a low proportion of the flowers will be nonrewarding females. Mimicry of males suggests that females are under selection for attracting pollinators, and that females are possibly also subject to sexual selection by competition over pollinators (Bawa 1980a). Female mimicry of males has been inferred in other dioecious and monoecious species, for example papaya *Carica papaya* (Baker 1976; also see Little 1983; Ågren et al. 1986).

SCENT

Scent, the other main long-distance signal to pollinators, may be under similar selection pressure as visual flower display (e.g., Proctor and Yeo 1973; N. H. Williams 1983). Our noses being what they are, sexual selection of scent in plants largely remains to be explored, but at least one study in the wild has found effects of variation in scent for pollination and reproductive success. The alpine sky pilot *Polemonium viscosum* has one scent morph with sweet, and another with skunky smell. Galen (1985, 1989) showed by hand pollination that seed set was limited in both morphs by the level of pollination. Plants with skunky scent had lower success in alpine habitats where bumblebees are the main pollinators, and higher success at lower altitudes, where flies are common pollinators.

NECTAR

The standing crop and rate of nectar production influence how long a pollinator stays at each plant, how many flowers it visits, and what travel pattern it follows. These aspects are likely to affect male as well as female fitness in complex ways, discussed by Stephenson and Bertin (1983), Waser (1983b), and Zimmerman (1988). For an ESS approach to the rate of nectar production, considering male and female contributions to fitness, see Pyke (1981).

Nectar production in flowers of *Lobelia cardinalis*, a herb pollinated by hummingbirds, was roughly twice as large per day during the active male phase as during the later female phase (Devlin and Stephenson 1985). The nectar reward was highest near the top of the inflorescence, where male-phase flowers are located. Hummingbirds preferred this upper part, apparently responding to its better nectar reward. This and other aspects of the visitation pattern may promote outcrossing, and leads to longer and more frequent pollinator visits to flowers in active male phase. The results are consistent with ideas on flower phase separation presented by Lloyd and Yates (1982; see Flower Phenology, below), but it remains to measure the strength of sexual selection of nectar production in *L. cardinalis* (Devlin and Stephenson 1985).

S. Sutherland (1987) analyzed fruit set and nectar production in *Agave mckelveyana*, another protandrous hermaphrodite. Fruit set was resource limited; most fruits are produced by flowers in the upper part of the inflorescence, lower flowers mainly making pollen. Most of the lower flowers abort before fruit production, even if they are pollinated. The plant produced relatively more nectar in upper than lower flowers during the female phase, suggesting that it allocated relatively more nectar to female function in flowers that might yield fruits than in flowers likely to become aborted. This also suggests that nectar production is important not only for male (pollen) success, but also for female success in being fertilized.

In *Asclepias quadrifolia*, nectar production is more strongly correlated with the level of pollinia removal than with insertion, suggesting that nectar production has a stronger effect on male than female function (Pleasants and Chaplin 1983). Similar results were obtained in *Ipomopsis aggregata* by Mitchell (1993) and Mitchell and Waser (1992).

Some dioecious plants show sex differences in nectar production; the reasons for the differences among species are not well understood. Female flowers in five of six species of Costa Rican trees produced several times more nectar than males. Bawa and Opler (1975) suggested that this may be a compensation for lack of pollen reward to pollinators in female flowers. In some other species, however, females offer no nectar (e.g., Baker 1976; Bawa 1980a).

Flower Phenology

Competition over pollination may explain several aspects of flowering phenology (e.g., Zimmerman 1980; 1984, 1988). For example, many plants release pollen gradually over several days. Using an ESS model, Lloyd and Yates (1982) explored temporal division of male and female function in cosexual plants where reproduction is limited by resources in females, and by pollination success under competition in males. The model indicated that it is then adaptive to spend longer time in male (pollen-releasing) phase than in female (receptive) phase. Sexual selection of males may therefore explain longer male than female phase in some plants (Lloyd and Yates 1982; Devlin and Stephenson 1985; Harder and Thomson 1989).

Separation between male and female phase in some species may be favored because it reduces selfing, but in self-incompatible species, the reason must be different. Lloyd and Yates (1982) suggested that separation is favored because it segregates pollen and stigmas in time and prevents mechanical interference between male and female function during pollination. (Interference might also select for unisexual flowers; Bawa and Opler 1975.) Separation in hermaphrodites usually takes the form of protandry, with pollen release preceding stigma exposure in the same flower (Bawa and Beach 1981). Lloyd and Yates (1982) point to reasons why separation should usually be most easily achieved by protandry.

In addition, protandry in plants may be sexually selected for the same reason as in some animals. Males are involved in scrambles to reach and fertilize females before competitors do so; this favors protandry rather than protogyny (e.g., Wiklund and Fagerström 1977; Bawa 1980a; Fagerström and Wiklund 1982; Bulmer 1983). There are many exceptions, particularly among monoecious plants, where female flowers are often earlier than males, so selective pressures other than pollen competition are also likely to be important (Stephenson and Bertin 1983). The factors affecting flowering time in the two sexes remain to be clarified in detail.

17.7 Male Scrambles and Female Choice in the Style?

Male Scrambles

Competition over pollination does not end when the pollen reaches the stigma: as grains from several individuals may arrive at the same time, quick pollen germination and tube growth should be crucial for male success (Mulcahy 1979; Lee 1984). Genetic variation in germination time or tube growth rate has been found in several species (see Bertin 1988). Sexual selection in the style was reviewed by, for example, Charlesworth et al. (1987), Queller (1987), Bertin (1988), Lyons et al. (1989), Marshall and Folsom (1991), and Walsh and Charlesworth (1992); this possibility and some of its fascinating consequences are discussed below.

In Haldane's (1932) words, "clearly a higher plant species is at the mercy of its pollen grains. A gene which greatly accelerates pollen tube growth will spread through a species even if it causes moderately disadvantageous changes in the adult plant. A gene producing changes which would be valuable in the adult will be unable to spread through a community if it slows down pollen tube growth." In accordance, pollen tubes are among the most rapidly growing tissues in the organismal world, in some cases growing 1 cm in 20 minutes (Mulcahy and Mulcahy 1987).

The mechanisms that lead to differences in success after pollen has reached the stigma are hard to identify. Among the many possibilities are differences in germination time, tube growth rate, interactions with other pollen, and with the style or ovule. Success therefore needs to be decomposed into a number of sources of variation, such as paternal genotype and environment, pollen genotype, number and genotype of other pollen present at the style, maternal genotype and environment, and interactions among these components (see Schemske and Fenster 1983; Lyons et al. 1989; Marshall and Folsom 1991; Walsh and Charlesworth 1992). This formidable task has not been carried out in full in any species. The problems of demonstrating genetic differences in competitive ability among pollen, and the evidence for such differences, are discussed by D. Charlesworth (1988), Stephenson et al. (1988), and Walsh and Charlesworth (1992).

There is evidence for pollen competition in cultivated species, but its importance under natural conditions has rarely been studied (see Bertin 1988; Lyons et al. 1989; Marshall and Folsom 1991; Walsh and Charlesworth 1992; for reviews and discussions). There is now evidence, however, that individual differences in the growth rates of pollen tubes lead to nonrandom differences in paternal success in wild rose mallow *Hibiscus moscheutos* (Snow and Spira 1991a,b). Experiments revealed consistent paternal differences in tube growth rate among individuals tested on fifteen unrelated recipients of pollen (figure 17.7.1). Experiments using Mendelian marker alleles showed that pollen tube growth rate was correlated with the number of seeds the plant sired when pollen mixtures were applied to the stigmas of recipients (Snow and Spira 1991a). These results strongly

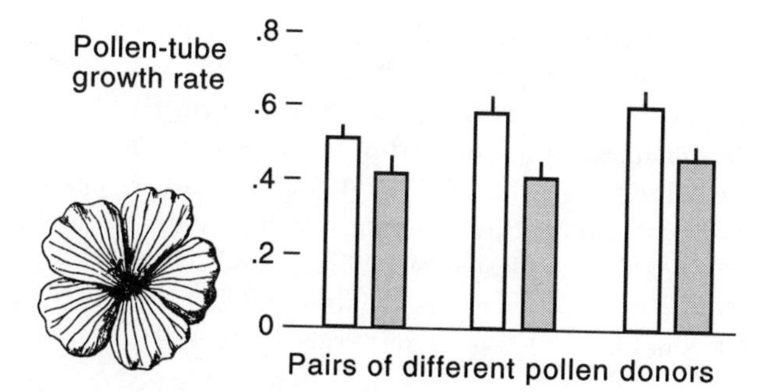

Figure 17.7.1 Experiments in wild rose mallow *Hibiscus moscheutos* show that individuals differ in the rate at which the tubes of their pollen grow (there are corresponding differences in pollination success). Relative growth rates for three pairs of pollen donors are shown in the figure. Each pair was tested on 14–15 unrelated recipients; open bars represent the donors with higher growth rate (vertical lines = s.e.). (After Snow and Spira 1991; flower from Gleason 1952)

suggest that sexual selection by pollen scrambles in the style may occur in the wild. It is not known, however, whether the differences in pollen tube growth rates are genetic or purely phenotypic. An example suggesting the latter possibility was presented by Young and Stanton (1990a; section 17.3 above).

Part of the haploid gene complement of pollen codes for traits that are also expressed in the diploid plant. A gene favorable at the pollen stage might therefore also be favorable at the sporophyte stage. Because pollen is haploid, both recessive and dominant mutations can be expressed and exposed to selection during pollen tube growth. Mulcahy (1979) suggested that there is more scope for pollen competition in angiosperms than in gymnosperms, and that adaptive evolution and radiation in angiosperms for these reasons can proceed faster than in gymnosperms (Mulcahy 1979; Mulcahy and Mulcahy 1987; but see Walsh and Charlesworth 1992). There are, however, several alternative hypotheses for angiosperm success, especially as regards their diversification in relation to insect pollination (reviewed, e.g., by Eriksson and Bremer 1992; section 9.3 above).

Female Choice?

Several authors have suggested that females to some extent can "choose" among the pollen grains competing to fertilize the ovules.[4] One example is self-incompatibility, which probably arose at least in part owing to advantages of inbreeding avoidance. Many incompatibility systems are based on genetic variability at one or a few highly polymorphic loci (reviewed, e.g., by Charlesworth et al. 1987; Barrett 1988). This special kind of mate choice does not favor any particular phenotypic trait.

Another example of female choice of pollen concerns effects of outcrossing distance (reviewed by Bertin 1988; Lovett Doust and Lovett Doust 1988; Waser 1993). Matings with close relatives often lead to inbreeding depression (Charlesworth and Charlesworth 1987b). At the other extreme, distant matings might disrupt local adaptations, leading to "coadaptive breakdown" (Endler 1977) and "outbreeding depression." These effects in combination can favor matings with partners from intermediate distances, an idea first suggested by Müller (1883). There is some support from work on animals as well as plants (e.g., Bateson 1978; Price and Waser 1979; Shields 1982; Waser and Price 1989, 1991; Waser 1993).

[4] E.g., Janzen 1977, Charnov 1979b, Stephenson and Bertin 1983, Willson and Burley 1983, Lee 1984, Waser et al. 1987, Galen and Rotenberry 1988. For reviews, see Charlesworth et al. 1987, Queller 1987, Lyons et al. 1989, Marshall and Folsom 1991. Other evidence that interactions between the sexes influence pollen success is provided, e.g., by Bertin 1982, Marshall and Ellstrand 1986, 1988, Snow and Mazer 1988, Schlichting and Devlin 1989, Marshall and Folsom 1991, Walsh and Charlesworth 1992.

In Nelson's larkspur *Delphinium nelsonii* (figure 17.6.1), seed set per flower was higher for medium mating distances (10 m) than for shorter or longer distances (1 m, 100 m, and 1,000 m; Waser and Price 1991). Seedling survival in the wild also seemed to be highest for offspring from medium mating distances, and so was the growth rate and number of pollen tubes reaching the ovary (Waser et al. 1987). This effect may have been caused by interactions between pollen and style favoring pollen from medium distances. Whether it arises by an adaptive, evolved discrimination mechanism in the female, or as an incidental, inevitable consequence of interactions between the paternal and maternal genomes, is not known. It should not be adaptive for a pollen grain to refrain from fertilization once it has reached the stigma (Charnov 1979b), at least not if it belongs to an unrelated female. In matings between close relatives, in particular for pollen and ovules from the same plant, kin selection might favor pollen "temperance," if this raises the chances that unrelated pollen instead fertilize the ovule and leads to sufficiently much higher success for the outbred zygote. Still, the female should usually have a greater genetic advantage of avoiding fertilization by a related pollen grain, suggesting that more discriminating maternal control of fertilization (rather than pollen control) is likely to evolve (also see Waser and Price 1993).

Female choice in *D. nelsonii* apparently concerned favorable combinations of genes from male and female. As Waser et al. (1987) pointed out, choice of a genetically complementary mate does not usually lead to sexual selection of any particular trait.

Mate choice in plants has sometimes been defined in ways that include choice among haploid pollen as well as embryos to mature (Stephenson and Bertin 1983; Willson and Burley 1983; also see, e.g., Charlesworth et al. 1987; Mazer 1987; Queller 1987; Lee 1988; Marshall and Folsom 1991). Also in some animals, females, after insemination, can influence which sperm will fertilize her eggs, and perhaps also which embryos will be nourished (reviewed by Eberhard 1985). Queller (1987) emphasized that selective maturation of embryos is not mate choice; females do not "choose" among embryos to get a good mate, but to get fitter offspring.

Embryo choice permits females to take advantage also of purely phenotypic variation among embryos, and of nonadditive genetic variation that arises from heterotic, epistatic, or other effects. A female that matures vigorous embryos is likely to leave more offspring than females that mature embryos at random (Queller 1987; also see Waser et al. 1987). The common abortion of embryos or fruits in many plants makes selective maturation a plausible possibility (e.g., Stephenson 1981; Lee 1988; Marshall and Folsom 1991). There is evidence that certain types of embryos are selectively aborted. For instance, in *Lotus corniculatus*, half the number of immature fruits are often aborted. Seeds were less successful from plants in which fruits were thinned randomly by hand than from plants that aborted fruits naturally (Stephenson and Winsor 1986). Mature fruits from the latter plants contained more seeds, which germinated and grew better, and produced more offspring. This suggests that fruits with lower than average numbers of seeds were aborted, and that offspring success as a consequence was raised, perhaps by elimination of embryos of low quality (Stephenson and Winsor 1986; also see Lee 1984).

Selective maturation of embryos seems to be less common in animals than plants (early embryo abortions are hard to document, however, so perhaps there is a bias here). Queller (1987) suggested the reason is partly that animals achieve much the same advantage simply by avoiding mating with close kin (see Bateson 1978).

17.8 Sexual Advertisement in Plants Compared with Animals

Why do plants usually show smaller secondary sex differences in advertising traits than animals? Lloyd and Webb (1977) suggested the reason is weaker sexual selection and less ecological niche differentiation between the sexes in plants. To these hypotheses can be added that females as well as males have to attract pollinators in order to reproduce; both sexes therefore need to display in most higher plants. In many animals this is not necessary, as one of the sexes searches for and visits the other. Selection for attractive display should therefore be more similar in the two sexes among plants than in many animals. This may be so even if there is strong sexual selection in plants, that is, competition over matings, mediated by pollinators (e.g., Bawa 1980a; D. Charlesworth 1984). Female flowers can probably not be much less conspicuous than males without having reduced pollination and reproductive success. In plants where females offer no pollinator reward such as pollen or nectar, females may be strongly selected to mimic males. This should increase the chances that pollinators rewarded by male flowers will also visit females (Bawa 1980a; section 17.6 above). Sexual selection of conspicuous male flowers therefore seems likely to lead to similarly conspicuous females. Moreover, there may be competition over pollination also in females of some plants. An additional reason for similarity of the sexes in animal-pollinated plants is that selection should favor flowers of each sex that resemble the other, as this will increase the chances that pollinators transfer pollen between individuals of the same species.

A need for females in dioecious plants pollinated by animals to resemble males perhaps explains why many such species have small flowers that should leave females with more resources for seed production. A radically different evolutionary response might be to avoid the need for resemblance between the sexes by evolving wind pollination, which is common in dioecious plants (Bawa 1980b).

To the extent that male and female flowers share the same genetic determination, selection of male flowers will lead to a correlated response in females. The minority of plants that are dioecious probably represent a derived condition, arisen many times from hermaphroditism (e.g., Bawa 1980b; Thomson and Brunet 1990). Dioecious plants may therefore have had less time than higher animals, where dioecy is ancient, to break up genetic correlations and evolve large secondary sex differences, even if they are adaptive. Such a process may perhaps require millions of generations (Lande 1980; Lande and Arnold 1985).

These hypothetical explanations for smaller secondary sex differences in plants than in animals should be testable with experiments and comparative analyses. Lloyd and Webb (1977) remarked that secondary sex characters in plants is an understudied subject that offers "an ideal opportunity to investigate differences in the reproductive strategies of males and females." In spite of exciting progress, this situation prevails today.

Colorful, conspicuous flowers are favored by advantages in pollinator attraction; sexual selection has probably often played a role in this process (section 17.6). Fisher (1930) and several later workers (e.g., O'Donald 1980; Lande 1981; Kirkpatrick 1982; Seger 1985) have suggested that conspicuous, attractive sex traits in animals may evolve by sexual runaway selection: self-reinforcing coevolution of trait and preference that arises because the genes for the preference and the preferred trait are combined in the offspring of the mating pair (chapter 2).

In plants, runaway selection by genetic coupling seems less likely for floral display, as the conspicuous flower and the attracted pollinator belong to different species. Even if an allele that confers improved pollinator attraction makes a plant reproduce better, a corresponding preference allele in the pollinator therefore cannot in general spread by genetic coupling with the new plant allele, in contrast with the situation in animals. This should usually prevent Fisherian runaway selection of trait and preference in plant-pollinator systems. Other processes may, however, lead to rapid evolution of floral display (West-Eberhard 1983). Some flowers seem as conspicuous and extreme as the most remarkable secondary set traits found in animals (figure 17.8.1).

Even if a Fisherian process may perhaps work in certain plant-pollinator systems (Kiester et al. 1984), the special conditions required make it unlikely to explain some of the most remarkable floral displays. For example, conspicuous, often fiery red flowers are typical of western American plants pollinated by migrant hummingbirds. A possible reason is that a convergent floral signal facilitates discovery of the flowers by vagrant pollinating

Figure 17.8.1 Some plants have extremely conspicuous flowers; in *Rafflesia arnoldii* they approach 1 meter in size. This genus consists of parasites on woody vines in SE Asian rainforests. The reason why their flowers are so large is poorly known. In *R. arnoldii*, the color and odor of the flower resemble those of rotting meat; *Rafflesia* flowers are pollinated by carrion flies. (Beaman et al. 1988; drawing by F. Bauer in Brown 1820)

birds passing through the area (Grant and Grant 1968). These flowers do not seem to fit the assumptions required for a runaway process. Perhaps some form of Fisherian process may sometimes lead to coevolution by genetic coupling between male and female parts of the flower (A. Pomiankowski, pers. comm.).

The extremely long tubes of some flowers (figure 17.8.2) likewise seem to have evolved at least in part by other than Fisherian processes. For instance, among plants pollinated by hummingbirds, flowers with long tubes typically occur in plants with widely scattered individuals and copious nectar secretion. The long tubes may be favored, for example, because they make it unprofitable for other than specialized pollinators to visit the flowers, hence reducing the risk of losing pollen to other species or becoming clogged by foreign pollen (see Feinsinger 1983 for review). The relationship between flowers with long tubes and their pollinators is often highly specialized. It enabled Darwin (1877a), upon seeing the 30 cm long corolla tube of the orchid *Angraecum sesquipedale* from Madagascar, to predict the existence of a pollinating hawkmoth with equally long tongue. In 1903 a candidate species was described and named *Xanthopan morgani praedicta* (see Nilsson 1988). Darwin (1877a) suggested that long corolla tubes often evolve because they force pollinators to contact the sexual organs of

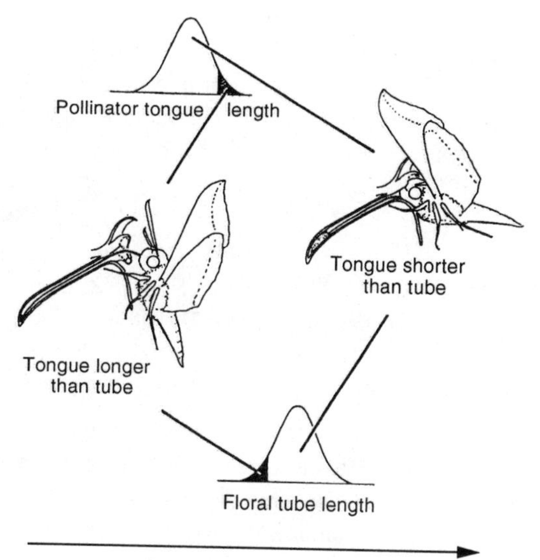

Figure 17.8.2 Possible mechanism of evolution of deep flowers. Long-tongued pollinator individuals are not compelled to contact the sex organs in short-tubed plant individuals, which therefore receive fewer pollinations than the long-tubed individuals that pollinate freely with each other through the pollinators. (After Nilsson 1988)

the flower. Long tubes in turn favor long-tongued individuals among the pollinator, which better reach the nectar. This idea was further developed and supported with field data from moth-pollinated orchids by Nilsson (1988; figure 17.8.2). The extreme corolla tubes, then, have probably not evolved by a Fisherian process (based on genetic coupling between trait and preference; chapter 2), but by a process more similar to other types of coevolution (see, e.g., Darwin 1877a; Dawkins and Krebs 1979; Nilsson 1988).

In sum, there is little empirical support for a Fisherian process in plants. Simpler forms of selection for efficiency in attracting pollinators from long distances may be responsible even for the most extreme and visually conspicuous flowers (e.g., figure 17.8.1), a hypothesis that should be testable by manipulating the size, color, and other aspects of flowers in the field. If this view is correct, conspicuous floral display in plants is favored by a process that resembles direct attraction to the strongest signal in animals (section 1.4). As in animals, competition over mates is likely to contribute to such a process.

Many benthic marine animals, like terrestrial plants, are sessile. In contrast with such plants, however, many of these animals reproduce by releasing female as well as male gametes in the surrounding medium (e.g.,

Strathman 1990; also see section 11.4 above). There are no known marine analogs of pollinators. Two possible reasons are that water, owing to its higher density and viscosity, is more favorable than air for passive transport of gametes, and that poorer visibility in water limits the effectiveness of visual advertisment for attraction of pollinator analogues (Strathman 1990).

17.9 Allocation to Male and Female Function in Cosexual Organisms

Cosexuality versus Separate Sexes

Cosexuality is common in plants and in some animals (with exception for arthropods and vertebrates). Resource use for male and female function in cosexual organisms is one of the main problems in sex allocation theory[5] (Charnov 1982). Cosexual plants that allocate resources to pollinator attraction are discussed here.

Whether cosexuals or separate sexes are favored by selection can be examined with a fitness set approach and ESS analysis (figure 17.9.1). (For an alternative quantitative genetic approach to these problems, see Morgan 1992.) In essence, hermaphrodites are favored if the relations between male and female fitness are such that the combined fitness from male and female function in a hermaphrodite is larger than the fitness of a pure male or female (figure 17.9.1; Charnov et al. 1976; Charnov 1982).

What conditions might lead to such a fitness set? It seems to require that the fitness gain curve shows strong enough diminishing returns (i.e., that the second derivative is always negative) from additional investment in male or female function, or both (figure 17.9.2). For instance, if the seeds of a plant do not disperse far from the parent, and hence to some extent compete among each other, the proportion of seeds that results in mature offspring decreases with increasing seed numbers. This is because additional seeds will raise competition among siblings over suitable growth sites. Also, pollen travels only limited distances, so pollen grains from the same plant may end up competing at the same stigma. The number of fertilizations in relation to pollen output will then also follow a law of diminishing returns (Charnov 1979b; D. G. Lloyd 1984; Charlesworth and Charlesworth 1987a; Thomson and Thomson 1989). Local pollen competition may therefore influence the relative allocations to male and female function among cosexual plants.

[5] See, e.g., Charnov et al. 1976, Maynard Smith 1978b, Charnov 1979b, Charlesworth and Charlesworth 1981, 1987, D. G. Lloyd 1984, 1988, Goldman and Willson 1986, Charlesworth and Morgan 1991, Brunet 1992, Morgan 1992, Emms 1993, and section 16.3 above.

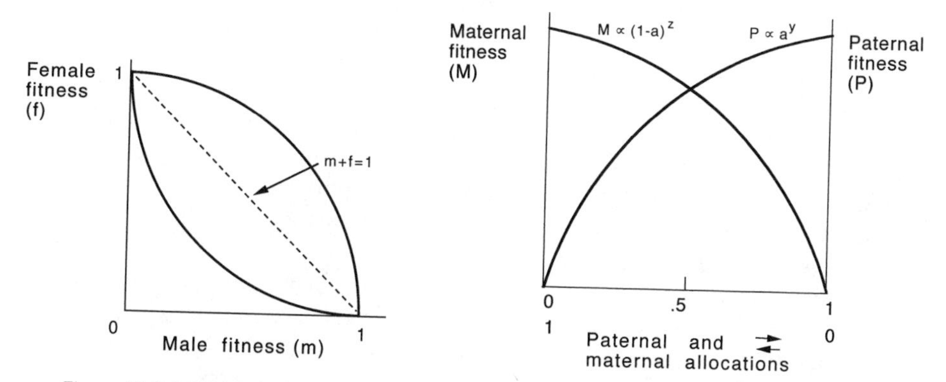

Figure 17.9.1 (*left*) Possible fitness sets for the trade-off between male and female reproduction in a simultaneous hermaphrodite. Pure males and females (end points of the fitness set) here have a relative fitness of one. The upper curve represents situations where hermaphrodites are favored, as the combined male and female fitness ($m + f$) of a hermaphrodite is larger than that of a pure male or female (with fitness = 1). The lower curve represents opposite situations, where pure sexes are favored. (After Charnov 1982)

Figure 17.9.2 (*right*) Graphs of paternal allocation (a) versus paternal fitness (P), and of maternal allocation ($1 - a$) versus maternal fitness (M). Hermaphroditism can be favored if the fitness gain curve shows strong enough diminishing returns from additional investment in male or female function, or both. The evolutionarily stable allocation to male function in this model is $a = y/(y + z)$. Note that allocation is measured on a relative (0–1) scale. (After D. G. Lloyd 1984)

In contrast, what Charnov (1982) termed a "concave" fitness set implies that separate sexes (dioecy) is the ESS, because the combined male and female fitness of a hermaphrodite is then smaller than that of a pure male or female (figure 17.9.1). Such a fitness set can arise if the fitness gain curve for male or female function follows a law of increasing returns from additional effort. This might come about, for example, if pollinator attraction increases more than linearly with the size of male flower display, or if the attraction of seed-dispersing animals increases more than linearly with the seed crop of a plant (Bawa 1980b; Beach 1981; Givnish 1980). In accordance with the latter idea, dioecy is more common among plants with fleshy, animal-dispersed propagules than among wind- or gravity-dispersed plants (Givnish 1980; but see Donoghue 1989; Charlesworth and Morgan 1991). Refined comparative analyses and, if possible, estimates of real gain curves should permit further tests of these hypotheses. Quantitative models of sex allocation and the evolution of dioecy and other types of unisexuality offer many ideas and predictions for empirical testing (e.g., Charlesworth and Charlesworth 1987a; Charlesworth and Morgan 1991; Morgan 1992).

In addition to the previous ideas, selection against selfing has been the traditional explanation for the evolution of separate sexes in plants. It is probably an important contributing factor, possibly the main one (e.g., Bawa 1980b; Givnish 1980; Lloyd 1982; D. Charlesworth 1984).

Reviewing male and female allocation in cosexual plants, D. G. Lloyd (1984) found that maternal investment is usually larger. He suggested the most likely reason is that there are upper limits to male allocation beyond which fitness does not increase. Such limits might arise, for example, if large pollen production attracts inefficient pollinators that eat or scatter pollen wastefully.

Empirical evidence is not yet sufficient to evaluate these possibilities; measuring resource allocation to various reproductive functions is problematic (e.g., D. G. Lloyd 1984; Goldman and Willson 1986; Reekie and Bazzaz 1987). For a thorough review and discussion of theoretical models and empirical tests of resource allocation in flowering plants, see Charlesworth and Morgan (1991).

Allocation to Pollinator Attraction

Reproducing cosexual plants invest not only in pollen and seeds, but also in attractive and other accessory structures, some of which cannot easily be ascribed to only one of the sexes, such as flower stems, petals, and nectar (e.g., D. G. Lloyd 1984, 1988; Charnov and Bull 1986; Charlesworth and Charlesworth 1987a; Brunet 1992). Lloyd (1988) found that such structures can make up a large part, in some species more than half, of the total dry weight of reproductive structures. Stamen mass was smaller than that of seeds and fruits.

How does allocation to accessory structures influence plant fitness? Of particular interest as regards sexual selection: Are such traits favored largely by male (pollen) competition, or is there also sexual selection of females competing over pollen? Such questions are now being examined by quantitative models (box 17.9.1) (e.g., Charnov and Bull 1986; Charlesworth and Charlesworth 1987a; Lloyd 1988; Morgan 1992; reviewed by Charlesworth and Morgan 1991).

Lloyd (1988) concluded that the ESS allocation of resources to pollinator attraction is proportional to the additive marginal benefits for female as well as male function. Female function will therefore influence the level of investment in attractive traits (unless male function favors an investment so large that no further female gain at all can be achieved from stronger attraction of pollinators; see Bell 1985). Charlesworth and Charlesworth (1987a) reached similar conclusions from another ESS model. They found that, in many cases, male fitness is much more sensitive to changes in the allocation to attractive structures than is female fitness. This is as expected if pollinator-attracting traits are more important for male than female func-

BOX 17.9.1 ESS ALLOCATION TO MALE, FEMALE, AND ACCESSORY FUNCTION IN COSEXUAL PLANTS

Using an ESS approach, Lloyd (1988) considered five types of reproductive allocation in cosexual plants:

direct paternal investment in pollen, with cost a_i
direct maternal investment in seeds, with cost b_i
accessory bilateral investment in petals, etc., with cost g_i
accessory paternal investment in petals, stamens, etc., with cost k_i
accessory maternal investment in petals, carpel walls, etc., with cost l_i

For an individual i, these relative allocations sum to 1. The paternal fitness P_i of the individual is related to its direct paternal investment as $p_i \propto (o\,a_i)^{\alpha}$. Here, o is a scaling factor representing the total number of pollen grains produced if the entire effort is allocated to pollen ($a_i = 1$). The exponent α determines whether the male fitness curve will positively accelerate ($\alpha > 1$) or decelerate ($\alpha < 1$) as the allocation to pollen increases. If corresponding expressions are used for the other investments, and if they have independent, multiplicative effects, the total maternal and paternal fitness for the plant can be written

$$m_i = (n\,b_i)^{\beta}(c_e\,g_i)^{\varepsilon}(c_i\,l_i)^{\lambda}$$

and

$$p_i \propto (o\,a_i)^{\alpha}(c_g\,g_i)^{\gamma}(c_k\,k_i)^{\kappa}.$$

The paternal fitness of the plant is the product of the seed fitness of its mates and the fraction of these ovules that it fertilizes. That fraction is given by the plant's own male competitiveness (right-hand side of p above), divided by that of all other plants competing to fertilize the same ovules. The total fitness of the individual is the sum of its maternal and paternal fitness.

The evolutionarily stable set of allocations is such that (Lloyd 1988)

$$a : b : g : k : l = \alpha : \beta : \gamma + \varepsilon : \kappa : \lambda.$$

where γ and ε are the exponents for male and female fitness in relation to accessory bilateral investment.

The allocation to each function is therefore proportional to its fitness exponent, reflecting the degree to which the fitness component increases with further allocation to that function. For bilateral accessory structures such as petals, the allocation is proportional to the sum of the exponents for male and female fitness, $\gamma + \varepsilon$. Note also that the ratio of two allocations does not depend on other allocations. This means, for example, that the ESS relation between direct male and female investment is not affected by investment in accessory structures. Total male and female allocations are related as $(\alpha + \gamma + \kappa) : (\beta + \varepsilon + \lambda)$. Since the exponents are independent, there is no

BOX 17.9.1 CONT.

intrinsic reason why total female allocation should be larger than male allocation in this model, although that may often be the case in practice (Lloyd 1988; Stanton and Preston 1988). Note also that it may not be realistic to ascribe all costs for pollinator-attracting devices such as petals and nectar to male function. Attraction of more pollinators may also raise female fitness, even if usually not to the same extent as in males (also see Charnov and Bull 1986; Charlesworth and Charlesworth 1987a; Charlesworth et al. 1987; Lloyd 1988). The occurrence of male-mimetic flowers in female plants of several dioecious species suggests that also females need to invest in attraction of pollinators (e.g. Baker 1976; Bawa 1980a; D. Charlesworth 1984).

tion in hermaphrodites. Under certain circumstances, however, especially if there is a nearly linear relation between the number of ovules fertilized and the number of seeds that mature, the sensitivity of female fitness to investment in attractive structures is not much less than that of male function.

Do cosexual plants have an advantage because costs of pollination are shared between male and female function? D. G. Lloyd (1988) suggested that this need not be the case. When traits that attract pollinators benefit female as well as male function, the ESS allocation to such traits can be higher in cosexuals than in plants with separate sexes. Pollination is therefore not necessarily more economical in hermaphrodites.

Few studies have separated the different kinds of reproductive allocations in sufficient detail for quantitative tests of these ideas. The importance of nectar, corolla, and other accessory structures can be estimated by experimental changes to see how they influence pollen removal and seed set. The strength of sexual selection through male and female function can also be tested by changes in the presence of competing neighbor plants (section 17.2). Experiments of the first kind have shown that larger investment in pollinator attraction increases fitness more for males than females (section 17.6). Some complicating factors, such as pollen itself being an attractant for pollinators that consume part of it, have not been included in the models (see D. G. Lloyd 1984, 1987, 1988; Charnov and Bull 1986; Charlesworth and Charlesworth 1987a). Testing and refining these ideas by comparison with field results is a challenging task, reviewed by Charlesworth and Morgan (1991).

Perhaps the clearest sex allocation trend in plants is that pollen:ovule ratios decrease with increasing self-fertilization (reviewed, e.g., by Cruden 1977; Bertin 1988; Brunet 1992). Data from several species suggest that the relationship may be linear, and several ESS models predict such a decrease (e.g., Maynard Smith 1978b; Charnov 1979b; 1982; Charlesworth and Charlesworth 1981; Lloyd 1987). The reasons for this pattern are debated. Charnov (1979b, 1982) suggested that local mate competition in inbred populations is responsible. Lloyd (1987, 1988) explained the trend by selfed seeds having twice the reproductive value of outcrossed seeds, with exception for inbreeding effects, because the parent contributes two sets of genes to each selfed seed (also see Charlesworth and Charlesworth 1981; Queller 1984a). The relative investment in seeds versus pollen should therefore increase with the degree of selfing. Cross-fertilizing species also tend to invest relatively more in male structures than do self-fertilizers (Lovett-Doust and Cavers 1982; Cruden and Lyon 1985; Brunet 1992). The models of Lloyd (1987) suggest that the consequences differ depending on the form of selfing, which therefore needs to be considered in tests of these ideas.

Improved methods for estimating resource allocation and its costs for different reproductive structures, and application of such methods to natural populations, should lead to further insights about sex allocation in cosexual plants. There is rapid development of theory and testing in this field (Charlesworth and Morgan 1991).

Sex Change in Plants

Most plants are cosexual, producing pollen as well as ovules. Many plants can modify their relative investment in male and female function in response to resource levels (reviewed, e.g., by Freeman et al. 1980; Charnov 1982; Lloyd and Bawa 1984; Schlessman 1988; Charlesworth and Morgan 1991; Brunet 1992). Some plants can change sex completely, for example Catasetinae orchids (Dodson 1962; Gregg 1975). The sex-changing plant studied in most detail is a perennial forest herb, jack-in-the-pulpit *Arisaema triphyllum*. As in many other members of this genus, young, small plants produce male flowers, whereas older plants with larger energy reserves are females. Females have higher reproductive costs, and males gain less than females from being large. Reduced nutrient reserves or partial damage can make a female change back to male sex (e.g., Policansky 1981; Lovett Doust and Cavers 1982; Bierzychudek 1984).

Reviewing sex change in plants, Schlessman (1988) concluded that it conforms to the general predictions of Ghiselin's (1969a) size advantage hypothesis (section 16.3), and that high cost of female reproduction explains why sex change is from male to female as the plant grows and stores more resources. In addition, Freeman et al. (1980) suggested that some plants may choose sex in relation to environmental patch quality; some patches might favor one sex function, other patches the other sex function. There seems to be no plant analogy to the size advantage in competition over mates that explains why some animals are first females, then males (section 16.3).

17.10 Summary

Reproductive effort in male plants is largely mating effort; in females it is parental effort. A major form of sexual selection in higher plants is scrambles among pollen to reach and fertilize ovules. Competition over pollination and fertilization may occur by attraction and reward of pollinators, and by rapid germination and growth of pollen tubes in the style. The evolution of pollinator-attracting floral display has traditionally been explained by advantages in cross-fertilization and inbreeding avoidance, but sexual selection is also likely to be involved. Pollen competition is suggested by many studies, but more direct tests are needed.

There is much observational and experimental evidence that floral display is crucial for success in attracting pollinators and exporting pollen. Selection via male function often favors more and larger flowers in hermaphrodites than needed from a female point of view. Nectar production and flowering phenology may also be subject to sexual selection mainly through male function. A direct competitive advantage in the attraction of pollinators is likely to play a role in the evolution of conspicuous floral traits and pollinator rewards. Complex genetic mechanisms such as Fisherian runaway selection, or indicator processes, seem less likely in plants.

Fruit set in many plants is limited by resources other than pollen, and fruit set sometimes shows no reduction even if many flowers are removed. In certain species, for example some orchids, there is pollen limitation of female reproduction. Sexual selection may then favor floral display also through female function.

There is indirect and some direct evidence that pollen scrambles in the style favor rapid pollen germination and tube growth. There is no demonstration that pollen scrambles or female choice in the style favors any other male trait by sexual selection. The possibility remains, however, that such selection plays a role in some cases.

Sexual selection may affect the allocation of resources to male and female function in cosexual plants and animals. The relative roles of male and female function in allocations to pollinator-attracting and other accessory reproductive structures in plants are likely to be influenced by competition over fertilizations. This possibility is now beginning to be explored.

18 Sexual Selection: Conclusions and Open Questions

Why are males and females so different in many species? What are the reasons for all the striking secondary sex traits? How can conspicuous ornamental traits that reduce survival evolve? The theory of sexual selection was Darwin's controversial answer to these questions. Massive modeling and empirical work has now tested and corroborated the theory in many respects, modified it in others, and extended it in new directions.

The basic reasons for sexual selection, and for the usually stronger competition over mates in males than females, have been clarified by theoretical advances and field work. The importance of sex differences in parental roles and relative investment in offspring has been added to Bateman's (1948) explanation based on anisogamy (larger eggs than sperm). Such aspects influence which sex will have the lowest potential reproductive rate, and the degree to which it limits the reproduction of members of the other sex (chapter 7; Williams 1966; Trivers 1972; Clutton-Brock and Parker 1992). Because females invest more resources in offspring, males will usually compete more strongly over females than vice versa. The theory has been tested in the field in a few species (e.g., Gwynne and Simmons 1990; Simmons 1992), but much more empirical testing is needed. Exceptions occur in some animals with male uniparental care. As predicted, their reversed parental roles and stronger female than male competition over mates has led to more pronounced secondary sex traits in females (e.g., Jenni and Collier 1972; Berglund et al. 1986b). The reasons for parental role reversal are, however, poorly understood (Clutton-Brock 1991). Until they have been explained, the causes for sex differences in parental and other roles, and in the strength of sexual selection, remain incompletely known.

A major advance is that female choice and mating preferences have been convincingly demonstrated for many secondary sex traits. Examples include a wide range of characters, such as pheromones in moths, song in crickets, frogs, and birds, and conspicuous colors and other visual ornaments in butterflies, fish, and birds (chapter 6, Ryan and Keddy-Hector 1992). In the main, Darwin was clearly correct: female choice often favors conspicuous secondary sex traits in males, also in monogamous species. He underestimated the importance of male contests, however: not only physical weapons such as horns, but also many male signals and ornaments may be favored by male contests, alone or in concert with female choice. In some some cases, contests seem to be the main selection factor.

18.1 Mechanisms of Sexual Selection

The theory has been enriched by many new ideas. Supported by results from genetic models (e.g., O'Donald 1980; Lande 1981; Kirkpatrick 1982; Pomiankowski et al. 1991), the Fisherian process based on genetic coupling between trait and preference long remained the main explanation for sexual selection of male ornaments, but strong empirical support is still lacking (but see Eberhard 1993).

Although other possibilities were suggested, for instance by Williams (1966), only recently have genetic and game theory models shown that entirely different mechanisms of female choice might be at work. For example, females may be attracted directly to the locally most conspicuous male, without detailed comparison, visitation, and discrimination among several males. There is evidence for such a simple mechanism in some cases, for example in the toads studied by Arak (1988c), and it may be more common than presently realized. Still, as females do not mate indiscriminately, the question remains how the female sensory system has come to have the properties that favor certain traits in conspecific males.

One possibility is genetic indicator mechanisms, proposed, for example, by Fisher (1915), Williams (1966), and Zahavi (1975). Results from formal models now suggest that such mechanisms might be important in the evolution of female preferences and conspicuous male traits, especially if the trait is condition-dependent and develops in proportion to the nutritional status and health of the individual (e.g., Andersson 1986a; Pomiankowski 1988; Grafen 1990a; Iwasa et al. 1991; Maynard Smith 1991a). The probably low heritability of fitness was previously assumed to make indicator processes unlikely, but this restriction now seems less serious. Empirical evidence shows that male secondary sex traits are often condition-dependent, but genetic aspects of the process have not been sufficiently tested.

One particular indicator model, Hamilton and Zuk's (1982) idea of host-parasite cycles maintaining the heritability of fitness, has inspired many empirical test. Comparative analyses have been difficult to interpret and may not allow critical tests. Some detailed studies of individual species (e.g., Møller 1990a) have supported the predictions, but many of them also follow from simpler hypotheses (e.g., Borgia and Collis 1989). Like other genetic indicator processes, the version based on resistance to parasites therefore remains to be convincingly demonstrated.

Indicator models based on direct phenotypic benefits have no problems with fitness heritability, as they work not through complex genetic mechanisms, but through direct material advantages to females or their offspring (e.g., Heywood 1989; Hoelzer 1989; Price et al. 1993). There is much evidence that such phenotypic benefits are often important, and may suffice to explain female preferences in many cases (chapter 8; e.g., Thornhill 1986; Reynolds and Gross 1990; Hill 1991; Kirkpatrick and Ryan 1991; Williams 1992), but genetic benefits may also be involved (Norris 1993).

A rather different explanation is also receiving empirical support. An attractive male trait may be favored because it corresponds to already existing features of the female sensory system, which have arisen independently of the male trait, perhaps in contexts other than mate choice (e.g., West-Eberhard 1984; Ryan 1985, 1992; Kirkpatrick 1987a; Endler and McLellan 1988). There is increasing evidence that a preexisting sensory bias makes females favor certain male secondary sex traits in some species. This process, like others, is debated; more empirical work is required.

In spite of substantial progress, the evolution of female preferences for male traits remains an outstanding empirical problem in sexual selection. By which genetic mechanisms do such systems evolve? How can critical tests be devised that distinguish among the alternatives? What are the relative roles of the different mechanisms, within and among species? During the heydays of the evolutionary synthesis, species isolation was the main explanation for secondary sex ornaments and mating preferences, but the picture has changed gradually, beginning in the 1960s. A decade ago, Fisherian runaway selection had become the most common explanation for female choice of male traits, although critical empirical evidence was lacking. More recently, indicator processes, especially host-parasite ideas, have received most attention. Is this because reality is becoming more accurately portrayed, or is there a new myth in the making? Complex genetic mechanisms have perhaps received attention out of proportion to their importance compared with other, simpler mechanisms, which also need more study. The alternatives are manifold, and include the possibility that some secondary sex ornaments are not at all, or at least not only, favored by female choice but by other mechanisms.

Male contests over females have long been obvious, and probably therefore less studied than female choice. In some cases they favor male ornaments, but the only trait often shown to be selected by sexual contest competition is large body size (chapter 6). Contest competition needs more attention in studies of sexual selection, particularly as regards signal traits. More work also needs to be done on scramble competition and endurance rivalry, the importance of which has become clear only recently. These aspects have received scant interest compared to other mechanisms of sexual selection, including sperm competition, the role of which is now well established (e.g., Parker 1970b; R. L. Smith 1984; Eberhard 1985; Birkhead and Møller 1992). Whereas scrambles may select (directly or indirectly) for reduced body size, endurance rivalry may favor large body size; these possibilities also need much more empirical testing.

The relative roles of the various mechanisms of sexual selection, which may occur together, have not yet been assessed in any case; this needs to be done in a range of species. Important theoretical developments probably also lie ahead, but the greatest obstacle to understanding now seems to be shortage of adequate empirical tests, not lack of theoretical ideas. Maynard Smith (1991) remarked that no topic in evolutionary biology has presented greater difficulties to theorists than sexual selection. One might add that critical empirical testing of some aspects of the theory seems even more difficult than formulating it. Many kinds of tests are needed, based on combinations of observations, experiments, and comparative analyses. Realistic and critical testing will probably require as much creative thinking as the development of the theory, plus detailed biological knowledge of suitable test organisms. In particular, decisive testing of genetic mechanisms of sexual selection remains a major challenge.

For example, a convincing demonstration of a genetic indicator mechanism should preferably show, among other things, that (1) females choosing the most ornamented males bear offspring with higher than average survival, and that (2) the higher survival has heritable genetic, not only phenotypic (direct, material), causes. This is no easy task. Until the roles of different selection mechanisms behind a range of secondary sex traits have been estimated in a variety of species, the relative importance of mate choice, contests, scrambles, endurance rivalry, and other mechanisms in the selection of secondary sex traits will remain unknown. Studies that mainly examine one mechanism and a single or a few traits will not suffice; more studies that simultaneously deal with several mechanisms and traits are needed from a wide spectrum of species.

In addition, there is need for further theoretical modeling of sexual selection. For example, most models have focused on equilibrium properties of traits and preferences. Rates of change also deserve attention, as they may be crucial for the effects of sexual selection in nonequilibrium situations, which are probably common. Scrambles and endurance rivalry and their consequences also need more modeling.

18.2 Traits Favored by Sexual Selection

Song and other behavioral display are the male traits most often shown to be sexually selected, followed by body size and visual ornaments (chapters 6, 11, 13, 14). Many studies have demonstrated sexual selection of territories and other material resources, such as food offered to females (chapter 8). It has also become clear that males often discriminate among potential mates, for example favoring large females when their fecundity increases with body size.

A variety of selection pressures influence body size in both sexes, in ways that differ strongly among groups. Large size is often selected by male contests, and probably also by endurance rivalry, if reproduction is based on resources built up over a long period. Larger female than male size is usually favored by fecundity advantages, and only rarely by stronger sexual selection in females. Sexual size dimorphism shows a bewildering variation among and within taxa, presenting many fascinating unsolved problems (chapter 11).

Sexually selected traits have usually been analyzed with univariate methods, but mating success often depends on several traits, which may be correlated with one another. Multivariate analyses are needed (Lande and Arnold 1983), such as the long-term study of Darwin's finches by Price (1984a,b), Grant (1986), and Grant and Grant (1989), and of sage grouse by Gibson et al. (1991). If possible, they might be combined with controlled experiments and estimation of the relative roles of the main mechanisms of sexual selection: scrambles, contests, mate choice, endurance rivalry, and sperm competition. Such studies need to be carried out over a sufficiently long period, as mating success can vary greatly with male age; two examples are red deer and black grouse (Clutton-Brock et al. 1982, 1988; Kruijt and de Vos 1988; Alatalo et al. 1991). Genetic long-term consequences of selection over generations also need to be clarified. This has been done in guppies (e.g., Endler 1983; Houde and Endler 1990), sticklebacks (Bakker 1993), and Darwin's finches (e.g., Price 1984a,b; Grant 1986; Grant and Grant 1989).

Whereas many secondary sex traits have now been shown to be sexually selected, much of their variation is poorly understood. For example, tail ornaments differ greatly in size and shape among birds, from the modestly sex-dimorphic outer tail streamers of swallows, to the enormous plumes of male long-tailed widowbirds and peacocks (e.g., Andersson 1982a; Møller 1988; Petrie et al. 1991). Part of the variation seems to be related to food and foraging. Species that hunt on the wing and need to be agile have relatively small tail ornaments that should not seriously impede flight. Large, unwieldy tails are found mainly in species that feed on the ground or in vegetation (Evans and Thomas 1992). Balmford et al. (1993) discuss the factors that influence the variation in elongated tails among species.

There is also much variation in coloration, song, and other secondary sex traits. For instance, two conspicuous members of the spring chorus of birds singing in marshes near my home in Göteborg are the grasshopper warbler *Locustella naevia* and the marsh warbler *Acrocephalus palustris*. As its name suggests, the song of the grasshopper warbler is a monotonous trill, which may go on for hours with brief pauses. The marsh warbler, on the other hand, has an extremely varied mimetic song, with sounds from over one hundred other birds, many of them African species from its winter quarter (e.g., Dowsett-Lemaire 1979; Kelsey 1989; Cramp et al. 1992). Why have these two sylvid warblers, neighbors in the same habitat, evolved the most contrasting songs one can imagine? This is only one example of the many puzzling problems that still surround the variation in form and expression of secondary sex traits. Relating it quantitatively to ecology, habitat, life history, sensory physiology, and other aspects of species' biology should help explain the variation. Some of it, for example in sexual size dimorphism, has been related to the mating system and other aspects (chapters 7 and 11), but most of the variation remains to be explained, a fascinating topic for future work.

Although much remains to be learned about male secondary sex traits, the characteristics and variation of female preferences within and among species have received much less study. Further progress and tests among alternative explanations require that mating preferences be explored and characterized in more detail. The nature of female sampling behavior and preferences, their genetic and evolutionary background, their biases and dependence on learning, copying, and imprinting, and their beneficial effects as well as costs in terms of, for example, predation risk and time loss, are crucial for tests among rival hypotheses.[1] This rich field awaits cultivation, which seems likely to yield important insights into sexual selection, perhaps even to change present views in major ways. The study of mating preferences has fertile connections with other vigorous research areas, such as speciation theory and animal communication (see below). Analysis of evolutionary interactions among sensory systems, environmental conditions such as light, predators, and food organisms, and the properties of sexually selected signals, also holds much promise for clarifying developments (e.g., Endler 1978, 1992b; Ryan 1990b; Ryan and Keddy-Hector 1992; Enquist and Arak 1993; Marchetti 1993).

More study is needed of the proximate mechanisms by which female choice favors male secondary sex traits, such as long-range detection and attraction, versus close-range and short-term stimulation of mating. The effects of signals can also be divided into "releasing" (short-term behavioral changes mediated directly by the central nervous system) and "priming" (hormonally mediated changes in reproductive condition and behavior). A priming effect of male signals on female reproduction has been

[1] See e.g., Kirkpatrick 1987a,b, ten Cate and Bateson 1988, 1989, Trail and Adams 1989, Real 1990, Ryan 1990b, Gerhardt 1991, Pomiankowski et al. 1991, Iwasa et al. 1991, Dugatkin 1992, Endler 1992b, Gibson 1992, Gibson and Höglund 1992, Pruett-Jones 1992, Ryan and Keddy-Hector 1992, Enquist and Arak 1993.

documented in many vertebrates (e.g., Bastock 1967; Hinde 1970; chapters 13–15 above). In some animals, male display is required for females to enter reproductive condition (e.g., Erickson 1970; Crews 1980; reviewed by Halliday 1983). The role of priming in sexual selection remains, however, to be explored. The consequences of sexual imprinting for sexual selection also presents several problems that need further attention (e.g., ten Cate and Bateson 1988, 1989; Bakker 1990; Barlow et al. 1990; Kruijt and Meeuwissen 1991).

18.3 Sexual Selection in Plants

Studies of higher plants have shown that there is male competition over pollination, favoring conspicuous flowers for attracting pollinators (e.g., Bell 1985), and rapid growth of pollen tubes down the style to the ovules (e.g., Snow and Spira 1991a,b). Sexual selection via male function may in part explain why (1) many cosexual plants have more flowers than needed for full seed set, (2) species with separate sexes usually have more and larger flowers in males than females, and (3) many plants are protandrous, with male function starting earlier in the season than female function. Most aspects have been studied in only few species, and there are many open questions. For example, seed set seems to be limited by insufficient pollination in some species, so competition over matings may occur also among females. The relative roles of the two sexes in the selection of traits that increase mating success remain to be clarified in most plants. So does the importance of "mate choice," for example by female discrimination among pollen grains in stigma and style (e.g., Willson and Burley 1983; Marshall and Folsom 1991; Walsh and Charlesworth 1992).

Central in the Fisherian process is the genetic coupling that arises between trait and preference. In plants that attract pollinating animals by conspicuous advertisement, such coupling usually is unlikely, as the two organisms do not share a common genome. Yet, the advertising flowers of many plants seem as conspicuous and extreme as the most remarkable secondary sex traits of animals. Simpler processes, such as direct attraction to the strongest signal, may often be responsible for the evolution of conspicuous advertising traits in plants, and perhaps also in animals.

Most flowering plants are hermaphrodites, with male and female function in the same flower. Resource allocation to male and female function in cosexual plants and animals is dealt with in sex allocation theory; so is sex change in sequential hermaphrodites (e.g., Ghiselin 1969a; Charnov 1979a, 1982; Lloyd 1988; Warner 1988c). Sexual selection is one of the factors that may affect sex allocation in hermaphrodites, another area rich in problems that await empirical testing.

18.4 Constraints

A variety of constraints on sexual selection have been identified. As long suspected, predation often works against sexually selected male traits, such as mobile search for females, acoustic and visual display, and contests among rivals. In mammals and birds where males are much larger than females, males are more susceptible to starvation or other hazards during the juvenile growth period. Sexual selection can also be self-limiting, as a trait may be advantageous in competition over mates in one respect, but disadvantageous in another (Searcy 1979d).

Energetic and physiological constraints can limit male display behavior and mating success. There is increasing evidence that display rate is important for mating success, but sexual display is often among the most energy-demanding activities of an animal, raising the metabolism 15–20 times. Sexual selection should therefore strongly favor good phenotypic condition and the ability of displaying at a high rate over extended periods, in many species based on resources stored over a long time.

Although several constraints on sexually selected traits have been demonstrated, no empirical study has yet shown quantitatively what compromise trait size is favored by the enhancing and opposing selection pressures that are likely to work in concert.

18.5 Relations with Some Adjacent Fields

Sexual selection partly overlaps with several other research fields, with which exchange of ideas and approaches is likely to provide new insights. The study of mating preferences will benefit from closer connections with sensory and brain physiology. There is increasing evidence that preferences in part depend on preexisting biases of the neurosensory system, the analysis of which should contribute to explaining mating preferences and preferred traits (e.g., Ryan and Keddy-Hector 1992; Enquist and Arak 1993).

Endocrinology is another field that seems likely to increase the understanding of aspects of sexual selection (e.g., Wingfield et al. 1990; Folstad and Karter 1992; Ketterson and Nolan 1992). There is hormonal control of many secondary sex traits, and testosterone levels strongly influence sexually selected male behavior, such as song and territorial aggression. Hormone implants may permit experimental changes in behavior and ornamentation that would otherwise be difficult to alter. Experimentally raised testosterone levels make some males gain several females in normally monogamous sparrows (Wingfield 1984). Raised levels of testosterone also may, however, reduce male parental behavior and success, and increase mortality (Silverin 1980; Hegner and Wingfield 1987; Marler and Moore 1988).

The development of secondary sex traits is part of the effort an animal expends in order to reproduce. Such traits can therefore be analyzed from a life history perspective. For example, there are indications that secondary sex traits, like other aspects of reproductive effort, are more extreme in organisms that reproduce only once in life than in forms that breed repeatedly (e.g., Williams 1966; Partridge and Endler 1987). Although a beginning has been made, probably much more remains to be learned from life history ideas applied to sexual selection, and vice versa (chapter 10).

Mating system evolution is another area with close relations to sexual selection; there are strong causal links in both directions (figure 17.2.2). There are many unsolved problems as regards both the effects of mating systems on the strength and consequences of sexual selection, and the role played by sexual selection in mating system evolution. Even the factors that lead to such contrasts in parental care and mating systems as lek polygyny versus polyandry are not well understood. Sexual selection clearly is important in both cases, but the detailed reasons for the evolution of such contrasting mating systems are not known (e.g., Emlen and Oring 1977; Bradbury 1981; Clutton-Brock 1991); some crucial aspects may still remain to be discovered.

Species recognition and speciation is a field with which sexual selection theory is likely to have fruitful interactions (e.g., West-Eberhard 1983; chapter 9 above). After a period of relative quiescence, speciation theory has been revitalized by new conceptual developments, and by exchange of ideas between different branches of biology (e.g., Otte and Endler 1989). New discoveries have shed light on the debated evolution of mating preferences and mate recognition in relation to speciation. An extensive analysis of pairs of *Drosophila* species suggests that there is selection of specific mate recognition in sympatry, making it evolve more rapidly there than in allopatry (Coyne and Orr 1989). There is also other suggestive evidence that mating preferences are sometimes strongly affected by the risk of hybridization. It therefore seems that costs of hybrid matings sometimes influence the evolution of mating preferences and secondary sex traits, a possibility that deserves much more study.

Speciation rates can probably be strongly influenced by sexual selection (Lande 1981; West-Eberhard 1983; Thornhill and Alcock 1983). Sensory characteristics and the scope for distinct variation in traits used in mate choice may be of crucial importance (chapter 9). Speciation rates seem particularly high in several taxa that have in common (1) much room for variation in signal traits important for mate recognition, and (2) fine discriminatory ability in the choosing sex. Some examples are frogs, where speciation rates increase with the range of frequencies to which the inner ear responds (Ryan 1986), and passerine birds with syringes capable of advanced sound production (Raikow 1986). Angiosperms may provide similar examples among plants; their profuse speciations often involve changes in visual, chemical, or tactile stimuli for pollinating insects (e.g., Crepet 1984; Eriksson and Bremer 1992).

Another developing field with which sexual selection theory shares much ground is animal communication. A debated problem concerns the information conveyed by signals used in conflicts: is the information "honest," or are there signals that mislead the receiver to the advantage of the sender?[2] Examples from sexual selection are signals used in contests over mates or in mate attraction. Emlen (1973) and Dawkins and Krebs (1978) suggested that signals in such situations are likely to be unreliable and manipulative. On the other hand, Zahavi (1975, 1987) argued that such situations will favor the evolution of costly honest signals that accurately reflect the fighting ability or other qualities of the sender (but see Borgia 1993). This idea has received support from formal models (see Enquist 1985; Grafen 1990b; Maynard Smith 1991b). On the other hand, Dawkins and Guilford (1991) pointed out that accurate assessment of signals may be costly, for example in terms of predation risk or time loss. Honest signals might then be replaced by conventional signals that are not reliable indicators of quality. Costs of assessment may influence the degree of signal reliability when the interests of sender and receiver differ. Costs of mate assessment and choice can have profound effects in sexual selection (chapter 2). The effects of such costs need to be further explored, and their occurrence tested in mate choice and other contexts. Johnstone and Grafen (1992) found that also error-prone signaling systems may be "honest" at equilibrium, provided that the more strongly a signaler advertises, the more likely he is to also be perceived as advertising strongly.

In sum, competition over mates is an important selective agent in animals as well as plants, with far-reaching consequences for morphology, behavior, life history, and other characteristics of many sexual organisms. Abundant empirical evidence shows that, as Darwin suggested, female choice often favors conspicuous secondary sex ornaments and display in males. It has become clear, however, that male contests also often favor such signal traits, and not only weapons that are directly used in fights. Other forms of

[2] For discussion, see, e.g., Maynard Smith 1958, 1991a,b, Zahavi 1975, 1977b, 1987, Dawkins and Krebs 1978, Andersson 1980, Enquist and Leimar 1983, Wiley 1983, Kodric-Brown and Brown 1984, Krebs and Dawkins 1984, Enquist 1985, Michod and Hasson 1990, Grafen 1991b, Dawkins and Guilford 1991, Guilford and Dawkins 1991, Harper 1991, Zuk 1991a, Johnstone and Grafen 1992, Borgia 1993.

sexual selection with important consequences are scrambles, endurance rivalry, and sperm competition.

Whereas sexual selection clearly is an important factor in evolution, many basic problems remain unsolved. The relative roles of the various mechanisms remain to be clarified. Although female choice has often been demonstrated, the evolution of mate choice itself remains unclear. In particular, Fisherian and indicator processes have not been tested as regards several crucial genetic aspects. There are alternative explanations, such as direct material advantages of mate choice and sensory biases, that have arisen in other contexts. The characteristics, variation, selection, and evolution of mating preferences largely remain to be clarified. Although the need for theoretical modeling is far from exhausted, the greatest obstacles to better understanding of sexual selection now seem to be empirical rather than theoretical.

Many kinds of constraints on sexually selected traits, such as predation risk and energy loss, have been identified, but quantitative demonstration of a balance between sexual selection and other selection pressures is still lacking. Most of the enormous variation among species in secondary sex traits remains to be explained.

There is much exciting progress in areas of overlap with other fields, such as sensory and neural physiology, animal communication, endocrinology, life history, mating systems, species recognition, and speciation theory. Growing research in these areas of common ground should also lead to better knowledge of sexual selection. A reasonably detailed understanding of the selection and evolution of secondary sex traits and mating preferences may be within reach during the next decades.

References

Ackerman, J. D., and A. M. Montalvo. 1990. Short- and long-term limitation to fruit production in a tropical orchid. *Ecology* 271: 263–272.

Adams, J., and P. J. Greenwood. 1983. Why are males bigger than females in pre-copula pairs of *Gammarus pulex*? *Behav. Ecol. Sociobiol.* 13: 239–241.

Ågren, J., T. Elmqvist, and A. Tunlid. 1986. Pollination by deceit, floral sex ratios and seed set in dioecious *Rubus chamaemorus* L. *Oecologia* 70: 332–338.

Ahearn, J. N.. and A. R. Templeton. 1989. Interspecific hybrids of *Drosophila heteroneura* and *D. silvestris*. I. Courtship success. *Evolution* 43: 347–361.

Ahlén, I. 1965. Studies on the red deer, *Cervus elaphus* L., in Scandinavia. III. Ecological investigations. *Viltrevy* 3: 177–376.

Aiken, R. B. 1982. Theories of sexual difference: The sexual selection hypothesis and its antecedents, 1786–1919. *Quaest. Entomol.* 18: 1–14.

Åkesson, B. 1972. Sex determination in *Ophryotrocha labronica*. *Fifth Eur. Marine Biol. Symp.*: 163–172.

Alatalo, R. V., and A. Lundberg. 1984. Polyterritorial polygyny in the pied flycatcher *Ficedula hypoleuca*—evidence for the deception hypothesis. *Ann. Zool. Fennici* 21: 217–228.

Alatalo, R. V., A. Carlson, A. Lundberg, and S. Ulfstrand. 1981. The conflict between male polygamy and female monogamy: The case of the pied flycatcher *Ficedula hypoleuca*. *Am. Nat.* 117: 285–291.

Alatalo, R. V., A. Lundberg, and K. Ståhlbrandt. 1984. Female mate choice in the pied flycatcher *Ficedula hypoleuca*. *Behav. Ecol. Sociobiol.* 14: 253–261.

Alatalo, R. V., A. Lundberg, and C. Glynn. 1986. Female pied flycatchers choose territory quality and not male characteristics. *Nature* 323: 152–153.

Alatalo, R. V., A. Carlson, and A. Lundberg. 1988a. The search cost in mate choice of the pied flycatcher. *Anim. Behav.* 36: 289–291.

Alatalo, R. V., J. Höglund, and A. Lundberg. 1988b. Patterns of variation in tail ornament size in birds. *Biol. J. Linn. Soc.* 34: 363–374.

Alatalo, R. V., D. Eriksson, L. Gustafsson, and A. Lundberg. 1990a. Hybridization between pied and collared flycatchers—sexual selection and speciation theory. *J. Evol. Biol.* 3: 375–389.

Alatalo, R. V., C. Glynn, and A. Lundberg. 1990b. Singing rate and female attraction in the pied flycatcher: An experiment. *Anim. Behav.* 39: 601–603.

Alatalo, R. V., A. Lundberg, and J. Sundberg. 1990c. Can female preference explain sexual dichromatism in the pied flycatcher, *Ficedula hypoleuca*? *Anim. Behav.* 39: 244–252.

Alatalo, R. V., A. Lundberg, and O. Rätti. 1990d. Male polyterritoriality and imperfect female choice in the pied flycatcher *Ficedula hypoleuca*. *Behav. Ecol.* 1: 171–177.

Alatalo, R. V., J. Höglund, and A. Lundberg. 1991. Lekking in the black grouse—a test of male viability. *Nature* 352: 155–156.

Alatalo, R. V., J. Höglund, A. Lundberg, and W. J. Sutherland. 1992. Evolution of black grouse leks: Female preferences benefit males in larger leks. *Behav. Ecol.* 3: 53–59.

Alberts, A. C. 1992. Constraints on the design of chemical communication systems in terrestrial vertebrates. *Am. Nat.* 139: s62–s89.

Alcock, J. 1979a. The evolution of intraspecific diversity in male reproductive strategies in some bees and wasps. In M. S. Blum and N. A. Blum, eds., *Sexual Selection and Reproductive Competition in Insects*, 381–402. Academic Press, New York.

Alcock, J. 1979b. Selective mate choice by females of *Harpobittacus australis* (Mecoptera: *Bittacidae*). *Psyche* 86: 213–217.

Alcock, J. 1984a. Long-term maintenance of size variation in populations of *Centris pallida* (Hymenoptera: Anthophoridae). *Evolution* 38: 220–223.

Alcock, J. 1984b. *Animal Behavior: An Evolutionary Approach, 3d ed.* Sinauer, Sunderland, Mass.

Alcock, J., and D. W. Pyle. 1979. The complex courtship behavior of *Physiphora demandata* (F.) (Diptera: Otitidae). *Z. Tierpsych.* 49: 352–362.

Alcock, J., C. E. Jones, and S. L. Buchmann. 1977. Male mating strategies in the bee *Centris pallida* Fox (Hymenoptera: Anthophoridae). *Am. Nat.* 111: 145–155.

Alexander, R. D. 1961. Aggressiveness, territoriality and sexual behavior in field crickets (Orthoptera: Gryllidae). *Behavior* 17: 130–223.

Alexander, R. D. 1962. Evolutionary change in cricket communication. *Evolution* 16: 443–467.

Alexander, R. D., and G. Borgia. 1979. On the origin and basis of the male-female phenomenon. In M. S. Blum and N. A. Blum, eds., *Sexual Selection and Reproductive Competition in Insects*, 417–440. Academic Press, New York.

Alexander, R. D., and T. E. Moore. 1962. The evolutionary relationships of 17-year and 13-year cicadas and three new species (Homoptera, Cicadidae, *Magicicada*). *Misc. Publ. Mus. Zool., Univ. Michigan* 121: 1–59.

Alexander, R. D., J. L. Hoogland, R. D. Howard, K. M. Noonan, and P. W. Sherman. 1979. Sexual dimorphisms and breeding systems in pinnipeds, ungulates, primates, and humans. In N. A. Chagnon and W. Irons, eds., *Evolutionary Biology and Human Social Behavior: An Anthropological Perspective*, 402–435. Duxbury Press, North Scituate, Mass.

Allen, G. A. 1986. Flowering pattern and fruit production in the dioecious shrub *Oemleria cerasiformis* (Rosaceae). *Can. J. Bot.* 64: 1216–1220.

Allen, G. A., and J. A. Antos. 1988. Relative reproductive effort in males and females of the dioecious shrub *Oemleria cerasiformis*. *Oecologia* 76: 111–118.

Altmann, S. A., S. S. Wagner, and S. Lenington. 1977. Two models for the evolution of polygyny. *Behav. Ecol. Sociobiol.* 2: 397–410.

Amadon, D. 1959. The significance of sexual differences in size among birds. *Proc. Am. Phil. Soc.* 103: 531–536.

Amadon, D. 1975. Why are female birds of prey larger than males? *Raptor Research* 9: 1–11.

Amadon, D. 1977. Further comments on sexual size dimorpism in birds. *Wilson Bull.* 89: 619–620.

Anderson, S. S., and M. F. Fedak. 1985. Grey seal males: Energetic and behavioural links between size and sexual success. *Anim. Behav.* 33: 829–838.

Andersson, M. 1971. Breeding behaviour of the long-tailed skua *Stercorarius longicaudus* (Viellot). *Ornis Scand.* 2: 35–54.

Andersson, M. 1976. Population ecology of the long-tailed skua (*Stercorarius longicaudus* Vieill.). *J. Anim. Ecol.* 45: 537–559.

Andersson, M. 1980. Why are there so many threat displays? *J. theor. Biol.* 86: 773–781.

Andersson, M. 1982a. Female choice selects for extreme tail length in a widowbird. *Nature* 299: 818–820.

Andersson, M. 1982b. Sexual selection, natural selection and quality advertisement. *Biol. J. Linn. Soc.* 17: 375–393.

Andersson, M. 1983. On the functions of conspicuous seasonal plumages in birds. *Anim. Behav.* 31: 1262–1264.

Andersson, M. 1986a. Evolution of condition-dependent sex ornaments and mating preferences: Sexual selection based on viability differences. *Evolution* 40: 804–816.

Andersson, M. 1986b. Sexual selection and the importance of viability differences: A reply. *J. theor. Biol.* 120: 251–254.

Andersson, M. 1987. Genetic models of sexual selection: Some aims, assumptions, and tests. In J. W. Bradbury and M. B. Andersson, eds., *Sexual Selection: Testing the Alternatives*, 41–53. Wiley, Chichester, U.K..

Andersson, M. 1994. Reversed sex roles, sexual size dimorphism, and mating system in coucals (Centropodidae, Aves). In prep.

Andersson, M., and R. Å. Norberg. 1981. Evolution of reversed sexual size dimorphism and role partitioning among predatory birds, with a size scaling of flight performance. *Biol. J. Linn. Soc.* 15: 105–130.

Andersson, S. 1988. Size-dependent pollination efficiency in *Anchusa officinalis* (Boraginaceae): Causes and consequences. *Oecologia* 76: 125–130.

Andersson, S. 1989. Sexual selection and cues for female choice in leks of Jackson's widowbird *Euplectes jacksoni*. *Behav. Ecol. Sociobiol.* 25: 403–410.

Andersson, S. 1991. Bowers on the savanna: Display courts and mate choice in a lekking widowbird. *Behav. Ecol.* 2: 210–218.

Andersson, S. 1992. Female preference for long tails in lekking Jackson's widowbirds: Experimental evidence. *Anim. Behav.* 43: 379–388.

Andersson, S. 1993. Sexual dimorphism and modes of sexual selection in lekking Jackson's widowbirds *Euplectes jacksoni*. *Biol. J. Linn. Soc.* 49: 1–17.

Andersson, S., and M. Andersson. 1994. Tail ornamentation, size dimorphism and wing length in the genus *Euplectes* (Ploceinae). *Auk* 111: in press.

Andersson, S., and C. G. Wiklund. 1987. Sex role partitioning in the rough-legged buzzard *Buteo lagopus*. *Ibis* 129: 103–107.

Andrén, C. 1986. Courtship, mating and agonistic behaviour in a free-living population of adders, *Vipera berus*. *Amphibia-Reptilia* 7: 353–383.

Angelstam, P. 1984. Sexual and seasonal differences in mortality of the Black Grouse *Tetrao tetrix* in boreal Sweden. *Ornis Scand.* 15: 123–134.

Anholt, B. R., J. H. Marden, and D. M. Jenkins. 1991. Patterns of mass gain and

sexual dimorphism in adult dragonflies (Insecta: Odonata). *Can. J. Zool.* 69: 1156–1163.
Antonovics, J. 1968. Evolution in closely adjacent plant populations. V. Evolution of self-fertility. *Heredity* 23: 219–238.
Aoki, K. 1989. A sexual-selection model for the evolution of imitative learning of song in polygynous birds. *Am. Nat.* 134: 599–612.
Apollonio, M., M. Festa-Bianchet, and F. Mari. 1989. Correlates of copulatory success in a fallow deer lek. *Behav. Ecol. Sociobiol.* 25: 89–97.
Apollonio, M., M. Festa-Bianchet, F. Mari, and M. Riva. 1990. Site-specific asymmetries in male copulatory success in a fallow deer lek. *Anim. Behav.* 39: 205–212.
Apollonio, M., M. Festa-Bianchet, F. Mari, S. Mattiolo, and B. Sarno. 1992. To lek or not to lek: Mating strategies of male fallow deer. *Behav. Ecol.* 3: 25–31.
Arak, A. 1983a. Sexual selection by male-male competition in natterjack toad choruses. *Nature* 306: 261–262.
Arak, A. 1983b. Male-male competition and mate choice in anuran amphibians. In P. Bateson, ed., *Mate Choice*, 181–210. Cambridge University Press, Cambridge, U.K.
Arak, A. 1984. Sneaky breeders. In C. J. Barnard, ed., *Producers and Scroungers: Strategies of Exploitation and Parasitism*, 154–194. Croom Helm, London.
Arak, A. 1988a. Sexual dimorphism in body size: A model and a test. *Evolution* 42: 820–825.
Arak, A. 1988b. Callers and satellites in the natterjack toad: Evolutionarily stable decision rules. *Anim. Behav.* 36: 416–432.
Arak, A. 1988c. Female mate selection in the natterjack toad: Active choice or passive attraction? *Behav. Ecol. Sociobiol.* 22: 317–327.
Arcese, P., P. K. Stoddard, and S. M. Hiebert. 1988. The form and function of song in female song sparrows. *Condor* 90: 44–50.
Armbruster, W. S. 1985. Patterns of character divergence and the evolution of reproductive ecotypes of *Dalechampia scandens* (Euphorbiaceae). *Evolution* 39: 733–752.
Armstrong, E. A. 1965. *Bird Display and Behaviour*. Dover, New York.
Arnold, S. J. 1977. The courtship behavior of North American salamanders with some comments on Old World salamandrids. In D. H. Taylor and S. I. Guttman, eds., *The Reproductive Biology of Amphibians*, 141-183. Plenum, New York.
Arnold, S. J. 1983. Sexual selection: The interface of theory and empiricism. In P. Bateson, ed., *Mate Choice*, 67–107. Cambridge University Press, Cambridge, U.K.
Arnold, S. J. 1987. Quantitative genetic models of sexual selection: A review. In S. C. Stearns, ed., *The Evolution of Sex and Its Consequences*, 283–316. Birkhäuser, Basel.
Arnold, S. J. 1988. Behavior, energy and fitness. *Am. Zool.* 28: 815–827.
Arnold, S. J., and L. D. Houck. 1982. Courtship pheromones: Evolution by natural and sexual selection. In M. Nitecki, ed., *Biochemical Aspects of Evolutionary Biology*, 173–221. University of Chicago Press, Chicago.
Arnold, S. J., and M. J. Wade. 1984a. On the measurement of natural and sexual selection: Theory. *Evolution* 38: 709–719.

Arnold, S. J., and M. J. Wade 1984b. On the measurement of natural and sexual selection: Applications. *Evolution* 38: 720–734.

Arnqvist, G. 1989. Sexual selection in a water strider: The function, mechanism of selection and heritability of a male grasping apparatus. *Oikos* 56: 344–350.

Arnqvist, G. 1992a. The effects of operational sex ratio on the relative mating success of extreme male phenotypes in the water strider *Gerris odontogaster*. *Anim. Behav.* 43: 681–683.

Arnqvist, G. 1992b. Pre-copulatory fighting in a water strider: Inter-sexual conflict or mate assessment? *Anim. Behav.* 43: 559–567.

Arnqvist, G. 1992c. Spatial variation in selective regimes: Sexual selection in the water strider, *Gerris odontogaster*. *Evolution* 46: 914–929.

Arrow, G. J. 1951. *Horned Beetles*. Junk, The Hague.

Arvidsson, B. L. 1992. Copulation and mate guarding in the willow warbler. *Anim. Behav.* 43: 501–509.

Arvidsson, B. L., and R. Neergaard. 1991. Mate choice in the willow warbler—a field experiment. *Behav. Ecol. Sociobiol.* 29: 225–229.

Askenmo, C. 1984. Polygyny and nest site selection in the pied flycatcher. *Anim. Behav.* 32: 972–980.

Askenmo, C., R. Neergaard, and B. L. Arvidsson. 1992. Pre-laying time budgets in rock pipits: Priority rules of males and females. *Anim. Behav.* 44: 957–965.

Atchley, W. R. 1987. Developmental quantitative genetics and the evolution of ontogenies. *Evolution* 41: 316–330.

Atkinson, D. 1991. Sexual showiness and parasite load: Correlations without parasite coevolutionary cycles. *J. theor. Biol.* 150: 251–260.

Austad, S. N. 1984. A classification of alternative reproductive behaviors and methods for field-testing of ESS models. *Am. Zool.* 24: 309–320.

Austad, S. N., and R. D. Howard. 1984. Introduction to the symposium: Alternative reproductive tactics. *Am. Zool.* 24: 307–308.

Ayre, D., and R. J. Whelan. 1989. Factors controlling fruit set in hermaphroditic plants: Studies with the Australian proteaceae. *TREE* 4: 267–272.

Baerends, G. P., R. Brouwer, and H. T. Waterbolk. 1955. Ethological studies on *Lebistes reticulatus* (Peters). I. An analysis of the male courtship pattern. *Behaviour* 8: 249–335.

Bailey, W. J., and D. R. Thiele. 1973. Male spacing behaviour in the Tettigoniidae: An experimental approach. In D. T. Gwynne and G. K. Morris, eds., *Orthopteran Mating Systems*, 163–184. Westview, Boulder, Colo.

Bailey, W. J., R. J. Cunningham, and L. Lebel. 1990. Song power, spectral distribution and female phonotaxis in the bushcricket, *Requena verticalis* (Tettigoniidae: Orthoptera): Active female choice or passive attraction? *Anim. Behav.* 40: 33–42.

Bajema, C. J. 1984. *Evolution by Sexual Selection Theory Prior to 1900*. Van Nostrand Reinhold, New York.

Baker, H. G. 1976. "Mistake" pollination as a reproductive system with special reference to the Caricaceae. In J. Burley and B. T. Styles, eds., *Tropical Trees: Variation, Breeding and Conservation*, 161–169. Linnean Society, London.

Baker, J.M.R. 1992. Body condition and tail height in great crested newts (*Triturus cristatus*). *Anim. Behav.* 43: 157–159.

Baker, M. C. 1983. The behavioral response of female Nuttall's white-crowned sparrows to male song of natal and alien dialects. *Behav. Ecol. Sociobiol.* 12: 309–315.

Baker, M. C., T. K. Bjerke, H. Lampe, and Y. Espmark. 1986. Sexual response of female great tits to variation in size of male's song repertoires. *Am. Nat.* 128: 491–498.

Baker, M. C., T. K. Bjerke, H. U. Lampe, and Y. O. Espmark. 1987a. Sexual response of female yellowhammers to differences in regional song dialects and repertoire sizes. *Anim. Behav.* 35: 395–401.

Baker, M. C., P. K. McGregor, and J. R. Krebs. 1987b. Sexual response of female great tits to local and distant songs. *Ornis. Scand.* 18: 186–188.

Baker, R. R. 1985. Bird coloration: In defence of unprofitable prey. *Anim. Behav.* 33: 1387–1388.

Baker, R. R., and C. J. Bibby. 1987. Merlin *Falco columbarius* predation and theories of the evolution of bird coloration. *Ibis* 129: 259–263.

Baker, R. R., and M. V. Hounsome. 1983. Bird coloration: Unprofitable prey model supported by ringing data. *Anim. Behav.* 31: 614–615.

Baker, R. R., and G. A. Parker. 1979. The evolution of bird coloration. *Phil. Trans. R. Soc. B.* 287: 63–130.

Baker, T. C. 1989. Sex pheromone communication in the Lepidoptera: New research progress. *Experientia* 45: 248–262.

Baker, T. C., and R. T. Cardé. 1979. Courtship behavior of the Oriental fruit moth (*Grapholitha molesta*): Experimental analysis and consideration of the role of sexual selection in the evolution of courtship pheromones in the *Lepidoptera*. *Ann. Entomol. Soc. Am.* 72: 173–188.

Baker, T. C., R. Nishida, and W. L. Roelofs. 1981. Close-range attraction of female Oriental fruit moths to herbal scent of male hairpencils. *Science* 214: 359–360.

Bakker, R. T. 1986. *The Dinosaur Heresies*. Morrow, New York.

Bakker, T.C.M. 1990. Genetic variation in female mating preferences. *Netherl. J. Zool.* 40: 617–642.

Bakker, T.C.M. 1993. Positive genetic correlation between female preference and preferred male ornament in sticklebacks. *Nature* 363: 255–257.

Bakker, T.C.M., and P. Sevenster. 1983. Determinants of dominance in male sticklebacks (*Gasterosteus aculeatus L.*). *Behaviour* 86: 55–71.

Balgooyen, T. G. 1976. Behavior and ecology of the American kestrel (*Falco sparverius*) in the Sierra Nevada of California. *Univ. Calif. Publ. Zool.* 103: 1–88.

Balinsky, B. I., and J. B. Balinsky. 1954. On the breeding habits of the South African bullfrog, *Pyxicephalus adspersus*. *S. Afr. J. Sci.51*: 55–58.

Balmford, A. 1991. Mate choice on leks. *TREE* 6: 87–91.

Balmford, A., and A. F. Read. 1991. Testing alternative models of sexual selection through female choice. *TREE* 6: 274–276.

Balmford, A., S. Albon, and S. Blakeman. 1992a. Correlates of male mating success and female choice in a lek-breeding antelope. *Behav. Ecol.* 3: 112–123.

Balmford, A., I. L. Jones, and A.L.R. Thomas. 1993. On avian asymmetry: Evidence of natural selection for symmetrical tails and wings in birds. *Proc. R. Soc. Lond.* B 252: 245–251.

Balmford, A., A. Thomas, G. Borgia, G. Wilkinson, and A. P. Møller. 1992b. Swallowing ornamental asymmetry. *Nature* 359: 487–488.

Balmford, A., A.L.R. Thomas, and I. L. Jones. 1993. Aerodynamics and the evolution of long tails in birds. *Nature* 361: 628–631.

Banks, E., R. J. Brooks, and J. Schnell. 1975. A radiotracking study of home range and activity of the brown lemming (*Lemmus trimucronatus*). *J. Mammal.* 56: 888–901.

Banks, M. J., and D. J. Thompson. 1985. Lifetime mating success in the damselfly *Coenagrion puella*. *Anim. Behav.* 33: 1175–1183.

Baptista, L. 1979. Territorial, courtship and duet songs of the Cuban grassquit (*Tiaris canora*). *J. Ornithol.* 119: 91–101.

Barlow, G. W. 1973. Competition between color morphs of the polychromatic Midas cichlid *Cichlasoma citrinellum*. *Science* 179: 806–807.

Barlow, G. W. 1974. Contrasts in social behavior between Central American cichlid fishes and coral-reef surgeon fishes. *Am. Zool.* 14: 9–34.

Barlow, G. W., R. C. Francis, and J. V. Baumgartner. 1990. Do the colours of parents, companions and self influence assortative mating in the polychromatic Midas cichlid? *Anim. Behav.* 40: 713–722.

Barnard, C. J., and J. M. Behnke. 1990. *Parasitism and Host Behaviour*. Taylor and Francis, London.

Barnard, C. J., and J. Fitzsimons. 1988. Kin recognition and mate choice in mice: The effects of kinship, familiarity and social interference on intersexual interaction. *Anim. Behav.* 36: 1078–1090.

Barnard, P. 1990. Male tail length, sexual display intensity and female sexual response in a parasitic African finch. *Anim. Behav.* 39: 652–656.

Barnard, P. 1991. Ornament and body size variation and their measurement in natural populations. *Biol. J. Linn. Soc.* 42: 379–388.

Barnes, R.F.W. 1982. Mate searching behaviour of elephant bulls in a semi-arid environment. *Anim. Behav.* 30: 1217–1223.

Barrett, S.C.H. 1988. The evolution, maintenance, and loss of self-incompatibility systems. In J. Lovett Doust and L. Lovett Doust, eds., *Plant Reproductive Ecology*, 98–124. Oxford University Press, New York.

Barrette, C. 1977. Fighting behavior of muntjac and the evolution of antlers. *Evolution* 31: 169–176.

Barrette, C., and D. Vandal. 1990. Sparring, relative antler size, and assessment in male caribou. *Behav. Ecol. Sociobiol.* 26: 383–387.

Bartholomew, G. A. 1952. Reproductive and social behaviour in the northern elephant seal. *Univ. Calif. Publ. Zool.* 47: 369–472.

Bartholomew, G. A. 1970. A model for the evolution of pinniped polygyny. *Evolution* 24: 546–559.

Barton, N. H., and M. Turelli. 1989. Evolutionary quantitative genetics: How little do we know ? *Ann. Rev. Genet.* 23: 337–370.

Barton, N. H., and M. Turelli. 1991. Natural and sexual selection on many loci. *Genetics* 127: 229–255.

Basolo, A. 1990a. Female preference for male sword length in the green swordtail, *Xiphophorus helleri* (Pisces: Poeciliidae). *Anim. Behav.* 40: 332–338.

Basolo, A. L. 1990b. Female preference predates the evolution of the sword in swordtail fish. *Science* 250: 808–810.

Bastock, M. 1967. *Courtship: An Ethological Study*. Aldine, Chicago.

Bateman, A. J. 1948. Intra-sexual selection in *Drosophila*. *Heredity* 2: 349–368.

Bateson, P. 1978. Sexual imprinting and optimal outbreeding. *Nature* 273: 659–660.

Bateson, P. 1980. Optimal outbreeding and the development of sexual preferences in Japanese quail. *Z. Tierpsychol.* 53: 231–244.

Bateson, P. 1982. Preferences for cousins in Japanese quail. *Nature* 295: 236–237.

Bateson, P. 1983. Optimal outbreeding. In P. Bateson, ed., *Mate Choice*, 257–278. Cambridge University Press, Cambridge, U.K.

Baughman, J. F. 1991. Do protandrous males have increased mating success? The case of *Euphydryas editha*. *Am. Nat.* 138: 536–542.

Bawa, K. S. 1980a. Mimicry of male by female flowers and intrasexual competition for pollinators in *Jacaratia dolichaula* (D. Smith) Woodson (Caricaceae). *Evolution* 34: 467–474.

Bawa, K. S. 1980b. Evolution of dioecy in flowering plants. *Ann. Rev. Ecol. Syst.* 11: 15–39.

Bawa, K. S., and J. H. Beach. 1981. Evolution of sexual systems in flowering plants. *Ann. Missouri Bot. Gard.* 68: 254–274.

Bawa, K. S., and P. A. Opler. 1975. Dioecism in tropical forest trees. *Evolution* 29: 167–179.

Bawa, K. S., and P. A. Opler. 1978. Why are pistillate inflorescences of *Simarouba glauca* eaten less than staminate inflorescences? *Evolution* 32: 673–676.

Bawa, K. S., and C. J. Webb. 1984. Flower, fruit and seed abortion in tropical forest trees: Implications for the evolution of paternal and maternal reproductive patterns. *Amer. J. Bot.* 71: 736–751.

Baylis, J. R. 1981. The evolution of parental care in fishes, with reference to Darwin's rule of male sexual selection. *Env. Biol. Fish.* 6: 223–251.

Beach, J. H. 1981. Pollinator foraging and the evolution of dioecy. *Am. Nat.* 118: 572–577.

Beacham, T. D., and C. B. Murray. 1988. A genetic analysis of body size in pink salmon *Oncorhynchus gorbuscha*. *Genome* 30: 31–35.

Beaman, R. S., P. J. Decker, and J. H. Beaman. 1988. Pollination in *Rafflesia*. *Am. J. Bot.* 75: 1148–1162.

Becker, P. H. 1982. The coding of species-specific characteristics in bird sounds. In D. E. Kroodsma and E. H. Miller, eds., *Acoustic Communication in Birds*, 1: 213–252. Academic Press, New York.

Beebe, W. 1944. The function of secondary sexual characters in two species of Dynastidae (Coleoptera). *Zoologica* 29: 53–58.

Beebe, W. 1947. Notes on the Hercules beetle, *Dynastes hercules* Linn., at Rancho Grande, Venezuela, with special reference to combat behavior. *Zoologica* 32: 109–116.

Beehler, B. 1983. Lek behavior of the lesser bird of paradise. *Auk* 100: 992–995.

Beehler, B. M. 1988. Lek behavior of the raggiana bird of paradise. *National Geographic Research* 4: 343–358.

Beehler, B. M., and M. S. Foster. 1988. Hotshots, hotspots, and female preference in the organization of lek mating systems. *Am. Nat.* 131: 203–219.

Beletsky, L. D. 1983. Aggressive and pair-bond maintenance songs of female red-winged blackbirds (*Agelaius phoeniceus*). *Z. Tierpsychol.* 62: 47–54.

Beletsky, L. D., and G. H. Orians. 1989. Territoriality among male red-winged blackbirds. III. Testing hypotheses of territorial dominance. *Behav. Ecol. Sociobiol.* 24: 333–339.

Bell, D. J. 1983. Mate choice in the European rabbit. In P. Bateson, ed., *Mate Choice*, 211–223. Cambridge University Press, Cambridge, U.K.

Bell, G. 1978. The handicap principle in sexual selection. *Evolution* 32: 872–885.

Bell, G. 1982. *The Masterpiece of Nature—The Evolution and Genetics of Sexuality.* University of California Press, Berkeley.

Bell, G. 1985. On the function of flowers. *Proc. R. Soc. Lond. B* 224: 223–265.

Bell, G., and V. Koufopanou. 1986. The cost of reproduction. In R. Dawkins and M. Ridley, eds., *Oxf. Surv. Evol. Biol.*, 83–131. Oxford University Press, Oxford.

Bell, P. D. 1979. Acoustic attraction of herons by crickets. *New York Entomol. Soc.* 87: 126–127.

Bellrose, F. C. 1980. *Ducks, Geese and Swans of North America.* 2d ed. Stackpole, Harrisburg, Penn.

Belt, T. 1874. *The Naturalist in Nicaragua.* John Murray, London.

Belwood, J. J.. and G. K. Morris. 1987. Bat predation and its influence on calling behavior in neotropical katydids. *Science* 238: 64–67.

Benedix, J. H., Jr., and D. J. Howard. 1991. Calling song displacement in a zone of overlap and hybridization. *Evolution* 45: 1751–1759.

Beninde, J. 1937. *Zur Naturgeschichte des Rothirsches.* Leipzig Monographien der Wildsäugetiere, Vol. 4. Paul Schöps, Leipzig.

Bennet-Clark, H. C., and A. W. Ewing. 1967. Stimuli provided by courtship of male *Drosophila melanogaster*. *Nature* 215: 669–671.

Bennet-Clark, H. C., and A. W. Ewing. 1969. Pulse interval as a critical parameter in the courtship song of *Drosophila melanogaster*. *Anim. Behav.* 17: 755–759.

Bensch, S., and D. Hasselquist. 1991. Nest predation lowers the polygyny threshold: A new compensation model. *Am. Nat.* 198: 1297–1306.

Bensch, S., and D. Hasselquist. 1992. Evidence for active female choice in a polygynous warbler. *Anim. Behav.* 44: 301–312.

Benzing, D. H. 1981. Why is orchidaceae so large, its seeds so small, and its seedlings mycotrophic? *Selbyana* 5: 241–242.

Berglund, A. 1986. Sex change by a polychaete: Effects of social and reproductive costs. *Ecology* 67: 837–845.

Berglund, A. 1991. Egg competition in a sex-role reversed pipefish: Subdominant females trade reproduction for growth. *Evolution* 45: 770–774.

Berglund, A., and G. Rosenqvist. 1990. Male limitation of female reproductive success in a pipefish: Effects of body-size differences. *Behav. Ecol. Sociobiol.* 27: 129–133.

Berglund, A., G. Rosenquist, and I. Svensson. 1986a. Reversed sex roles and parental energy investment in zygotes of two pipefish (*Syngnathidae*) species. *Mar. Ecol. Prog. Ser.* 29: 209–215.

Berglund, A., G. Rosenqvist, and I. Svensson. 1986b. Mate choice, fecundity and sexual dimorphism in two pipefish species (*Syngnathidae*). *Behav. Ecol. Sociobiol.* 19: 301–307.

Berglund, A., G. Rosenqvist, and I. Svensson. 1989. Reproductive success of females limited by males in two pipefish species. *Am. Nat.* 133: 506–516.

Bergman, G. 1965. Der sexuelle Grössendimorphismus der Anatiden als Anpassung an das Höhlenbruten. *Comm. Biol.* 28: 1–10.

Berry, J. F., and R. Shine. 1980. Sexual dimorphism and sexual selection in turtles (order Testudines). *Oecologia* 44: 185–191.

Bertelsen, E. 1951. *The Ceratioid Fishes. Ontogeny, Taxonomy, Distribution and Biology*. Dana-Report No. 39. Carlsberg Foundation, Copenhagen.

Bertin, R. I. 1982. Floral biology, hummingbird pollination and fruit production of trumpet creeper (*Campsis radicans*, Bignoniaceae). *Am. J. Bot.* 69: 122–134.

Bertin, R. I. 1988. Paternity in plants. In J. Lovett Doust and L. Lovett Doust, eds., *Plant Reproductive Ecology*, 30–59. Oxford University Press, New York.

Bertin, R. I., and A. G. Stephenson. 1983. Towards a definition of sexual selection. *Evol. Theor.* 6: 293–295.

Bertram, B.C.R. 1992. *The Ostrich Communal Nesting System*. Princeton University Press, Princeton, N.J.

Berven, K. A. 1981. Mate choice in the wood frog, *Rana sylvatica*. *Evolution* 35: 707–722.

Berven, K. A. 1987. The heritable basis of variation in larval developmental patterns within populations of the wood frog (*Rana sylvatica*). *Evolution* 41: 1088–1097.

Best, P. B. 1979. Social organization in sperm whales. In H. E. Winn and B. L. Olla, eds., *Behavior of Marine Animals*, Vol. 3., *Cetacea*, 227–290. Plenum, New York.

Bierzychudek, P. 1981. Pollinator limitation of plant reproductive effort. *Am. Nat.* 117: 838–840.

Bierzychudek, P. 1984. Determinants of gender in Jack-in-the-pulpit: The influence of plant size and reproductive history. *Oecologia* 65: 14–18.

Bierzychudek, P. 1987. Pollinators increase the cost of sex by avoiding female flowers. *Ecology* 68: 444–447.

Bildstein, K. L., S. G. McDowell, and I. L. Brisbin. 1989. Consequences of sexual dimorphism in sand fiddler crabs, *Uca pugilator*: Differential vulnerability to avian predation. *Anim. Behav.* 37: 133–139.

Birch, M. C. 1974. Aphrodisiac pheromones in insects. In M. C. Birch, ed., *Pheromones*, 115–134. North-Holland, Amsterdam.

Birkhead, T. R., and K. Clarkson. 1980. Mate selection and precopulatory guarding in *Gammarus pulex*. *Z. Tierpsychol.* 52: 365–380.

Birkhead, T. R., and F. M. Hunter. 1990. Mechanisms of sperm competition. *TREE* 5: 48–52.

Birkhead, T. R., and A. P. Møller. 1992. *Sperm Competition in Birds: Evolutionary Causes and Consequences*. Academic Press, London.

Birkhead, T. R., K. E. Lee, and P. Young. 1988. Sexual cannibalism in the praying mantis *Hierodula membranacea*. *Behaviour* 106: 112–118.

Bisazza, A., and A. Marconato. 1988. Female mate choice, male-male competition and parental care in the river bullhead, *Cottus gobio L.* (Pisces, Cottidae). *Anim. Behav.* 36: 1352–1360.

Bisazza, A., A. Marconato, and G. Marin. 1989. Male competition and female choice in *Padogobius martensi* (Pisces, Gobidae). *Anim. Behav.* 38: 406–413.

Bischoff, R. J., J. L. Gould, and D. I. Rubenstein. 1985. Tail size and female choice in the guppy. *Behav. Ecol. Sociobiol.* 17: 253–255.

Björklund, M. 1984. The adaptive significance of sexual indistinguishability in birds: A critique of a recent hypothesis. *Oikos* 43: 414–416.
Björklund, M. 1990. A phylogenetic interpretation of sexual dimorphism in body size and ornament in relation to mating system in birds. *J. Evol. Biol.* 3: 171–183.
Björklund, M. 1991a. Evolution, phylogeny, sexual dimorphism and mating system in the grackles (*Quiscalus* spp.: Icterinae). *Evolution* 45: 608–621.
Björklund, M. 1991b. Coming of age in fringillid birds: Heterochrony in the ontogeny of secondary sexual characters. *J. Evol. Biol.* 4: 83–92.
Björklund, M., B. Westman, and K. Allander. 1989. Song in Swedish great tits: Intra- or intersexual communication. *Behaviour* 111: 257–261.
Blair, W. F. 1964. Isolating mechanisms and interspecies interactions in anuran amphibians. *Q. Rev. Biol.* 39: 334–344.
Blaustein, A. R. 1981. Sexual selection and mammalian olfaction. *Am. Nat.* 117: 1006–1010.
Bleiweiss, R. 1992. Reversed plumage ontogeny in a female hummingbird: Implications for the evolution of iridescent colours and sexual dichromatism. *Biol. J. Linn. Soc.* 47: 183–195.
Boag, P. T. 1983. The heritability of external morphology in Darwin's ground finches (*Geospiza*) on Isla Daphne Major, Galápagos. *Evolution* 37: 877–894.
Boag, P. T., and A. J. van Noordwijk. 1987. Quantitative genetics. In F. Cooke and P. A. Buckley, eds., *Avian Genetics, A Population and Ecological Approach*, 45–78. Academic Press, London.
Boake, C.R.B. 1984. Male displays and female preferences in the courtship of a gregarious cricket. *Anim. Behav.* 32: 690–697.
Boake, C.R.B. 1985. Genetic consequences of mate choice: A quantitative genetic method for testing sexual selection theory. *Science* 227: 1061–1063.
Boake, C.R.B. 1986. A method for testing adaptive hypotheses of mate choice. *Am. Nat.* 127: 654–666.
Boake, C.R.B. 1991. Coevolution of senders and receivers of sexual signals: Genetic coupling and genetic correlations. *TREE* 6: 225–227.
Boake, C.R.B., and R. R. Capranica. 1982. Aggressive signal in "courtship" chirps of a gregarious cricket. *Science* 218: 580–582.
Bock, W. J., and J. Farrand, Jr. 1980. The number of species and genera of recent birds: A contribution to comparative systematics. *Am. Mus. Novit.* 2703: 1–29.
Boggs, C. L., and L. E. Gilbert. 1979. Male contribution to egg production in butterflies: Evidence for transfer of nutrients at mating. *Science* 206: 83–84.
Boggs, C. L., and W. Watt. 1981. Population structure of pierid butterflies. IV. Genetic and physiological investment in offspring by male *Colias*. *Oecologia* 50: 320–324.
Bohlin, T., C. Dellefors, and U. Faremo. 1986. Early sexual maturation of male sea trout and salmon—an evolutionary model and some practical implications. *Rep. Inst. Freshwat. Res. Drottningholm* 63: 17–25.
Bohlin, T., C. Dellefors, and U. Faremo. 1990. Large or small at maturity—theories on the choice of alternative male strategies in anadromous salmonids. *Ann. Zool. Fennici* 27: 139–147.
Bond, W. J., and J. Midgley. 1988. Allometry and sexual differences in leaf size. *Am. Nat.* 131: 901–910.

Bonner, J. T., and R. M. May. 1981. Introduction, in reprint of Darwin 1871. *The Descent of Man, and Selection in Relation to Sex*, vii–xli. Princeton University Press, Princeton, N.J.

Booth, J., and J. A. Peters. 1972. Behavioural studies on the green turtle (*Chelonia mydas*) in the sea. *Anim. Behav.* 20: 808–812.

Boppré, M. 1984. Chemically mediated interactions between butterflies. In R. I. Vane-Wright and P. R. Ackery, eds., *The Biology of Butterflies*, 259–275. Academic Press, London.

Borgerhoff-Mulder, M. 1990. Kipsigis women prefer wealthy men: Evidence for female choice in mammals? *Behav. Ecol. Sociobiol.* 27: 255–264.

Borgerhoff-Mulder, M. 1991. Human behavioural ecology. In J. R. Krebs and N. B. Davies, eds., *Behavioural Ecology: An Evolutionary Approach*, 3d ed., 69–98. Blackwell, Oxford.

Borgia, G. 1979. Sexual selection and the evolution of mating systems. In M. S. Blum and N. A. Blum, eds., *Sexual Selection and Reproductive Competition in Insects*, 19–80. Academic Press, New York.

Borgia, G. 1980. Sexual competition in *Scatophaga stercoraria*: Size- and density-related changes in male ability to capture females. *Behaviour* 75: 185–205.

Borgia, G. 1981. Mate selection in the fly *Scatophaga stercoraria*: Female choice in a male-controlled system. *Anim. Behav.* 29: 71–80.

Borgia, G. 1982. Experimental changes in resource structure and male density: Size-related differences in mating success among male *Scatophaga stercoraria*. *Evolution* 36: 307–315.

Borgia, G. 1985a. Bower destruction and sexual competition in the satin bowerbird (*Ptilonorhynchus violaceus*). *Behav. Ecol. Sociobiol.* 18: 91–100.

Borgia, G. 1985b. Bower quality, number of decorations and mating success of male satin bowerbirds (*Ptilonorhynchus violaceus*): An experimental analysis. *Anim. Behav.* 33: 266–271.

Borgia, G. 1986. Satin bowerbird parasites: A test of the bright male hypothesis. *Behav. Ecol. Sociobiol.* 19: 355–358.

Borgia, G. 1987. A critical review of sexual selection models. In J. W. Bradbury and M. B. Andersson, eds., *Sexual Selection: Testing the Alternatives*, 55–66. Wiley, Chichester, U.K.

Borgia, G. 1993. The cost of display in the non-resource-based mating system of the satin bowerbird. *Am. Nat.* 141: 729–743.

Borgia, G., and K. Collis. 1989. Female choice for parasite-free male satin bowerbirds and the evolution of bright male plumage. *Behav. Ecol. Sociobiol.* 25: 445–454.

Borgia, G., and K. Collis. 1990. Parasites and bright male plumage in the satin bowerbird (*Ptilonorhynchus violaceus*). *Am. Zool.* 30: 279–285.

Borgia, G., and M. A. Gore. 1986. Feather stealing in the satin bowerbird: Male competition and the quality of display. *Anim. Behav.* 34: 727–738.

Borgia, G., and J. C. Wingfield. 1991. Hormonal correlates of bower decoration and sexual display in the satin bowerbird (*Ptilonorhynchus violaceus*). *Condor* 93: 935–942.

Borgia, G., S. G. Pruett-Jones, and M. A. Pruett-Jones. 1985. The evolution of bower-building and the assessment of male quality. *Z. Tierpsychol.* 67: 225–236.

Bowman, R. 1987. Size dimorphism in mated pairs of American kestrels. *Wilson Bull.* 99: 465–467.

Box, G. E. P., W. G. Hunter, and J. S. Hunter. 1978. *Statistics for Experimenters.* Wiley, New York.

Boyce, M. S. 1990. The red queen visits sage grouse leks. *Am. Zool.* 30: 263–270.

Boyd, R., and P. Richerson. 1981. *Culture and the Evolutionary Process.* University of Chicago Press, Chicago.

Bradbury, J. W. 1981. The evolution of leks. In R. D. Alexander and D. W. Tinkle, eds., *Natural Selection and Social Behaviour: Recent Research and New Theory*, 138–173. Chiron Press, New York.

Bradbury, J. W. 1985. Contrasts between insects and vertebrates in the evolution of male display, female choice, and lek mating. *Fortschr. Zool.* 31: 273–289.

Bradbury, J. W., and M. B. Andersson, eds. 1987. *Sexual Selection: Testing the Alternatives.* Wiley, Chichester, U.K.

Bradbury, J. W., and N. B. Davies. 1987. Relative roles of intra- and intersexual selection. In J. W. Bradbury and M. B. Andersson, eds., *Sexual Selection: Testing the Alternatives.* Wiley, Chichester, U.K.

Bradbury, J. W., and R. M. Gibson. 1983. Leks and mate choice. In P. Bateson, ed., *Mate Choice*, 109–138. Cambridge University Press, Cambridge, U.K.

Bradbury, J. W., and S. L. Vehrencamp. 1977. Social organization and foraging in emballonurid bats. III. Mating systems. *Behav. Ecol. Sociobiol.* 2: 1–17.

Bradbury, J. W., S. Vehrencamp, and R. Gibson. 1985. Leks and the unanimity of female mate choice. In P. J. Greenwood, P. H. Harvey, and M. Slatkin, eds., *Evolution: Essays in the Honour of John Maynard Smith*, 301–314. Cambridge University Press, Cambridge, U.K.

Bradbury, J. W., R. Gibson, and I. M. Tsai. 1986. Hotspots and the dispersion of leks. *Anim. Behav.* 34: 1694–1709.

Bradbury, J. W., S. L. Vehrencamp, and R. M. Gibson. 1989a. Dispersion of displaying male sage grouse. 1. Patterns of temporal variation. *Behav. Ecol. Sociobiol.* 24: 1–14.

Bradbury, J. W., R. M. Gibson, C. E. McCarthy, and S. L. Vehrencamp. 1989b. Dispersion of displaying male sage grouse. 2. The role of female dispersion. *Behav. Ecol. Sociobiol.* 24: 15–24.

Bradshaw-Hawkins, V. I., and F. Sander. 1981. Notes on the reproductive biology and behavior of the West Indian fighting conch, *Strombus pugilis* Linnaeus in Barbados, with evidence of male guarding. *Veliger* 24: 159–164.

Brandt, C. A. 1989. Mate choice and reproductive success of pikas. *Anim. Behav.* 37: 118–132.

Bratton, B. O., and B. Kramer. 1989. Patterns of the electric organ discharge during courtship and spawning in the mormyrid fish, *Pollimyrus isidori*. *Behav. Ecol. Sociobiol.* 24: 349–368.

Breden, F. 1988. Sexual selection and predation risk in guppies. *Nature* 332: 593–594.

Breden, F., and G. Stoner. 1987. Male predation risk determines female preference in the Trinidad guppy. *Nature* 329: 831–833.

Breden, F., M. Scott, and E. Michel. 1987. Genetic differentiation for anti-predator behaviour in the Trinidad guppy (*Poecilia reticulata*). *Anim. Behav.* 35: 618–620.

Breed, M. D., S. K. Smith, and B. G. Gall. 1980. Systems of mate choice in a cockroach species with male dominance hierarchies. *Anim. Behav.* 28: 130–134.

Breiehagen, T., and T. Slagvold. 1988. Male polyterritoriality and female-female aggression in pied flycatchers *Ficedula hypoleuca*. *Anim. Behav.* 36: 604–606.

Bristowe, W. S. 1929. The mating habits of spiders, with special reference to the problems surrounding sex dimorphism. *Proc. Zool. Soc. London* 21: 309–358.

Brockman, H. J. 1973. The function of poster coloration in the beaugregory, *Eupomacentrus leucostictus* (Pomacentridae, Pisces). *Z. Tierpsychol.* 33: 13–34.

Brockway, B. F. 1965. Stimulation of ovarian development and egglaying by male courtship vocalizations in budgerigars (*Melopsittacus undulatus*). *Anim. Behav.* 13: 575–578.

Brodie, P. F. 1975. Cetacean energetics, an overview of intraspecific size variation. *Ecology* 56: 152–161.

Brodsky, L. 1988. Ornament size influences mating success in male rock ptarmigan. *Anim. Behav.* 36: 662–667.

Brodsky, L. 1985. Mating tactics of male rock ptarmigan, *Lagopus mutus*: A conditional mating strategy. *Anim. Behav.* 36: 335–342.

Brooks, D. R., and D. A. McLennan. 1990. *Phylogeny, Ecology, and Behavior: A Research Program in Comparative Biology*. University of Chicago Press, Chicago.

Brower, L. P., J.V.Z. Brower, and F. P. Cranston. 1965. Courtship behavior of the queen butterfly, *Danaus gilippus berenice* (Cramer). *Zoologica* 50: 1–39.

Brown, D. 1988. Components of lifetime reproductive success. In T. H. Clutton-Brock, ed., *Reproductive Success*, 439–453. University of Chicago Press, Chicago.

Brown, J. H., M. A. Cantrell, and S. M. Evans. 1973. Observations on the behaviour and colouration of some coral reef fish (Family Pomacentridae). *Mar. Behav. Physiol.* 2: 63–71.

Brown, J. L. 1975. *The Evolution of Behavior*. Norton, New York.

Brown, J. L. 1983. Intersexual selection. *Nature* 302: 472.

Brown, J. L. 1987. *Helping and Communal Breeding in Birds: Ecology and Evolution*. Princeton University Press, Princeton, N.J.

Brown, L. 1981. Patterns of female choice in mottled sculpins (*Cottidae, Teleostei*). *Anim. Behav.* 29: 375–382.

Brown, L., and J. Bartalon. 1986. Behavioral correlates of male morphology in a horned beetle. *Am. Nat.* 127: 565–570.

Brown, L., and L. L. Rockwood. 1986. On the dilemma of horns. *Nat. Hist.* 7: 54–61.

Brown, L., J. McDonell, and J. Fitzgerald. 1985. Courtship and female choice in the horned beetle, *Bolitotherus cornutus* (*Panzer*) (Coleoptera: Tenebrionidae). *Ann. Entomol. Soc. Am.* 78: 423–427.

Brown, L. H., E. K. Urban, and K. Newman 1982. *The Birds of Africa*. Vol. 1. Academic Press, London.

Brown, R. 1820. An account of a new genus of plants, named *Rafflesia*. *Trans. Linn. Soc. Lond.* 13:201–234.

Brown, R. D., ed. 1983. *Antler Development in Cervidae*. C. Kleiberg Wildlife Research Institute, Kingsville, Texas.

Brown, R. E. 1985. The rodents. I. Effects of odours on reproductive physiology (primer effects). In R. E. Brown and D. W. Macdonald, eds., *Social Odours in Mammals*, 245–344. Clarendon, Oxford.

Brown, R. E., and D. W. Macdonald, eds. 1985. *Social Odours in Mammals*. Clarendon, Oxford.

Brown, W. D. 1990a. Constraints on size-assortative mating in the blister beetle (*Tegrodera aloga*) (Coleoptera: Meloidae). *Ethology* 86: 146–160.

Brown, W. D. 1990b. Size-assortative mating in the blister beetle (*Lytta magister*) (Coleoptera: Meloidae) is due to male and female preference for lager mates. *Anim. Behav.* 40: 901–909.

Brown, W. L., Jr., and E. O. Wilson. 1956. Character displacement. *Syst. Zool.* 5: 49–64.

Bruce, H. M. 1959. An exteroceptive block to pregnancy in the mouse. *Nature* 184: 105.

Brüll, H. 1937. *Das Leben deutscher Greifvögel*. Gustav Fisher, Jena.

Brunet, J. 1992. Sex allocation in hermaphroditic plants. *TREE* 7: 79–84.

Bruning, D. F. 1974. Social structure and reproductive behaviour of the greater rhea. *Living Bird* 13: 251–294.

Brush, A. H. 1978. Avian pigmentation. In A. H. Brush, ed., *Chemical Zoology*, 141–161. Academic Press, New York.

Bryant, D. M. 1988. Lifetime reproductive success of house martins. In T. H. Clutton-Brock, ed., *Reproductive Success*, 173–188. University of Chicago Press, Chicago.

Bryant, D. M., and P. Tatner. 1991. Intraspecies variation in avian energy expenditure: Correlates and constraints. *Ibis* 133: 236–245.

Bryden, M. M. 1972. Growth and development in marine mammals. In R. J. Harrison, ed., *Functional Anatomy of Marine Mammals*, 1–79. Academic Press, New York.

Buchholz, R. 1991. Older males have bigger knobs: Correlates of ornamentation in two species of curassow. *Auk* 108: 153–160.

Bull, J., and E. Charnov 1988. How fundamental are Fisherian sex ratios? *Oxf. Surv. Evol. Biol.* 5: 96–135.

Bulmer, M. G. 1980. *The Mathematical Theory of Quantitative Genetics*. Oxford University Press, Oxford.

Bulmer, M. G. 1983. Models for the evolution of protandry in insects. *Theor. Pop. Biol.* 23: 314–322.

Bulmer, M.G. 1989a. Structural instability of models of sexual selection. *Theor. Pop. Biol.* 35: 195–206.

Bulmer, M. G. 1989b. Maintenance of genetic variability by mutation-selection balance: A child's guide through the jungle. *Genome* 23: 761–767.

Bürger, R. 1989. Linkage and the maintenance of heritable variation by mutation-selection balance. *Genetics* 121: 175–184.

Burk, T. 1982. Evolutionary significance of predation on sexually signalling males. *Florida Entomol.* 65: 90–104.

Burk, T. 1983. Male aggression and female choice in a field cricket (*Teleogryllus oceanicus*): The importance of courtship song. In D. T. Gwynne and G. K. Morris, eds., *Orthopteran Mating Systems*, 97–119. Westview, Boulder, Colo.

Burk, T. 1984. Male-male interactions in Caribbean fruit flies, *Anastrepha suspensa* (Loew) (Diptera: Tephritidae): territorial fights and signalling stimulation. *Florida Entomol.* 67: 542–547.

Burk, T., and J. C. Webb. 1983. Effect of male size on calling propensity, song parameters, and mating success in Caribbean fruit flies, *Anastrepha suspensa* (Loew) (Diptera: Tephritidae). *Ann. Entomol. Soc. Am.* 76: 678–682.

Burke, T. 1989. DNA fingerprinting and other methods for the study of mating success. *TREE* 4: 139–144.

Burke, T., N. B. Davies, M. W. Bruford, and B. J. Hatchwell. 1989. Parental care and mating behaviour of polyandrous dunnocks (*Prunella modularis*) related to paternity by DNA fingerprinting. *Nature* 338: 249–251.

Burkhardt, D., and I. de la Motte. 1987. Physiological, behavioural, and morphometric data elucidate the evolutive significance of stalked eyes in Diopsidae (Diptera). *Entomol. Gener.* 12: 221–233.

Burkhardt, D., and I. de la Motte. 1988. Big 'antlers' are favoured: Female choice in stalk-eyed flies (Diptera, Insecta), field collected harems and laboratory experiments. *J. Comp. Physiol. A* 162: 649–652.

Burkhardt, R. W., Jr. 1985. Darwin on animal behavior and evolution. In D. Kohn, ed., *The Darwinian Heritage*, 327–365. Princeton University Press, Princeton, N.J.

Burley, N. 1977. Parental investment, mate choice and mate quality. *Proc. Natl. Acad. Sci. USA* 74: 3476–3479.

Burley, N. 1981a. Mate choice by multiple criteria in a monogamous species. *Am. Nat.* 117: 515–528.

Burley, N. 1981b. Sex ratio manipulation and selection for attractiveness. *Science* 211: 721–722.

Burley, N. 1981c. The evolution of sexual indistinguishability. In R. D. Alexander and D. W. Tinkle, eds., *Natural Selection and Social Behaviour: Recent Research and New Theory*, 121–137. Chiron Press, New York.

Burley, N. 1985. The organization of behavior and the evolution of sexually selected traits. *Orn. Monogr.* 37: 22–44.

Burley, N. 1986a. Comparison of the band-colour preferences of two species of estrildid finches. *Anim. Behav.* 34: 1732–1741.

Burley, N. 1986b. Sex-ratio manipulation in color-banded populations of zebra finches. *Evolution* 40: 1191–1206.

Burley, N. 1986c. Sexual selection for aesthetic traits in species with biparental care. *Am. Nat.* 127: 415–445.

Burley, N. 1988a. Wild zebra finches have band-colour preferences. *Anim. Behav.* 36: 1235–1237.

Burley, N. 1988b. The differential-allocation hypothesis: An experimental test. *Am. Nat.* 132: 611–628.

Burley, N., and C. B. Coopersmith. 1987. Bill colour preferences of zebra finches. *Ethology* 76: 133–151.

Burley, N., and N. Moran. 1979. The significance of age and reproductive experience in the mate preferences of feral pigeons, *Columba livia*. *Anim. Behav.* 27: 686–698.

Burley, N., G. Krantzberg, and P. Radman. 1982. Influence of colour-banding on the conspecific preferences of zebra finches. *Anim. Behav.* 30: 444–455.

Burley, N., S. C. Tidemann, and K. Halupka. 1991. Bill colour and parasite levels of zebra finches. In J. E. Loye and M. Zuk, eds., *Bird-Parasite Interactions*, 359–376. Oxford University Press, Oxford.

Burtt, E. H., Jr., ed. 1979. *The Behavioral Significance of Color*. Garland STPM Press, New York.

Burtt, E. H., Jr. 1986. An analysis of physical, physiological and optical aspects of avian coloration with emphasis on wood warblers. *Orn. Monogr.* 38: 1–126.

Buskirk, R. E., C. Frohlich, and K. G. Ross. 1984. The natural selection of sexual cannibalism. *Am. Nat.* 123: 612–625.

Butcher, G. S. 1991. Mate choice in female northern orioles with a consideration of the role of the black male coloration in female choice. *Condor* 93: 82–88.

Butcher, G. S., and S. Rohwer. 1988. The evolution of conspicuous and distinctive coloration for communication in birds. *Curr. Ornithol.* 6: 51–108.

Butlin, R. 1987. Speciation by reinforcement. *TREE* 2: 8–13.

Butlin, R. 1989. Reinforcement of premating isolation. In D. Otte and J. A. Endler, eds., *Speciation and Its Consequences*, 158–179. Sinauer, Sunderland, Mass.

Butlin, R., and G. M. Hewitt. 1986. The response of female grasshoppers to male song. *Anim. Behav.* 34: 1896–1899.

Butlin, R. K., and M. G. Ritchie. 1989. Genetic coupling in mate recognition systems: What is the evidence? *Biol. J. Linn. Soc.* 37: 237–246.

Butlin, R. K., P. M. Collins, and T. H. Day. 1984. The effect of larval density on an inversion polymorphism in the seaweed fly, *Coelopa frigida*. *Heredity* 52: 415–423.

Butlin, R. K., G. M. Hewitt, and S. F. Webb. 1985. Sexual selection for intermediate optimum in *Chorthippus brunneus* (Orthoptera: Acrididae). *Anim. Behav.* 33: 1281–1292.

Butlin, R. K., C. W. Woodhatch, and G. M. Hewitt. 1987. Male spermatophore investment increases female fecundity in a grasshopper. *Evolution* 41: 221–225.

Bützler, W. 1974. Kampf und Paarungsverhalten, soziale Rangordnung und Aktivitätsperiodik beim Rothirsch (*Cervus elaphus* L.). *Z. Tierpsychol. Suppl.* 16: 1–80.

Cade, T. J. 1960. Ecology of the peregrine and gyrfalcon populations in Alaska. *Univ. Calif. Publ. Zool.* 63: 151–290.

Cade, T. J. 1982. *The Falcons of the World*. Cornell University Press, Ithaca, N.Y.

Cade, W. 1975. Acoustically orienting parasitoids: Fly phonotaxis to cricket song. *Science* 190: 1312–1313.

Cade, W. 1979. The evolution of alternative male reproductive strategies in field crickets. In M. S. Blum and N. A. Blum, eds., *Sexual Selection and Reproductive Competition in Insects*, 343–380. Academic Press, New York.

Cade, W. H. 1981a. Alternative male strategies: Genetic differences in crickets. *Science* 212: 563–564.

Cade, W. 1981b. Field cricket spacing, and the phonotaxis of crickets and parasitoid flies to clumped and isolated cricket songs. *Z. Tierpsychol.* 55: 365–375.

Cade, W. H. 1984. Genetic variation underlying sexual behavior and reproduction. *Am. Zool.* 24: 355–366.

Calder, W. A., III. 1984. *Size, Function and Life History*. Harvard University Press, Cambridge, Mass.

Callahan, J. R. 1981. Vocal solicitation and parental investment in female *Eutamias*. *Am. Nat.* 118: 872–875.

Calvo, R. N., and C. C. Horvitz. 1990. Pollinator limitation, cost of reproduction, and fitness in plants: A transition-matrix demographic approach. *Am. Nat.* 136: 499–516.

Campagna, C., and B. J. Le Boeuf. 1988. Reproductive behaviour of southern sea lions. *Behaviour* 104: 233–261.

Campanella, P. J., and L. L. Wolf. 1974. Temporal leks as a mating system in a temperate zone dragonfly (Odonata: Anisoptera). I. *Plathemis lydia* (Drury). *Behaviour* 51: 49–87.

Campbell, D. R. 1989a. Inflorescence size: Test of the male function hypothesis. *Am. J. Bot.* 76: 730–738.

Campbell, D. R. 1989b. Measurements of selection in a hermaphroditic plant: Variation in male and female pollination success. *Evolution* 43: 318–334.

Campbell, D. R. 1991. Effects of floral traits on sequential components of fitness in *Ipomopsis aggregata*. *Am. Nat.* 137: 713–737.

Campbell, D. R., and N. M. Waser. 1987. The evolution of plant mating systems: Multilocus simulations of pollen dispersal. *Am. Nat.* 129: 593–609.

Campbell, D. R., N. M. Waser, M. V. Price, E. A. Lynch, and R. J. Mitchell. 1991. Components of phenotypic selection: Pollen export and flower corolla width in *Ipomopsis aggregata*. *Evolution* 45: 1458–1467.

Capp, M. S., and W. A. Searcy 1991a. An experimental study of song type function in the bobolink (*Dolichonyx oryzivorus*). *Behav. Ecol. Sociobiol.* 28: 179–186.

Capp, M. S., and W. A. Searcy. 1991b. Acoustical communication of aggressive intentions by territorial male bobolinks. *Behav. Ecol.* 2: 319–326.

Capranica, R. R., L. S. Frischkopf, and E. Nevo. 1973. Encoding of geographic dialects in the auditory system of the cricket frog. *Science* 182: 1272–1275.

Caravello, H. E., and G. N. Cameron. 1987. The effects of sexual selection on the foraging behaviour of the Gulf coast fiddler crab, *Uca panacea*. *Anim. Behav.* 35: 1864–1874.

Cardé, R. T. 1986. The role of pheromones in reproductive isolation and speciation of insects. In M. D. Muettel, ed., *Evolutionary Genetics of Invertebrate Behavior*, 303–315. Plenum, New York.

Carlson, A. 1989. Courtship feeding and clutch size in red-backed shrikes (*Lanius collurio*). *Am. Nat.* 133: 454–457.

Caro, T. M. 1985. Intersexual selection and cooption. *J. theor. Biol.* 112: 275–277.

Caro, T. M., and P. Bateson. 1986. Organization and ontogeny of alternative tactics. *Anim. Behav.* 34: 1483–1499.

Carothers, J. H. 1984. Sexual selection and sexual dimorphism in some herbivorous lizards. *Am. Nat.* 124: 244–254.

Carpenter, C. C., J. C. Gillingham, and J. B. Murphy. 1976. The combat ritual of the rock rattlesnake (*Crotalus lepidus*). *Copeia* 1976: 764–780.

Carr, D. E. 1991. Sexual dimorphism and fruit production in a dioecious understory tree, *Ilex opaca* Ait. *Oecologia* 85: 381–388.

Carson, H. L. 1978. Speciation and sexual selection in Hawaiian *Drosophila*. In P. F. Brussard, ed., *Ecological Genetics: The Interface*, 93–107. Springer, New York.

Carson, H. L. 1986. Sexual selection and speciation. In S. Karlin and E. Nevo, eds., *Evolutionary Processes and Theory*, 391–409. Academic Press, London.

Carson, H. L., and R. Lande. 1984. Inheritance of a secondary sexual character in *Drosophila silvestris. Proc. Natl. Acad. Sci. USA* 81: 6904–6907.

Carson, H. L., K. Y. Kaneshiro, and F. C. Val. 1989. Natural hybridization between the sympatric Hawaiian species *Drosophila silvestris* and *Drosophila heteroneura. Evolution* 43: 190–203.

Catchpole, C. K. 1973. The functions of advertising song in the sedge warbler (*Acrocephalus schoenobaenus*) and the reed warbler (*A. scirpaceus*). *Behaviour* 46: 300–320.

Catchpole, C. K. 1980. Sexual selection and the evolution of complex songs among European warblers of the genus *Acrocephalus. Behaviour* 74: 149–166.

Catchpole, C. K. 1982. The evolution of bird sounds in relation to mating and spacing behavior. In D. E. Kroodsma and E. H. Miller, eds., *Acoustic Communication in Birds*, 1: 297–319. Academic Press, London.

Catchpole, C. K. 1983. Variation in the song of the great reed warbler (*Acrocephalus arundinaceus*) in relation to mate attraction and territorial defence. *Anim. Behav.* 31: 1217–1225.

Catchpole, C. K. 1986. Song repertoires and reproductive success in the great reed warbler (*Acrocephalus arundinaceus*). *Behav. Ecol. Sociobiol.* 19: 439–445.

Catchpole, C. K. 1987. Bird song, sexual selection and female choice. *TREE* 2: 94–97.

Catchpole, C. K. 1988. Sexual selection and the song of the great reed warbler. *Acta XIX Congr. Int. Ornithol.* 2: 1366–1372.

Catchpole, C. K. 1989. Pseudoreplication and external validity: Playback experiments in avian bioacoustics. *TREE* 4: 286.

Catchpole, C. K., and B. Leisler. 1989. Variation in the song of the aquatic warbler (*Acrocephalus paludicola*) in response to playback of different song structures. *Behaviour* 108: 125–138.

Catchpole, C. K., and P. K. McGregor. 1985. Sexual selection, song complexity and plumage dimorphism in European buntings of the genus *Emberiza. Anim. Behav.* 33: 1378–1379.

Catchpole, C. K., J. Dittami, and B. Leisler. 1984. Differential responses to male song repertoires in female songbirds implanted with oestradiol. *Nature* 312: 563–564.

Catchpole, C. K., B. Leisler, and H. Winkler. 1985. Polygyny in the great reed warbler *Acrocephalus arundinaceus*: A possible case of deception. *Behav. Ecol. Sociobiol.* 16: 285–291.

Catchpole, C. K., B. Leisler, and J. Dittami. 1986. Sexual differences in the responses of captive great reed warblers *Acrocephalus arundinaceus* to variation in song structure and repertoire size. *Ethology* 73: 69–77.

Chandler, M., and G. Cabana. 1991. Sexual dichromatism in North American freshwater fish: Do parasites play a role? *Oikos* 60: 322–328.

Charlesworth, B. 1978. The population genetics of anisogamy. *J. theor. Biol.* 73: 347–357.

Charlesworth, B. 1980. *Evolution in Age-Structured Populations*. Cambridge University Press, Cambridge, U.K.

Charlesworth, B. 1984. The cost of phenotypic evolution. *Paleobiology* 10: 319–327.

Charlesworth, B. 1987. The heritability of fitness. In J. W. Bradbury and M. B. Andersson, eds., *Sexual Selection: Testing the Alternatives*, 21–40. Wiley, Chichester, U.K.

Charlesworth, B. 1988. The evolution of mate choice in a fluctuating environment. *J. theor. Biol.* 130: 191–204.

Charlesworth, B. 1989. The evolution of sex and recombination. *TREE* 4: 264–267.

Charlesworth, B. 1990. Optimization models, quantitative genetics, and mutation. *Evolution* 44: 520–38.

Charlesworth, D. 1984. Androdioecy and the evolution of dioecy. *Biol. J. Linn. Soc.* 23: 333–348.

Charlesworth, D. 1988. Evidence for pollen competition in plants and its relationship to progeny fitness: A comment. *Am. Nat.* 132: 298–302.

Charlesworth, D. 1989. Evolution of low female fertility in plants: Pollen limitation, resource allocation and genetic load. *TREE* 4: 289–292.

Charlesworth, D. 1993. Why are unisexual flowers associated with wind pollination and unspecialized pollinators? *Am. Nat.* 141: 481–490.

Charlesworth, D., and B. Charlesworth. 1981. Allocation of resources to male and female functions in hermaphrodites. *Biol. J. Linn. Soc.* 15: 57–74.

Charlesworth, D., and B. Charlesworth. 1987a. The effect of investment in attractive structures on allocation to male and female functions in plants. *Evolution* 41: 948–968.

Charlesworth, D., and B. Charlesworth. 1987b. Inbreeding depression and its evolutionary consequences. *Ann. Rev. Ecol. Syst.* 18: 237–268.

Charlesworth, D., and M. T. Morgan. 1991. Allocation of resources to sex functions in flowering plants. *Phil. Trans. R. Soc. London B* 332: 91–102.

Charlesworth, D., D. W. Schemske, and V. L. Sork. 1987. The evolution of plant reproductive characters: Sexual versus natural selection. In S. C. Stearns, ed., *The Evolution of Sex and Its Consequences*, 317–336. Birkhäuser, Basel.

Charnov, E. L. 1979a. Natural selection and sex change in Pandalid shrimp: Test of a life history theory. *Am. Nat.* 113: 715–734.

Charnov, E. L. 1979b. Simultaneous hermaphroditism and sexual selection. *Proc. Natl. Acad. Sci. USA* 76: 2480–2484.

Charnov, E. L. 1982. *The Theory of Sex Allocation.* Princeton University Press, Princeton, N.J.

Charnov, E. L. 1984. Behavioural ecology of plants. In J. R. Krebs and N. B. Davies, eds., *Behavioural Ecology: An Evolutionary Approach*, 2d ed., 362–379. Blackwell, Oxford.

Charnov, E. L., and D. Berrigan. 1991. Evolution of life history parameters in animals with indeterminate growth, particularly fish. *Evol. Ecol.* 5: 63–68.

Charnov, E. L., and J. J. Bull. 1986. Sex allocation, pollinator attraction and fruit dispersal in cosexual plants. *J. theor. Biol.* 118: 321–325.

Charnov, E. L., J. Maynard Smith, and J. J. Bull. 1976. Why be an hermaphrodite? *Nature* 263: 125–126.

Cherry, M. I. 1990. Tail length and female choice. *TREE* 5: 349–350.

Cheverud, J. M., M. M. Dow, and W. Leutenegger. 1985. The quantitative assessment of phylogenetic constraints in comparative analyses: Sexual dimorphism in body weight among primates. *Evolution* 39: 1335–1351.

Christy, J. H. 1983. Female choice in the resource-defense mating system of the sand fiddler crab, *Uca pugilator*. *Behav. Ecol. Sociobiol.* 12: 169–180.

Christy, J. H. 1987. Female choice and the breeding behavior of the fiddler crab *Uca beebei*. *J. Crust. Biol.* 7: 624–635.

Christy, J. H. 1988. Pillar function in the fiddler crab *Uca beebei*. II. Competitive courtship signaling. *Ethology* 78: 113–128.

Christy, J. H., and M. Salmon. 1991. Comparative studies of reproductive behavior in mantis shrimps and fiddler crabs. *Am. Zool.* 31: 329–337.

Cinat-Tomson, H. 1926. Die geschlechtliche Zuchtwahl beim Wellensittich (*Melopsittacus undulatus* Shaw.). *Biol. Zentralbl.* 46: 545–552.

Claridge, M. F. 1985. Acoustic signals in the Homoptera. *Ann. Rev. Entomol.* 30: 297–317.

Claridge, M. F. 1990. Acoustic recognition signals: Barriers to hybridization in *Homoptera auchenorrhyncha*. *Can. J. Zool.* 68: 1741–1746.

Clayton, D. H. 1990. Mate choice in experimentally parasitized rock doves: Lousy males lose. *Am. Zool.* 30: 251–262.

Clayton, D. H. 1991. The influence of parasites on host sexual selection. *Paras. Today* 7: 329–334.

Clayton, D. H., S. G. Pruett-Jones, and R. Lande. 1992. Reappraisal of the interspecific prediction of parasite-mediated sexual selection: Opportunity knocks. *J. theor. Biol.* 157: 95–108.

Clayton, N. S. 1990a. Assortative mating in zebra finch subspecies, *Taeniopygia guttata guttata* and *T. g. castanotis*. *Phil. Trans. Roy. Soc. London* 330: 351–370.

Clayton, N. S. 1990b. Mate choice and pair formation in Timor and Australian mainland zebra finches. *Anim. Behav.* 39: 474–480.

Clutton-Brock, T. H. 1982. The functions of antlers. *Behaviour* 70: 108–125.

Clutton-Brock, T. H. 1983. Selection in relation to sex. In D. S. Bendall, ed., *Evolution from Molecules to Men*, 457–481. Cambridge University Press, Cambridge, U.K.

Clutton-Brock, T. H. 1988a. Reproductive success. In T. H. Clutton-Brock, ed., *Reproductive Success*, 472–485. University of Chicago Press, Chicago.

Clutton-Brock, T. H., ed. 1988b. *Reproductive Success*. University of Chicago Press, Chicago.

Clutton-Brock, T. H. 1989. Mammalian mating systems. *Proc. Roy. Soc. London* B 236: 339–372.

Clutton-Brock, T. H. 1991. *The Evolution of Parental Care*. Princeton University Press, Princeton, N.J.

Clutton-Brock, T. H., and S. D. Albon. 1979. The roaring of red deer and the evolution of honest advertisement. *Behaviour* 69: 145–170.

Clutton-Brock, T. H., and C. Godfray. 1991. Parental investment. In J. R. Krebs and N. B. Davies, eds., *Behavioural Ecology: An Evolutionary Approach*, 3d ed., 234–262. Blackwell, Oxford.

Clutton-Brock, T. H., and P. H. Harvey. 1977. Primate ecology and social organization. *J. Zool. Lond.* 183: 1–39.

Clutton-Brock, T. H., and P. H. Harvey. 1983. The functional significance of variation in body size among mammals. In J. F. Eisenberg and D. G. Kleiman, eds., *Advances in the Study of Mammalian Behavior*, 632–663. Spec. Publ. Amer. Soc. Mamm. 7.

Clutton-Brock, T. H., and P. H. Harvey. 1984. Comparative approaches to investigating adaptation. In J. R. Krebs and N. B. Davies, eds., *Behavioural Ecology: An Evolutionary Approach*, 2d ed., 7–29. Blackwell, Oxford.

Clutton-Brock, T. H., and G. A. Parker. 1992. Potential reproductive rates and the operation of sexual selection. *Quart. Rev. Biol.* 67: 437–456.

Clutton-Brock, T. H., and A.C.J. Vincent. 1991. Sexual selection and the potential reproductive rates of males and females. *Nature* 351: 58–60.

Clutton-Brock, T. H., P. H. Harvey, and B. Rudder. 1977. Sexual dimorphism, socionomic sex ratio and body weight in primates. *Nature* 269: 797–800.

Clutton-Brock, T. H., S. D. Albon, R. M. Gibson, and F. E. Guinness. 1979. The logical stag: Adaptive aspects of fighting in red deer (*Cervus elaphus* L.). *Anim. Behav.* 27: 211–225.

Clutton-Brock, T. H., S. D. Albon, and P. H. Harvey. 1980. Antlers, body size and breeding group size in the Cervidae. *Nature* 285: 565–566.

Clutton-Brock, T. H., F. E. Guinness, and S. D. Albon. 1982. *Red Deer: Behavior and Ecology of Two Sexes*. University of Chicago Press, Chicago.

Clutton-Brock, T. H., S. D. Albon, and F. E. Guinness 1985. Parental investment and sex differences in juvenile mortality in birds and mammals. *Nature* 313: 131–133.

Clutton-Brock, T. H., S. D. Albon, and F. E. Guinness. 1988a. Reproductive success in male and female red deer. In T. H. Clutton-Brock, ed., *Reproductive Success*, 325–343. University of Chicago Press, Chicago.

Clutton-Brock, T. H., D. Green, M. Hiraiwa-Hasegawa, and S. D. Albon. 1988b. Passing the buck: Resource defence, lek breeding and mate choice in fallow deer. *Behav. Ecol. Sociobiol.* 23: 281–296.

Clutton-Brock, T. H., M. Hiraiwa-Hasegawa, and A. Robertson. 1989. Mate choice on fallow deer leks. *Nature* 340: 463–465.

Coates, D. 1982. Some observations on the sexuality of humbug damselfish, *Dascyllus aruanus* (Pisces, Pomacentridae) in the field. *Z. Tierpsychol.* 59: 7–18.

Cockburn, A. 1991. *An Introduction to Evolutionary Ecology*. Blackwell, Oxford.

Cockburn, A., M. P. Scott, and D. J. Scotts. 1985. Inbreeding avoidance and male-biased natal dispersal in *Antechinus* spp. (Marsupialia: Dasyuridae). *Anim. Behav.* 33: 908–915.

Cohan, F. M., and A. A. Hoffman. 1989. Uniform selection as a diversifying force in evolution: Evidence from *Drosophila*. *Am. Nat.* 134: 613–637.

Cohen, J. A. 1984. Sexual selection and the psychophysics of female choice. *Z. Tierpsychol.* 64: 1–8.

Colbert, E. H. 1955. *The Evolution of the Vertebrates*. Wiley, New York.

Colgan, P. 1983. *Comparative Social Recognition*. Wiley, New York.

Collias, E. C., N. E. Collias, C. H. Jacobs, F. McAlary, and J. T. Fujimoto. 1979. Experimental evidence for facilitation of pair formation by bright color in weaverbirds. *Condor* 81: 91–93.

Collias, N. E., and E. C. Collias. 1984. *Nest Building and Bird Behavior*. Harvard University Press, Cambridge, Mass.

Collias, N. E., and J. K. Victoria. 1978. Nest and mate selection in the village weaverbird *Ploceus cucullatus*. *Anim. Behav.* 26: 470–479.

Conner, J. 1988. Field measurements of natural and sexual selection in the fungus beetle, *Bolitotherus cornutus*. *Evolution* 42: 735–749.

Conner, J. 1989a. Older males have higher insemination success in a beetle. *Anim. Behav.* 38: 503–509.

Conner, J. 1989b. Density-dependent sexual selection in the fungus beetle *Bolitotherus cornutus*. *Evolution* 43: 1378–1386.

Conner, W. E., T. Eisner, R. K. Vander Meer, A. Guerrero, and J. Meinwald. 1981. Precopulatory sexual interaction in an arctiid moth (*Utetheisa ornatrix*): Role of a pheromone derived from dietary alkaloids. *Behav. Ecol. Sociobiol.* 9: 227–235.

Conner, W. E., B. Roach, E. Benedict, J. Meinwald, and T. Eisner. 1990. Courtship pheromone production and body size as correlates of larval diet in males of the arctiid moth, *Utetheisa ornatrix*. *J. Chem. Ecol.* 16: 543–552.

Conrad, K. F., and G. Pritchard. 1989. Female dimorpism and physiological colour change in the damselfly *Argia vivida* Hagen (Odonata: Coenagrionidae). *Can. J. Zool.* 67: 298–304.

Conrad, K. F., and G. Pritchard. 1992. An ecological classification of odonate mating systems: The relative influence of natural, inter- and intra-sexual selection on males. *Biol. J. Linn. Soc.* 45: 255–269.

Constantz, G. D. 1985. Alloparental care in the tessellated darter, *Etheostoma olmstedi* (Pisces: Percidae). *Env. Biol. Fish.* 14: 175–183.

Convey, P. 1989. Influences on the choice between territorial and satellite behaviour in male *Libellula quadrimaculata* Linn. (Odonata: Libellulidae). *Behaviour* 109: 125–141.

Cooper, W. E., Jr. 1984. Female secondary sexual coloration and sex recognition in the keeled earless lizard, *Holbrookia propinqua*. *Anim. Behav.* 32: 1142–1150.

Cooper, W. E., Jr., and N. Burns. 1987. Social significance of ventrolateral coloration in the fence lizard, *Sceloporus undulatus*. *Anim. Behav.* 35: 526–532.

Cooper, W. E., Jr., and D. Crews. 1987. Hormonal induction of secondary sexual coloration and rejection behaviour in female keeled earless lizards, *Holbrookia propinqua*. *Anim. Behav.* 35: 1177–1187.

Cooper, W. E., Jr., and L. J. Vitt. 1986. Lizard pheromones: Behavioural responses and adaptive significance in skinks of the genus *Eumeces*. In D. Duvall, D. Müller-Schwarze and R. M. Silverstein, eds., *Chemical Signals in Vertebrates*, 323–340. Plenum, New York.

Cooper, W. E., Jr., and L. J. Vitt. 1988. Orange head coloration of the male broad-headed skink *Eumeces laticeps*, a sexually selected social cue. *Copeia* 1988: 1–6.

Cooper, W. T., and J. M. Forshaw. 1977. *The Birds of Paradise and Bowerbirds*. Collins, Sydney.

Cosmides, L. M., and J. Tooby. 1981. Cytoplasmic inheritance and intragenomic conflict. *J. theor. Biol.* 89: 83–129.

Côte, I. M., and W. Hunte. 1989. Male and female mate choice in the redlip blenny: Why bigger is better. *Anim. Behav.* 38: 78–88.

Cott, H. B. 1957. *Adaptive Coloration in Animals*. 2d ed. Methuen, London.

Cott, H. B., and C. W. Benson. 1970. The palatability of birds, mainly based on observations of a tasting panel in Zambia. *Ostrich* Suppl. 8: 357–384.

Coulson, J. C., and C. S. Thomas. 1983. Mate choice in the kittiwake gull. In

P. Bateson, ed., *Mate Choice*, 361–376. Cambridge University Press, Cambridge, U.K.

Cowan, D. P. 1981. Parental investment in two solitary wasps *Ancistrocerus adiabatus* and *Euodynerus foraminatus*. *Behav. Ecol. Sociobiol.* 9: 95–102.

Cox, C. R. 1981. Agonistic encounters among male elephant seals: Frequency, context, and the role of female preference. *Am. Zool.* 21: 197–209.

Cox, C. R., and B. J. Le Boeuf. 1977. Female incitation of male competition: A mechanism in sexual selection. *Am. Nat.* 111: 317–335.

Cox, P. A. 1983. Search theory, random motion, and the convergent evolution of pollen and spore morphologies in aquatic plants. *Am. Nat.* 121: 9–31.

Cox, P. A. 1988a. Hydrophilus pollination. *Ann. Rev. Ecol. System.* 19: 261–280.

Cox, P. A. 1988b. Monomorphic and dimorphic sexual strategies: A modular approach. In J. Lovett Doust and L. Lovett Doust, eds., *Plant Reproductive Ecology*, 80–97. Oxford University Press, New York.

Cox, P. A., and R. B. Knox. 1988. Pollination postulates and two-dimensional pollination in hydrophilous monocotyledons. *Ann. Miss. Bot. Gard.* 75: 811–818.

Coyne, J. A. 1983. Genetic basis of differences in genital morphology among three sibling species of *Drosophila*. *Evolution* 37: 1101–1118.

Coyne, J. A. 1989. Genetics of sexual isolation between two sibling species, *Drosophila simulans* and *Drosophila mauritiana*. *Proc. Nat. Acad. Sci. USA* 86: 5464–5468.

Coyne, J. A., and H. A. Orr. 1989. Patterns of speciation in *Drosophila*. *Evolution* 43: 362–381.

Craig, A.J.F.K. 1980. Behaviour and evolution in the genus *Euplectes*. *J. Ornithol.* 121: 144–161.

Craig, A.J.F.K. 1989. Tail length and sexual selection in the polygynous longtailed widow (*Euplectes progne*): A cautionary tale. *S. Afr. J. Sci.* 85: 523–524.

Cramp, S., et al., eds. 1977, 1980, 1983, 1988. *Handbook of the Birds of Europe, the Middle East and North Africa*. Vols. 1, 2, 3, 5. Oxford University Press, Oxford.

Crane, J. 1975. *Fiddler Crabs of the World*. Princeton University Press, Princeton, N.J.

Crankshaw, O. S. 1979. Female choice in relation to calling and courtship songs in *Acheta domesticus*. *Anim. Behav.* 27: 1274–1275.

Crawford, J. D. 1986. Acoustic communication in an electric fish, *Pollimyrys isidori* (Mormyridae). *J. Comp. Physiol.* A 159: 297–310.

Crepet, W. L. 1984. Advanced (constant) insect pollination mechanisms: pattern of evolution and implications vis-à-vis angiosperm diversity. *Ann. Missouri Bot. Gard.* 71: 607–630.

Crespi, B. J. 1986a. Size assessment and alternative fighting tactics in *Elaphrothrips tuberculatus* (Insecta: Thysanoptera). *Anim. Behav.* 34: 1324–1335.

Crespi, B. J. 1986b. Territoriality and fighting in a colonial thrips, *Hoplothrips pedicularius*, and sexual dimorphism in Thysanoptera. *Ecol. Ent.* 11: 119–130.

Crespi, B. J. 1988. Risks and benefits of lethal male fighting in the colonial, polygynous thrips *Hoplothrips karnyi* (Insecta: Thysanoptera). *Behav. Ecol. Sociobiol.* 22: 293–301.

Crespi, B. J. 1989. Causes of assortative mating in arthropods. *Anim. Behav.* 38: 980–1000.

Crews, D. 1975. Effects of different components of male courtship behaviour on environmentally induced ovarian recrudescence and mating preferences in the lizard, *Anolis carolinensis*. *Anim. Behav.* 23: 349–356.

Crocker, G., and T. Day. 1987. An advantage to mate choice in the seaweed fly (*Coelopa frigida*). *Behav. Ecol. Sociobiol.* 20: 295–301.

Cronin, E. W., and P. W. Sherman. 1977. A resource-based mating system: The orange-rumped honeyguide. *Living Bird* 15: 5–32.

Cronin, H. 1991. *The Ant and the Peacock*. Cambridge University Press, Cambridge, U.K.

Cronly-Dillon, J., and S. C. Sharma. 1968. Effect of season and sex on the photopic spectral sensitivity of the three-spined stickleback. *J. Exp. Biol.* 49: 679–687.

Crook, J. H. 1965. The adaptive significance of avian social organization. *Symp. Zool. Soc. London* 14: 181–218.

Crook, J. H. 1972. Sexual selection, dimorphism, and social organization in the primates. In B. Campbell, ed., *Sexual Selection and the Descent of Man, 1871–1971*, 231–281. Heinemann, London.

Crossley, S. A. 1974. Changes in mating behavior produced by selection for ethological isolation between ebony and vestigial mutants of *Drosophila melanogaster*. *Evolution* 28: 631–647.

Crow, J. F., and M. J. Simmons. 1983. The mutation load in *Drosophila*. In M. Ashburner, H. L. Carson. and J. N. Thompson, eds., *The Genetics and Biology of Drosophila*, vol. 3c, 2–35. Academic Press, New York.

Crozier, R. H. 1987. Genetic aspects of kin recognition: Concepts, models, and synthesis. In D.J.C. Fletcher and C. D. Michener, eds., *Kin Recognition in Animals*, 55–74. Wiley, Chichester, U.K.

Cruden, R. W. 1977. Pollen-ovule ratios: A conservative indicator of breeding systems in flowering plants. *Evolution* 31: 32–46.

Cruden, R. W., and D. Lyon. 1985. Patterns of biomass allocation to male and female function in plants with different mating systems. *Oecologia* 66: 299–306.

Curtsinger, J. W., and I. L. Heisler. 1988. A diploid "sexy son" model. *Am. Nat.* 132: 439–453.

Curtsinger, J. W., and I. L. Heisler. 1989. On the consistency of sexy-son models: A reply to Kirkpatrick. *Am. Nat.* 134: 978–981.

Cuthill, I., and A. Hindmarsh. 1985. Increase in starling song activity with removal of mate. *Anim. Behav.* 33: 326–328.

Dahlgren, J. 1990. Females choose vigilant males: An experiment with the monogamous grey partridge, *Perdix perdix*. *Anim. Behav.* 39: 646–651.

Dale, S., and T. Slagsvold. 1990. Random settlement of female pied flycatchers, *Ficedula hypoleuca*: Significance of male territory size. *Anim. Behav.* 39: 231–243.

Dale, S., T. Amundsen, J. T. Lifjeld, and T. Slagsvold. 1990. Mate sampling behaviour of female pied flycatchers: Evidence for active mate choice. *Behav. Ecol. Sociobiol.* 27: 87–91.

Daly, M., and M. Wilson. 1983. *Sex, Evolution, and Behavior*. 2d ed. Wadsworth, Belmont, Mass.

Danforth, C. H. 1950. Evolution and plumage traits in pheasant hybrids, *Phasianus* x *Chrysolophus*. *Evolution* 4: 301–315.

Darling, F. F. 1937. *A Herd of Red Deer*. Oxford University Press, London.

Darwin, C. 1859. *On the Origin of Species by Means of Natural Selection*. Murray, London.

Darwin, C. 1871. *The Descent of Man, and Selection in Relation to Sex*. Murray, London.

Darwin, C. 1874. *The Descent of Man, and Selection in Relation to Sex*. 2d ed. Murray, London.

Darwin, C. 1876. Sexual selection in relation to monkeys. *Nature* 15: 18–19.

Darwin, C. 1877a. *The Various Contrivances by which Orchids are Fertilized by Insects*. 2d ed. Murray, London.

Darwin, C. 1877b. *The Different Forms of Flowers on Plants of the Same Species*. Murray, London.

Darwin, C. 1882. Preliminary Notice. *Proc. Zool. Soc. London* 25: 367–369.

Darwin, E. 1794. *Zoonomia*. J. Johnson, London.

Davidson, D. W. 1982. Sexual selection in Harvester ants (Hymenoptera: Formicidae: *Pogonomyrmex*). *Behav. Ecol. Sociobiol.* 10: 245–250.

Davies, N. B. 1978a. Ecological questions about territorial behaviour. In J. R. Krebs and N. B. Davies, eds., *Behavioural Ecology, an Evolutionary Approach*, 317–351. Blackwell, London.

Davies, N. B. 1978b. Territorial defence in the speckled wood butterfly (*Pararge aegeria*): The resident always wins. *Anim. Behav.* 26: 138–147.

Davies, N. B. 1983. Polyandry, cloaca-pecking and sperm competition in dunnocks. *Nature* 302: 334–336.

Davies, N. B. 1985. Cooperation and conflict among dunnocks, *Prunella modularis*, in a variable mating system. *Anim. Behav.* 33: 628–648.

Davies, N. B. 1989. Sexual conflict and the polygamy threshold. *Anim. Behav.* 38: 226–234.

Davies, N. B. 1991. Mating systems. In J. R. Krebs and N. B. Davies, eds., *Behavioural Ecology: An Evolutionary Approach*, 3d ed., 263–294. Blackwell, Oxford.

Davies, N. B. 1992. *Dunnock Behaviour and Social Evolution*. Oxford University Press, Oxford.

Davies, N. B., and T. R. Halliday. 1977. Optimal mate selection in the toad *Bufo bufo*. *Nature* 269: 56–58.

Davies, N. B., and T. R. Halliday. 1978. Deep croaks and fighting assessment in toads (*Bufo bufo*). *Nature* 274: 683–685.

Davies, N. B., and T. R. Halliday. 1979. Competitive mate searching in common toads, *Bufo bufo*. *Anim. Behav.* 27: 1253–1267.

Davies, N. B., and A. I. Houston. 1986. Reproductive success of dunnocks, *Prunella modularis*, in a variable mating system. II. Conflicts of interest among breeding adults. *J. Anim. Ecol.* 55: 139–154.

Davies, N. B., and A. Lundberg. 1984. Food distribution and a variable mating system in the dunnock (*Prunella modularis*). *J. Anim. Ecol.* 53: 895–912.

Davis, J.W.F., and P. O'Donald. 1975. Sexual selection for a handicap: A critical analysis of Zahavi's model. *J. theor. Biol.* 57: 345–354.

Davis, M. A. 1981. The effect of pollinators, predators, and energy constraints on the floral ecology and evolution of *Trillium erectum*. *Oecologia* 48: 400–406.

Davison, G.W.H. 1981. Sexual selection and the mating system of *Argusianus argus* (Aves: Phasianidae). *Biol. J. Linn. Soc.* 15: 91–104.

Davison, G.W.H. 1983. The eyes have it: Ocelli in a rainforest pheasant. *Anim. Behav.* 31: 1037–1042.

Davison, G.W.H. 1985. Avian spurs. *J. Zool. London* 206: 353–366.

Davitashvili, L. 1961. *The Theory of Sexual Selection*. Acad. Sci. USSR, Moscow. (In Russian.)

Dawkins, M. S., and T. Guilford. 1991. The corruption of honest signalling. *Anim. Behav.* 41: 865–873.

Dawkins, R. 1980. Good strategy or evolutionarily stable strategy? In G. W. Barlow and S. Silverberg, eds., *Sociobiology: Beyond Nature/Nurture?*, 331–367. Westview, Boulder, Colo.

Dawkins, R., and J. R. Krebs. 1978. Animal signals: Information or manipulation? In J. R. Krebs and N. B. Davies, eds., *Behavioural Ecology: An Evolutionary Approach*, 282–309. Blackwell, Oxford.

Dawkins, R., and J. R. Krebs. 1979. Arms races between and within species. *Proc. Roy. Soc., London* B 205: 489–511.

Day, T. H., and R. K. Butlin. 1987. Non-random mating in natural populations of the seaweed fly, *Coelopa frigida. Heredity* 58: 213–220.

Day, T. H., S. Miles, M. D. Pilkington, and R. K. Butlin. 1987. Differential mating success in populations of seaweed flies (*Coelopa frigida*). *Heredity* 58: 203–212.

de Kiriline, L. 1954. The voluble singer of the treetops. *Audubon* 56: 109–111.

Delacour, J. 1951. *The Pheasants of the World*. Country Life, London.

de la Motte, I., and D. Burkhardt. 1983. Portrait of an Asian stalk-eyed fly. *Naturwissenschaften* 70: 451–461.

Dellefors, C., and U. Faremo. 1988. Early sexual maturation in males of wild sea trout, *Salmo trutta* L., inhibits smoltification. *J. Fish. Biol.* 33: 741–749.

DeMartini, E. E. 1987. Paternal defence, cannibalism and polygamy: Factors influencing the reproductive success of painted greeling (Pisces, Hexagrammidae). *Anim. Behav.* 35: 1145–1158.

Deutsch, C. J., M. P. Haley, and B. J. Le Boeuf. 1990. Reproductive effort of male northern elephant seals: Estimates from mass loss. *Can. J. Zool.* 68: 2580–2593.

Devlin, B., and A. G. Stephenson. 1985. Sex differential floral longevity, nectar secretion, and pollinator foraging in a protandrous species. *Am. J. Bot.* 72: 303–310.

DeVore, I., and K.R.L. Hall. 1965. Baboon ecology. In I. DeVore, ed., *Primate Behavior*, 20–52. Holt, New York.

de Vos, G. J. 1979. Adaptedness of arena behaviour in black grouse *Tetrao tetrix* and other grouse species (Tetraoninae). *Behaviour* 68: 277–314.

de Vos, G. J. 1983. Social behaviour of black grouse: An observational and experimental field study. *Ardea* 71: 1–103.

Dewar, D., and F. Finn. 1909. *The Making of Species*. Bodley Head, London.

DeWinter, A. J., and T. Rollenhagen. 1990. The importance of male and female acoustic behaviour for reproductive isolation in *Ribautodelphax* planthoppers (Homoptera: Delphacidae). *Biol. J. Linn. Soc.* 40: 191–206.

Dewsbury, D. A. 1982a. Dominance rank, copulatory behavior, and differential reproduction. *Q. Rev. Biol.* 57: 135–159.

Dewsbury, D. A. 1982b. Ejaculate cost and male choice. *Am. Nat.* 119: 601–610.

Diamond, J. 1986a. Animal art: Variation in bower decorating style among male bowerbirds *Amblyornis inornatus*. *Proc. Natl. Acad. Sci. USA* 83: 3042–3046.

Diamond, J. 1986b. Biology of birds of paradise and bowerbirds. *Ann. Rev. Ecol. Syst.* 17: 17–37.

Diamond, J. 1987. Bower building and decoration by the bowerbird *Amblyornis inornatus*. *Ethol.* 74: 177–204.

Diamond, J. 1988. Experimental study of bower decoration by the bowerbird *Amblyornis inornatus*, using colored poker chips. *Am. Nat.* 131: 631–653.

Diamond, J. M. 1991a. Borrowed sexual ornaments. *Nature* 349: 105.

Diamond, J. M. 1991b. *The Rise and Fall of the Third Chimpanzee*. Radius, London.

Diamond, J. M. 1992. Rubbish birds are poisonous. *Nature* 360: 19–20.

Dick, J.T.A., and R. W. Elwood. 1990. Symmetrical assessment of female quality by male *Gammarus pulex* (Amphipoda) during struggles over precopula females. *Anim. Behav.* 40: 877–883.

Dobzhansky, T. 1937. *Genetics and the Origin of Species*. Columbia University Press, New York.

Dobzhansky, T. 1940. Speciation as a stage in evolutionary divergence. *Am. Nat.* 74: 312–332.

Dodson, C. H. 1962. Pollination and variation in the subtribe Catasetinae (Orchidaceae). *Ann. Missouri Bot. Gard.* 49: 35–56.

Doflein, F. 1914. *Das Tier als Glied des Naturganzen*. Teubner, Leipzig.

Doherty, J. A., and H. C. Gerhardt. 1984. Evolutionary and neurobiological implications of selective phonotaxis in the spring peeper (*Hyla crucifer*). *Anim. Behav.* 32: 875–881.

Doherty, J., and R. Hoy. 1985. Communication in insects. III. The auditory behavior of crickets: Some views of genetic coupling, song recognition, and predator detection. *Q. Rev. Biol.* 60: 457–472.

Dominey, W. J. 1980. Female mimicry in male bluegill sunfish—a genetic polymorphism? *Nature* 284: 546–548.

Dominey, W. J. 1983. Sexual selection, additive genetic variance and the "phenotypic handicap." *J. theor. Biol.* 101: 495–502.

Dominey, W. J. 1984a. Effects of sexual selection and life history on speciation: Species flocks in African cichlids and Hawaiian *Drosophila*. In A. A. Echelle and I. Kornfield, eds., *Evolution of Fish Species Flocks*, 231–249. University of Maine Press, Orono.

Dominey, W. J. 1984b. Alternative mating tactics and evolutionarily stable strategies. *Am. Zool.* 24: 385–396.

Donoghue, M. J. 1989. Phylogenies and the analysis of evolutionary sequences, with examples from seed plants. *Evolution* 43: 1137–1156.

Doty, R. L., ed. 1976. *Mammalian Olfaction, Reproductive Processes, and Behavior*. Academic Press, New York.

Dowling, T. E., and B. D. De Marais. 1993. Evolutionary significance of introgressive hybridization in cyprinid fishes. *Nature* 362: 444–446.

Downhower, J. F. 1976. Darwin's finches and the evolution of sexual dimorphism in body size. *Nature* 263: 558–563.

Downhower, J. F., and L. Brown. 1980. Mate preferences of female mottled sculpins, *Cottus bairdi*. *Anim. Behav.* 28: 728–734.

Downhower, J. F., and L. Brown. 1981. The timing of reproduction and its behavioral consequences for mottled sculpins, *Cottus bairdi*. In R. D. Alexander and D. W. Tinkle, eds., *Natural Selection and Social Behavior*, 78–95. Chiron, New York.

Downhower, J. F., L. Brown, R. Pederson, and G. Staples. 1983. Sexual selection and sexual dimorphism in mottled sculpins. *Evolution* 37: 96–103.

Downhower, J. F., L. S. Blumer, and L. Brown. 1987. Opportunity for selection: An appropriate measure for evaluating variation in the potential for selection? *Evolution* 41: 1395–1400.

Dowsett-Lemaire, F. 1979. The imitative range of the song of the marsh warbler *Acrocephalus palustris*, with special reference to imitations of African birds. *Ibis* 121: 453–468.

Draper, N., and H. Smith. 1981. *Applied Regression Analysis*. 2d ed. Wiley, New York.

Drent, R. H., and S. Daan. 1981. The prudent parent: Energetic adjustment in avian breeding. *Ardea* 68: 225–252.

Dressler, R. L. 1981. *The Orchids*. Harvard University Press, Cambridge, Mass.

Duellman, W. E., and A. H. Savitzky. 1976. Aggressive behavior in a centrolenid frog, with comments on territoriality in anurans. *Herpetologica* 32: 401–404.

Duellman, W. E., and L. Trueb. 1986. *Biology of Amphibians*. McGraw-Hill, New York.

Dugan, B. 1982. The mating behavior of the green iguana, *Iguana iguana*. In G. M. Burghardt and A. S. Rand, eds., *Iguanas of the World: Their Behavior, Ecology, and Conservation*, 320–341. Noyes, Park Ridge, N.J.

Dugatkin, L. A. 1992. Sexual selection and imitation: Females copy the mate choice of others. *Am. Nat.* 139: 1384–1389.

Dugatkin, L. A., and J.-G. J. Godin. 1992. Reversal of female mate choice by copying in the guppy (*Poecilia reticulata*). *Proc. Roy. Soc. London* B 249: 179–184.

Dumbacher, J. P., B. M. Beehler, T. F. Spande, H. M. Garraffo, and J. W. Daly. 1992. Homobatrachotoxin in the genus *Pitohui*: Chemical defense in birds? *Science* 258: 799–801.

Dunbar, R.I.M. 1988. *Primate Social Systems*. Croom Helm, London.

Duvall, D., R. Herskowitz, and J. Trupiano-Duvall. 1980. Responses of five-lined skinks (*Eumeces fasciatus*) and ground skinks (*Scincella lateralis*) to conspecific and interspecific chemical cues. *J. Herpetol.* 14: 121–127.

Duvall, D., D. Müller-Schwarze, and R. M. Silverstein, eds. 1986. *Chemical Signals in Vertebrates*. Vol. 4. Plenum, New York.

Duvall, D., S. J. Arnold, and G. W. Schuett. 1992. Pitviper mating systems: Ecological potential, sexual selection, and microevolution. In J. A. Campbell and E. D. Brodie, eds., *Biology of the Pitvipers*, 321–336. Selva, Tyler, Texas.

Dyson, M. L., and N. I. Passmore. 1988a. Two-choice phonotaxis in *Hyperolius marmoratus* (Anura: Hyperoliidae): The effect of temporal variation in presented stimuli. *Anim. Behav.* 36: 648–652.

Dyson, M. L., and N. I. Passmore. 1988b. The combined effect of intensity and the temporal relationship of stimuli on phonotaxis in female painted reed frogs *Hyperolius marmoratus*. *Anim. Behav.* 36: 1555–1556.

Earhart, C. M., and N. K. Johnson. 1970. Size dimorphism and food habits of North American owls. *Condor* 72: 251–264.

East, M. L. 1982. Time-budgeting by European robins *Erithacus rubecula*: Inter- and intra-sexual comparisons during autumn, winter and early spring. *Ornis Scand.* 13: 85–93.

East, M. L., and H. Hofer. 1991. Loud calling in a female-dominated mammalian society. II. Behavioural contexts and functions of whooping of spotted hyaenas, *Crocuta crocuta*. *Anim. Behav.* 42: 651–669.

Eberhard, W. G. 1977. Aggressive chemical mimicry by a bolas spider. *Science* 198: 1173–1175.

Eberhard, W. G. 1979. The function of horns in *Podischnus agenor* (Dynastinae) and other beetles. In M. S. Blum and N. A. Blum, eds., *Sexual Selection and Reproductive Competition in Insects*, 231–258. Academic Press, New York.

Eberhard, W. G. 1980a. The natural history and behavior of the bolas spider *Mastophora dizzydeani* sp. n. (*Araneidae*). *Psyche* 87: 143–169.

Eberhard, W. G. 1980b. Horned beetles. *Sci. Amer.* 242: 166–182.

Eberhard, W. G. 1982. Beetle horn dimorphism: Making the best of a bad lot. *Am. Nat.* 119: 420–426.

Eberhard, W. G. 1985. *Sexual Selection and Animal Genitalia*. Harvard University Press, Cambridge, Mass.

Eberhard, W. G. 1991. Copulatory courtship and cryptic female choice in insects. *Biol. Rev.* 66: 1–31.

Eberhard, W. G. 1993. Evaluating models of sexual selection: Genitalia as a test case. *Am. Nat.* 142: 564–571.

Eberhard, W. G., and E. E. Gutiérrez. 1991. Male dimorphisms in beetles and earwigs and the question of developmental constraints. *Evolution* 45: 18–28.

Eckert, C. G., and P. J. Weatherhead. 1987a. Owners, floaters and competitive asymmetries among territorial red-winged blackbirds. *Anim. Behav.* 35: 1317–1323.

Eckert, C. G., and P. J. Weatherhead. 1987b. Ideal dominance distributions: A test using red-winged blackbirds (*Agelaius phoeniceus*). *Behav. Ecol. Sociobiol.* 20: 43–52.

Eckert, C. G., and P. J. Weatherhead. 1987c. Male characteristics, parental quality and the study of mate choice in the red-winged blackbird (*Agelaius phoeniceus*). *Behav. Ecol. Sociobiol.* 20: 35–42.

Eckert, C. G., and P. J. Weatherhead. 1987d. Competition for territories in red-winged blackbirds: Is resource-holding potential realized? *Behav. Ecol. Sociobiol.* 20: 369–376.

Edmunds, M. 1974. *Defence in Animals: A Survey of Anti-Predator Defenses*. Longmans, London.

Eff, D. 1962. A little about the little-known *Papilio indra minori*. *J. Lepid. Soc.* 16: 137–142.

Egid, K., and J. L. Brown. 1989. The major histocompatibility complex and female mating preferences in mice. *Anim. Behav.* 38: 548–550.

Ehrlich, P. R., F. H. Talbot, B. C. Russell, and G.R.V. Anderson. 1977. The behaviour of chaetodontid fishes with special reference to Lorenz's "poster colouration" hypothesis. *J. Zool. London* 183: 213–228.

Eisen, E. J., and J. P. Hanrahan. 1972. Selection for sexual dimorphism in body weight of mice. *Aust. J. Biol. Sci.* 25: 1015–1024.

Elgar, M. A. 1991. Sexual cannibalism, size dimorphism and courtship behavior in orb-weaving spiders (Aranae). *Evolution* 45: 444–448.

Elgar, M. A. 1992. Sexual cannibalism in spiders and other invertebrates. In M. A. Elgar and B. J. Crespi, eds., *Cannibalism: Ecology and Evolution among Diverse Taxa*, 128–155. Oxford University Press, Oxford.

Elgar, M. A., and D. R. Nash 1988. Sexual cannibalism in the garden spider *Araneus diadematus*. *Anim. Behav.* 36: 1511–1517.

Elgar, M. A., N. Ghaffar, and A. F. Read. 1990. Sexual dimorphism in leg length among orb-weaving spiders: a possible role for sexual cannibalism. *J. Zool.* 222: 455–470.

Ellstrand, N. C. 1984. Multiple paternity within the fruits of the wild radish, *Raphanus sativus*. *Am. Nat.* 123: 819–828.

Ellstrand, N. C., and D. L. Marshall. 1986. Patterns of multiple paternity in populations of *Raphanus sativus*. *Evolution* 40: 837–842.

Elwood, R., J. Gibson, and S. Neil. 1987. The amorous *Gammarus*: Size assortative mating in *G. pulex*. *Anim. Behav.* 35: 1–6.

Emlen, J. M. 1973. *Ecology: An Evolutionary Approach*. Addison-Wesley, Reading, Mass.

Emlen, J. T., Jr. 1957. Display and mate selection in the whydahs and bishop birds. *Ostrich* 28: 202–213.

Emlen, S. T. 1968. Territoriality in the bullfrog, *Rana catesbeiana*. *Copeia* 2: 240–243.

Emlen, S. T. 1976. Lek organization and mating strategies in the bullfrog. *Behav. Ecol. Sociobiol.* 1: 283–313.

Emlen, S. T., and L. W. Oring. 1977. Ecology, sexual selection, and the evolution of mating systems. *Science* 197: 215–223.

Emlen, S. T., N. J. Demong, and D. J. Emlen. 1989. Experimental induction of infanticide in female wattled jacanas. *Auk* 106: 1–7.

Emms, S. K. 1993. On measuring fitness gain curves in plants. *Ecology* 74: 1750–1756.

Endler, J. A. 1977. *Geographic Variation, Speciation, and Clines*. Princeton University Press, Princeton, N.J.

Endler, J. 1978. A predator's view of animal colour patterns. *Evol. Biol.* 11: 319–364.

Endler, J. A. 1980. Natural selection on color patterns in *Poecilia reticulata*. *Evolution* 34: 76–91.

Endler, J. A. 1983. Natural and sexual selection on color patterns in poeciliid fishes. *Envir. Biol. Fishes* 9: 173–190.

Endler, J. A. 1986a. *Natural Selection in the Wild*. Princeton University Press, Princeton, N.J.

Endler, J. A. 1986b. Defense against predators. In M. E. Feder and G. V. Lauder, eds., *Predator—Prey Relationships*, 109–134. University of Chicago Press, Chicago.

Endler, J. A. 1987. Predation, light intensity and courtship behaviour in *Poecilia reticulata* (Pisces: Poeciliidae). *Anim. Behav.* 35: 1376–1385.

Endler, J. A. 1988a. Frequency-dependent predation, crypsis and aposematic coloration. *Phil. Trans. Roy. Soc. London* B 319: 505–523.

Endler, J. A. 1988b. Sexual selection and predation risk in guppies. *Nature* 332: 593–594.

Endler, J. A. 1989. Conceptual and other problems in speciation. In D. Otte and J. A. Endler, eds., *Speciation and Its Consequences*, 626–648. Sinauer, Sunderland, Mass.

Endler, J. A. 1990. On the measurement and classification of colour in studies of animal colour patterns. *Biol. J. Linn. Soc.* 41: 315–352.

Endler, J. A. 1991. Variation in the appearance of guppy color patterns to guppies and their predators under different visual conditions. *Vision Res.* 31: 587–608.

Endler, J. A. 1992a. Introduction to the symposium. *Am. Nat.* 139: s1–s3.

Endler, J. A. 1992b. Signals, signal conditions, and the direction of evolution. *Am. Nat.* 139: s125-s153.

Endler, J. A., and A. M. Lyles. 1989. Bright ideas about parasites. *TREE* 4: 246–248.

Endler, J. A., and T. McLellan. 1988. The processes of evolution: Towards a newer synthesis. *Ann. Rev. Ecol. Syst.* 19: 395–421.

Engelhard, G., S. P. Foster, and T. H. Day. 1989. Genetic differences in mating success and female choice in seaweed flies (*Coelopa frigida*). *Heredity* 62: 123–131.

Engen, S., and B.-E. Saether. 1985. The evolutionary significance of sexual selection. *J. theor. Biol.* 117: 277–289.

Enquist, M. 1985. Communication during aggressive interactions with particular reference to variation in choice of behaviour. *Anim. Behav.* 33: 1152–1161.

Enquist, M., and A. Arak. 1993. Selection of exaggerated male traits by female aesthetic senses. *Nature* 361: 446–448.

Enquist, M., and O. Leimar. 1983. Evolution of fighting behaviour: Decision rules and assessment of relative strength. *J. theor. Biol.* 102: 387–410.

Enström, D. A. 1992. Breeding season communication hypotheses for delayed plumage maturation in passerines: Tests in the orchard oriole (*Icterus spurius*). *Anim. Behav.* 43: 463–472.

Erckmann, W. J. 1981. The Evolution of Sex Role Reversal and Monogamy in Shorebirds. Ph.D. diss., University of Washington, Seattle.

Erckmann, W. J. 1983. The evolution of polyandry in shorebirds: An evaluation of hypotheses. In S. K. Wasser, ed., *Social Behavior of Female Vertebrates*, 114–168. Academic Press, New York.

Erickson, C. J. 1970. Induction of ovarian activity in female ring doves by androgen treatment of castrated males. *J. Comp. Phys. Psych.* 71: 210–215.

Eriksson, D. 1991. The Significance of Song for Species Recognition and Mate Choice in the Pied Flycatcher *Ficedula hypoleuca*. Ph.D. diss., University of Uppsala, Sweden.

Eriksson, D., and L. Wallin. 1986. Male bird song attracts females—a field experiment. *Behav. Ecol. Sociobiol.* 19: 297–299.

Eriksson, O., and B. Bremer 1992. Pollination systems, dispersal modes, life forms, and diversification rates in angiosperm families. *Evolution* 46: 258–266.

Erlinge, S. 1977. Agonistic behaviour and dominance in stoats (*Mustela erminea*, L.). *Z. Tierpsychol.* 44: 375–388.

Erlinge, S. 1979. Adaptive significance of sexual dimorphism in weasels. *Oikos* 33: 233–245.

Erlinge, S., M. Sandell, and C. Brinck. 1982. Scent-marking and its territorial significance in stoats, *Mustela erminea*. *Anim. Behav.* 30: 811–818.

Eshel, I. 1978. On the handicap principle—a critical defence. *J. theor. Biol.* 70: 245–250.

Eshel, I., and W. D. Hamilton. 1984. Parent-offspring correlation in fitness under fluctuating selection. *Proc. Roy. Soc. London* B 222: 1–14.

Espmark, Y. 1964. Studies in dominance-subordination relationship in a group of semi-domestic reindeer (*Rangifer tarandus L.*). *Anim. Behav.* 12: 420–426.

Evans, M. R. 1991. The size of adornments of male scarlet-tufted malachite sunbirds varies with environmental conditions, as predicted by handicap theories. *Anim. Behav.* 42: 797–803.

Evans, M. R., and B. J. Hatchwell. 1992a. An experimental study of male adornment in the scarlet-tufted malachite sunbird. I. The role of pectoral tufts in territorial defence. *Behav. Ecol. Sociobiol.* 29: 413–420.

Evans, M. R., and B. J. Hatchwell. 1992b. An experimental study of male adornment in the scarlet-tufted malachite sunbird. II. The role of the elongated tail in mate choice and experimental evidence for a handicap. *Behav. Ecol. Sociobiol.* 29: 421–428.

Evans, M. R., and B. J. Hatchwell. 1993. New slants on ornament asymmetry. *Proc. Roy. Soc. London* B 251: 171–177.

Evans, M. R., and A.L.R. Thomas. 1992. The aerodynamic and mechanical effects of elongated tails in the scarlet-tufted malachite sunbird: Measuring the cost of a handicap. *Anim. Behav.* 43: 337–347.

Ewald, P. W., and S. Rohwer. 1982. Effects of supplemental feeding on time of breeding, clutch-size and polygyny in red-winged blackbirds *Agelaius phoeniceus*. *J. Anim. Ecol.* 51: 429–450.

Ewing, A. W. 1989. *Arthropod Bioacoustics: Neurobiology and Behaviour*. Edinburgh University Press, Edinburgh, U.K.

Ewing, E. P. 1979. Genetic variation in a heterogeneous environment. VII. Temporal and spatial heterogeneity in infinite populations. *Am. Nat.* 114: 197–212.

Faegri, K., and L. van der Pijl. 1979. *The Principles of Pollination Ecology*. 3d ed. Pergamon, Oxford.

Fagerström, T., and C. Wiklund. 1982. Why do males emerge before females? Protandry as a mating strategy in male and female butterflies. *Oecologia 52* : 164–166.

Fairbairn, D. J. 1990. Factors influencing sexual size dimorphism in temperate waterstriders. *Am. Nat.* 130: 61–86.

Fairchild, L. 1981. Mate selection and behavioral thermoregulation in Fowler's toads. *Science* 212: 950–951.

Falconer, D. S. 1989. *Introduction to Quantitative Genetics*. 3d ed. Longman, Harlow, U.K.

Falls, J. B. 1988. Does song deter territorial intrusion in white-throated sparrows (*Zonotrichia albicollis*)? *Can. J. Zool.* 66: 206–211.

Falls, J. B., J. R. Krebs, and P. K. McGregor. 1982. Song matching in the great tit

Parus major: The effect of similarity and familiarity. *Anim. Behav.* 30: 997–1009.

Farabaugh, S. M. 1982. The ecological and social significance of duetting. In D. E. Kroodsma and E. H. Miller, eds., *Acoustic Communication in Birds*, 85–124. Academic Press, New York.

Farr, J. A. 1975. The role of predation in the evolution of social behavior of natural populations of the guppy, *Poecilia reticulata* (Pisces: Poeciliidae). *Evolution* 29: 151–158.

Farr, J. A. 1980. Social behavior patterns as determinants of reproductive success in the guppy, *Poecilia reticulata* Peters (Pisces: Poeciliidae). *Behaviour* 74: 38–91.

Farr, J. A. 1983. The inheritance of quantitative fitness traits in guppies, *Poecilia reticulata* (Pisces: Poeciliidae). *Evolution* 37: 1193–1209.

Fay, F. H., G. C. Ray, and A. A. Kibalchich. 1984. Time and location of mating and associated behavior of the Pacific walrus, *Odobenus rusmarus divergens* Illiger. In *Society of American Cooperative Research on Marine Mammals*, Vol. 1, *Pinnipedi*. NOAA Tech. Rep. NMFS 12.

Feh, C. 1990. Long-term paternity data in relation to different aspects of rank for Camargue stallions, *Equus caballus*. *Anim. Behav.* 40: 995–996.

Feinsinger, P. 1983. Coevolution and pollination. In D. J. Futuyma and M. Slatkin, eds., *Coevolution*, 282–310. Sinauer, Sunderland, Mass.

Feinsinger, P., and H. M. Tiebout. 1991. Competition among plants sharing hummingbird pollinators: Laboratory experiments on a mechanism. *Ecology* 72: 1946–1952.

Feinsinger, P., H. M. Tiebout, and B. E. Young. 1991. Do tropical bird-pollinated plants exhibit density-dependent interactions? Field experiments. *Ecology* 72: 1953–1963.

Fellers, G. M. 1979. Aggression, territoriality, and mating behaviour in North American treefrogs. *Anim. Behav.* 27: 107–19.

Felsenstein, J. 1976. The theoretical population genetics of variable selection and migration. *Ann. Rev. Genet.* 10: 253–280.

Felsenstein, J. 1985. Phylogenies and the comparative method. *Am. Nat.* 125: 1–15.

Felsenstein, J. 1988. Phylogenies and quantitative methods. *Ann. Rev. Ecol. Syst.* 19: 445–471.

Ferguson, G. W. 1976. Colour change and reproductive cycling in female collared lizards (*Crotaphytus collaris*). *Copeia* 1976: 491–494.

Ferguson, G. W. 1977. Display and communication in reptiles: An historical perspective. *Amer. Zool.* 17: 167–176.

ffrench, R. P. 1980. *A Guide to the Birds of Trinidad and Tobago*. Harrowood, Newton Square, Penn.

Fiedler, K. 1954. Vergleichende Verhaltensstudien an Seenadeln, Schlangennadeln und Seepferdchen (Syngnathidae). *Z. Tierpsychol.* 11: 358–416.

Fincke, O. M. 1984. Giant damselflies in a tropical forest: Reproductive biology of *Megaloprepus coerulatus* with notes on *Mecistogaster* (Odonata: Pseudostigmatidae). *Adv. Odonatol.* 2: 13–27.

Fisher, R. A. 1915. The evolution of sexual preference. *Eugenics Review* 7: 184–192.

Fisher, R. A. 1930. *The Genetical Theory of Natural Selection.* Clarendon Press, Oxford.

Fisher, R. A. 1958. *The Genetical Theory of Natural Selection.* 2d ed. Dover, New York.

Fisher, R. A., and C. S. Stock. 1915. Cuénot on preadaptation. *Eugenics Review* 7: 46–61.

Fitch, H. S. 1970. Reproductive cycles in lizards and snakes. *Univ. Kans. Mus. Nat. Hist. Misc. Publ.* 52: 1–247.

Fitch, H. S., and D. M. Hillis. 1984. The *Anolis* dewlap: Interspecific variability and morphological associations with habitat. *Copeia* 1984: 315–323.

Fitzpatrick, J. W. 1988. Why so many passerine birds? A response to Raikow. *Syst. Zool.* 37: 71–76.

Fleishman, L. J. 1992. The influence of the sensory system and the environment on motion patterns in the visual displays of anoline lizards and other vertebrates. *Am. Nat.* 139: s36-s61.

Flood, N. J. 1984. Adaptive significance of delayed plumage maturation in male northern orioles. *Evolution* 38: 267–279.

Flood, N. J. 1989. Coloration in New World orioles. 1. Tests of predation-related hypotheses. *Behav. Ecol. Sociobiol.* 25: 49–56.

Foelix, R., ed. 1982. *Biology of Spiders.* Harvard University Press, Cambridge, Mass.

Folstad, I., and A. J. Karter. 1992. Parasites, bright males, and the immunocompetence handicap. *Am. Nat.* 139: 603–622.

Foltz, D. W., and P. L. Schwagmeyer. 1989. Sperm competition in the thirteen-lined ground squirrel: Differential fertilization success under field conditions. *Am. Nat.* 133: 257–265.

Foote, C. J. 1988. Male mate choice dependent on male size in salmon. *Behaviour* 106: 63–80.

Foote, C. J. 1989. Female mate preference in Pacific salmon. *Anim. Behav.* 38: 721–723.

Foote, C. J. 1990. An experimental comparison of male and female spawning territoriality in a pacific salmon. *Behaviour* 115: 283–314.

Foote, C. J., and P. A. Larkin. 1988. The role of male choice in the assortative mating of anadromous and non-anadromous sockeye salmon (*Oncorhynchus nerka*). *Behaviour* 106: 43–62.

Ford, N. B. 1986. The role of pheromone trails in the sociobiology of snakes. In D. Duvall, D. Müller-Schwarze, and R. M. Silverstein, eds., *Chemical Signals in Vertebrates*, 261–278. Plenum, New York.

Forester, D. C., and R. Czarnowsky. 1985. Sexual selection in the spring peeper, *Hyla crucifer* (Amphibia, Anura): Role of the advertisement call. *Behaviour* 92: 112–128.

Forester, D. C., D. V. Lykens, and W. K. Harrison. 1989. The significance of persistent vocalisation by the spring peeper, *Pseudacris crucifer* (Anura: Hylidae). *Behaviour* 108: 197–208.

Forrest, T. G. 1980. Phonotaxis in mole crickets: Its reproductive significance. *Fla. Entomol.* 63: 45–53.

Forrest, T. G. 1983. Calling songs and mate choice in mole crickets. In D. T.

Gwynne and G. K. Morris, eds., *Orthopteran Mating Systems*, 185–204. Westview, Boulder, Colo.
Forsberg, J. 1987. A model for male mate discrimination in butterflies. *Oikos* 49: 46–54.
Forsman, A. 1991. Variation in sexual size dimorphism and maximum body size among adder populations: Effects of prey size. *J. Anim. Ecol.* 60: 253–267.
Foster, M. S. 1977. Odd couples in manakins: A study of social organization and cooperative breeding in the swallow-tailed manakin *Chiroxiphia linearis*. *Am. Nat.* 111: 845–853.
Foster, M. S. 1981. Cooperative behavior and social organization in the swallow-tailed manakin (*Chiroxiphia caudata*). *Behav. Ecol. Sociobiol.* 9: 167–177.
Foster, M. S. 1983. Disruption, dispersion and dominance in lek-breeding birds. *Am. Nat.* 122: 53–72.
Fouquette, M. J., Jr. 1975. Speciation in chorus frogs. I. Reproductive character displacement in the *Pseudacris nigrita* complex. *Syst. Zool.* 24: 16–23.
Frankham, R. 1966. Sex and selection for a quantitative character in *Drosophila*. *Aust. J. Biol. Sci.* 21: 1215–1223.
Frankie, G. W., S. B. Vinson, and R. E. Colville. 1980. Territorial behavior of *Centris adani* and its reproductive function in the Costa Rican dry forest (Hymenoptera: Anthophoridae). *J. Kansas Ent. Soc.* 53: 837–857.
Freed, L. 1987. Prospective infanticide and protection of genetic paternity in tropical house wrens. *Am. Nat.* 130: 948–954.
Freeland, W. J. 1976. Pathogens and the evolution of primate sociality. *Biotropica* 8: 12–24.
Freeman, D. C., K. T. Harper, and E. L. Charnov. 1980. Sex change in plants: Old and new observations and new hypotheses. *Oecologia* 47: 222–232.
Freeman, S., and W. M. Jackson. 1990. Univariate metrics are not adequate to measure avian body size. *Auk* 107: 69–74.
Fricke, H. W. 1966. Attrappenversuche mit einigen plakatfarbigen Korallenfischen im Roten Meer. *Z. Tierpsychol.* 23: 4–7.
Fricke, H. W., and S. Fricke. 1977. Monogamy and sex change by aggressive dominance in coral reef fish. *Nature* 266: 830–832.
Fricke, H. W., and S. Holzberg. 1974. Social units and hermaphroditism in a pomacentrid fish. *Naturwissensch.* 61: 367–368.
Fritzsch, B., M. J. Ryan, W. Wilczynski, T. E. Hetherington, and W. Walkowiak, eds. 1988. *The Evolution of the Amphibian Auditory System*. Wiley, New York.
Fry, C. H., S. Keith, and E. K. Urban, eds. 1988. *The Birds of* Africa, vol. 3. Academic Press, London.
Futuyma, D. J. 1986. *Evolutionary Biology*. 2d ed. Sinauer, Sunderland, Mass.
Gadgil, M. 1972. Male dimorphism as a consequence of sexual selection. *Am. Nat.* 106: 574–580.
Gadgil, M., and W. Bossert. 1970. Life history consequences of natural selection. *Am. Nat.* 104: 1–24.
Galen, C. 1985. Regulation of seed-set in *Polemonium viscosum*: Floral scents, pollination, and resources. *Ecology* 66: 792–797.
Galen, C. 1989. Measuring pollinator-mediated selection on morphometric floral

traits: Bumblebees and the alpine sky pilot, *Polemonium viscosum*. *Evolution* 43: 882–890.

Galen, C., and J. T. Rotenberry. 1988. Variance in pollen carryover in animal-pollinated plants: Implications for mate choice. *J. theor. Biol.* 135: 419–429.

Gardner, R. 1990. Mating calls. *Nature* 344: 495–496.

Garstka, W. R., and D. Crews. 1986. Pheromones and reproduction in garter snakes. In D. Duvall, D. Müller-Schwarze, and R. M. Silverstein, eds., *Chemical Signals in Vertebrates*, 243–260. Plenum, New York.

Gartside, D. F. 1980. Analysis of a hybrid zone between chorus frogs of the *Pseudacris nigrita* complex in the Southern United States. *Copeia* 1980: 56–66.

Gaskin, D. E. 1982. *The Ecology of Whales and Dolphins*. Heinemann, London.

Gatz, A. J. 1981a. Size selective mating in *Hyla versicolor* and *Hyla crucifer*. *J. Herpetol.* 15: 114–116.

Gatz, A. J. 1981b. Non-random mating by size in American toads, *Bufo americanus*. *Anim. Behav.* 29: 1004–1012.

Gaulin, S. J., and L. D. Sailer. 1984. Sexual dimorphism in weight among the primates: The relative impact of allometry and sexual selection. *Int. J. Primatol.* 5: 525–535.

Gaulin, S. J., and L. D. Sailer. 1985. Are females the ecological sex? *American Anthropologist* 87: 111–119.

Gaulin, S.J.C., and R. W. FitzGerald. 1986. Sex differences in spatial ability: An evolutionary hypothesis and test. *Am. Nat.* 127: 74–88.

Gaulin, S.J.C., and R. W. FitzGerald. 1989. Sexual selection for spatial-learning ability. *Anim. Behav.* 37: 322–331.

Geist, V. 1964. On the rutting behaviour of the mountain goat (*Oreamnos americanus*). *J. Mamm.* 45: 551–568.

Geist, V. 1966. The evolution of horn-like organs. *Behaviour* 27: 175–214.

Geist, V. 1971a. The relation of social evolution and dispersal in ungulates during the Pleistocene, with emphasis on the Old World deer and the genus *Bison*. *Quat. Res.* 1: 283–315.

Geist, V. 1971b. *Mountain Sheep: A Study in Behavior and Evolution*. University of Chicago Press, Chicago.

Geist, V. 1978. On weapons, combat, and ecology. In L. Kramer, P. Pliner, and T. Alloway, eds., *Aggression, Dominance and Individual Spacing*. Plenum, New York.

Geist, V. 1986. The paradox of the great Irish stags. *Nat. Hist.* (3): 54–64.

Gerhardt, H. C. 1978. Mating call recognition in the green treefrog (*Hyla cinerea*): The significance of some fine-temporal properties. *J. Exp. Biol.* 74: 59–73.

Gerhardt, H. C. 1981. Mating call recognition in barking treefrog (*Hyla gratiosa*): Responses to synthetic calls and comparisons with the green treefrog (*Hyla cinerea*). *J. Comp. Physiol.* 144: 17–25.

Gerhardt, H. C. 1982. Sound pattern recognition in some North American treefrogs (Anura: Hylidae). Implications for mate choice. *Am. Zool.* 22: 581–595.

Gerhardt, H. C. 1987. Mating behaviour and male mating success in the green treefrog. *Anim. Behav.* 35: 1490–1503.

Gerhardt, H. C. 1988. Acoustic properties used in call recognition by frogs and toads. In B. Fritzsch, M. Ryan, W. Wilczynski, T. E. Hetherington, and W. Walkowiak, eds., *The Evolution of the Amphibian Auditory System*, 455–483. Wiley, New York.

Gerhardt, H. C. 1991. Female mate choice in treefrogs: Static and dynamic acoustic criteria. *Anim. Behav.* 42: 615–635.

Gerhardt, H. C. 1993. Reproductive character displacement on female mate choice in the gray tree frog *Hyla chrysoscelis*. *Anim. Behav.*: in press.

Gerhardt, H. C., and J. A. Doherty. 1988. Acoustic communication in the gray treefrog, *Hyla versicolor*: Evolutionary and neurobiological implications. *J. Comp. Physiol.* A 162: 261–278.

Gerhardt, H. C., R. E. Daniel, S. A. Perrill, and S. Schramm. 1987. Mating behaviour and male mating success in the green treefrog. *Anim. Behav.* 35: 1490–1503.

Gertsch, W. J. 1979. *American Spiders*. 2d ed. Van Nostrand Reinhold, New York.

Ghiselin, M. T. 1969a. The evolution of hermaphroditism among animals. *Q. Rev. Biol.* 44: 189–208.

Ghiselin, M. T. 1969b. *The Triumph of the Darwinian Method*. University of California Press, Berkeley.

Ghiselin, M. T. 1974. *The Economy of Nature and the Evolution of Sex*. University of California Press, Berkeley.

Gibbons, B. 1986. *Dragonflies and Damselflies of Britain and Northern Europe*. Country Life, London.

Gibbons, D. W. 1989. Seasonal reproductive success of moorhen (*Gallinula chloropus*): The importance of male weight. *Ibis* 131: 57–68.

Gibbons, M. M., and T. K. McCarthy. 1986. The reprodutive output of frogs (*Rana temporaria L.)* with particular reference to body size and age. *J. Zool. London* 209: 579–593.

Gibbs, H. L., P. J. Weatherhead, P. T. Boag, B. N. White, L. M. Tabak, and D. J. Hoysak. 1990. Realized reproductive success of polygynous red-winged blackbirds revealed by DNA markers. *Science* 250: 1394–1397.

Gibson, R. M. 1987. Bivariate versus multivariate analyses of sexual selection in red deer. *Anim. Behav.* 35: 292–305.

Gibson, R. M. 1989. Field playback of male display attracts females in lek breeding sage grouse. *Behav. Ecol. Sociobiol.* 24: 439–443.

Gibson, R. M. 1990. Relationships between blood parasites, mating success and phenotypic cues in male sage grouse (*Centrocercus urophasianus). Amer. Zool.* 30: 271–278.

Gibson, R. M., and G. C. Bachman. 1992. The costs of female choice in a lekking bird. *Behav. Ecol.* 3: 300–309.

Gibson, R. M., and J. W. Bradbury. 1985. Sexual selection in lekking sage grouse: Phenotypic correlates of male mating success. *Behav. Ecol. Sociobiol.* 18: 117–123.

Gibson, R. M., and F. E. Guinness. 1980a. Differential reproductive success in red deer stags. *J. Anim. Ecol.* 49: 199–208.

Gibson, R. M., and F. E. Guinness. 1980b. Behavioural factors affecting male reproductive success in red deer (*Cervus elaphus* L.). *Anim. Behav.* 28: 1163–1174.

Gibson, R. M., and J. Höglund. 1992. Copying and sexual selection. *TREE* 7: 229–232.

Gibson, R. M., J. W. Bradbury, and S. Vehrencamp. 1991. Mate choice in lekking sage grouse revisited: The roles of vocal display, female site fidelity, and copying. *Behav. Ecol.* 2: 165–180.

Giese, A. C., L. Greenfield, H. Huang, A. Farmanfarmaian, R. Boolootian, and R. Lasker. 1959. Organic productivity in the reproductive cycle of the purple sea urchin. *Biol. Bull.* 116: 49–58.

Gilbert, J. J., and C. E. Williamson. 1983. Sexual dimorphism in zooplankton (Copepoda, Cladocera, and Rotifera). *Ann. Rev. Ecol. Syst.* 14: 1–34.

Gilbert, L. E. 1975. Ecological consequences of a coevolved mutualism between butterflies and plants. In L. E. Gilbert and P. H. Raven, eds., *Coevolution of Animals and Plants*, 210–240. University of Texas Press, Austin.

Gilbert, L. E. 1976. Postmating female odor in *Heliconius* butterflies: A male-contributed antiaphrodisiac? *Science* 193: 419–420.

Gill, D. E. 1989. Fruiting failure, pollination inefficiency, and speciation in orchids. In D. Otte and J. A. Endler, eds., *Speciation and Its Consequences*, 458–481. Sinauer, Sunderland, Mass.

Gill, F. B. 1990. *Ornithology*. Freeman, New York.

Gill, F. B., and B. G. Murray, Jr. 1972. Discrimination behavior and hybridization of the blue-winged and golden warblers. *Evolution* 26: 282–293.

Gillespie, J. 1973. Polymorphism in random environments. *Theor. Pop. Biol.* 4: 193–195.

Gilliard, E. T. 1969. *Birds of Paradise and Bowerbirds*. Weidenfeld and Nicholson, London.

Gittleman, J. L. 1981. The phylogeny of parental care in fishes. *Anim. Behav.* 29: 936–941.

Gittleman, J. L. 1989. The comparative approach in ethology: Aims and limitations. In P. G. Bateson and P. H. Klopfer, eds., *Perspectives in Ethology*, 55–83. Plenum, New York.

Gittleman, J. L., and M. Kot. 1990. Adaptation: Statistics and a null model for estimating phylogenetic effects. *Syst. Zool.* 39: 227–241.

Given, M. F. 1988. Growth rate and the cost of calling activity in male carpenter frogs, *Rana virgatipes*. *Behav. Ecol. Sociobiol.* 22: 153–160.

Givnish, T. J. 1980. Ecological constraints on the evolution of breeding systems in seed plants: Dioecy and dispersal in gymnosperms. *Evolution* 34: 959–972.

Gleason, H. A. 1952. *Illustrated Flora of the Northeastern United States and Adjacent Canada*. Vols. 2, 3. Lancaster Press, Lancaster, Pa.

Goldman, A. D., and M. F. Willson. 1986. Sex allocation in functionally hermaphroditic plants: A review and critique. *Bot. Rev.* 52: 158–194.

Gomulkiewicz, R. S., and A. Hastings. 1990. Ploidy and evolution by sexual selection: A comparison of haploid and diploid female choice models near fixation equlibria. *Evolution* 44: 757–770.

Goodwin, G. J., and S. M. Roble. 1983. Mating success in male treefrogs, *Hyla chrysoscelis* (Anura: Hylidae). *Herpetologica* 39: 141–146.

Göransson, G., T. von Schantz, I. Fröberg, A. Helgée, and H. Wittzell. 1990. Male characteristics, viability and harem size in the pheasant, *Phasianus colchicus*. *Anim. Behav.* 40: 89–104.

Gosling, L. M. 1982. A reassessment of the function of scent marking in territories. *Z. Tierpsychol.* 60: 89–118.

Gosling, L. M. 1986. Economic consequences of scent marking in mammalian territoriality. In D. Duvall, D. Müller-Schwarze, and R. M. Silverstein, eds., *Chemical Signals in Vertebrates*, vol. 4, 385–395. Plenum, New York.

Gosling, L. M. 1990. Scent-marking by resource holders: Alternative mechanisms for advertising the costs of competition. In D. W. MacDonald, D. Müller-Schwarze, and S. Natymezuk, eds., *Chemical Signals in Vertebrates*, vol. 5, 315–328. Oxford University Press, Oxford.

Gosling, L. M., and H. V. McKay. 1990. Competitor assessment by scent matching: An experimental test. *Behav. Ecol. Sociobiol.* 26: 415–420.

Gosling, L. M., and M. Petrie. 1990. Lekking in topi: A consequence of satellite behaviour by small males at hotspots. *Anim. Behav.* 40: 272–287.

Goss, R. J. 1983. *Deer Antlers: Regeneration, Function and Evolution*. Academic Press, New York.

Götmark, F. 1992. Anti-predator effect of conspicuous plumage in a male bird. *Anim. Behav.* 44: 51–55.

Götmark, F. 1993a. Conspicuous coloration in male birds: Favoured by predation in some species, disfavoured in others. *Proc. Roy. Soc. London* B 253: 143–146.

Götmark, F. 1993b. An experimental study of the importance of plumage coloration in breeding males of the white-crowned sparrow. *Ornis Scand.* 24: 149–154.

Götmark, F. 1994a. Are bright birds distasteful? A reanalysis of H. B. Cott's data on the edibility of birds. *J. Avian Biol.*, in press.

Götmark, F. 1994b. Does a novel bright color patch increase or decrease predation? Red wings reduce predation risk in European blackbirds. *Proc. R. Soc. Lond.* B, in press.

Götmark, F., and U. Unger. 1994. Are conspicuous birds unprofitable prey? Field experiments with hawks and stuffed prey species. *Auk* 111: in press.

Gottlander, K. 1987. Variation in the song rate of the male pied flycatcher *Ficedula hypoleuca*: Causes and consequences. *Anim. Behav.* 35: 1037–1043.

Gould, S. J. 1974. The origin and function of "Bizarre" structures: Antler size and skull size in the "Irish elk," *Megaloceros giganteus*. *Evolution* 28: 191–220.

Gould, S. J. 1983. *Hen's Teeth and Horse's Toes*. Norton, New York.

Gould, S. J. 1984. Only his wings remained. *Nat. Hist.* 93(9): 10–18.

Gould, S. J. 1989. A developmental constraint in *Cerion*, with comments on the definition and interpretation of constraint in evolution. *Evolution* 43: 516–539.

Gould, S. J., and R. C. Lewontin. 1979. The spandrels of San Marco and the Panglossian paradigm: A critique of the adaptationist programme. *Proc. Roy. Soc. London* B 205: 581–598.

Gowaty, P. A. 1985. Multiple parentage and apparent monogamy in birds. *Ornithol. Monogr.* 37: 11–21.

Gowaty, P. A., and H. L. Gibbs 1993. DNA fingerprinting in avian behavioral ecology: Two cultures arise. *Auk* 110: 152–155.

Gowaty, P. A., and A. A. Karlin. 1984. Multiple paternity and maternity in single broods of apparently monogamous Eastern bluebirds. *Behav. Ecol. Sociobiol.* 15: 91–95.

Grafen, A. 1987. Measuring sexual selection: Why bother? In J. W. Bradbury and M. B. Andersson, eds., *Sexual Selection: Testing the Alternatives*, 221–233. Wiley, Chichester, U.K.

Grafen, A. 1988. On the uses of data on lifetime reproductive success. In T. H. Clutton-Brock, ed., *Reproductive Success*, 454–471. University of Chicago Press, Chicago.

Grafen, A. 1990a. Sexual selection unhandicapped by the Fisher process. *J. theor. Biol.* 144: 473–516.

Grafen, A. 1990b. Biological signals as handicaps. *J. theor. Biol.* 144: 517–546.

Grafen, A., and M. Ridley. 1983. A model of mate guarding. *J. theor. Biol.* 102: 549–567.

Grant, B. R. 1984. The significance of song variation in a population of Darwin's finches. *Behaviour* 89: 90–116.

Grant, B. R. 1985. Selection on bill characters in a population of Darwin's finches: *Geospiza conirostris* on Isla Genovesa, Galápagos. *Evolution* 39: 523–532.

Grant, B. R. 1990. The significance of subadult plumage in Darwin's finches, *Geospiza fortis*. *Behav. Ecol.* 1: 161–170.

Grant, B. R., and P. R. Grant. 1987. Mate choice in Darwin's finches. *Biol. J. Linn. Soc.* 32: 247–270.

Grant, B. R., and P. R. Grant. 1989. *Evolutionary Dynamics of a Natural Population*. University of Chicago Press, Chicago.

Grant, J.W.A., and P. W. Colgan. 1983. Reproductive success and mate choice in the johnny darter, *Etheostoma nigrum* (Pisces: Percidae). *Can. J. Zool.* 61: 437–445.

Grant, K. A., and V. Grant. 1968. *Hummingbirds and Their Flowers*. Columbia University Press, New York.

Grant, P. R. 1972. Convergent and divergent character displacement. *Biol. J. Linn. Soc.* 4: 39–68.

Grant, P. R. 1975. The classical case of character displacement. *Evol. Biol.* 8: 237–337.

Grant, P. R. 1986. *Ecology and Evolution of Darwin's Finches*. Princeton University Press, Princeton, N.J.

Grant, P. R., and B. R. Grant. 1992. Hybridization of bird species. *Science* 256: 193–197.

Green, A. J. 1989. The sexual behaviour of the great crested newt, *Triturus cristatus* (Amphibia: Salamandridae). *Ethology* 83: 129–153.

Green, A. J. 1990. Determinants of chorus participation and the effects of size, weight and competition on advertisement calling in the tungara frog, *Physalaemus pustulosus* (Leptodactulidae). *Anim. Behav.* 39: 620–638.

Green, A. J. 1991. Large male crests, an honest indicator of condition, are preferred by female smooth newts, *Triturus vulgaris* (Salamandridae) at the spermatophore transfer stage. *Anim. Behav.* 41: 367–369.

Green, A. J. 1992. Positive allometry is likely with mate choice, competitive display and other functions. *Anim. Behav.* 43: 170–172.

Greenberg, B., and G. K. Noble. 1944. Social behavior of the American chameleon, *Anolis carolinensis*. *Physiol. Zool.* 17: 392–439.

Greenewalt, C. H. 1975. The flight of birds. *Trans. Am. Phil. Soc.* 65: 244–292.

Greenfield, M. D. 1981. Moth sex pheromones: An evolutionary perspective. *Florida Entomol.* 64: 4–17.

Greenfield, M. D., and T. E. Shelly. 1985. Alternative mating strategies in a desert grasshopper: Evidence of density-dependence. *Anim. Behav.* 33: 1192–1210.

Greenspan, B. N. 1980. Male size and reproductive success in the communal courtship system of the fiddler crab *Uca rapax*. *Anim. Behav.* 28: 387–392.

Greenwood, P. J. 1980. Mating systems, philopatry and dispersal in birds and mammals. *Anim. Behav.* 28: 1140–1162.

Greenwood, P. J., and J. Adams. 1987. *The Ecology of Sex*. Edward Arnold, London.

Greenwood, P. J., and P. H. Harvey. 1982. The natal and breeding dispersal of birds. *Ann. Rev. Ecol. Syst.* 13: 1–21.

Greer, B. J., and K. D. Wells. 1980. Territorial and reproductive behavior of the Tropical American frog *Centrolenella fleischmanni*. *Herpetologica* 36: 318–326.

Gregg, K. B. 1975. The effects of light intensity on sex expression in species of *Cycnoches* and *Catasetum* (Orchidaceae). *Selbyana* 1: 101–112.

Greig-Smith, P. W. 1982. Song-rates and parental care by individual male stonechats (*Saxicola torquata*). *Anim. Behav.* 30: 245–252.

Gribbin, S. D., and K. J. Thompson. 1991. The effects of size and residency on territorial disputes and short-term mating success in the damselfy, *Pyrrhosoma nymphula* (Sulzer) (Zygoptera: Coenagrionidae). *Anim. Behav.* 41: 689–696.

Gronell, A. M. 1984. Courtship, spawning and social organization of the pipefish, *Corythoichthys intestinalis* (Pisces: Syngnathidae) with notes on two congeneric species. *Z. Tierpsychol.* 65: 1–24.

Gronell, A. M. 1989. Visiting behaviour by females of the sexually dichromatic damselfish, *Chrysiptera cyanea* (Teleostei: Pomacentridae): A probable method of assessing male quality. *Ethology* 81: 89–122.

Gross, M. R. 1979. Cuckoldry in sunfishes (Lepomis: Centrarchidae). *Can. J. Zool.* 57: 1507–1509.

Gross, M. R. 1982. Sneakers, satellites and parentals: Polymorphic mating strategies in North American sunfishes. *Z. Tierpsychol.* 60: 1–26.

Gross, M. R. 1984. Sunfish, salmon, and the evolution of alternative reproductive strategies and tactics in fishes. In G. W. Potts and R. J. Wootton, eds., *Fish Reproduction: Strategies and Tactics*, 55–75. Academic Press, London.

Gross, M. R. 1985. Disruptive selection for alternative life histories in salmon. *Nature* 313: 47–48.

Gross, M. R. 1991a. Evolution of alternative reproductive strategy: Frequency-dependent sexual selection in male bluegill sunfish. *Phil. Trans. Roy. Soc. London* B 332: 59–66.

Gross, M. R. 1991b. Salmon breeding behavior and life history evolution in changing environments. *Ecology* 72: 1180–1186.

Gross, M. R., and E. L. Charnov. 1980. Alternative male life histories in bluegill sunfish. *Proc. Natl. Acad. Sci. USA* 77: 6937–6940.

Gross, M. R., and R. C. Sargent. 1985. The evolution of male and female parental care in fishes. *Amer. Zool.* 25: 807–822.

Gross, M. R., and R. Shine. 1981. Parental care and mode of fertilization in ectothermic vertebrates. *Evolution* 35: 775–795.

Grula, J. W., and O. R. Taylor, Jr. 1980. The effect of x-chromosome inheritance on mate-selection behavior in the sulfur butterflies, *Colias eurytheme* and *C. philodice*. *Evolution* 34: 688–695.

Guilford, T., and M. S. Dawkins. 1991. Receiver psychology and the evolution of animal signals. *Anim. Behav.* 42: 1–14.

Gulick, J. T. 1888. Divergent evolution through cumulative segregation. *J. Linn. Soc.* 20: 189–274, 312–380.

Gunnarsson, B., and J. Johnsson. 1990. Protandry and moulting to maturity in the spider *Pityohyphantes phrygianus*. *Oikos* 59: 205–212.

Gustafsson, L. 1986. Lifetime reproductive success and heritability: Empirical support for Fisher's fundamental theorem. *Am. Nat.* 128: 761–764.

Guthrie, R. D. 1970. Evolution of human threat display organs. In T. Dobzhansky, M. K. Hecht, and W. C. Steere, eds., *Evolutionary Biology*, 257–302. Appleton Century, New York.

Gwynne, D. T. 1981. Sexual difference theory: Mormon crickets show role reversal in mate choice. *Science* 213: 779–780.

Gwynne, D. T. 1982. Mate selection by female katydids (Orthoptera: Tettigoniidae, *Conocephalus nigropleurum*). *Anim. Behav.* 30: 734–738.

Gwynne, D. T. 1984a. Sexual selection and sexual differences in mormon crickets (Orthoptera: Tettigoniidae, *Anabrus simplex*). *Evolution* 38: 1011–1022.

Gwynne, D. T. 1984b. Courtship feeding increases female reproductive success in bushcrickets. *Nature* 307: 361–363.

Gwynne, D. T. 1984c. Nuptial feeding behavior and female choice of mates in *Harpobittacus similis* (Mecoptera: Bittacidae). *J. Aust. Ent. Soc.* 23: 271–276.

Gwynne, D. T. 1985. Role-reversal in katydids: Habitat influences reproductive behaviour (Orthoptera: Tettigonidae, *Metaballus* sp.). *Behav. Ecol. Sociobiol.* 16: 355–361.

Gwynne, D. T. 1986. Courtship feeding in katydids (Orthoptera: Tettigoniidae). Investment in offspring or in obtaining fertilizations? *Am. Nat.* 128: 342–352.

Gwynne, D. T. 1987. Sex-biased predation and the risky mate-locating behaviour of male tick-tock cicadas (Homoptera: Cicadidae). *Anim. Behav.* 35: 571–576.

Gwynne, D. T. 1988a. Courtship feeding in katydids benefits the mating male's offspring. *Behav. Ecol. Sociobiol.* 23: 373–377.

Gwynne, D. T. 1988b. Courtship feeding and the fitness of female katydids (Orthoptera: Tettigoniidae). *Evolution* 42: 545–555.

Gwynne, D. T. 1989. Does copulation increase the risk of predation? *TREE* 4: 54–56.

Gwynne, D. T. 1990. Testing parental investment and the control of sexual selection in katydids: The operational sex ratio. *Am. Nat.* 136: 474–484.

Gwynne, D. T. 1991. Sexual competition among females: What causes courtship-role reversal? *TREE* 6: 118–121.

Gwynne, D. T., and W. J. Bailey. 1988. Mating system, mate choice and ultrasonic calling in a zaprochiline katydid (Orthoptera: Tettigoniidae). *Behaviour* 105: 202–223.

Gwynne, D. T., and G. K. Morris, eds. 1983. *Orthopteran Mating Systems: Sexual Competition in a Diverse Group of Insects*. Westview, Boulder, Colo.

Gwynne, D. T., and G. K. Morris. 1986. Heterospecific recognition and behavioral isolation in acoustic orthoptera (Insecta). *Evol. Theory* 8: 33–38.

Gwynne, D. T., and K. M. O'Neill. 1980. Territoriality in digger wasps results in sex biased predation on males (Hymenoptera: Sphecidae, *Philantus*). *J. Kansas Ent. Soc.* 53: 220–224.

Gwynne, D. T., and L. W. Simmons. 1990. Experimental reversal of courtship roles in an insect. *Nature* 346: 171–174.

Gyllensten, U. B., S. Jakobsson, and H. Temrin. 1990. No evidence for illegitimate young in monogamous and polygynous warbles. *Nature* 343: 168–170.

Haas, R. 1976. Sexual selection in *Nothobranchius guentheri* (Pisces: Cyprinodontidae). *Evolution* 30: 614–622.

Hafernik, J. E., Jr., and R. W. Garrison. 1986. Mating success and survival rate in a population of damselfies: Results at variance with theory? *Am. Nat.* 128: 353–365.

Hagen, D. W., G.E.E. Moodie, and P. F. Moodie. 1980. Polymorphism for breeding colors in *Gasterosteus aculeatus*. II. Reproductive success as a result of convergence for threat display. *Evolution* 34: 1050–1059.

Hagen, J. M., and J. M. Reed. 1988. Red color bands reduce fledging success in red-cockaded woodpeckers. *Auk* 105: 498–503.

Haig, D., and M. Westoby. 1988. On limits to seed production. *Am. Nat.* 131: 757–759.

Haigh, J., and M. R. Rose. 1980. Evolutionary game auctions. *J. theor. Biol.* 85: 381–397.

Hailman, J. P. 1977. *Optical Signals: Animal Communication and Light*. Indiana University Press, Bloomington.

Hainsworth, F. R., L. L. Wolf, and T. Mercier. 1984. Pollination and pre-dispersal seed predation: Net effects on reproduction and inflorescence characteristics in *Ipomopsis aggregata*. *Oecologia* 63: 405–409.

Hakkarainen, H., and E. Korpimäki. 1991. Reversed sexual size dimorphism in Tengmalm's owl: Is small size adaptive?. *Oikos* 61: 337–346.

Haldane, J.B.S., 1932. *The Causes of Evolution*. Longmans, London.

Haldane, J.B.S., and S. D. Jayakar. 1963. Polymorphism due to selection of varying direction. *J. Genet.* 58: 237–242.

Halliday, T. R. 1976. The libidinous newt. An analysis of variations in the sexual behaviour of the male smooth newt, *Triturus vulgaris*. *Anim. Behav.* 24: 398–414.

Halliday, T. R. 1977a. The courtship of European newts: An evolutionary perspective. In D. H. Taylor and S. I. Guttman, eds., *The Reproductive Biology of Amphibians*, 185–232. Plenum, New York.

Halliday, T. R. 1977b. The effect of experimental manipulations of breathing behaviour on the sexual behaviour of the smooth newt, *Triturus vulgaris*. *Anim. Behav.* 25: 39–45.

Halliday, T. R. 1978. Sexual selection and mate choice. In J. R. Krebs and N. B.

Davies, eds., *Behavioural Ecology: An Evolutionary Approach*, 180–213. Blackwell, Oxford.

Halliday, T. R. 1983. The study of mate choice. In P. Bateson, ed., *Mate Choice*, 3–32. Cambridge University Press, Cambridge, U.K.

Halliday, T. R. 1987. Physiological constraints on sexual selection. In J. W. Bradbury and M. B. Andersson, eds., *Sexual Selection: Testing the Alternatives*, 247–264. Wiley, Chichester, U.K.

Halliday, T. R., and S. J. Arnold. 1987. Multiple mating by females: A perspective from quantitative genetics. *Anim. Behav.* 35: 939–941.

Halliday, T. R., and A. I. Houston. 1978. The newt as an honest salesman. *Anim. Behav.* 26: 1273–1274.

Halliday, T. R., and P. A. Verrell. 1986. Sexual selection and body size in Amphibians. *Herpet. J.* 1: 86–92.

Hamilton, W. D. 1967. Extraordinary sex ratios. *Science* 156: 477–487.

Hamilton, W. D. 1979. Wingless and fighting males in fig wasps and other insects. In M. S. Blum and N. A. Blum, eds., *Sexual Selection and Reproductive Competition in Insects*, 167–220. Academic Press, New York.

Hamilton, W. D. 1982. Pathogens as causes of genetic diversity in their host populations. In R. M. Anderson and R. M. May, eds., *Population Biology of Infectious Diseases*, 269–296. Springer, Berlin.

Hamilton, W. D. 1990. Mate choice near or far. *Amer. Zool.* 30: 341- 352.

Hamilton, W. D., and M. Zuk. 1982. Heritable true fitness and bright birds: A role for parasites? *Science* 218: 384–387.

Hamilton, W. D., and M. Zuk. 1989. Parasites and sexual selection: A reply. *Nature* 341: 289–290.

Hamilton, W. D., R. Axelrod, and R. Tanese. 1990. Sexual reproduction as an adaptation to resist parasites (a review). *Proc. Natl. Acad. Sci. USA* 87: 3566–3573.

Hammerstein, P., and G. A. Parker. 1987. Sexual selection: Games between the sexes. In J. W. Bradbury and M. B. Andersson, eds., *Sexual Selection: Testing the Alternatives*, 119–142. Wiley, Chichester, U.K.

Handford, P., and M. A. Mares. 1985. The mating systems of ratites and tinamous: An evolutionary perspective. *Biol. J. Linn. Soc.* 25: 77–104.

Hansen, A. J., and S. Rohwer. 1986. Coverable badges and resource defence in birds. *Anim. Behav.* 34: 69–76.

Hansson, B. S., C. Löfstedt, and W. L. Roelofs. 1987. Inheritance of olfactory response to sex pheromone components in the European corn borer, *Ostrinia nubilalis*. *Naturwissensch.* 74: 497–499.

Happ, G. M. 1969. Multiple sex pheromones of the mealworm beetle *Tenebrio molitor*. *Nature* 222: 180–181.

Harcourt, A. H., P. H. Harvey, S. G. Larson, and R. V. Short. 1981. Testis weight, body weight and breeding system in primates. *Nature* 293: 55–57.

Harder, L. D., and J. D. Thomson. 1989. Evolutionary options for maximizing pollen dispersal of animal-pollinated plants. *Am. Nat.* 133: 323–344.

Hardy, J. W., and R. W. Dickerman. 1965. Relationships between two forms of the red-winged blackbird in Mexico. *Living Bird* 4: 107–129.

Haresign, T. W., and S. E. Schumway. 1981. Permeability of the marsupium of the pipefish *Syngnathus fuscus* to ^{14}C-alpha amino isobutyric acid. *Comp. Biochem. Physiol. A, Comp. Physiol.* 69: 603–604.

Harper, D. 1985. Pairing strategies and mate choice in female robins *Erithacus rubecula*. *Anim. Behav.* 33: 862–875.

Harper, D. 1991. Communication. In J. R. Krebs and N. B. Davies, eds., *Behavioural Ecology: An Evolutionary Approach*, 3d ed., 374–397. Blackwell, Oxford.

Harris, V. A. 1964. *The Life of the Rainbow Lizard*. Hutchinson, London.

Harris, V. E., and J. W. Todd. 1980. Male-mediated aggregation of male, female and 5th-instar southern green stink bugs and concomitant attraction of a tachinid parasite, *Trichopoda pennipes*. *Ent. Exp. Appl.* 27: 117–126.

Harvey, A. W. 1990. Sexual differences in contemporary selection acting on size in the hermit crab *Clibanarius digueti*. *Am. Nat.* 136: 292–304.

Harvey, P. H., and P. M. Bennett. 1985. Sexual dimorphism and reproductive strategies. In J. Ghesquiere, R. D. Martin, and F. Newcombe, eds., *Human Sexual Dimorphism*, 43–59. Taylor and Francis, London.

Harvey, P. H., and J. W. Bradbury. 1991. Sexual selection. In J. R. Krebs and N. B. Davies, eds., *Behavioural Ecology: An Evolutionary Approach*, 3d ed., 203–233. Blackwell, Oxford.

Harvey, P. H., and A. H. Harcourt. 1984. Sperm competition, testes size, and breeding systems in primates. In R. L. Smith, ed., *Sperm Competition and the Evolution of Animal Mating Systems*, 589–600. Academic Press, Orlando, Florida.

Harvey, P. H., and M. D. Pagel. 1991. *The Comparative Method in Evolutionary Biology*. Oxford University Press, Oxford.

Harvey, P., and L. Partridge. 1982. Bird coloration and parasites—a task for the future? *Nature* 300: 480–481.

Harvey, P. H., M. J. Kavanagh, and T. H. Clutton-Brock. 1978. Sexual dimorphism in primate teeth. *J. Zool. London* 186: 475–485.

Harvey, P. H., A. F. Read, J. L. John, R. D. Gregory, and A. E. Keymer. 1991. An evolutionary perspective. In C. A. Toft and A. Aeschlimann, eds., *Parasitism: Coexistence or Conflict. Ecological, Physiological and Immunological Aspects*, 344–355. Oxford University Press, Oxford.

Haskins, C. P., E. F. Haskins, J.J.A. McLaughlin, and R. E. Hewitt. 1961. Polymorphism and population structure in *Lebistes reticulatus*, a population study. In W. F. Blair, ed., *Vertebrate Speciation*, 320–395. University of Texas Press, Austin.

Hasselquist, D., and S. Bensch. 1991. Trade-off between mate guarding and mate attraction in the polygynous great reed warbler. *Behav. Ecol. Sociobiol.* 28: 187–193.

Hasson, O. 1989. Amplifiers and the handicap principle in sexual selection: A different emphasis. *Proc. Roy. Soc. London* 235: 383–406.

Hasson, O. 1990. The role of amplifiers in sexual selection: An integration of the amplifying and the Fisherian mechanisms. *Evol. Ecol.* 4: 277–289.

Hasson, O. 1991. Sexual displays as amplifiers: Practical examples with an emphasis on feather decorations. *Behav. Ecol.* 2: 189–197.

Hastings, P. A. 1988a. Correlates of male reproductive success in the browncheek blenny, *Acanthemblemaria crockeri* (Blennioidea: Chaenopsidae). *Behav. Ecol. Sociobiol.* 22: 95–102.

Hastings, P. A. 1988b. Female choice and male reproductive success in the angel blenny, *Coralliozetus angelica* (Teleostei: Chaenopsidae). *Anim. Behav.* 36: 115–124.

Hatziolos, M. E., and R. L. Caldwell. 1983. Role reversal in courtship in the stomatopod *Pseudosquilla ciliata* (Crustacea). *Anim. Behav.* 31: 1077–1087.

Hausfater, G., and S. B. Hrdy, eds. 1984. *Infanticide: Comparative and Evolutionary Perspectives*. Aldine, New York.

Hausfater, G., H. C. Gerhardt, and G. M. Klump. 1990. Parasites and mate choice in gray treefrogs, *Hyla versicolor*. *Am. Zool.* 30: 299–311.

Hedlund, L. 1990. Factors affecting differential mating success in male crested newts, *Triturus cristatus*. *J. Zool. London* 220: 33–40.

Hedrick, A. V. 1986. Female preferences for male calling bout duration in a field cricket. *Behav. Ecol. Sociobiol.* 19: 73–77.

Hedrick, A. V. 1988. Female choice and the heritability of attractive male traits: An empirical study. *Am. Nat.* 132: 267–276.

Hedrick, A. V., and E. J. Temeles. 1989. The evolution of sexual dimorphism in animals: Hypotheses and tests. *TREE* 4: 136–138.

Hegner, R. E., and J. C. Wingfield. 1987. Effects of experimental manipulation of testosterone levels on parental investment and breeding success in male house sparrows. *Auk* 104: 462–469.

Heisler, I. L. 1984a. Inheritance of female mating propensities for yellow locus genotypes in *Drosophila melonogaster*. *Genet. Res. Camb.* 44: 133–149.

Heisler, I. L. 1984b. A quantitative genetic model for the origin of mating preferences. *Evolution* 38: 1283–1295.

Heisler, I. L. 1985. Quantitative genetic models of female choice based on "arbitrary" male characters. *Heredity* 55: 187–198.

Heisler, I. L., and J. W. Curtsinger. 1990. Dynamics of sexual selection in diploid populations. *Evolution* 44: 1164–1176.

Heisler, I. L., et al. 1987. The evolution of mating preferences and sexually selected traits: Group report. In J. W. Bradbury and M. B. Andersson, eds., *Sexual Selection: Testing the Alternatives*, 96–118. Wiley, Chichester, U.K.

Henry, C. S. 1985. Sibling species, call differences, and speciation in green lacewings (Neuroptera: Chrysopidae, *Chrysoperla*). *Evolution* 39: 965–984.

Heptner, W. A., A. A. Nasimovitsch, and A. G. Bannikov. 1961. *Mammals of the Soviet Union*. Fischer, Jena, Germany.

Hews, D. K. 1990. Examining hypotheses generated by field measures of sexual selection on male lizards, *Uta palmeri*. *Evolution* 44: 1956–1966.

Heywood, J. S. 1989. Sexual selection by the handicap mechanism. *Evolution* 43: 1387–1397.

Hieber, C. S., and J. A. Cohen. 1983. Sexual selection in the lovebug, *Plecia nearctica*: The role of male choice. *Evolution* 37: 987–992.

Hill, G. E. 1989. Late spring arrival and dull nuptial plumage: Aggression avoidance by yearling males? *Anim. Behav.* 37: 665–673.

Hill, G. E. 1990. Female house finches prefer colourful males: Sexual selection for a condition-dependent trait. *Anim. Behav.* 40: 563–572.

Hill, G. E. 1991. Plumage coloration is a sexually selected indicator of male quality. *Nature* 350: 337–339.

Hill, G. E. 1992. Proximate basis of variation in carotenoid pigmentation in male house finches. *Auk* 109: 1–12.

Hill, G. E. 1993. Geographic variation in male ornamentation and female mate preference in the house finch: A comparative test of models of sexual selection. *Behav. Ecol.*, in press.

Hill, W. L. 1991. Correlates of male mating success in the ruff (*Philomachus pugnax*), a lekking shorebird. *Behav. Ecol. Sociobiol.* 29: 367–372.

Hillgarth, N. 1990a. Parasites and female choice in the ring-necked pheasant. *Am. Zool.* 30: 227–233.

Hillgarth, N. 1990b. Pheasant spurs out of fashion. *Nature* 345: 119–120.

Hinde, R. A. 1970. *Animal Behaviour*. 2d ed. McGraw-Hill, New York.

Hingston, R.W.G. 1933. *The Meaning of Animal Colour and Adornment*. Edward Arnold, London.

Hitchcock, C. L., A. Cronquist, M. Ownbey, and J. W. Thompson. 1961. *Vascular Plants of the Pacific Northwest*. University of Washington Press, Seattle.

Hjorth, I. 1970. Reproductive behavior in Tetraonidae. *Viltrevy* 7: 184–596.

Hobson, K. A., and S. G. Sealy. 1990. Female song in the yellow warbler. *Condor* 92: 259–261.

Hoekstra, R. F. 1987. The evolution of sexes. In S. C. Stearns, ed., *The Evolution of Sex and Its Consequences*, 59–92. Birkhäuser, Basel.

Hoelzer, G. A. 1989. The good parent process of sexual selection. *Anim. Behav.* 38: 1067–1078.

Hoelzer, G. A. 1990. Male-male competition and female choice in the Cortez damselfish, *Stegastes rectifraenum*. *Anim. Behav.* 40: 339–349.

Hoffman, J. A., L. L. Getz, and L. Gavish. 1987. Effect of multiple short-term exposures of pregnant *Microtus ochrogaster* to strange males. *J. Mammal.* 68: 166–169.

Hoffman, S. G., M. P. Schildhauer, and R. R. Warner. 1985. The costs of changing sex and the ontogeny of males under contest competition for mates. *Evolution* 39: 915–927.

Hogan-Warburg, A. J. 1966. Social behaviour of the ruff, *Philomachus pugnax* (L.). *Ardea* 54: 109–229.

Höglund, J. 1989a. Pairing and spawning patterns in the common toad, *Bufo bufo*: The effects of sex ratios and the time available for male-male competition. *Anim. Behav.* 38: 423–429.

Höglund, J. 1989b. Size and plumage dimorphism in lek-breeding birds: A comparative analysis. *Am. Nat.* 134: 72–87.

Höglund, J., and A. Lundberg. 1987. Sexual selection in a monomorphic lek-breeding bird: Correlates of male mating success in the great snipe *Gallinago media*. *Behav. Ecol. Sociobiol.* 21: 211–216.

Höglund, J., and A. Lundberg. 1989. Plumage color correlates with body size in the ruff (*Philomachus pugnax*). *Auk* 106: 336–337.

Höglund, J., and J.G.M. Robertson. 1988. Chorusing behaviour, a density-dependent alternative mating strategy in male common toads (*Bufo bufo*). *Ethology* 79: 324–332.

Höglund, J., and J.G.M. Robertson. 1990. Female preferences, male decision rules and the evolution of leks in the great snipe *Gallinago media*. *Anim. Behav.* 40: 15–22.

Höglund, J., M. Eriksson, and L. E. Lindell. 1990. Females of the lek-breeding great snipe, *Gallinago media*, prefer males with white tails. *Anim. Behav.* 40: 23–32.

Höglund, J., R. V. Alatalo, and A. Lundberg. 1992a. The effects of parasites on male ornaments and female choice in the lek-breeding black grouse (*Tetrao tetrix*). *Behav. Ecol. Sociobiol.* 30: 71–76.

Höglund, J., J. A. Kålås, and P. Fiske. 1992b. The costs of secondary sexual characters in the lekking great snipe (*Gallinago media*). *Behav. Ecol. Sociobiol.* 30: 309–316.

Hölldobler, B. 1976. The behavioral ecology of mating in harvester ants (Hymenoptera: Formicidae: *Pogonomyrmex*). *Behav. Ecol. Sociobiol.* 1: 405–423.

Hölldobler, B., and E. O. Wilson. 1990. *The Ants*. Springer, Berlin.

Holm, C. H. 1973. Breeding sex ratios, territoriality and reproductive success in the red-winged blackbird (*Agelaius phoeniceus*). *Ecology* 54: 356–365.

Holmberg, K., L. Edsman, and T. Klint. 1989. Female mate preferences and male attributes in mallard ducks, *Anas platyrhynchos*. *Anim. Behav.* 38: 1–7.

Honěk, A. 1993. Intraspecific variation in body size and fecundity in insects: A general relationship. *Oikos* 66: 483–492.

Horvitz, C. C., and D. W. Schemske. 1988. A test of the pollinator limitation hypothesis for a neotropical herb. *Ecology* 69: 200–206.

Houck, L. D. 1986. The evolution of salamander courtship pheromones. In D. Duvall, D. Müller-Schwarze, and R. M. Silverstein, eds., *Chemical Signals in Vertebrates*, 173–190. Plenum, New York.

Houck, L. D., and N. L. Reagan. 1990. Male courtship pheromones increase female receptivity in a plethodontid salamander. *Anim. Behav.* 39: 729–734.

Houck, L. D., S. J. Arnold, and R. A. Thisted. 1985. A statistical study of mate choice: Sexual selection in a plethodontid salamander (*Desmognathus ochrophaeus*). *Evolution* 39: 370–386.

Houde, A. E. 1987. Mate choice based upon naturally occurring color-pattern variation in a guppy population. *Evolution* 41: 1–10.

Houde, A. E. 1988a. Genetic difference in female choice between two guppy populations. *Anim. Behav.* 36: 510–516.

Houde, A. E. 1988b. The effects of female choice and male-male competition on the mating success of male guppies. *Anim. Behav.* 36: 888–896.

Houde, A. E. 1988c. Sexual selection in guppies called into question. *Nature* 333: 711.

Houde, A. E. 1992. Sex-linked heritability of a sexually selected character in a natural population of *Poecilia reticulata* (Pisces: Poeciliidae) (guppies). *Heredity* 69: 229–235.

Houde, A. E., and J. A. Endler. 1990. Correlated evolution of female mating preferences and male color patterns in the guppy *Poecilia reticulata*. *Science* 248: 1405–1408.

Houston, A. I., and N. B. Davies. 1985. The evolution of cooperation and life history in the dunnock *Prunella modularis*. In R. M. Sibley and R. H. Smith, eds., *Behavioural Ecology*. Blackwell, Oxford.

Howard, R. D. 1974. The influence of sexual selection and interspecific competition on mockingbird song (*Mimus polyglottus*). *Evolution* 28: 428–438.

Howard, R. D. 1978a. The evolution of mating strategies in bullfrogs, *Rana catesbeiana*. *Evolution* 32: 850–871.

Howard, R. D. 1978b. The influence of male-defended oviposition sites on early embryo mortality in bullfrogs. *Ecology* 59: 789–798.

Howard, R. D. 1980. Mating behaviour and mating success in woodfrogs, *Rana sylvatica*. *Anim. Behav.* 28: 705–716.

Howard, R. D. 1981a. Sexual dimorphism in bullfrogs. *Ecology* 62: 303–310.

Howard, R. D. 1981b. Male age-size distribution and male mating success in bullfrogs. In R. D. Alexander and D. W. Tinkle, eds., *Natural Selection and Social Behavior*. Chiron Press, New York.

Howard, R. D. 1983. Sexual selection and variation in reproductive success in a long-lived organism. *Am. Nat.* 122: 301–325.

Howard, R. D. 1984. Alternative mating behaviors of young male bullfrogs. *Am. Zool.* 24: 397–406.

Howard, R. D. 1988a. Sexual selection on male body size and mating behaviour in American toads, *Bufo americanus*. *Anim. Behav.* 36: 1796–1808.

Howard, R. D. 1988b. Reproductive success in two species of anurans. In T. H. Clutton-Brock, ed., *Reproductive Success*, 99–117. University of Chicago Press, Chicago.

Howard, R. D., and A. G. Kluge. 1985. Proximate mechanisms of sexual selection in wood frogs. *Evolution* 39: 260–277.

Hoy, R. R., J. Hahn, and R. C. Paul. 1977. Hybrid cricket auditory behavior: Evidence for genetic coupling in animal communication. *Science* 195: 82–83.

Hoy, R. R., A. Hoikkala, and K. Kaneshiro. 1988. Hawaiian courtship songs: Evolutionary innovation in communication signals of *Drosophila*. *Science* 240: 217–219.

Hrdy, S. B. 1979. Infanticide among animals: A review, classification, and examination of the implications for the reproductive strategies of females. *Ethol. Sociobiol.* 1: 13–40.

Hubbell, S. P., and L. K. Johnson. 1987. Environmental variance in lifetime mating success, mate choice and sexual selection. *Am. Nat.* 130: 91–112.

Huck, W., and E. M. Banks. 1982. Male dominance status, female choice and mating success in the brown lemming, *Lemmus trimucronatus*. *Anim. Behav.* 30: 665–675.

Huck, U. W., E. M. Banks, and S.-C. Wang. 1981. Olfactory discrimination of social status in the brown lemming. *Behav. Neur. Biol.* 33: 364–371.

Huey, R. B., E. R. Pianka, and T. W. Schoener. 1983. *Lizard Ecology: Studies of a Model Organism*. Harvard University Press, Cambridge, Mass.

Hughes, A. L., and M. K. Hughes. 1985. Female choice of mates in a polygynous insect, the whitespotted sawyer *Monochamus scutellatus*. *Behav. Ecol. Sociobiol.* 17: 385–387.

Hull, D. L. 1973. *Darwin and His Critics*. University of Chicago Press, Chicago.

Hurlbert, S. H. 1984. Pseudoreplication and the design of ecological field experiments. *Ecol. Monogr.* 54: 187–211.

Hurst, G.D.D., L. D. Hurst, and R. A. Johnstone. 1992. Intranuclear conflict and its role in evolution. *TREE* 7: 373–378.
Hurst, L. D. 1990. Parasite diversity and the evolution of diploidy, multicellularity and anisogamy. *J. theor. Biol.* 144: 429–443.
Hurst, L. D. 1992. Intragenomic conflict as an evolutionary force. *Proc. Roy. Soc. London* B 248: 135–140.
Hurst, L. D., and W. D. Hamilton. 1992. Cytoplasmic fusion and the nature of sexes. *Proc. Roy. Soc. London* B 247: 189–194.
Huxley, J. S. 1914. The courtship-habits of the great crested grebe (*Podiceps cristatus*); with an addition to the theory of sexual selection. *Proc. Zool. Soc. London* 35: 491–562.
Huxley, J. S. 1923. Courtship activities in the red-throated diver. *J. Linn. Soc. London (Zool.)* 35: 253–292.
Huxley, J. S. 1932. *Problems of Relative Growth.* Methuen, London.
Huxley, J. S. 1938a. Threat and warning coloration in birds, with a general discussion of the biological functions of colour. *Proc. 8th Internat. Orn. Congr.*, 430–455. Oxford.
Huxley, J. S. 1938b. Darwin's theory of sexual selection and the data subsumed by it, in the light of recent research. *Am. Nat.* 72: 416–433.
Huxley, J. S. 1938c. The present standing of the theory of sexual selection. In G. R. De Beer, ed., *Evolution (Essays on Aspects of Evolutionary Biology)*, 11–41. Oxford University Press, Oxford.
Huxley, J. S. 1942. *Evolution, the Modern Synthesis.* Allen and Unwin, London.
Hyvärinen, H., R.N.B. Kay, and W. J. Hamilton. 1977. Variation in the weight, specific gravity and composition of the antlers of red deer (*Cervus elaphus*). *Br. J. Nutr.* 38: 301–311.
Ims, R. A. 1987. Male spacing systems in microtine rodents. *Am. Nat.* 130: 475–484.
Ims, R. A. 1988a. Spatial clumping of sexually receptive females induces space sharing among male voles. *Nature* 335: 541–543.
Ims, R. A. 1988b. The potential for sexual selection in males: Effect of sex ratio and spatiotemporal distribution of receptive females. *Evol. Ecol.* 2: 338–352.
Ims, R. A. 1990. Mate detection success of male *Clethrionomys rufocanus* in relation to the spatial distribution of sexually receptive females. *Evol. Ecol.* 4: 57–61.
Ince, S. A., and P.J.B. Slater. 1985. Versatility and continuity in the song of thrushes *Turdus* spp. *Ibis* 127: 355–364.
Iwasa, Y., F. J. Odendaal, D. D. Murphy, P. R. Ehrlich, and A. E. Launer. 1983. Emergence patterns of male butterflies: A hypothesis and a test. *Theor. Pop. Biol.* 23: 363–379.
Iwasa, Y., A. Pomiankowski, and S. Nee. 1991. The evolution of costly mate preferences. II. The "handicap" principle. *Evolution* 45: 1431–1442.
Jacobs, G. H. 1981. *Comparative Color Vision.* Academic Press, New York.
Jacobs, L. F., S.J.C. Gaulin, D. F. Sherry, and G. E. Hoffman. 1990. Evolution of spatial cognition: Sex-specific patterns of spatial behavior predict hippocampal size. *Proc. Natl. Acad. Sci. USA* 87: 6349–6352.
Jacobs, M. E. 1955. Studies on territorialism and sexual selection in dragonflies. *Ecology* 36: 566–586.

Jacobson, M. 1972. *Insect Sex Pheromones*. Academic Press, New York.

Jacobson, S. K. 1985. Reproductive behavior and male success in two species of glass frogs (Centrolenidae). *Herpetologica* 41: 396–404.

Jaeger, R. G. 1986. Pheromonal markers as territorial advertisement by terrestrial salamanders. In D. Duvall, D. Müller-Schwarze, and R. M. Silverstein, eds., *Chemical Signals in Vertebrates*, 191–204. Plenum, New York.

Jägerskiöld, L. A. 1908. *Om Spel och Parningslekar Bland Djuren*. Gebers, Stockholm.

Jain, S. K., and A. D. Bradshaw. 1966. Evolutionary divergence among adjacent plant populations. I. The evidence and its theoretical analysis. *Heredity* 21: 407–441.

James, F. C., and C. E. McCulloch. 1985. Data analysis and the design of experiments in ornithology. In R. F. Johnston, ed., *Current Ornithology*, 1–63. Plenum, New York.

Jamieson, I. G., and P. W. Colgan. 1989. Eggs in the nests of males and their effect on mate choice in the three-spined stickleback. *Anim. Behav.* 38: 859–865.

Janetos, A. C. 1980. Strategies of female mate choice: A theoretical analysis. *Behav. Ecol. Sociobiol.* 7: 107–112.

Jannett, F. J., Jr. 1986. Morphometric patterns among microtine rodents. I. Sexual selection suggested by relative scent gland development in representative voles (*Microtus*). In D. Duvall, D. Müller-Schwarze, and R. M. Silverstein, eds., *Chemical Signals in Vertebrates*, 541–550. Plenum, New York.

Janzen, D. H. 1977. A note on optimal mate selection by plants. *Am. Nat.* 111: 365–371.

Janzen, D. H., P. DeVries, D. E. Gladstone, M. L. Higgins, and T. M. Lewinsohn. 1980. Self- and cross-pollination of *Encyclia cordigera* (Orchidaceae) in Santa Rosa National Park, Costa Rica. *Biotropica* 12: 72–74.

Jarman, P. J. 1974. The social organisation of antelope in relation to their ecology. *Behaviour* : 215–266.

Jarman, P. J. 1983. Mating system and sexual dimorphism in large, terrestrial, mammalian herbivores. *Biol. Rev.* 58: 485–520.

Jarman, P. J. 1989. On being thick-skinned: Dermal shields in large mammalian herbivores. *Biol. J. Linn. Soc.* 36: 169–191.

Järvi, T. 1983. The evolution of song versatility in the willow warbler *Phylloscopus trochilus*: A case of evolution by intersexual selection explained by the "female's choice of the best mate." *Ornis Scand.* 14: 123–128.

Järvi, T. 1990. The effects of male dominance, secondary sexual characteristics and female mate choice on the mating success of male Atlantic salmon (*Salmo salar*). *Ethology* 84: 123–132.

Järvi, T., and M. Bakken. 1984. The function of the variation in the breast-stripe of the great tit (*Parus major*). *Anim. Behav.* 32: 590–596.

Järvi, T., T. Radesäter, and S. Jakobsson. 1980. The song of the willow warbler *Phylloscopus trochilus* with special reference to singing behaviour in agonistic situations. *Ornis Scand.* 11: 236–242.

Järvi, T., E. Røskaft, M. Bakken, and B. Zumsteg. 1987. Evolution of variation in male secondary sexual characteristics. A test of eight hypotheses applied to pied flycatchers. *Behav. Ecol. Sociobiol.* 20: 161–169.

Jehl, J. R., Jr. 1970. Sexual selection for size differences in two species of sandpipers. *Evolution* 24: 311–319.

Jehl, J. R., Jr., and B. G. Murray, Jr. 1986. The evolution of normal and reverse sexual size dimorphism in shorebirds and other birds. In R. F. Johnston, ed., *Current Ornithology*, vol. 3, 1–86. Plenum, New York.

Jenni, D. A. 1974. Evolution of polyandry in birds. *Am. Zool.* 14: 129–144.

Jenni, D. A., and G. Collier. 1972. Polyandry in the American jacana. *Auk* 89: 743–765.

Johns, J. E. 1964. Testosterone-induced nuptial feathers in Phalaropes. *Condor* 66: 449–455.

Johnsgard, P. A. 1983. *The Grouse of the World*. Croom Helm, London.

Johnson, D. H., and A. B. Sargeant. 1977. Impact of red fox predation on the sex ratio of prairie mallards. *U.S. Dept. Interior, Wildl. Res. Rep.* 6: 1–56.

Johnson, K. 1988a. Sexual selection in pinyon jays. I. female choice and male-male competition. *Anim. Behav.* 36: 1038–1047.

Johnson, K. 1988b. Sexual selection in pinyon jays. II. male choice and female-female competition. *Anim. Behav.* 36: 1048–1053.

Johnson, K., and J. M. Marzluff. 1990. Some problems and approaches in avian mate choice. *Auk* 107: 296–304.

Johnson, L. K. 1982. Sexual selection in a brentid weevil. *Evolution* 36: 251–262.

Johnson, L. K., and S. P. Hubbell. 1984. Male choice. Experimental demonstration in a brentid weevil. *Behav. Ecol. Sociobiol.* 15: 183–188.

Johnson, L. L., and M. S. Boyce. 1991. Female choice of males with low parasite loads in sage grouse. In J. E. Loye and M. Zuk, eds., *Bird-Parasite Interactions*, 377–388. Oxford University Press, Oxford.

Johnson, S. G. 1991. Effects of predation, parasites, and phylogeny on the evolution of bright coloration in North American male passerines. *Evol. Ecol.* 5: 52–62.

Johnston, M. O. 1991a. Natural selection on floral traits in two species of *Lobelia* with different pollinators. *Evolution* 45: 1468–1479.

Johnston, M. O. 1991b. Pollen limitation of female reproduction in *Lobelia cardinalis* and *L. siphilitica*. *Ecology* 72: 1500–1503.

Johnston, R. F. 1966. The adaptive basis of geographic variation in color of the purple martin. *Condor* 68: 219–228.

Johnstone, R. A., and A. Grafen. 1992. Error-prone signalling. *Proc. Roy. Soc. London* B 248: 229–233.

Jones, G. P. 1981. Spawning-site choice by female *Pseudolabrus celidotus* (Pisces: Labridae) and its influence on the mating system. *Behav. Ecol. Sociobiol.* 8: 129–142.

Jones, I. L. 1992. Sexual selection and the evolution of extravagant traits in birds: problems with testing good-genes models of sexual selection. *Auk* 109: 197–199.

Jones, I. L., and F. M. Hunter. 1993. Mutual sexual selection in a monogamous seabird. *Nature* 362: 238–239.

Jones, I. L., and R. D. Montgomerie. 1991. Mating and remating of least auklets relative to ornamental traits. *Behav. Ecol.* 2: 249–257.

Jones, I. L., and R. D. Montgomerie. 1992. Least auklet ornaments: Do they function as quality indicators? *Behav. Ecol. Sociobiol.* 30: 43–52.

Jones, K. N., F. J. Odendaal, and P. R. Ehrlich. 1986. Evidence against the spermatophore as paternal investment in the checker-spot butterflies (*Euphydryas*: Nymphalidae). *Am. Midl. Nat.* 116: 1–6.

Jönsson, P. E. 1987. Sexual size dimorphism and disassortative mating in the Dunlin (*Calidris alpina schinzii*) in southern Sweden. *Ornis. Scand.* 18: 257–264.

Jönsson, P. E., and T. Alerstam. 1990. The adaptive significance of parental role division and sexual size dimorphism in breeding shorebirds. *Biol. J. Linn. Soc.* 41: 301–314.

Jormalainen, V., and J. Tuomi. 1989. Sexual differences in habitat selection and activity of the colour polymorphic isopod *Idotea baltica*. *Anim. Behav.* 38: 576–585.

Jumper, G. Y., Jr., and R. C. Baird. 1991. Location by olfaction: A model and application to the mating problem in the deep-sea hatchetfish *Argyropelecus hemigymnus*. *Am. Nat.* 138: 1431–1458.

Källander, H. 1983. Aspects of the Breeding Biology, Migratory Movements, Winter Survival, and Population Fluctuations in the Great Tit *Parus major* and the Blue Tit *P. caeruleus*. Ph.D. diss., University of Lund, Sweden.

Kaneshiro, K. Y. 1983. Sexual selection and direction of evolution in the biosystematics of Hawaiian Drosophilidae. *Ann. Rev. Entomol.* 28: 161–178.

Kaneshiro, K. Y. 1988. Speciation in the Hawaiian *Drosophila*: Sexual selection appears to play an important role. *BioScience* 38: 258–263.

Kaneshiro, K. Y., and C.R.B. Boake. 1987. Sexual selection and speciation: Issues raised by Hawaiian *Drosophila*. *TREE* 2: 207–212.

Kaufman, G. W., D. B. Siniff, and R. Reichle. 1975. Colony behavior of Weddell seals, *Leptonychotes weddelli*, at Hutton Cliffs, Antarctica. *Rapp. Proc.-Verb. Réun. Cons. Int. Expl. Mer* 169: 228–246.

Keane, B. 1990. The effect of relatedness on reproductive success and mate choice in the white-footed mouse, *Peromyscus leucopus*. *Anim. Behav.* 39: 264–273.

Kearns, P.W.E., I.P.M. Tomlinson, C. J. Veltman, and P. O'Donald. 1992. Nonrandom mating in *Adalia bipunctata* (the two-spot ladybird). Further tests for female mating preference. *Heredity* 68: 385–389.

Keenleyside, M.H.A., R. W. Rangeley, and B. U. Kuppers. 1985. Female mate choice and parental defense behaviour in the cichlid fish *Cichlasoma nigrofasciatum*. *Can. J. Zool.* 63: 2489–2493.

Kellogg, V. L. 1907. *Darwinism To-day*. Holt, New York.

Kelsey, M. G. 1989. A comparison of the song and territorial behaviour of a long-distance migrant, the marsh warbler (*Acrocephalus palustris*), in summer and winter. *Ibis* 131: 403–414.

Kempenaers, B., G. R. Verheyen, M. Van den Broeck, T. Burke, C. V. Van Broeckhoven, and A. Dhondt. 1992. Extra-pair paternity results from female preference for high-quality males in the blue tit. *Nature* 357: 494–496.

Kennedy, C.E.J., J. A. Endler, S. L. Poynton, and H. McMinn. 1987. Parasite load predicts mate choice in guppies. *Behav. Ecol. Sociobiol.* 21: 291–295.

Ketterson, E. D., and V. Nolan, Jr. 1992. Hormones and life histories: An integrative approach. *Am. Nat.* 140: s33-s62.

Kiester, A. R., R. Lande, and D. W. Schemske. 1984. Models of coevolution and speciation in plants and their pollinators. *Am. Nat.* 124: 220–243.

Kiltie, R. A. 1985. Evolution and function of horns and hornlike organs in female ungulates. *Biol. J. Linn. Soc.* 24: 299–320.

King, A. P., M. J. West, and D. H. Eastzer. 1980. Song structure and song development as potential contributors to reproductive isolation in cowbirds (*Molothrus ater*). *J. Comp. Physiol. Psychol.* 94: 1028–1039.

King, C. M., and P. J. Moors. 1979. On co-existence, foraging strategy and the biogeography of weasels and stoats (*Mustela nivalis* and *M. erminea*) in Britain. *Oecologia* 39: 129–150.

Kirkpatrick, C. E., S. K. Robinson, and U. D. Kitron. 1991. Phenotypic correlates of blood parasitism in the common grackle. In J. E. Loye and M. Zuk, eds., *Bird-Parasite Interactions*, 349–358. Oxford University Press, Oxford.

Kirkpatrick, M. 1982. Sexual selection and the evolution of female choice. *Evolution* 36: 1–12.

Kirkpatrick, M. 1985. Evolution of female choice and male parental investment in polygynous species: The demise of the "sexy son." *Am. Nat.* 125: 788–810.

Kirkpatrick, M. 1986a. Sexual selection and cycling parasites: A simulation study of Hamilton's hypothesis. *J. theor. Biol.* 119: 263–271.

Kirkpatrick, M. 1986b. The handicap mechanism of sexual selection does not work. *Am. Nat.* 127: 222–240.

Kirkpatrick, M. 1987a. Sexual selection by female choice in polygynous animals. *Ann. Rev. Ecol. Syst.* 18: 43–70.

Kirkpatrick, M. 1987b. The evolutionary forces acting on female mating preferences in polygynous animals. In J. W. Bradbury and M. B. Andersson, eds., *Sexual Selection: Testing the Alternatives*, 67–82. Wiley, Chichester, U.K.

Kirkpatrick, M. 1992. Direct selection of female mating preferences: Comments on Grafen's models. *J. theor. Biol.* 154: 127–129.

Kirkpatrick, M., and M. J. Ryan. 1991. The evolution of mating preferences and the paradox of the lek. *Nature* 350: 33–38.

Kirkpatrick, M., T. Price, and S. J. Arnold. 1990. The Darwin-Fisher theory of sexual selection in monogamous birds. *Evolution* 44: 180–193.

Kitchen, D. W. 1974. Social behavior and ecology of the pronghorn. *Wildl. Monogr.* 38: 1–96.

Kleiman, D. 1977. Monogamy in mammals. *Q. Rev. Biol.* 52: 39–69.

Klint, T. 1980. Influence of male nuptial plumage on mate selection in the female mallard (*Anas platyrhynchos* L.). *Anim. Behav.* 28: 1230–1238.

Kluge, A. G. 1981. The life history, social organization, and parental behavior of *Hyla rosenbergi* Boulenger, a nest-building gladiator frog. *Misc. Publ. Mus. Zool. Univ. Michigan* 160: 1–170.

Klump, G. M., and H. C. Gerhardt. 1987. Use of non-arbitrary acoustic criteria in mate choice by female gray tree frogs. *Nature* 326: 286–288.

Klun, J. A., and S. Maini. 1979. Genetic basis of an insect chemical communication system: The European cornborer. *Environ. Entomol.* 8: 423–426.

Knapp, R. A., and J. T. Kovach. 1991. Courtship as an honest indicator of male parental quality in the bicolor damselfish, *Stegastes partitus*. *Behav. Ecol.* 2: 295–300.

Knowlton, N. 1974. A note on the evolution of gamete dimorphism. *J. theor. Biol.* 46: 283–285.

Knowlton, N. 1980. Sexual selection and dimorphism in two demes of a symbiotic, pair-bonding snapping shrimp. *Evolution* 34: 161–173.

Knuth, P. 1906. *Handbook of Flower Pollination*. Clarendon, Oxford.

Kochmer, J. P., and R. H. Wagner. 1988. Why are there so many kinds of passerine birds? Because they are small. A reply to Raikow. *Syst. Zool.* 37: 68–69.

Kodric-Brown, A. 1983. Determinants of male reproductive success in pupfish (*Cyprinodon pecosensis*). *Anim. Behav.* 31: 128–137.

Kodric-Brown, A. 1985. Female preferences and sexual selection for male coloration in the guppy. *Behav. Ecol. Sociobiol.* 17: 199–205.

Kodric-Brown, A. 1989. Dietary carotenoids and male mating success in the guppy: An environmental component to female choice. *Behav. Ecol. Sociobiol.* 25: 393–401.

Kodric-Brown, A. 1992. Male dominance can enhance mating success in guppies. *Anim. Behav.* 44: 165–167.

Kodric-Brown, A., and J. H. Brown. 1984. Truth in advertising: The kinds of traits favoured by sexual selection. *Am. Nat.* 124: 309–323.

Koenig, W. D., and S. S. Albano. 1986. On the measurement of sexual selection. *Am. Nat.* 127: 403–409.

Koepfer, H. R. 1987. Selection for sexual isolation between geographic forms of *Drosophila mojavensis*. II. Effects of selection on mating preference and propensity. *Evolution* 41: 1409–1413.

Kondrashov, A. S. 1988. Deleterious mutations as an evolutionary factor. III. Mating preference and some general remarks. *J. theor. Biol.* 131: 487–496.

Koopman, P., J. Gubbay, N. Vivian, P. Goodfellow, and R. Lovell-Badge. 1991. Male development of chromosomally female mice transgenic for *Sry*. *Nature* 351: 117–121.

Korpimäki, E. 1987a. Reversed size dimorphism in birds of prey, especially in Tengmalm's owl *Aegolius funereus*: A test of the "starvation hypothesis." *Ornis Scand.* 17: 309–315.

Korpimäki, E. 1987b. Sexual size dimorphism and life-history traits of Tengmalm's owl: A review. In R. W. Nero, R. J. Clark, R. J. Knapton, and R. H. Hamre, eds., *Biology and Conservation of Northern Forest Owls*, 157–161. U.S. Dept. of Agriculture, Fort Collins, Colo.

Kottler, M. J. 1980. Darwin, Wallace, and the origin of sexual dimorphism. *Proc. Am. Phil. Soc.* 124: 203–226.

Kottler, M. J. 1985. Charles Darwin and Alfred Russel Wallace: Two decades of debate over natural selection. In D. Kohn, ed., *The Darwinian Heritage*, 367–432. Princeton University Press, Princeton, N.J.

Kozlowski, J. 1989. Sexual size dimorphism: A life history perpective. *Oikos* 54: 253–255.

Král, M., T. Järvi, and V. Bicík. 1988. Inter-specific aggression between the collared flycatcher and the pied flycatcher: The selective agent for the evolution of light-coloured male pied flycatcher populations? *Ornis Scand.* 19: 287–289.

Kramer, B. 1990. Sexual signals in electric fishes. *TREE* 5: 247–250.

Krebs, J. R. 1976. Habituation and song repertoires in the great tit. *Behav. Ecol. Sociobiol.* 1: 215–227.

Krebs, J. R. 1977. The significance of song repertoires: The Beau Geste hypothesis. *Anim. Behav.* 25: 475–478.

Krebs, J. R. 1979. Bird colours. *Nature* 282: 14–16.

Krebs, J. R., and N. B. Davies. 1987. *An Introduction to Behavioural Ecology*. 2d ed. Blackwell, Oxford.

Krebs, J. R., and R. Dawkins. 1984. Animal signals: Mind-reading and manipulation. In J. R. Krebs and N. B. Davies, eds., *Behavioural Ecology: An Evolutionary Approach*, 2d ed., 380–402. Blackwell, Oxford.

Krebs, R. A., and D. A. West. 1988. Female mate preference and the evolution of female-limited batesian mimicry. *Evolution* 42: 1101–1104.

Krebs, J. R., R. Ashcroft, and M. I. Webber. 1978. Song repertoires and territory defence in the great tit. *Nature* 271: 539–542.

Kronester-Frei, A. 1975. Licht-und Elektronenmikroskopische Untersuchungen am Brutephitel des Männchens von *Nerophis lumbriciformis* (Pennant 1976), Syngnathidae. *Forma Funct.* 8: 419–462.

Kroodsma, D. E. 1976. Reproductive development in a female songbird: Differential stimulation by quality of male song. *Science* 192: 574–575.

Kroodsma, D. E. 1977. Correlates of song organisation among North American wrens. *Am. Nat.* 111: 995–1008.

Kroodsma, D. E. 1989a. Pseudoreplication, external validity and the design of playback experiments. *Anim. Behav.* 37: 715–719.

Kroodsma, D. E. 1989b. Suggested experimental designs for song playbacks. *Anim. Behav.* 37: 600–609.

Kroodsma, D. E. 1990. Using appropriate experimental designs for intended hypotheses in "song" playbacks, with examples for testing effects of song repertoire sizes. *Anim. Behav.* 40: 1138–1150.

Kroodsma, D. E., and B. E. Byers. 1991. The function(s) of bird song. *Am. Zool.* 31: 318–328.

Kroodsma, D. E., R. C. Bereson, B. E. Byers, and E. Minear. 1989. Use of song types by the chestnut-sided warbler: Evidence for both intra- and inter-sexual functions. *Can. J. Zool.* 67: 447–456.

Kruijt, J. P., and G. J. de Vos. 1988. Individual variation in reproductive success in male black grouse, *Tetrao tetrix* L. In T. H. Clutton-Brock, ed., *Reproductive Success*, 279–290. University of Chicago Press, Chicago.

Kruijt, J. P., and G. B. Meeuwissen. 1991. Sexual preferences of male zebra finches, effects of early and adult experience. *Anim. Behav.* 42: 91–102.

Kruijt, J. P., G. J. de Vos, and I. Bossema. 1972. The arena system of black grouse. *Proc. XV Int. Ornithol. Congr.*: 399–423.

Krupa, J. J. 1989. Alternative mating tactics in great plains toads. *Anim. Behav.* 37: 1035–1043.

Kruse, K. C. 1990. Male back space availability in the giant waterbug (*Belostoma flumineum* Say). *Behav. Ecol. Sociobiol.* 26: 281–290.

Kunin, W. E. 1993. Sex and the single mustard: Population density and pollinator behavior effects on seed set. *Ecology* 74: 2145–2160.

Kuwamura, T. 1985. Social and reproductive behavior of three mouthbrooding cardinalfishes, *Apogon doederleini*, *A. niger* and *A. notatus*. *Env. Biol. Fish.* 13: 17–24.

Kyriacou, C. P., and J. C. Hall. 1982. The function of courtship song rhythms in *Drosophila*. *Anim. Behav.* 30: 794–801.

Labove, J. B. 1981. Pregnancy blocking in rodents: Adaptive advantages for females. *Am. Nat.* 118: 361–371.

Lack, D. 1940. Courtship feeding in birds. *Auk* 57: 169–178.

Lack, D. 1968. *Ecological Adaptations for Breeding in Birds*. Methuen, London.

Lambrechts, M., and A. A. Dhondt. 1986. Male quality, reproduction, and survival in the great tit (*Parus major*). *Behav. Ecol. Sociobiol.* 19: 57–63.

Lambrechts, M., and A. A. Dhondt. 1988. The anti-exhaustion hypothesis: A new hypothesis to explain song performance and song switching in the great tit. *Anim. Behav.* 36: 327–334.

Lancaster, D. A. 1964. Life history of the Boucard Tinamou in British Honduras. I. Distribution and general behavior. *Condor* 66: 165–181.

Lande, R. 1976. The maintenance of genetic variability by mutation in a polygenic character with linked loci. *Genet. Res.* 26: 221–235.

Lande, R. 1979. Quantitative genetic analysis of multivariate evolution, applied to brain:body size allometry. *Evolution* 33: 402–416.

Lande, R. 1980. Sexual dimorphism, sexual selection, and adaptation in polygenic characters. *Evolution* 34: 292–305.

Lande, R. 1981. Models of speciation by sexual selection on polygenic traits. *Proc. Natl. Acad. Sci. USA* 78: 3721–3725.

Lande, R. 1982a. Rapid origin of sexual isolation and character divergence in a cline. *Evolution* 36: 213–223.

Lande, R. 1982b. A quantitative genetic theory of life history evolution. *Ecology* 63: 607–615.

Lande, R. 1987. Genetic correlations between the sexes in the evolution of sexual dimorphism and mating preferences. In J. W. Bradbury and M. B. Andersson, eds., *Sexual Selection: Testing the Alternatives.*, 83–94. Wiley, Chichester, U.K.

Lande, R., and S. J. Arnold. 1983. The measurement of selection on correlated characters. *Evolution* 37: 1210–1226.

Lande, R., and S. J. Arnold. 1985. Evolution of mating preference and sexual dimorphism. *J. theor. Biol.* 117: 651–664.

Lande, R., and M. Kirkpatrick. 1988. Ecological speciation by sexual selection. *J. theor. Biol.* 133: 85–98.

Landolt, P. J., and R. R. Heath. 1990. Sexual role reversal in mate-finding strategies of the cabbage looper moth. *Science* 249: 1026–1028.

Langston, N. E., S. Freeman, S. Rohwer, and D. Gori. 1990. The evolution of female body size in red-winged blackbirds: The effects of timing of breeding, social competition, and reproductive energetics. *Evolution* 44: 1764–1779.

Lank, D. B., and C. M. Smith. 1987. Conditional lekking in ruff (*Philomachus pugnax*). *Behav. Ecol. Sociobiol.* 20: 137–145.

Lank, D. B., and C. M. Smith. 1992. Females prefer larger leks: Field experiments with ruffs (*Philomachus pugnax*). *Behav. Ecol. Sociobiol.* 30: 323–329.

Lank, D. B., L. W. Oring, and S. J. Maxson. 1985. Mate and nutrient limitation of egg-laying in a polyandrous shorebird. *Ecology* 66: 1513–1524.

Larson, K. S., and R. J. Larson. 1990. Lure of the locks: Showiest ladies-tresses orchids, *Spiranthes romanzoffiana*, affect bumblebee, *Bombus* spp., foraging behavior. *Can. Field-Nat.* 104: 519–525.

Latimer, W., and M. Sippel. 1987. Acoustic cues for female choice and male competition in *Tettigonia cantans*. *Anim. Behav.* 35: 887–900.

Lawrence, W. S. 1986. Male choice and competition in *Tetraopes tetraophthalmus*: Effects of local sex ratio variation. *Behav. Ecol. Sociobiol.* 18: 289–296.

Laws, R. M. 1956. The elephant seal (*Mirounga leonina* Linn.). II. General, social, and reproductive behaviour. *Falkland Is. Depend. Surv., Sci. Rep.* 15: 1–66.

Leary, R. F., and F. W. Allendorf 1989. Fluctuating asymmetry as an indicator of stress: Implications for conservation biology. *TREE* 4: 214–217.

Le Boeuf, B. J. 1974. Male-male competition and reproductive success in elephant seals. *Am. Zool.* 14: 163–176.

Le Boeuf, B. J. 1986. Sexual strategies of seals and walruses. *New Sci.* 16: 36–39.

Le Boeuf, B. J., and S. Mesnick. 1990. Sexual behavior of male Northern elephant seals. I. Lethal injuries to adult females. *Behaviour* 116: 143–162.

Le Boeuf, B. J., and R. S. Peterson. 1969. Social status and mating activity in elephant seals. *Science* 163: 91–93.

Le Boeuf, B. J., and J. Reiter. 1988. Lifetime reproductive success in northern elephant seals. In T. H. Clutton-Brock, ed., *Reproductive Success*, 344–362. University of Chicago Press, Chicago.

LeCroy, M. 1981. The genus *Paradisaea*—display and evolution. *Am. Mus. Nov.* 2714: 1–52.

LeCroy, M., A. Kulupi, and W. S. Peckover. 1980. Goldie's bird of paradise: Display, natural history and traditional relationships of people to the bird. *Wilson Bull.* 92: 289–301.

Lee, A. K., and A. Cockburn. 1985. *Evolutionary Ecology of Marsupials*. Cambridge University Press, Cambridge, U.K.

Lee, T. D. 1984. Patterns of fruit maturation: A gametophyte competition hypothesis. *Am. Nat.* 123: 427–432.

Lee, T. D. 1988. Patterns of fruit and seed production. In J. Lovett-Doust and L. Lovett-Doust, eds., *Plant Reproductive Ecology: Patterns and Strategies*, 179–202. Oxford University Press, New York.

Lefcort, H., and A. R. Blaustein. 1991. Parasite load and brightness in lizards: An interspecific test of the Hamilton and Zuk hypothesis. *J. Zool., London* 224: 491–499.

Lehrman, D., and M. Friedman. 1969. Auditory stimulation of ovarian activity in the ring dove (*Streptopelia risoria*). *Anim. Behav.* 17: 494–497.

Leigh, E. G., E. L. Charnov, and R. R. Warner. 1976. Sex ratio, sex change, and natural selection. *Proc. Natl. Acad. Sci. USA* 73: 3655–3660.

Lemel, J. 1993. Evolutionary and Ecological Perspectives of Status Signalling in the Great Tit (*Parus major L.*). Ph.D. diss., University of Göteborg, Sweden.

Lemnell, P. A. 1978. Social behaviour of the great snipe *Capella media* at the arena display. *Ornis Scand.* 9: 146–163.

Lemon, R. E., S. Monette, and D. Roff. 1987. Song repertoires of American warblers (*Parulinae*): Honest advertising or assessment? *Ethology* 74: 265–284.

Lenington, S. 1980. Female choice and polygyny in redwinged blackbirds. *Anim. Behav.* 28: 347–361.

Lenington, S. 1984. The evolution of polyandry in shorebirds. In J. Burger and B. L. Olla, eds., *Behavior of Marine Animals*, vol. 5, 149–167. Plenum, New York.

Lenington, S., and K. Egid. 1985. Female discrimination of male odors correlated with male genotype at the T locus: A response to T-locus or H-2-locus variability? *Behav. Genet.* 15: 53–67.

Leonard, M. L., and R. Picman. 1988. Mate choice by marsh wrens: The influence of male and territory quality. *Anim. Behav.* 36: 517–528.

Leutenegger, W. 1978. Scaling of sexual dimorphism in body size and breeding system in primates. *Nature* 272: 610–611.

Leutenegger, W., and J. Cheverud. 1982. Correlates of sexual dimorphism in primates: Ecological and size variables. *Int. J. Primatol.* 3: 387–402.

Leutenegger, W., and J. T. Kelly. 1977. Relationship of sexual dimorphism in canine size and body size to social, behavioral, and ecological correlates in anthropoid primates. *Primates* 18: 117–136.

Levin, D. A. 1978. The origin of isolating mechanisms in flowering plants. *Evol. Biol.* 11: 185–317.

Levin, D. A. 1985. Reproductive character displacement in *Phlox*. *Evolution* 39: 1275–1281.

Levin, D. A., and H. W. Kerster. 1967. Natural selection for reproductive isolation in *Phlox*. *Evolution* 21: 679–687.

Levin, D. A., and B. A. Schaal. 1970. Corolla color as an inhibitor of interspecific hybridization in *Phlox*. *Am. Nat.* 104: 273–283.

Levins, R. 1966. The strategy of model building in population biology. *Am. Scient.* 54: 421–431.

Lewis, D. M. 1972. Importance of face-mask in sexual recognition and territorial behavior in the yellowthroat. *Jack-Pine Warbler* 50: 98–109.

Lewis, E. R. 1981. Suggested evolution of tonotopic organization in the frog amphibian papilla. *Neurosci. Lett.* 21: 131–136.

Lewontin, R. C. 1965. Selection for colonizing ability. In H. G. Baker and G. L. Stebbins, eds., *The Genetics of Colonizing Species*, 77–94. Academic Press, New York.

Licht, L. E. 1976. Sexual selection in toads (*Bufo americanus*). *Can. J. Zool.* 54: 1277–1284.

Lifjeld, J. T., and T. Slagsvold. 1988. Female pied flycatchers *Ficedula hypoleuca* choose male characteristics in homogeneous habitats. *Behav. Ecol. Sociobiol.* 22: 27–36.

Lightbody, J. P., and P. J. Weatherhead. 1987. Polygyny in yellow-headed blackbirds: Female choice versus male competition. *Anim. Behav.* 35: 1670–1684.

Lightbody, J. P., and P. J. Weatherhead. 1988. Female settling patterns and polygyny: Tests of a neutral mate-choice hypothesis. *Am. Nat.* 132: 20–33.

Ligon, J. D. 1968. Sexual differences in foraging behavior in two species of *Dendrocopus* woodpeckers. *Auk* 85: 203–215.

Ligon, J. D., R. Thornhill, M. Zuk, and K. Johnson. 1990. Male-male competition, ornamentation and the role of testosterone in sexual selection in red jungle fowl. *Anim. Behav.* 40: 357–373.

Liley, R. N. 1966. Ethological isolating mechanisms in four sympatric species of Poecilid fishes. *Behaviour* (suppl.) 13: 1–197.

Liley, R. N., and B. H. Seghers. 1975. Factors affecting the morphology and behavior of guppies in Trinidad. In G. P. Baerends, C. Beer, and A. Manning, eds., *Function and Evolution in Behaviour*, 92–118. Oxford University Press, Oxford.

Lill, A. 1974. Sexual behavior of the lek-forming white-bearded manakin. *Z. Tierpsychol.* 36: 1–36.

Lill, A. 1976. Lek behavior in the golden-headed manakin, *Pipra erythrocephala* in Trinidad (West Indies). *Z. Tierpsychol.* (suppl.) 18: 1–84.

Lill, A. 1979. An assessment of male parental investment and pair bonding in the polygamous superb lyrebird. *Auk* 96: 489–498.

Lincoln, G. A. 1972. The role of antlers in the behaviour of red deer. *J. Exp. Zool.* 182: 233–250.

Lincoln, G. A. 1992. Biology of antlers. *J. Zool., London* 226: 517–528.

Lincoln, G. A., and J. Fletcher. 1977. History of a hummel. 5. Offspring from father/daughter matings. *Deer* 4: 86–87.

Lincoln, G. A., R. W. Youngson, and R. V. Short. 1970. The social and sexual behaviour of the red deer stag. *J. Reprod. Fert.* (suppl.) 11: 71–103.

Lindberg, P. 1983. Relations between the diet of Fennoscandian peregrines *Falco peregrinus* and organochlorines and mercury in their feathers, with a comparison with the gyrfalcon *Falco rusticolus*. Ph.D. diss., University of Göteborg, Sweden.

Lindén, M. 1991. Divorce in great tits—chance or choice? An experimental approach. *Am. Nat.* 138: 1039–1048.

Linnaeus, C. 1758. *Systema Naturae*. 10th ed. Stockholm.

Liske, E., and W. J. Davis. 1987. Courtship and mating behaviour of the Chinese praying mantis *Tenodera aridfolia sinensis*. *Anim. Behav.* 35: 1524–1538.

Litte, M. 1979. *Mischocyttarus flavitarsis* in Arizona: Social and nesting biology of a polistine wasp. *Z. Tierpsychol.* 50: 282–312.

Little, R. J. 1983. A review of floral food deception mimicries with comments on floral mutualism. In C. E. Jones and R. J. Little, eds., *Handbook of Experimental Pollination Biology*, 294–309. Van Nostrand Reinhold, New York.

Littlejohn, M. J. 1965. Premating isolation in the *Hyla ewingi* complex (Anura: Hylidae). *Evolution* 19: 234–243.

Littlejohn, M. J. 1981. Reproductive isolation: A critical review. In W. R. Atchley and D. S. Woodruff, eds., *Evolution and Speciation*, 298–334. Cambridge University Press, Cambridge, U.K.

Littlejohn, M. J., and J. J. Loftus-Hills. 1968. An experimental evaluation of premating isolation in the *Hyla ewingi* complex (Anura: Hylidae). *Evolution* 22: 659–663.

Lloyd, D. G. 1973. Sex ratios in sexually dimorphic umbelliferae. *Heredity* 31: 239–249.

Lloyd, D. G. 1980. Sexual strategies in plants. III. A quantitative method for describing the gender of plants. *New Zeal. J. Bot.* 18: 103–108.

Lloyd, D. G. 1982. Selection of combined versus separate sexes in seed plants. *Am. Nat.* 120: 571–585.

Lloyd, D. G. 1984. Gender allocations in outcrossing cosexual plants. In R. Dirzo and J. Sarukhan, eds., *Perspectives on Plant Population Ecology*, 277–300. Sinauer, Sunderland, Mass.

Lloyd, D. G. 1987. Allocations to pollen, seeds and pollination mechanisms in self-fertilizing plants. *Func. Ecol.* 1: 83–89.

Lloyd, D. G. 1988. Benefits and costs of biparental and uniparental reproduction in plants. In R. E. Michod and B. R. Levin, eds., *The Evolution of Sex*, 233–252. Sinauer, Sunderland, Mass.

Lloyd, D. G., and K. S. Bawa. 1984. Modification of the gender of seed plants in varying conditions. *Evol. Biol.* 17: 255–338.

Lloyd, D. G., and C. J. Webb. 1977. Secondary sex characters in plants. *Bot. Rev.* 43: 177–215.

Lloyd, D. G., and J. M. Yates. 1982. Intrasexual selection and the segregation of pollen and stigmas in hermaphrodite plants, exemplified by *Wahlenbergia albomarginata* (Campanulaceae). *Evolution* 36: 903–913.

Lloyd, J. E. 1975. Aggressive mimicry in *Photuris* fireflies: Signal repertoires in femmes fatales. *Science* 187: 452–453.

Lloyd, J. E. 1979. Sexual selection in luminescent beetles. In M. S. Blum and N. A. Blum, eds., *Sexual Selection and Reproductive Competition in Insects*, 293–342. Academic Press, New York.

Lloyd, J. E. 1980. Male *Photuris* fireflies mimic sexual signals of their female's prey. *Science* 210: 669–671.

Lloyd, J. E. 1984. On deception, a way of all flesh, and firefly signaling and systematics. *Oxford Surv. Evol. Biol.* 1: 48–84.

Lloyd, L. 1867. *Game Birds and Wild Fowl of Sweden and Norway*. Warne, London.

Lloyd, M., and H. S. Dybas. 1966. The periodical cicada problem. II. Evolution. *Evolution* 20: 466–505.

Lockie, J. D. 1966. Territory in small carnivores. *Symp. Zool. Soc. London* 18: 143–165.

Lockyer, C. 1976. Body weights of some species of large whales. *J. Cons. Int. Explor. Mer.* 36: 259–273.

Loffredo, C. A., and G. Borgia. 1986. Sexual selection, mating systems, and the evolution of avian acoustical displays. *Am. Nat.* 128: 773–794.

Löfstedt, C. 1990. Population variation and genetic control of pheromone communication systems in moths. *Entomol. Exp. Appl.* 54: 199–218.

Löfstedt, C., N. J. Vickers, W. L. Roelofs, and T. C. Baker. 1989. Diet related courtship success in the Oriental fruit moth, *Grapholita molesta* (Tortricidae). *Oikos* 55: 402–408.

Loftus-Hills, J. J., and M. J. Littlejohn. 1971. Pulse repetition rate as the basis for mating call discrimination by two sympatric species of *Hyla*. *Copeia*: 154–156.

Loiselle, P. V. 1982. Male spawning-partner preference in an arena-breeding teleost *Cyprinodon macularius californiensis* Girard (Atherinomorpha: Cyprinodontidae). *Am. Nat.* 120: 721–732.

Loiselle, P. V., and G. W. Barlow. 1978. Do fishes lek like birds? In E. S. Reese and F. J. Lighter, eds., *Contrasts in Behavior*, 31–76. Wiley, New York.

Lombardi, J. R., and J. G. Vandenbergh. 1977. Pheromonally induced sexual maturation in females: Regulation by the social environment of the male. *Science* 196: 545–546.

Long, K. D., and A. E. Houde. 1989. Orange spots as a visual cue for female mate choice in the guppy (*Poecilia reticulata*). *Ethology* 82: 316–324.

Lorenz, K. 1962. The function of colour in coral reef fishes. *Proc. Roy. Inst. Great Brit.* 39: 282–296.

Losey, G. S., F. G. Stanton, T. M. Telecky, and W. A. Tyler. 1986. Copying others, an evolutionarily stable strategy for mate choice: A model. *Am. Nat.* 128: 653–664.

Lovett Doust, J. 1989. Plant reproductive strategies and resource allocation. *TREE* 4: 230–233.

Lovett Doust, J., and P. B. Cavers. 1982. Sex and gender dynamics in Jack-in-the-Pulpit, *Arisaema triphyllum* (Araceae). *Ecology* 63: 797–808.

Lovett Doust, J., and L. Lovett Doust. 1988. Sociobiology of plants: An emerging systhesis. In J. Lovett Doust and L. Lovett Doust, eds., *Plant Reproductive Ecology: Patterns and Strategies*, 5–29. Oxford University Press, New York.

Lovett-Doust, J., and L. Lovett-Doust, eds. 1988. *Plant Reproductive Ecology: Patterns and Strategies*. Oxford University Press, New York.

Low, B. S. 1978. Environmental uncertainty and the parental strategies of marsupials and placentals. *Am. Nat.* 112: 197–213.

Lundberg, A. 1986. Adaptive significance of reversed sexual size dimorphism in European owls. *Ornis Scand.* 17: 133–140.

Lundberg, A., and R. V. Alatalo. 1992. *The Pied Flycatcher*. Poyser, London.

Lutz, B. 1960. Fighting and incipient notion of territoriality in male tree frogs. *Copeia* 1960: 61–63.

Luyten, P. H., and N. R. Liley. 1985. Geographic variation in the sexual behaviour of the guppy, *Poecilia reticulata* (Peters). *Behaviour* 95: 164–179.

Luyten, P. H., and N. R. Liley. 1991. Sexual selection and competitive mating success of males guppies (*Poecilia reticulata*) from four Trinidad populations. *Behav. Ecol. Sociobiol.* 28: 329–336.

Lyon, B. E., and R. D. Montgomerie. 1985. Conspicuous plumage of birds: Sexual selection or unprofitable prey? *Anim. Behav.* 33: 1038–1040.

Lyon, B. E., and R. D. Montgomerie. 1986. Delayed plumage maturation in passerine birds: Reliable signaling by subordinate males? *Evolution* 40: 605–615.

Lyons, E. E., N. M. Waser, M. V. Price, J. Antonovics, and A. F. Motten. 1989. Sources of variation in plant reproductive success and implications for concepts of sexual selection. *Am. Nat.* 134: 409–433.

McAlpine, D. K. 1979. Agnostic behavior in *Achias australis* (Diptera, Platystomatidae) and the significance of eyestalks. In M. S. Blum and N. A. Blum, eds., *Sexual Selection and Reproductive Competition in Insects*, 221–230. Academic Press, New York.

McCann, T. S. 1981. Aggression and sexual activity of male Southern elephant seals, *Mirounga leonina*. *J. Zool., London* 195: 295–310.

McCauley, D. E. 1982. The behavioural components of sexual selection in the milkweed beetle *Tetraopes tetraophtalmus*. *Anim. Behav.* 30: 23–28.

McCauley, D. E. 1983. An estimate of the relative opportunities for natural and sexual selection in a population of milkweed beetles. *Evolution* 37: 701–707.

McComb, K. 1987. Roaring by red deer stags advances the date of oestrus in hinds. *Nature* 330: 648–649.

McComb, K. E. 1991. Female choice for high roaring rates in red deer, *Cervus elaphus*. *Anim. Behav.* 41: 79–88.

McDiarmid, R. W., and K. Adler. 1974. Notes on territorial and vocal behavior of Neotropical frogs of the genus *Centrolenella*. *Herpetologica* 30: 75–78.

McDonald, D. B. 1989a. Correlates of male mating success in a lekking bird with male-male cooperation. *Anim. Behav.* 37: 1007–1022.

McDonald, D. B. 1989b. Cooperation under sexual selection: Age-graded changes in a lekking bird. *Am. Nat.* 134: 709–730.

Macdonald, D. W., D. Müller-Schwarze, and S. E. Natynczuk, eds. 1990. *Chemical Signals in Vertebrates*. Vol. 5. Oxford University Press, Oxford.

McDonald, M. V. 1989. Function of song in Scott's seaside sparrow, *Ammodramus maritimus peninsulae*. *Anim. Behav.* 38: 468–485.

McGregor, P. K., and J. R. Krebs. 1982. Mating and song types in the great tit. *Nature* 297: 60–61.

McGregor, P. K., and J. R. Krebs. 1989. Song learning in adult great tits (*Parus major*): Effects of neighbours. *Behaviour* 103: 139–159.

McGregor, P. K., J. R. Krebs, and C. M. Perrins. 1981. Song repertoires and lifetime reproductive success in the great tit (*Parus major*). *Am. Nat.* 118: 149–159.

McGregor, P. K., et al. 1992. Design of playback experiments: The Thornbridge Hall Nato ARW consensus. In P. K. McGregor, ed., *Playback Studies and Animal Communication*, 1–9. Plenum, New York.

McKaye, K. R., S. M. Louda, and J.R.J. Stauffer. 1990. Bower size and male reproductive success in a cichlid fish lek. *Am. Nat.* 135: 597–613.

McKinney, F. 1986. Ecological factors influencing the social systems of migratory dabbling ducks. In D. Rubenstein and R. W. Wrangham, eds., *Ecological Aspects of Social Evolution*, 153–171. Princeton University Press, Princeton, N.J.

McKinney, F., S. R. Derrickson, and P. Mineau. 1983. Forced copulation in waterfowl. *Behaviour* 86: 250–294.

MacKinnon, J. 1981. The structure and function of the tusks of babirusa. *Mammal Rev.* 11: 37–40.

McLachlan, A. J. 1986. Sexual dimorphism in midges: Strategies for flight in the rain-pool dweller *Chironomus imicola* (Diptera: Chironomidae). *J. Anim. Ecol.* 261–267.

McLachlan, A. J., and D. F. Allen. 1987. Male mating success in Diptera: Advantages of small size. *Oikos* 48: 11–14.

McLain, D. K. 1980. Female choice and the adaptive significance of prolonged copulation in *Nezara viridula*. *Psyche* 87: 325–336.

McLain, D. K. 1982. Density dependent sexual selection and positive phenotypic assortative mating in natural populations of the soldier beetle, *Chauliognathus pennsylvanicus*. *Evolution* 36: 1227–1235.

McLain, D. K. 1987. Heritability of size, a sexually selected character, and the response to sexual selection in a natural population of the southern green stink bug, *Nezara viridula* (Hemiptera: Pentatomidae). *Heredity* 59: 391–395.

McLain, D. K. 1988. Male mating preferences and assortative mating in the soldier beetle. *Evolution* 42: 729–735.

McLain, D. K., and R. D. Boromisa. 1987. Male choice, fighting ability, assortative mating and the intensity of sexual selection in the milkweed longhorn beetle, *Tetraopes tetraophthalmus* (Coleoptera, Cerambycidae). *Behav. Ecol. Sociobiol.* 20: 239–246.

McLaren, A. 1988. Somatic and germ-cell sex in mammals. *Phil. Trans. Roy. Soc. London* B 322: 3–9.

McLennan, D. A., and J. D. McPhail. 1989. Experimental investigations of the evolutionary significance of sexually dimorphic nuptial colouration in *Gasterosteus aculeatus* (L.): Temporal changes in the structure of the male mosaic signal. *Can. J. Zool.* 67: 1767–1777.

McMahon, T. A., and J. T. Bonner. 1983. *On Size and Life*. Scientific American Books, New York.

McMinn, H. 1990. Effects of the nematode parasite *Camallanus cotti* on sexual and non-sexual behaviors in the guppy *Poecilia reticulata*. *Am. Zool.* 30: 245–249.

MacNally, R., and D. Young. 1981. Song energetics of the bladder cicada, *Cystosoma saundersii*. *J. Exp. Biol.* 90: 185–196.

McPhail, J. D. 1969. Predation and the evolution of a stickleback (*Gasterosteus*). *J. Fish. Res. Board Can.* 26: 3183–3208.

Madden, J. R. 1974. Female territoriality in a Suffolk County, Long Island, population of *Glaucomys volans*. *J. Mammal.* 55: 647–652.

Maddison, W. P. 1990. A method for testing the correlated evolution of two binary characters: Are gains or losses concentrated on certain branches of a phylogenetic tree? *Evolution* 44: 539–557.

Madhavi, M., and R. M. Anderson. 1985. Variability in the susceptibility of the fish host, *Poecilia reticulata*, to infection with *Gyrodactylus bullatarudis* (Monogenea). *Parasitology* 91: 531–544.

Madsen, J. D., and D. M. Waller. 1983. A note on the evolution of gamete dimorphism in algae. *Am. Nat.* 121: 443–447.

Madsen, T. 1988. Reproductive success, mortality and sexual size dimorphism in the adder, *Vipera berus*. *Holarctic Ecol.* 11: 77–80.

Madsen, T., and J. Loman. 1987. On the role of colour display in the social and spatial organization of male rainbow lizards, *Agama agama*. *Amphibia-Reptilia* 8: 365–372.

Madsen, T., and R. Shine. 1992. Determinants of reproductive success in female adders. *Vipera berus*. *Oecologia* 92: 40–48.

Madsen, T., and R. Shine. 1993. A rapid, sexually-selected shift in mean body size in a population of snakes. *Evolution*, in press.

Madsen, T., R. Shine, J. Loman, and T. Håkansson. 1992. Why do female adders copulate so frequently? *Nature* 355: 440–441.

Magnhagen, C. 1991. Predation risk as a cost of reproduction. *TREE* 6: 183–186.

Majerus, M.E.N., P. O'Donald, and J. Weir. 1982. Female mating preference is genetic. *Nature* 300: 521–523.

Majerus, M.E.N., P. O'Donald, P.W.E. Kearns, and H. Ireland. 1986. Genetics and evolution of female choice. *Nature* 321: 164–167.

Malacarne, G. 1984. Female attractiveness and male choice in the crested newt, *Triturus cristatus carnifex* (Laur.). *Monit. Zool. Ital.* 18: 33–39.

Malacarne, G., and R. Cortassa. 1983. Sexual selection in the crested newt. *Anim. Behav.* 31: 1256–1257.

Malacarne, G., and C. Vellano. 1987. Behavioral evidence of a courtship pheromone in the crested newt, *Triturus cristatus carnifex*. *Copeia*: 245–247.

Malacarne, G., L. Bottini, R. Massa, and C. Vellano. 1984. The abdominal gland of the crested newt: A possible source of courtship pheromones. *Monit. Zool. Ital.* (n.s.) 18: 33–39.

Mallory, F. F., and R. J. Brooks. 1978. Infanticide and other reproductive strategies in the collared lemming, *Dicrostonyx groenlandicus*. *Nature* 273: 144–146.

Manly, B.F.J. 1985. *The Statistics of Natural Selection*. Chapman and Hall, London.

Manning, J. T. 1975. Male discrimination and investment in *Asellus aquaticus* L. and *A. meridianus* Racovitsza (Crustacea : Isopoda). *Behaviour* 55: 1–14.

Manning, J. T. 1984. Males and the advantage of sex. *J. theor. Biol.* 108: 215–220.

Manning, J. T. 1985. Choosy females and correlates of male age. *J. theor. Biol.* 116: 349–354.

Manning, J. T. 1987. The peacock's train and the age-dependency model of female choice. *J. World Pheasant Assoc.* 12: 44–56.

Manning, J. T. 1989. Age-advertisement and the evolution of the peacock's train. *J. Evol. Biol.* 2: 379–384.

Manning, J. T., and M. A. Hartley. 1991. Symmetry and ornamentation are correlated in the peacock's train. *Anim. Behav.* 42: 1020–1021.

Marchetti, K. 1993. Dark habitats and bright birds illustrate the role of the environment in species divergence. *Nature* 362: 149–152.

Marchetti, K., and T. Price. 1989. Differences in the foraging of juvenile and adult birds: The importance of developmental constraints. *Biol. Rev.* 64: 51–70.

Marconato, A., and A. Bisazza. 1986. Males whose nests contain eggs are preferred by female *Cottus gobio* L. (Pisces, Cottidae). *Anim. Behav.* 34: 1580–1582.

Marden, J. H., and J. K. Waage. 1990. Escalated damselfly territorial contests are energetic wars of attrition. *Anim. Behav.* 39: 954–959.

Markow, T. A. 1988. *Drosophila* males provide a material contribution to offspring sired by other males. *Func. Ecol.* 2: 77–79.

Markow, T. A., M. Quaid, and S. Kerr. 1978. Male mating experience and competitive courtship success in *Drosophila melanogaster*. *Nature* 276: 821–822.

Marler, C. A., and M. C. Moore. 1988. Evolutionary costs of aggression revealed by testosterone manipulations in free-living male lizards. *Behav. Ecol. Sociobiol.* 23: 21–26.

Marler, P. 1955. Studies of fighting in chaffinches. 3. The effect on dominance relations of disguising females as males. *Brit. J. Anim. Behav.* 3: 137–146.

Marler, P. 1956. The voice of the chaffinch and its function as a language. *Ibis* 98: 231–261.

Marshall, D. L., and D. C. Ellstrand. 1986. Sexual selection in *Raphanus sativus*: Experimental data on non-random fertilization, maternal choice, and consequences of multiple paternity. *Am. Nat.* 127: 446–461.

Marshall, D. L., and N. C. Ellstrand. 1988. Effective mate choice in wild radish: Evidence for selective seed abortion and its mechanism. *Am. Nat.* 131: 736–759.

Marshall, D. L., and M. W. Folsom. 1991. Mate choice in plants: An anatomical to population perspective. *Ann. Rev. Ecol. Syst.* 22: 37–63.

Marshall, F.H.A. 1936. Sexual periodicity and the causes which determine it. *Phil. Trans. Roy. Soc. London* B 226: 443–449.

Marshall, L. D. 1982. Male nutrient investment in the Lepidoptera: What nutrients should males invest? *Am. Nat.* 120: 273–279.

Marti, C. D. 1990. Sex and age dimorphism in the barn owl and a test of mate choice. *Auk* 107: 246–254.

Martin, W. C., and C. R. Hutchins. 1980. *A Flora of New Mexico*. A. R. Gantner Verlag, Vaduz.

Marzluff, J. M., and R. P. Balda. 1988. Pairing patterns and fitness in a free-ranging population of pinyon jays: What do they reveal about mate choice? *Condor* 90: 201–213.

Mason, L. G. 1969. Mating selection in the Califonia oak moth (*Lepidoptera: Dioptidae*). *Evolution* 23: 55–58.

Mathieu, J. M. 1969. Mating behavior of five species of Lucanidae (Coleoptera: Insecta). *Can. Entomol.* 101: 1054–1062.

Mathis, A. 1990. Territorial salamanders assess sexual and competitive information using chemical signals. *Anim. Behav.* 40: 953–962.

Maxson, S. J., and L. W. Oring. 1980. Breeding season time and energy budgets of the polyandrous spotted sandpiper. *Behaviour* 74: 200–263.

Maynard Smith, J. 1956. Fertility, mating behaviour and sexual selection in *Drosophila subobscura*. *J. Genet.* 54: 261–279.

Maynard Smith, J. 1958. Sexual selection. In S. A. Barnett, ed., *A Century of Darwin*, 231–244. Heinemann, London.

Maynard Smith, J. 1976. Sexual selection and the handicap principle. *J. theor. Biol.* 57: 239–242.

Maynard Smith, J. 1977. Parental investment—a prospective analysis. *Anim. Behav.* 25: 1–9.

Maynard Smith, J. 1978a. The handicap principle—a comment. *J. theor. Biol.* 70: 251–252.

Maynard Smith, J. 1978b. *The Evolution of Sex*. Cambridge University Press, Cambridge, U.K.

Maynard Smith, J. 1982. *Evolution and the Theory of Games*. Cambridge University Press, Cambridge, U.K.

Maynard Smith, J. 1985. Sexual selection, handicaps and true fitness. *J. theor. Biol.* 115: 1–8.

Maynard Smith, J. 1987. Sexual selection—a classification of models. In J. W. Bradbury and M. B. Andersson, eds., *Sexual Selection: Testing the Alternatives*, 9–20. Wiley, Chichester, U.K.

Maynard Smith, J. 1988. Can a mixed strategy be stable in a finite population? *J. theor. Biol.* 130: 247–251.

Maynard Smith, J. 1989. *Evolutionary Genetics*. Oxford University Press, Oxford.

Maynard Smith, J. 1991a. Theories of sexual selection. *TREE* 6: 146–151.

Maynard Smith, J. 1991b. Honest signalling: The Philip Sidney game. *Anim. Behav.* 42: 1034–1035.

Maynard Smith, J., and R.L.W. Brown. 1986. Competition and body size. *Theor. Pop. Biol.* 30: 166–179.

Maynard Smith, J., and D.G.C. Harper. 1988. The evolution of aggression: Can selection generate variability? *Phil. Trans. Roy. Soc. London* B 319: 557–570.

Maynard Smith, J., and G. R. Price. 1973. The logic of animal conflict. *Nature* 246: 15–18.

Mayr, E. 1942. *Systematics and the Origin of Species*. Columbia University Press, New York.

Mayr, E. 1963. *Animal Species and Evolution.* Harvard University Press, Cambridge, Mass.

Mayr, E. 1972. Sexual selection and natural selection. In B. Cambpell, ed., *Sexual Selection and the Descent of Man, 1871–1971*, 87–104. Aldine, Chicago.

Mayr, E. 1982. *The Growth of Biological Thought: Diversity, Evolution, and Inheritance.* Harvard University Press, Cambridge, Mass.

Mazer, S. J. 1987. Maternal investment and male reproductive success in angiosperms: Parent-offspring conflict or sexual selection? *Biol. J. Linn. Soc.* 30: 115–131.

Meagher, T. R. 1986. Analysis of paternity within a natural population of *Chamaelirium luteum.* Identification of most-likely male parents. *Am. Nat.* 128: 199–215.

Meagher, T. R. 1991. Analysis of paternity within a natural population of *Chamaelirium luteum.* II. Patterns of male reproductive success. *Am. Nat.* 137: 738–752.

Mech, L. D. 1966. *The Wolves of Isle Royal.* U.S. Dept. of Interior, National Park Service, Washington, D.C.

Merrell, D. 1949. Selective mating in *Drosophila melanogaster. Genetics* 34: 370–389.

Metz, K. J., and P. J. Weatherhead. 1991. Color bands function as secondary sexual traits in male red-winged blackbirds. *Behav. Ecol. Sociobiol.* 28: 23–27.

Metz, K. J., and P. J. Weatherhead. 1992. Seeing red: Uncovering coverable badges in red-winged blackbirds. *Anim. Behav.* 43: 223–229.

Michaud, T. C. 1962. Call discrimination by females of the chorus frogs, *Pseudacris clarki* and *Pseudacris nigrita. Copeia* 1962: 213–215.

Michod, R. E., and O. Hasson. 1990. On the evolution of reliable indicators of fitness. *Am. Nat.* 135: 788–808.

Michod, R. E., and B. R. Levin, eds. 1988. *The Evolution of Sex.* Sinauer, Sunderland, Mass.

Mikkola, H. 1983. *Owls of Europe.* Poyser, Calton, U.K.

Milinski, M., and T.C.M. Bakker. 1990. Female sticklebacks use male coloration in mate choice and hence avoid parasitized males. *Nature* 344: 330–333.

Millar, J. S., and G. J. Hickling. 1990. Fasting endurance and the evolution of mammalian body size. *Funct. Ecol.* 4: 5–12.

Millar, J. S., and G. J. Hickling. 1992. Evolution of animal body size: A cautionary note on assessments of the role of energetics. *Func. Ecol.* 6: 495–498.

Milstein, P.L.S. 1979. The evolutionary significance of wild hybridization in South African highveld ducks. *Ostrich* 13 (suppl.): 1–48.

Mitani, J. 1985. Gibbon song duets and intergroup spacing. *Behaviour* 92: 59–96.

Mitani, J. C. 1988. Male gibbon (*Hylobates agilis*) singing behavior: Natural history, song variations and function. *Ethology* 79: 177–194.

Mitchell, B., B. W. Staines, and D. Welch. 1977. *Ecology of Red Deer: A Research Review Relevant to Their Management.* Institute of Terrestrial Ecology, Cambridge, U.K.

Mitchell, F. J. 1973. Studies on the ecology of the agamid lizard *Amphibolurus maculosus* (Mitchell). *Trans. Roy. Soc. S. Aust.* 97: 47–76.

Mitchell, R. J. 1993. Adaptive significance of *Ipomopsis aggregata* nectar production: Observation and experiment in the field. *Evolution* 47: 25–35.

Mitchell, R. J., and N. M. Waser. 1992. Adaptive significance of *Ipomopsis aggregata* nectar production. *Ecology* 73: 633–638.

Mitchell, S. L. 1990. The mating system genetically affects offspring performance in Woodhouse's toad (*Bufo woodhousei*). *Evolution* 44: 502–519.

Mitchell, W. C., and R.F.L. Mau. 1971. Response of the female southern green stink bug and its parasite, *Trichopoda pennipes*, to male stink bug pheromones. *J. Econ. Ent.* 64: 856–859.

Mitchell-Olds, T., and R. G. Shaw. 1987. Regression analysis of natural selection: Statistical inference and biological interpretation. *Evolution* 41: 1149–1161.

Mock, D. W. 1985. An introduction to the neglected mating system. *Ornithol. Monogr.* 37: 1–10.

Molau, U., B. Eriksen, and J. Teilmann Knudsen. 1989. Predispersal seed predation in *Bartsia alpina*. *Oecologia* 81: 181–185.

Moll, E. O., K. E. Matson, and E. B. Krehbiel. 1981. Sexual and seasonal dichromatism in the Asian river turtle (*Callagur borneoensis*). *Herpetologica* 37: 181–194.

Møller, A. P. 1987. Variation in badge size in male house sparrows *Passer domesticus*: Evidence for status signalling. *Anim. Behav.* 35: 1637–1644.

Møller, A. P. 1988a. Female choice selects for male sexual tail ornaments in the monogamous swallow. *Nature* 332: 640–642.

Møller, A. P. 1988b. Badge size in the house sparrow *Passer domesticus*: Effects of intra- and intersexual selection. *Behav. Ecol. Sociobiol.* 22: 373–378.

Møller, A. P. 1988c. Ejaculate quality, testes size, and sperm production in mammals. *J. Human Evol.* 17: 479–488.

Møller, A. P. 1989. Viability costs of male tail ornaments in a swallow. *Nature* 339: 132–135.

Møller, A. P. 1990a. Effects of a haematophagous mite on the Barn Swallow (*Hirundo rustica*): A test of the Hamilton and Zuk hypothesis. *Evolution* 44: 771–784.

Møller, A. P. 1990b. Fluctuating asymmetry in male sexual ornaments may reliably reveal male quality. *Anim. Behav.* 40: 1185–1187.

Møller, A. P. 1991a. Viability is positively related to degree of ornamentation in male swallows. *Proc. Roy. Soc. London* B 243: 145–148.

Møller, A. P. 1991b. Preferred males acquire mates of higher phenotypic quality. *Phil. Trans. Roy. Soc. London* B 245: 179–182.

Møller, A. P. 1991c. Sexual ornament size and the cost of fluctuating asymmetry. *Proc. Roy. Soc. London* B 243: 59–62.

Møller, A. P. 1991d. Sexual selection in the monogamous barn swallow (*Hirundo rustica*). I. Determinants of tail ornament size. *Evolution* 45: 1823–1836.

Møller, A. P. 1991e. Influence of wing and morphology on the duration of song flight in skylarks. *Behav. Ecol. Sociobiol.* 28: 309–314.

Møller, A. P. 1991f. Parasite load reduces song output in a passerine bird. *Anim. Behav.* 41: 723–730.

Møller, A. P. 1992a. Female swallow preference for symmetrical male sexual ornaments. *Nature* 357: 238–240.

Møller, A. P. 1992b. Frequency of female copulations with multiple males and sexual selection. *Am. Nat.* 139: 1089–1101.

Møller, A. P., and J. Höglund. 1991. Patterns of fluctuating asymmetry in avian feather ornaments: Implications for models of sexual selection. *Proc. Roy. Soc. London* 245: 1–5.

Møller, A. P.; P. H. Harvey; S. Nee; A. F. Read; I. C. Cuthill; J. P. Swaddle; and M. S. Witter. 1993. Fluctuating asymmetry. *Nature* 363: 217–218.

Montgomerie, R., and R. Thornhill. 1989. Fertility advertisement in birds: A means of inciting male-male competition? *Ethology* 81: 209–220.

Moodie, G.E.E. 1972. Predation, natural selection and adaptation in an unusual threespine stickleback. *Heredity* 28: 155–167.

Moore, A. J. 1988. Female preferences, male social status, and sexual selection in *Nauphoeta cinerea*. *Anim. Behav.* 36: 303–305.

Moore, A. J. 1989. Sexual selection in *Nauphoeta cinerea*: Inherited mating preference? *Behavior Genetics* 19: 717–724.

Moore, A. J. 1990a. The evolution of sexual dimorphism by sexual selection: The separate effects of intrasexual selection and intersexual selection. *Evolution* 44: 315–331.

Moore, A. J. 1990b. The inheritance of social dominance, mating behaviour and attractiveness to mates in male *Nauphoeta cinerea*. *Anim. Behav.* 39: 388–397.

Moore, A. J. 1991. Genetics, inheritance and social behaviour. *Anim. Behav.* 43: 497–498.

Moore, A. J., and M. D. Breed. 1986. Mate assessment in a cockroach, *Nauphoeta cinerea*. *Anim. Behav.* 34: 1160–1165.

Moore, A. J., and P. J. Moore. 1988. Female strategy during mate choice threshold assessment. *Evolution* 42: 387–391.

Moore, P. D. 1984. Pollination renewed. *Nature* 312: 84.

Moore, S. D. 1987. Male-biased mortality in the butterfly *Euphydryas editha*: A novel cost of mate acquisition. *Am. Nat.* 130: 306–309.

Moore, W. S. 1987. Random mating in the northern flicker hybrid zone: Implications for the evolution of bright and contrasting plumage patterns in birds. *Evolution* 41: 539–546.

Moors, P. J. 1980. Sexual dimorphism in the body size of mustelids (Mammalia: Carnivora): The role of food habits and breeding systems. *Oikos* 34: 147–158.

Morgan, M. T. 1992. The evolution of traits influencing male and female fertility in outcrossing plants. *Am. Nat.* 139: 1022–1051.

Morgan, T. H. 1903. *Evolution and Adaptation*. Macmillan, New York.

Morgan, T. H. 1919. The genetic and the operative evidence relating to secondary sexual characters. *Carnegie Inst. of Washington*, publ. no. 285, 1–108.

Morgan, T. H. 1932. *The Scientific Basis of Evolution*. Norton, New York.

Morris, G. K. 1979. Mating systems, paternal investment and aggressive behavior of acoustic Orthoptera. *Florida Entomol.* 62: 9–17.

Morris, G. K. 1980. Calling display and mating behaviour of *Copiphora rhinoceros* Pictet (Orthoptera: Tettigoniidae). *Anim. Behav.* 28: 42–51.

Morris, G. K., and J. H. Fullard. 1983. Random noise and congeneric discrimination in *Conocephalus* (Orthoptera: Tettigoniidae). In D. G. Gwynne and G. K. Morris, eds., *Orthopteran Mating Systems: Sexual Competition in a Diverse Group of Insects*, 73–96. Westview Press, Boulder, Colo.

Morris, L. 1975. Effect of blackened epaulets on the territorial behavior and breeding success of male redwinged blackbirds, *Agelaius phoeniceus*. *Ohio J. Sci.* 75: 168–176.

Morris, M. R. 1989. Female choice of large males in the treefrog *Hyla chrysoscelis*: The importance of identifying the scale of choice. *Behav. Ecol. Sociobiol.* 25: 275–281.

Morris, M. R., and S. L. Yoon. 1989. A mechanism for female choice of large males in the treefrog *Hyla chrysoscelis*. *Behav. Ecol. Sociobiol.* 25: 65–71.

Morse, D. H. 1970. Territorial and courtship songs of birds. *Nature* 226: 659–661.

Morse, D. H. 1989. *American Warblers: An Ecological and Behavioral Perspective*. Harvard University Press, Cambridge, Mass.

Morton, E. 1977. On the occurrence and significance of motivation-structural rules in some bird and mammal sounds. *Am. Nat.* 111: 855–869.

Morton, E. 1982. Grading, discreteness, redundancy, and motivation-structural rules. In D. E. Kroodsma and E. H. Miller, eds., *Acoustic Communication in Birds*, 183–212. Academic Press, New York.

Mosher, J. A., and P. F. Matray. 1974. Size dimorphism: A factor in energy savings for broad-winged hawks. *Auk* 91: 325–341.

Moss, C. J. 1983. Oestrus behaviour and female choice in the African elephant. *Behaviour* 86: 167–196.

Moss, R., H. H. Kolb, M. Marquiss, A. Watson, B. Treca, D. Watt, and W. Glennie. 1979. Aggressivenes and dominance in captive cock red grouse. *Aggress. Behav.* 5: 59–84.

Motro, U. 1982. The courtship handicap-phenotypic effect. *J. theor. Biol.* 97: 319–324.

Motro, U. 1991. Avoiding inbreeding and sibling competition: The evolution of sexual dimorphism for dispersal. *Amer. Natur.* 137: 108–115.

Mousseau, T. A., and D. A. Roff. 1987. Natural selection and the heritability of fitness components. *Heredity* 59: 181–197.

Moynihan, M. H., and A. F. Rodaniche. 1977. Communication, crypsis and mimicry among Cephalopods. In T. A. Sebeok, ed., *How Animals Communicate*, 293–302. Indiana University Press, Bloomington.

Mueller, H. C. 1986. The evolution of reversed sexual dimorphism in owls: An empirical analysis of possible selective factors. *Wilson Bull.* 98: 387–406.

Mueller, H. C. 1989a. Aerial agility and the evolution of reversed sexual dimorphism (RSD) in shorebirds. *Auk* 106: 154–157.

Mueller, H. C. 1989b. Evolution of reversed sexual size dimorphism: Sex or starvation? *Ornis Scand.* 20: 265–272.

Mueller, H. C. 1990. The evolution of reversed sexual dimorphism in size in monogamous species of birds. *Biol. Rev.* 65: 553–585.

Mueller, H. C., and K. Meyer. 1985. The evolution of reversed sexual dimorphism in size: A comparative analysis of the Falconiformes of the Western Palearctic. In R. F. Johnston, ed., *Current Ornithology*, 65–101. Plenum, New York.

Mukai, T., and O. Yamaguchi. 1974. The genetic structure of natural populations of *Drosophila melanogaster*. XI. Genetic variability in a local population. *Genetics* 76: 339–366.

Mulcahy, D. L. 1979. The rise of the angiosperms: A genecological factor. *Science* 206: 20–23.

Mulcahy, D. L., and G. B. Mulcahy. 1987. The effects of pollen competition. *Am. Sci.* 75: 44–50.

Muldal, A. M., J. D. Moffatt, and R. J. Robertson. 1986. Parental care of nestlings by male red-winged blackbirds. *Behav. Ecol. Sociobiol.* 19: 105–114.

Müller, F. 1877. Über Haarpinsel, Filzflecke, und ähnliche Gebilde auf den Flügeln männlicher Schmetterlinge. *Jena Z. Naturw.* 5: 99–114.

Müller, H. 1873. *Die Befruchtung der Blumen durch Insekten.* Engelmann, Leipzig.

Müller-Using, D., and R. Schloeth. 1967. Das Verhalten der Hirsche. In W. Kükenthal, ed., *Handbuch der Zoologie* 10 (28): 1–60.

Mullins, D. E., and C. B. Keil. 1980. Paternal investment of urates in cockroaches. *Nature* 283: 567–569.

Muma, K. E., and P. J. Weatherhead. 1989. Male traits expressed in females: Direct or indirect sexual selection? *Behav. Ecol. Sociobiol.* 25: 23–31.

Murray, B. G., Jr. 1984. A demographic theory on the evolution of mating systems as exemplified by birds. *Evol. Biol.* 18: 71–140.

Murray, M. G. 1990. Comparative morphology and mate competition of flightless male fig wasps. *Anim. Behav.* 39: 434–443.

Myers, C. W., and J. W. Daly. 1983. Dart-poison frogs. *Scient. Am.* 248(2): 96–105.

Myers, J., and L. P. Brower. 1969. A behavioural analysis of the courtship pheromone receptors of the queen butterfly, *Danaus gilippus berenice*. *J. Insect Physiol.* 15: 2117–2130.

Myers, P. 1978. Sexual dimorphism in size of vespertilionid bats. *Am. Nat.* 112: 701–711.

Myrberg, A. A., M. Mohler, and J. D. Catala. 1986. Sound production by males of a coral reef fish (*Pomacentrus partitus*): Its significance to females. *Anim. Behav.* 34: 913–923.

Nakatsuru, K., and D. L. Kramer. 1982. Is sperm cheap? Limited male fertility and female choice in the lemon tetra (Pisces, Characidae). *Science* 216: 753–755.

Narins, P. M. 1992. Evolution of anuran chorus behavior: Neural and behavioral constraints. *Am. Nat.* 139: s90-s104.

Neems, R. M., A. J. McLachlan, and R. Chambers. 1990. Body size and lifetime mating success of male midges (Diptera: Chironomidae). *Anim. Behav.* 40: 648–652.

Nei, M. 1987. *Molecular Evolutionary Genetics.* Columbia University Press, New York.

Nei, M., T. Maruyama, and C. I. Wu. 1983. Models of evolution of reproductive isolation. *Genetics* 103: 557–579.

Nelson, K. 1964. Behavior and morphology in the Glandulocaudine fishes (Ostariophysi, Characidae). *Univ. Calif. Publ. Zool.* 75: 59–152.

Neudecker, S. 1989. Eye camouflage and false eyespots: Chaetodontid responses to predators. *Env. Biol. Fish.* 25: 143–157.

Nevo, E., and R. R. Capranica. 1985. Evolutionary origin of ethological reproductive isolation in cricket frogs, *Acris*. *Evol. Biol.* 19: 147–214.

Newman, J. A., and M. A. Elgar 1991. Sexual cannibalism in orb-weaving spiders: An economic model. *Am. Nat.* 138: 1372–1395.

Newton, I. 1979. *Population Ecology of Raptors.* Poyser, Calton, U.K.

Newton, I. 1986. *The Sparrowhawk*. Poyser, Calton, U.K.

Newton, I. 1988 Individual performance in sparrowhawks: The ecology of two sexes. *Acta XIX Congr. Int. Orn.* 125–154.

Newton, I., ed. 1989. *Lifetime Reproduction in Birds*. Academic Press, London.

Nicoletto, P. F. 1991. The relationship between male ornamentation and swimming performance in the guppy, *Poecilia reticulata*. *Behav. Ecol. Sociobiol.* 28: 365–370.

Nilsson, L. A. 1980. The pollination ecology of *Dactylorhiza sambucina* (Orchidaceae). *Bot. Notiser* 133: 367–385.

Nilsson, L. A. 1988. The evolution of flowers with deep corolla tubes. *Nature* 334: 147–149.

Nilsson, L. A. 1992. Orchid pollination biology. *TREE* 7: 255–259.

Nilsson, L. A., E. Rabakonandrianina, and B. Petterson. 1992. Exact tracking of pollen transfer and mating in plants. *Nature* 360: 666–668.

Nisbet, I.C.T. 1973. Courtship-feeding, egg-size and breeding success in common terns. *Nature* 241: 141–142.

Nisbet, I.C.T. 1977. Courtship-feeding and clutch size in common terns *Sterna hirundo*. In B. Stonehouse and C. M. Perrins, eds., *Evolutionary Ecology*, 101–109. Macmillan, London.

Nishida, R., H. Fukami, T. C. Baker, W. L. Roelofs, and T. E. Acree. 1985. Oriental fruit moth pheromone: Attraction of females by an herbal essence. In T. E. Acree and D. M. Soderlund, eds., *Semiochemistry. Flavors and Pheromones*, 47–63. Walter de Gruyter & Co, Berlin.

Noble, G. K. 1934. Experimenting with the courtship of lizards. *Nat. Hist.* 34: 5–15.

Noble, G. K. 1936. Courtship and sexual selection of the flicker (*Colaptes auratus luteus*). *Auk* 53: 269–282.

Noble, G. K. 1938. Sexual selection among fishes. *Biol. Rev.* 13: 133–158.

Noble, G. K., and H. T. Bradley. 1933. The mating behavior of lizards; its bearing on the theory of sexual selection. *Ann. N.Y. Acad. Sci.* 35: 35–100.

Noble, G. K., and B. Curtis. 1939. The social behavior of the jewel fish, *Hemichromis bimaculatus* Gill. *Bull. Am. Mus. Nat. Hist.* 76: 1–47.

Noble, G. K., and W. Vogt. 1935. An experimental study of sex recognition in birds. *Auk* 34: 278–286.

Noonan, K. C. 1983. Female mate choice in the cichlid fish *Cichlasoma nigrofasciatum*. *Anim. Behav.* 31: 1005–1010.

Norberg, R. Å. 1981. Temporary weight decrease in breeding birds may result in more fledged young. *Am. Nat.* 118: 838–850.

Norberg, R. Å. 1987. Evolution, structure, and ecology of Northern forest owls. In R. W. Nero, R. J. Clark, R. J. Knapton, and R. H. Hamre, eds., *Biology and Conservation of Northern Forest Owls*, 9–43. U.S. Dept. of Agriculture, Fort Collins, Colo.

Norberg, R. Å. 1991. The flappet lark *Mirafra rufocinnamomea* doubles its wingbeat rate to 24 Hz in wing-clap display flight: A sexually selected feat. *J. Exp. Biol.* 159: 515–523.

Norberg, U. M. 1990. *Vertebrate Flight*. Springer, Berlin.

Norris, K. J. 1990a. Female choice and the evolution of the conspicuous plumage coloration of monogamous male great tits. *Behav. Ecol. Sociobiol.* 26: 129–138.

Norris, K. J. 1990b. Female choice and the quality of parental care in the great tit *Parus major*. *Behav. Ecol. Sociobiol.* 27: 275–281.

Norris, K. 1993. Heritable variation in a plumage indicator of viability in male great tits *Parus major*. *Nature* 362: 537–539.

Nottebohm, F. 1972. The origins of vocal learning. *Am. Nat.* 106: 116–140.

Nottebohm, F., and M. E. Nottebohm. 1971. Vocalizations and breeding behaviour of surgically deafened ring doves (*Streptopelia risoria*). *Anim. Behav.* 19: 311–327.

Nur, N., and O. Hasson. 1984. Phenotypic plasticity and the handicap principle. *J. theor. Biol.* 110: 275–297.

Oakes, E. J. 1992. Lekking and the evolution of sexual dimorphism in birds: Comparative approaches. *Am. Nat.* 140: 665–684.

Oberhauser, K. S. 1989. Effects of spermatophores on male and female monarch butterfly reproductive success. *Behav. Ecol. Sociobiol.* 25: 237–246.

O'Donald, P. 1962. The theory of sexual selection. *Heredity* 17: 541–552.

O'Donald, P. 1967. A general model of sexual and natural selection. *Heredity* 22: 499–518.

O'Donald, P. 1977. Theoretical aspects of sexual selection. *Theor. Pop. Biol.* 12: 298–334.

O'Donald, P. 1980a. *Genetic Models of Sexual Selection*. Cambridge University Press, Cambridge, U.K.

O'Donald, P. 1980b. Genetic models of sexual and natural selection in monogamous organisms. *Heredity* : 391–415.

O'Donald, P. 1983a. Sexual selection by female choice. In P. Bateson, ed., *Mate Choice*, 53–66. Cambridge University Press, Cambridge, U.K.

O'Donald, P. 1983b. *The Arctic Skua*. Cambridge University Press, Cambridge, U.K.

O'Donald, P. 1987. Polymorphism and sexual selection in the arctic skua. In F. Cooke and P. A. Buckley, eds., *Avian Genetics: A Population and Ecological Approach*, 433–452. Academic Press, London.

O'Donald, P., and M.E.N. Majerus. 1985. Sexual selection and the evolution of preferential mating in ladybirds. I. Selection for high and low lines of female preference. *Heredity* 55: 401–412.

O'Donald, P., and M.E.N. Majerus. 1992. Non-random mating in *Adalia bipunctata* (the two-spot ladybird). III. New evidence of genetic preference. *Heredity* 69: 521–526.

Ohde, B. R., R. A. Bishop, and J. J. Dinsmore. 1983. Mallard reproduction in relation to sex ratios. *J. Wildl. Mgmt.* 47: 118–126.

Oldham, R. S., and H. C. Gerhardt. 1975. Behavioral isolation of the treefrogs *Hyla cinerea* and *Hyla gratiosa*. *Copeia* : 223–231.

Olsen, J., and P. Olsen. 1984. Review of "Falcons of the World." *Aust. Wildl. Res.* 11: 205–206.

Olsen, P., and J. Olsen. 1987. Sexual size dimorphism in raptors: Intrasexual competition in the larger sex for a scarce breeding resource, the smaller sex. *Emu* 87: 59–62.

Olson, E. C. 1969. Note: Sexual dimorphism in extinct amphibians and reptiles. In

G. E. G. Westermann, ed., *Sexual Dimorphism in Fossil Metazoa and Taxonomic Implications*, 223–225. E. Schweizerbart'sche Verlagsbuchhandlung, Stuttgart.
Olsson, M. 1992a. Sexual Selection and Reproductive Strategies in the Sand Lizard *(Lacerta agilis)*. Ph.D. diss., University of Göteborg, Sweden.
Olsson, M. 1992b. Contest success in relation to size and residency in male sand lizards, *Lacerta agilis*. *Anim. Behav.* 44: 386–388.
Olsson, M. 1993a. Male preference for large females and assortative mating for body size in the sand lizard (*Lacerta agilis*). *Behav. Ecol. Sociobiol.*: in press.
Olsson, M. 1993b. Nuptial coloration in the sand lizard (*Lacerta agilis*): An intrasexually selected cue to fighting ability. *Anim. Behav.*: in press.
Olsson, M. 1993c. Nuptial coloration and predation risk in model sand lizards. *Lacerta agilis*. *Anim. Behav.* 45: 410–412.
Olsson, M. 1994. The function of conspicuous coloration in reproductive female lizards. *Ctenophorus maculosus*. In prep.
Orians, G. 1961. The ecology of blackbird (*Agelaius*) social systems. *Ecol. Monogr.* 31: 285–312.
Orians, G. H. 1969. On the evolution of mating systems in birds and mammals. *Am. Nat.* 103: 589–603.
Orians, G. H., and G. M. Christman. 1968. A comparative study of the behavior of red-winged, tricolored, and yellow-headed blackbirds. *Univ. Calif. Publ. Zool.* 84: 1–81.
Oring, L. W. 1982. Avian mating systems. In D. S. Farner, J. R. King, and K. C. Parkes, eds., *Avian Biology*, 1–92. Academic Press, New York.
Oring, L. W. 1986. Avian polyandry. In R. F. Johnston, ed., *Current Ornithology*, vol. 3, 309–351. Plenum, New York.
Oring, L. W., and D. B. Lank. 1982. Sexual selection, arrival times, philopatry and site fidelity in the polyandrous spotted sandpiper. *Behav. Ecol. Sociobiol.* 10: 185–191.
Oring, L. W., and D. B. Lank. 1984. Breeding area fidelity, natal philopatry, and the social systems of sandpipers. In J. Burger and B. L. Olla, eds., *Behavior of Marine Animals*, 125–147. Plenum Press, New York.
Oring, L. W., and D. B. Lank. 1986. Polyandry in spotted sandpipers: The impact of environment and experience. In D. R. Rubenstein and R. W. Wrangham, eds., *Ecological Aspects of Social Evolution*, 21–42. Princeton University Press, Princeton, N.J.
Oring, L. W., M. A. Colwell, and J. M. Reed. 1991a. Lifetime reproductive success in the spotted sandpiper (*Actitis macularia*): Sex differences and variance components. *Behav. Ecol. Sociobiol.* 28: 425–432.
Oring, L. W., J. M. Reed, M. A. Colwell, D. B. Lank, and S. J. Maxson. 1991b. Factors regulating annual mating success and reproductive success in spotted sandpipers (*Actitis macularia*). *Behav. Ecol. Sociobiol.* 28: 433–442.
Ornduff, R. 1969. Reproductive biology in relation to systematics. *Taxon* 18: 121–133.
Osborn, H. F. 1942. *Proboscidea: A Monograph of the Discovery, Evolution, Migration and Extinction of the Mastodonts and Elephants of the World.* Vol. 2, *Stegodontoidea, Elephantoidea.* American Museum Press, New York.

Ostrom, J. H. 1986. Social and unsocial behavior in dinosaurs. In M. H. Nitecki and J. A. Kitchell, eds., *Evolution of Animal Behavior. Paleontological and Field Approaches*, 41–61. Oxford University Press, Oxford.

Otronen, M. 1984a. Male contests for territories and females in the fly *Dryomyza anilis*. *Anim. Behav.* 32: 891–898.

Otronen, M. 1984b. The effect of differences in body size on the male territorial system of the fly *Dryomyza anilis*. *Anim. Behav.* 32: 882–890.

Otronen, M. 1988a. The effect of body size on the outcome of fights in burying beetles (*Nicrophorus*). *Ann. Zool. Fenn.* 25: 191–201.

Otronen, M. 1988b. Intra- and intersexual interactions at breeding burrows in the horned beetle, *Coprophanaeus ensifer*. *Anim. Behav.* 36: 741–748.

Otte, D. 1974. Effects and functions in the evolution of signaling systems. *Ann. Rev. Ecol. Syst.* 5: 385–417.

Otte, D. 1979. Historical development of sexual selection theory. In M. S. Blum and N. A. Blum, eds., *Sexual Selection and Reproductive Competition in Insects*, 1–18. Academic Press, New York.

Otte, D. 1989. Speciation in Hawaiian crickets. In D. Otte and J. A. Endler, eds., *Speciation and Its Consequences*, 482–526. Sinauer, Sunderland, Mass.

Otte, D., and J. A. Endler, eds. 1989. *Speciation and Its Consequences*. Sinauer, Sunderland, Mass.

Otte, D., and K. Stayman. 1979. Beetle horns: Some patterns in functional morphology. In M. S. Blum and N. A. Blum, eds., *Sexual Selection and Reproductive Competition in Insects*, 259–292. Academic Press, New York.

Owen-Smith, N. 1993. Comparative mortality rates of male and female kudus: The costs of sexual size dimorphism. *J. Anim. Ecol.* 62: 428–440.

Packer, C. 1983. Sexual dimorphism: The horns of African antelopes. *Science* 221: 1191–1193.

Pagel, M. D., and P. H. Harvey. 1988. Recent developments in the analysis of comparative data. *Q. Rev. Biol.* 63: 413–440.

Palmer, T. J. 1978. A horned beetle which fights. *Nature* 274: 583–584.

Palokangas, P., R. V. Alatalo, and E. Korpimäki. 1992. Female choice in the kestrel under different availability of mating options. *Anim. Behav.* 43: 659–665.

Parker, G. A. 1970a. The repoductive behaviour and the nature of sexual selection in *Scatophaga stercoria* L. (Diptera: Scatophagidae). *J. Anim. Ecol.* 39: 205–228.

Parker, G. A. 1970b. Sperm competition and its evolutionary consequences in the insects. *Biol. Rev.* 45: 525–567.

Parker, G. A. 1970c. The reproductive behaviour and the nature of sexual selection in *Scatophaga stercoraria* L. IV. Epigamic recognition and competition between males for the possession of females. *Behaviour* 37: 113–139.

Parker, G. A. 1974. Courtship persistence and female guarding as male time investment strategies. *Behaviour* 48: 157–184.

Parker, G. A. 1978. Evolution of competitive mate searching. *Ann. Rev. Entomol.* 23: 173–196.

Parker, G. A. 1979. Sexual selection and sexual conflict. In M. S. Blum and N. A. Blum, eds., *Sexual Selection and Reproductive Competition in Insects*, 123–166. Academic Press, New York.

Parker, G. A. 1982. Phenotype-limited evolutionarily stable strategies. In King's

College Sociobiology Group, eds., *Current Problems in Sociobiology*, 173–202. Cambridge University Press, Cambridge, U.K.

Parker, G. A. 1983a. Arms races in evolution—an ESS to the opponent-independent costs game. *J. theor. Biol.* 101: 619–648.

Parker, G. A. 1983b. Mate quality and mating decisions. In P. Bateson, ed., *Mate Choice*, 141–166. Cambridge University Press, Cambridge, U.K.

Parker, G. A. 1984a. Sperm competition and the evolution of animal mating strategies. In R. L. Smith, ed., *Sperm Competition and the Evolution of Animal Maiing Systems*, 1–60. Academic Press, Orlando, Florida.

Parker, G. A. 1984b. Evolutionarily stable strategies. In J. R. Krebs and N. B. Davies, eds., *Behavioural Ecology: An Evolutionary Approach*, 2d ed., 30–61. Blackwell, Oxford.

Parker, G. A. 1992. Snakes and female sexuality. *Nature* 355: 395–396.

Parker, G. A., and L. W. Simmons. 1989. Nuptial feeding in insects: Theoretical models of male and female interests. *Ethology* 82: 3–26.

Parker, G. A., R. R. Baker, and V.G.F. Smith. 1972. The origin and evolution of gamete dimorphism and the male-female phenomenon. *J. theor. Biol.* 36: 529–533.

Parkes, K. C. 1978. Still another parulid intergeneric hybrid (*Mniotilta* x *Dendroica*) and its taxonomic and evolutionary implications. *Auk* 95: 682–690.

Parsons, P. A. 1992. Fluctuating asymmetry: A biological monitor of environmental and genomic stress. *Heredity* 68: 361–368.

Pärt, T. 1991. Philopatry pays: A comparison between collared flycatcher sisters. *Am. Nat.* 138: 790–796.

Partridge, L. 1980. Mate choice increases a component of offspring fitness in fruitflies. *Nature* 283: 290–291.

Partridge, L. 1983. Non-random mating and offspring fitness. In P. Bateson, ed., *Mate Choice*, 227–256. Cambridge University Press, Cambridge, U.K.

Partridge, L. 1989a. An experimentalist's approach to the role of costs of reproduction in the evolution of life-histories. In P. J. Grubb and I. Whittaker, eds., *Towards a More Exact Ecology*, 231–246. Blackwell, Oxford.

Partridge, L. 1989b. Lifetime reproductive success and life-history evolution. In I. Newton, ed., *Lifetime Reproduction in Birds*, 421–439. Academic Press, London.

Partridge, L., and J. A. Endler. 1987. Life history constraints on sexual selection. In J. W. Bradbury and M. B. Andersson, eds., *Sexual Selection: Testing the Alternatives*, 265–277. Wiley, Chichester, U.K.

Partridge, L., and M. Farquhar. 1983. Lifetime mating success of male fruitflies (*Drosophila melanogaster*) is related to their size. *Anim. Behav.* 31: 871–877.

Partridge, L., and T. Halliday. 1984. Mating patterns and mate choice. In J. R. Krebs and N. B. Davies, eds., *Behavioural Ecology: An Evolutionary Approach*, 2d ed., 222–250. Blackwell, Oxford.

Partridge, L., and P. Harvey. 1985. Costs of reproduction. *Nature* 316: 20–21.

Partridge, L., and W. G. Hill. 1984. Mechanisms for frequency-dependent mating success. *Biol. J. Linn. Soc.* 23: 113–132.

Partridge, L., and R. Sibly. 1991. Constraints in the evolution of life histories. *Phil. Trans. Roy. Soc. London* B 332: 3–13.

Partridge, L., A. Ewing, and A. Chandler. 1987a. Male size and mating success in *Drosophila melanogaster*: The roles of male and female behaviour. *Anim. Behav.* 35: 555–562.

Partridge, L., A. Hoffmann, and J. S. Jones. 1987b. Male size and mating success in *Drosophila melanogaster* and *D. pseudoobscura* under field conditions. *Anim. Behav.* 35: 468–476.

Paterson, H.E.H. 1978. More evidence against speciation by reinforcement. *S. Afr. J. Sci.* 74: 369–371.

Paterson, H.E.H. 1982. Perspective on speciation by reinforcement. *S. Afr. J. Sci.* 78: 53–57.

Paterson, H.E.H. 1985. The recognition concept of species. In E. S. Vrba, ed., *Species and Speciation*, 21–29. Transvaal Museum, Pretoria.

Payne, R. B. 1973. Vocal mimicry of the paradise whydahs (*Vidua*) and response of female whydas to songs of their hosts (*Pytilia*) and their mimics. *Anim. Behav.* 21: 762–771.

Payne, R. B. 1979. Song structure, behaviour, and sequence of song types in a population of village indigobirds, *Vidua chalybeata*. *Anim. Behav.* 27: 997–1013.

Payne, R. B. 1983. Bird song, sexual selection, and female mating strategies. In S. K. Wasser, ed., *Social Behavior of Female Vertebrates*, 55–90. Academic Press, New York.

Payne, R. B. 1984. Sexual selection, lek and arena behavior, and sexual size dimorphism in birds. *Orn. Monogr.* 33: 1–53.

Payne, R. B., and K. Payne. 1977. Social organization and mating success in local song populations of village indigobirds, *Vidua chalybeata*. *Z. Tierpsychol.* 45: 113–173.

Payne, R. B., and D. F. Westneat. 1988. A genetic and behavioral analysis of mate choice and song neighborhoods in indigo buntings. *Evolution* 42: 935–947.

Peckham, D. J., and A. W. Hook. 1980. Behavioral observations on *Oxybelus* in Southeastern North America. *Ann. Entomol. Soc. Am.* 73: 557–567.

Peckham, G. W., and E. G. Peckham. 1889. Observations on sexual selection in spiders of the family Attidae. *Occas. Pap. Nat. Hist. Soc. Wisconsin* 1: 1–60.

Peek, F. W. 1972. An experimental study of the territorial function of vocal and visual display in the male red-winged blackbird (*Agelaius phoeniceus*). *Anim. Behav.* 20: 112–118.

Pelkwijk, J.J.T., and N. Tinbergen. 1937. Eine reizbiologische Analyse einiger Verhaltensweisen von *Gasterosteus aculeatus* L. *Z. Tierpsychol.* 1: 193–204.

Pellmyr, O. 1992. Evolution of insect pollination and angiosperm diversification. *TREE* 7: 46–49.

Pemberton, J. M., S. D. Albon, F. E. Guinness, T. H. Clutton-Brock, and G. A. Dover. 1992. Behavioural estimates of male mating success tested by DNA fingerprinting in a polygynous mammal. *Behav. Ecol.* 3: 66–75.

Perrill, S. A., H. C. Gerhardt, and R. Daniel. 1978. Sexual parasitism in the green treefrog (*Hyla cinerea*). *Science* 200: 1179–1180.

Perrill, S. A., H. C. Gerhardt, and R. Daniel. 1982. Mating strategy shifts in male green treefrogs (*Hyla cinerea*): An experimental study. *Anim. Behav.* 30: 43–48.

Perrone, M. J. 1978. Mate size and breeding success in a monogamous cichlid fish. *Env. Biol. Fish.* 3: 193–201.

Peterman, R. M. 1971. A possible function of coloration in coral reef fishes. *Copeia*, 330–331.

Peters, R. H. 1983. *The Ecological Implications of Body Size*. Cambridge University Press, Cambridge, U.K.

Petersen, C. W. 1988. Male mating success, sexual size dimorphism, and site fidelity in two species of *Malacoctenus* (Labrisomidae). *Env. Biol. Fish.* 21: 173–183.

Petit, C. 1954. L'isolement sexual chez *Drosophila melanogaster*. Etude du mutant *white* et de son allélomorph *sauvage*. *Biol. Bull. France et Belgique* 88: 435–443.

Petrie, M. 1983a. Female moorhens compete for small fat males. *Science* 220: 413–415.

Petrie, M. 1983b. Mate choice in role-reversed species. In P. Bateson, ed., *Mate Choice*, 167–180. Cambridge University Press, Cambridge, U.K.

Petrie, M. 1992a. Are all secondary sexual display structures positively allometric and, if so, why? *Anim. Behav.* 43: 173–175.

Petrie, M. 1992b. Peacocks with low mating success are more likely to suffer predation. *Anim. Behav.* 44: 585–586.

Petrie, M., and A. Williams. 1993. Peahens lay more eggs for peacocks with larger trains. *Proc. Roy. Soc. London* B 251: 127–131.

Petrie, M., T. Halliday, and C. Sanders. 1991. Peahens prefer peacocks with elaborate trains. *Anim. Behav.* 41: 323–331.

Petrie, M., M. Hall, T. Halliday, H. Budgey, and C. Pierpoint. 1992. Multiple mating in a lekking bird: Why do peahens mate with more than one male and with the same male more than once? *Behav. Ecol. Sociobiol.* 31: 349–358.

Phelan, P. L., and T. C. Baker. 1986. Male-size-related courtship success and intersexual selection in the tobacco moth, *Ephestia elutella*. *Experientia* 42: 1291–1293.

Phelan, P. L., and T. C. Baker. 1987. Evolution of male pheromones in moths: Reproductive isolation through sexual selection? *Science* 235: 205–207.

Phelan, P. L., and T. C. Baker. 1990a. Comparative study of courtship in twelve phycitine moths (Lepidoptera: Pyralidae). *J. Insect Behav.* 3: 303–326.

Phelan, P. L., and T. C. Baker. 1990b. Information transmission during intra- and interspecific courtship in *Ephestia elutella* and *Cadra figulilella*. *J. Insect Behav.* 3: 589–601.

Pianka, E. 1978. *Evolutionary Ecology*. 2d ed. Harper and Row, New York.

Picman, J. 1980. Impact of marsh wrens on reproductive strategy of red-winged blackbirds. *Can. J. Zool.* 58: 337–350.

Picozzi, N. 1984. Breeding biology of polygynous hen harriers *Circus c. cyaneus* in Orkney. *Ornis Scand.* 15: 1–10.

Pietsch, T. W. 1976. Dimorphism, parasitism and sex: Reproductive strategies among deep-sea ceratioid anglerfishes. *Copeia* 1976: 781–793.

Pinto, J. D. 1980. Behavior and taxonomy of the *Epicauta maculata* group (Coleoptera: Meloidae). *Univ. Calif. Publ. Entomol.* 89: 1–111.

Pitelka, F. 1957. Some aspects of population structure in the short-term cycle of the brown lemming in northern Alaska. *Cold Spring Harbor Symp. Quant. Biol.* 22: 237–251.

Platt, J. R. 1964. Strong inference. *Science* 146: 347–353.

Pleasants, J. M., and S. J. Chaplin. 1983. Nectar production rates of *Asclepias quadrifolia*: Causes and consequences of individual variation. *Oecologia* 59: 232–238.

Pleszczynska, W. K. 1978. Microgeographic prediction of polygyny in the lark bunting. *Science* 201: 935–937.

Pliske, T. E., and T. Eisner. 1969. Sex pheromone of the queen butterfly: Biology. *Science* 164: 1170–1172.

Policansky, D. 1981. Sex choice and the size advantage model in jack-in-the-pulpit (*Arisaema triphyllum*). *Proc. Natl. Acad. Sci. USA* 78: 1306–1308.

Policansky, D. 1982. Sex change in plants and animals. *Ann. Rev. Ecol. Syst.* 13: 471–495.

Polis, G. 1981. The evolution and dynamics of intraspecific predation. *Ann. Rev. Ecol. Syst.* 12: 225–251.

Polis, G. A., and W. D. Sisson. 1990. Life history. In G. A. Polis, ed., *The Biology of Scorpions*, 161–223. Stanford University Press, Stanford, Calif.

Pomiankowski, A. 1987a. Sexual selection: The handicap principle does work—sometimes. *Proc. Roy. Soc. London* B 231: 123–145.

Pomiankowski, A. 1987b. The costs of choice in sexual selection. *J. theor. Biol.* 128: 195–218.

Pomiankowski, A. 1988. The evolution of female mate preferences for male genetic quality. *Oxford Surv. Evol. Biol.* 5: 136–184. Oxford University Press, Oxford.

Pomiankowski, A., and T. Guilford. 1990. Mating calls. *Nature* 344: 495–496.

Pomiankowski, A., Y. Iwasa, and S. Nee. 1991. The evolution of costly mate preferences. I. Fisher and biased mutation. *Evolution* 45: 1422–1430.

Poole, J. H. 1987. Rutting behaviour in African elephants: The phenomenon of musth. *Behaviour* 102: 283–316.

Poole, J. H. 1989a. Mate guarding, reproductive success and female choice in African elephants. *Anim. Behav.* 37: 842–849.

Poole, J. H. 1989b. Announcing intent: The aggressive state of musth in African elephants. *Anim. Behav.* 37: 140–152.

Poole, J. H., and C. J. Moss. 1981. Musth in the African elephant, *Loxodonta africana*. *Nature* 292: 830–831.

Poole, J. H., K. Payne, W. R. Langbauer, and C. J. Moss. 1988. The social contexts of some very low frequency calls of African elephants. *Behav. Ecol. Sociobiol.* 22: 385–392.

Pope, C. H. 1931. Notes on amphibians from Fukien, Hainan and other parts of China. *Bull. Am. Mus. Nat. Hist.* 61: 397–612.

Popp, J. W., and J. A. Reinartz. 1988. Sexual dimorphism in biomass allocation and clonal growth of *Xanthoxylum americanum*. *Am. J. Bot.* 75: 1732–1741.

Popper, K. R. 1969. *Conjectures and Refutations: The Growth of Scientific Knowledge*. 3d ed. Routledge and Kegan Paul, London.

Potti, J., and S. Montalvo. 1991. Male arrival and female mate choice in pied flycatchers *Ficedula hypoleuca* in Central Spain. *Ornis Scand.* 22: 45–54.

Poulton, E. B. 1890. *The Colours of Animals*. 2d ed. Kegan Paul, London.

Poulton, M. J., and D. J. Thompson. 1987. The effects of the acanthocephalan parasite *Pomphorhynchus laevis* on mate choice in *Gammarus pulex*. *Anim. Behav.* 35: 1577–1579.

Power, H. W. 1980. The foraging behavior of mountain bluebirds with special emphasis on sexual differences. *Orn. Monogr.* 28: 1–72.

Prestwich, K. N., K. E. Brugger, and M. Topping. 1989. Energy and communication in three species of Hylid frogs: Power input, power output and efficiency. *J. Exp. Biol.* 144: 53–80.

Price, M. V., and N. M. Waser. 1979. Pollen dispersal and optimal outcrossing in *Delphinium nelsoni*. *Nature* 277: 294–296.

Price, P. W., and M. F. Willson. 1976. Some consequences for a parasitic herbivore, the milkweed longhorn beetle, *Tetraopes tetraophtalmus*, of a host-plant shift from *Asclepias syriaca* to *A. verticillata*. *Oecologia* 25: 331–340.

Price, T. D. 1984a. The evolution of sexual size dimorphism in Darwin's finches. *Am. Nat.* 123: 500–518.

Price, T. D. 1984b. Sexual selection on body size, territory and plumage variables in a population of Darwin's finches. *Evolution* 38: 327–341.

Price, T. D., and P. T. Boag. 1987. Selection in natural populations of birds. In F. Cook and P. A. Buckley, eds., *Avian Genetics: A Population and Ecological Approach*, 257–287. Academic Press, London.

Price, T. D., and P. R. Grant. 1984. Life history traits and natural selection for small body size in a population of Darwin's finches. *Evolution* 38: 483–494.

Price, T., et. al. 1987. Constraints on the effects of sexual selection: Group report. In J. W. Bradbury and M. B. Andersson, eds., *Sexual Selection: Testing the Alternatives*, 279–294. Wiley, Chichester, U.K.

Price, T. D., M. Kirkpatrick, and S. Arnold. 1988. Directional selection and the evolution of breeding date in birds. *Science* 240: 798–799.

Price, T. D., D. Schluter, and N. E. Heckman. 1993. Sexual selection when the female directly benefits. *Biol. J. Linn. Soc.* 48: 187–211.

Primack, R. B., and P. Hall. 1990. Costs of reproduction in the pink lady's slipper orchid: A four-year experimental study. *Am. Nat.* 136: 638–656.

Proctor, H. C. 1991. Courtship in the water mite *Neumania papillator*: Males capitalize on female adaptations for predation. *Anim. Behav.* 42: 589–598.

Proctor, M., and P. Yeo. 1973. *The Pollination of Flowers*. Collins, London.

Promislow, D.E.L. 1992. Costs of sexual selection in natural populations of mammals. *Proc. Roy. Soc. London* B 247: 203–210.

Promislow, D.E.L., R. Montgomerie, and T. E. Martin. 1992. Mortality costs of sexual dimorphism in birds. *Proc. Roy. Soc. London* B 250: 143–150.

Provine, W. B. 1971. *The Origins of Theoretical Population Genetics*. University of Chicago Press, Chicago.

Prowse, D. L., J. S. Trilling, and J. R. Luick. 1980. Effects of antler removal on mating behavior of reindeer. In E. Reimers, E. Gaare, and S. Skjenneberg, eds., *Proceedings of the Second International Reindeer/Caribou Symposium*, 528–536. Direktoratet for Vilt og Ferskvannsfisk, Trondheim, Norway.

Pruett-Jones, S. 1992. Independent versus nonindependent mate choice: Do females copy each other? *Am. Nat.* 140: 1000–1009.

Pruett-Jones, S. G., and M. A. Pruett-Jones. 1990. Sexual selection through female choice in Lawes' Parotia, a lek-mating bird of paradise. *Evolution* 44: 486–501.

Pruett-Jones, S. G., M. A. Pruett-Jones, and H. I. Jones. 1990. Parasites and sexual selection in birds of paradise. *Am. Zool.* 30: 287–298.

Pruett-Jones, S. G., M. A. Pruett-Jones, and H. I. Jones. 1991. Parasites and sexual selection in a New Guinea avifauna. *Curr. Ornithol.* 8: 213–245.

Pyke, G. 1981. Optimal nectar production in a hummingbird-pollinated plant. *Theor. Pop. Biol.* 20: 326–343.

Pyke, G. H. 1984. Optimal foraging theory: A critical review. *Ann. Rev. Ecol. Syst.* 15: 523–575.

Pyke, G. H. 1991. What does it cost a plant to produce floral nectar ? *Nature* 350: 58–59.

Queller, D. C. 1983. Sexual selection in a hermaphroditic plant. *Nature* 305: 706–707.

Queller, D. C. 1984a. Pollen-ovule ratios and hermaphrodite sexual allocation strategies. *Evolution* 38: 1148–1151.

Queller, D. C. 1984b. Models of kin selection on seed provisioning. *Heredity* 53: 151–165.

Queller, D. C. 1985. Proximate and ultimate causes of low fruit production in *Asclepias exaltata. Oikos* 44: 373–381.

Queller, D. C. 1987. Sexual selection in flowering plants. In J. W. Bradbury and M. B. Andersson, eds., *Sexual Selection: Testing the Alternatives*, 165–179. Wiley, Chichester, U.K.

Quinn, J. S., and S. K. Sakaluk. 1986. Prezygotic male reproductive effort in insects: Why do males provide more than sperm? *Florida Entomol.* 69: 84–94.

Quinn, T. W., J. S. Quinn, F. Cooke, and B. N. White. 1987. DNA marker analysis detects multiple maternity and paternity in single broods of the lesser snow goose. *Nature* 326: 392–394.

Radesäter, T., and S. Jakobsson. 1988. Intra- and intersexual functions of song in the willow warbler (*Phylloscopus trochilus*). *Acta XIX Congr. Int. Orn.* (2): 1382–1390.

Radesäter, T., and S. Jakobsson. 1989. Song rate correlations of replacement territorial willow warblers *Phylloscopus trochilus. Ornis Scand.* 20: 71–73.

Radesäter, T., S. Jakobsson, N. Andbjer, A. Bylin, and K. Nyström. 1987. Song rate and pair formation in the willow warbler, *Phylloscopus trochilus. Anim. Behav.* 35: 1645–1651.

Raikow, R. J. 1986. Why are there so many kinds of passerine birds ? *Syst. Zool.* 35: 255–259.

Raikow, R. J. 1987. Hindlimb myology and evolution of the old world suboscine passerine birds (Acanthisittidae, Pittidae, Philepittidae, Eurylaimidae). *Orn. Monogr.* 41: 1–81.

Raikow, R. J. 1988. The analysis of evolutionary success. *Syst. Zool.* 37: 76–79.

Ralls, K. 1971. Mammalian scent marking. *Science* 171: 443–449.

Ralls, K. 1976. Mammals in which females are larger than males. *Q. Rev. Biol.* 51: 245–276.

Ralls, K. 1977. Sexual dimorphism in mammals: Avian models and unanswered questions. *Am. Nat.* 111: 917–937.

Ralls, K., and P. H. Harvey. 1985. Geographic variation in size and sexual dimorphism of North American weasels. *Biol. J. Linn. Soc.* 25: 119–167.

Ralph, C. L. 1969. The control of color in birds. *Am. Zool.* 9: 521–530.

Rand, A. S., and M. J. Ryan. 1981. The adaptive significance of a compex vocal repertoire in a Neotropical frog. *Z. Tierpsychol.* 57: 209–214.

Ratcliffe, L. M., and P. T. Boag. 1987. Effects of colour bands on male competition and sexual attractiveness in zebra finches (*Poephila guttata*). *Can. J. Zool.* 65: 333–338.

Ratcliffe, L. M., and P. R. Grant. 1983. Species recognition in Darwin's finches (*Geospiza*, Gould). II. Geographic variation in mate preference. *Anim. Behav.* 31: 1154–1165.

Rathke, B. 1983. Competition and facilitation among plants for pollination. In L. Real, ed., *Pollination Biology*, 305–329. Academic Press, Orlando, Florida.

Raxworthy, C. J. 1989. Courtship, fighting and sexual dimorphism of the banded newt, *Triturus vittatus ophryticus*. *Ethology* 81: 148–170.

Read, A. F. 1987. Comparative evidence supports the Hamilton and Zuk hypothesis on parasites and sexual selection. *Nature* 328: 68–70.

Read, A. F. 1988. Sexual selection and the role of parasites. *TREE* 3: 97–101.

Read, A. F. 1990. Parasites and the evolution of host sexual behaviour. In C. J. Barnard and J. M. Behnke, eds., *Parasitism and Host Behaviour*, 117–157. Taylor and Francis, London.

Read, A. F. 1991. Passerine polygyny: A role for parasites. *Am. Nat.* 138: 434–459.

Read, A. F., and P. H. Harvey. 1989a. Reassessment of comparative evidence for Hamilton and Zuk theory on the evolution of secondary sexual characters. *Nature* 339: 618–620.

Read, A. F., and P. H. Harvey. 1989b. Validity of sexual selection in birds. *Nature* 340: 105.

Read, A. F., and D. M. Weary. 1990. Sexual selection and the evolution of bird song: A test of the Hamilton-Zuk hypothesis. *Behav. Ecol. Sociobiol.* 26: 47–56.

Real, L., ed. 1983. *Pollination Biology*. Academic Press, New York.

Real, L. 1990. Search theory and mate choice. I. Models of single-sex discrimination. *Am. Nat.* 136: 376–405.

Real, L. A. 1991. Animal choice behavior and the evolution of cognitive architecture. *Science* 253: 980–986.

Reece-Engel, C. 1988. Female choice of resident male rabbits *Oryctolagus cuniculus*. *Anim. Behav.* 36: 1241–1242.

Reed, S. C., and E. W. Reed. 1950. Natural selection in laboratory populations of *Drosophila*. II. Competition between a white eye gene and its white allele. *Evolution* 4: 34–42.

Reekie, E. G., and F. A. Bazzaz. 1987. Reproductive effort in plants. 1. Carbon allocation to reproduction. *Am. Nat.* 129: 876–896.

Reese, E. S. 1975. A comparative field study of the social behavior and related ecology of reef fishes of the family Chaetodontidae. *Z. Tierpsychol.* 37: 37–61.

Reeve, H. K. and P. W. Sherman. 1993. Adaptation and the goals of evolutionary research. *Q. Rev. Biol.* 68: 1–32.

Reid, J. B. 1984. Bird coloration: Predation, conspicuousness and the unprofitable prey model. *Anim. Behav.* 32: 294–295.

Reid, M. L., and P. J. Weatherhead. 1990. Mate-choice criteria of Ipswich sparrows: The importance of variability. *Anim. Behav.* 40: 538–544.

Reimchen, T. E. 1989. Loss of nuptial color in threespine sticklebacks (*Gasterosteus aculeatus*). *Evolution* 43: 450–460.

Reiss, M. J. 1989. *The Allometry of Growth and Reproduction*. Cambridge University Press, Cambridge.

Renfree, M. B., and R. V. Short. 1988. Sex determination in marsupials: Evidence for a marsupial-eutherian dichotomy. *Phil. Trans. Roy. Soc. London* 322: 41–51.

Rensch, B. 1950. Die Abhängigkeit der relativen Sexualdifferenz von der Körpergrösse. *Bonn. Zool. Beitr.* 1: 58–69.

Rensch, B. 1959. *Evolution above the Species Level.* Columbia University Press, New York.

Ressel, S., and J. J. Schall. 1989. Parasites and showy males: Malarial infection and color variation in fence lizards. *Oecologia* 78: 158–164.

Reynolds, J. D. 1993. Should attractive individuals court more? Theory and a test. *Amer. Nat.* 141: 914–927.

Reynolds, J. D., and M. R. Gross. 1990. Costs and benefits of female mate choice: Is there a lek paradox? *Am. Nat.* 136: 230–243.

Reynolds, J. D., and M. R. Gross. 1992. Female mate preference enhances offspring growth and reproduction in a fish, *Poecilia reticulata. Proc. Roy. Soc. London* B 250: 57–62.

Reynolds, J. D., M. A. Colwell, and F. Cooke. 1986. Sexual selection and spring arrival times of red-necked and Wilsons phalaropes. *Behav. Ecol. Sociobiol.* 18: 303–310.

Reynolds, J. D., M. R. Gross, and M. J. Coombs. 1993. Environmental conditions and male morphology determine alternative mating behavior in Trinidadian guppies. *Anim. Behav.* 45: 145–152.

Reynolds, R. T. 1972. Sexual dimorphism in accipiter hawks: A new hypothesis. *Condor* 74: 191–197.

Reznick, D. 1985. Costs of reproduction: An evaluation of the empirical evidence. *Oikos* 44: 257–267.

Reznick, D. A., H. Bryga, and J. A. Endler. 1990. Experimentally induced life-history evolution in a natural population. *Nature* 346: 357–359.

Reznik, D. N., and J. A. Endler. 1982. The impact of predation on life history evolution in Trinidadian guppies (*Poecilia reticulata*). *Evolution* 36: 160–177.

Rice, W. 1988. Heritable variation in fitness as a prerequisite for adaptive female choice: The effect of mutation-selection balance. *Evolution* 42: 817–820.

Ridley, M. 1978. Paternal care. *Anim. Behav.* 26: 904–932.

Ridley, M. 1981. How the peacock got his tail. *New Sci.* 91: 398–401.

Ridley, M. 1983. *The Explanation of Organic Diversity: The Comparative Method and Adaptations for Mating*. Clarendon, Oxford.

Ridley, M. 1988. Mating frequency and fecundity in insects. *Biol. Rev.* 63: 509–549.

Ridley, M., and K. Rechten. 1981. Female sticklebacks prefer to spawn with males whose nests contain eggs. *Behaviour* 76: 152–161.

Ridley, M., and D. J. Thompson. 1979. Size and mating in *Asellus aquaticus* (Crustacea: Isopoda). *Z. Tierpsychol.* 51: 380–397.

Ridley, M., and D. J. Thompson. 1985. Sexual selection of population dynamics in aquatic crustacea. In R. M. Sibly and R. H. Smith, eds., *Behavioural Ecology: Ecological Consequences of Adaptive Behaviour*, 409–422. Blackwell, Oxford.

Ritchie, M. G. 1992. Setbacks in the search for mate-preference genes. *TREE* 7: 328–329.

Robertson, D. R. 1972. Social control of sex reversal in a coral reef fish. *Science* 177: 1007–1009.

Robertson, D. R., and S. G. Hoffman. 1977. The roles of female mate choice and predation in the mating systems of some tropical labroid fishes. *Z. Tierpsychol.* 45: 298–320.

Robertson, D. R., and R. R. Warner. 1978. Sexual patterns in the labroid fishes of the western Carribbean. II. The parrotfishes (Scaridae). *Smiths. Contrib. Zool.* 255: 1–26.

Robertson, J.G.M. 1984. Acoustic spacing by breeding males of *Uperoleia rugosa* (Anura: Leptodactylidae). *Z. Tierpsychol.* 64: 283–297.

Robertson, J.G.M. 1986a. Female choice, male strategies and the role of vocalizations in the Australian frog *Uperoleia rugosa. Anim. Behav.* 34: 773–784.

Robertson, J.G.M. 1986b. Male territoriality, fighting and assessment of fighting ability in the Australian frog *Uperoleia rugosa. Anim. Behav.* 34: 763–777.

Robertson, J.G.M. 1990. Female choice increases fertilization success in the Australian frog *Uperoleia laevigata. Anim. Behav.* 39: 639–645.

Robinson, B. W., and R. W. Doyle. 1985. Trade-off between male reproduction (amplexus) and growth in the Amphipod *Gammarus lawrencianus. Biol. Bull.* 168: 482–488.

Robinson, S. K. 1986a. Competitive and mutualistic interactions among females in a neotropical oriole. *Anim. Behav.* 34: 113–122.

Robinson, S. K. 1986b. Benefits, costs, and determinants of dominance in a polygynous oriole. *Anim. Behav.* 34: 241–255.

Roeder, K. D. 1935. An experimental analysis of the sexual behavior of the praying mantis (*Mantis religiosa*). *Biol. Bull.* 69: 203–220.

Roelofs, W. L., T. Glover, X. H. Tang, I. Sreng, P. Robbins, C. Eckenrode, C. Löfstedt, B. S. Hansson, and B. O. Bengtsson. 1987. Sex pheromone production and perception in European cornborer moths is determined by both autosomal and sex-linked genes. *Proc. Natl. Acad. Sci. USA* 84: 7585–7589.

Roff, D. A. 1986. Predicting body size with life history models. *Bioscience* 36: 316–323.

Roff, D. A. 1992. *The Evolution of Life Histories*. Chapman and Hall, London.

Roff, D. A., and T. A. Mousseau. 1987. Quantitative genetics and fitness: Lessons from *Drosophila. Heredity* 58: 103–118.

Rogers, A. R., and A. Mukherjee. 1992. Quantitative genetics of sexual dimorphism in human body size. *Evolution* 46: 226–234.

Rogoff, J. L. 1927. The hedonic glands of *Triturus viridescens*; a structural and functional study. *Anat. Rec.* 34: 132.

Rohwer, S. 1975. The social significance of avian winter plumage variability. *Evolution* 29: 593–610.

Rohwer, S. 1977. Status signaling in Harris' sparrows: Some experiments in deception. *Behaviour* 61: 107–129.

Rohwer, S. 1978. Parent cannibalism of offspring and egg raiding as a courtship strategy. *Am. Nat.* 112: 429–440.

Rohwer, S. 1982. The evolution of reliable and unreliable badges of fighting ability. *Am. Zool.* 22: 531–546.

Rohwer, S., and G. S. Butcher. 1988. Winter versus summer explanations of delayed plumage maturation in temperate passerine birds. *Am. Nat.* 131: 556–572.

Rohwer, S., and E. Røskaft. 1989. Results of dyeing male yellow-headed blackbirds solid black: Implications for the arbitrary identity badge hypothesis. *Behav. Ecol. Sociobiol.* 25: 39–48.

Rohwer, S., S. D. Fretwell, and D. M. Niles. 1980. Delayed maturation in passerine plumages and the deceptive acquisition of resources. *Am. Nat.* 115: 400–437.

Romer, A. S. 1966. *Vertebrate Paleontology*. 3d ed. University of Chicago Press, Chicago.

Romero, G. A., and C. E. Nelson. 1986. Sexual dimorphism in *Catasetum* orchids: Forcible pollen emplacement and male flower competition. *Science* 232: 1538–1540.

Rørvik, K. A., H. C. Pedersen, and J. B. Steen. 1990. Genetic variation and territoriality in willow ptarmigan (*Lagopus lagopus lagopus*). *Evolution* 44: 1490–1497.

Rose, M. R. 1984. Laboratory evolution of postponed senescence in *Drosophila melanogaster*. *Evolution* 38: 1004–1010.

Rose, M., and B. Charlesworth. 1981. Genetics of life history in *Drosophila melanogaster*. I. Sib analysis of adult females. *Genetics* 97: 173–186.

Rosenqvist, G. 1990. Male mate choice and female-female competition for mates in the pipefish *Nerophis ophidion*. *Anim. Behav.* 39: 1110–1115.

Røskaft, E., and T. Järvi. 1983. Male plumage colour and mate choice of female pied flycatchers *Ficedula hypoleuca*. *Ibis* 125: 396–400.

Røskaft, E., and T. Järvi. 1992. Interspecific competition and the evolution of plumage colour variation in three closely related old world flycatchers *Ficedula* spp. *J. Zool.* 228: 521–532.

Røskaft, E., and S. Rohwer. 1987. An experimental study of the function of the red epaulettes and the black body colour of male red-winged blackbirds. *Anim Behav.* 35: 1070–1077.

Røskaft, E., T. Järvi, N. E. Nyholm, M. Virolainen, W. Winkel, and H. Zang. 1986. Geographic variation in secondary sexual plumage colour characteristics of the male pied flycatcher. *Ornis Scand.* 17: 293–298.

Ross, H. A. 1979. Multiple clutches and shorebird egg and body weight. *Am. Nat.* 113: 618–622.

Roughgarden, J. 1979. *Theory of Population Genetics and Evolutionary Ecology: An Introduction*. Macmillan, New York.

Rowland, W. J. 1979. The use of colour in intraspecific communication. In E. M. Burtt Jr., ed., *The Behavioral Significance of Color*, 381–421. Garland STPM Press, New York.

Rowland, W. J. 1982a. Mate choice by male sticklebacks, *Gasterosteus aculeatus*. *Anim. Behav.* 30: 1093–1098.

Rowland, W. J. 1982b. The effects of male nuptial coloration on stickleback aggression: A reexamination. *Behaviour* 80: 118–126.

Rowland, W. J. 1989a. Mate choice and the supernormality effect in female sticklebacks (*Gasterosteus aculeatus*). *Behav. Ecol. Sociobiol.* 24: 433–438.

Rowland, W. J. 1989b. The effects of body size, aggression and nuptial coloration on competition for territories in male threespine sticklebacks, *Gasterosteus aculeatus*. *Anim. Behav.* 37: 282–289.

Rowland, W. J. 1989c. The ethological basis of mate choice in male threespine sticklebacks, *Gasterosteues aculeatus*. *Anim. Behav.* 38: 112–120.

Rowley, I. 1975. *Bird Life*. Collins, Sydney.

Rowley, I. 1983. Re-mating in birds. In P. Bateson, ed., *Mate Choice*, 331–360. Cambridge University Press, Cambridge, U.K.

Rubenstein, D. I. 1984. Resource acquisition and alternative mating strategies in water striders. *Amer. Zool.* 24: 345–353.

Rubenstein, D. I., and R. W. Wrangham, eds. 1986. *Ecological Aspects of Social Evolution*. Princeton University Press, Princeton, N.J.

Ruby, D. E. 1978. Seasonal changes in the territorial behaviour of the iguanid lizard *Sceloporus jarrovi*. *Copeia*, 431–438.

Ruby, D. E. 1981. Phenotypic correlates of male reproductive success in the lizard, *Sceloporus jarrovi*. In R. D. Alexander and D. W. Tinkle, eds., *Natural Selection and Social Behaviour: Recent Research and New Theory*, 96–107. Chiron Press, New York.

Ruse, M. 1979. *The Darwinian Revolution*. University of Chicago Press, Chicago.

Rütimeyer, L. 1881. Beiträge zu einer natürlichen Geschichte der Hirsche. *Abh. Schweiz. Paläontol. Ges.* 8: 3–97.

Rutowski, R. L. 1979. Courtship behavior of the checkered white, *Pieris protodice* (Pieridae). *J. Lepid. Soc.* 33: 42–49.

Rutowski, R. L. 1982a. Mate choice and lepidopteran mating behavior. *Florida Entomol.* 65: 72–82.

Rutowski, R. L. 1982b. Epigamic selection by males as evidenced by courtship partner preferences in the checkered white butterfly (*Pieris protodice*). *Anim. Behav.* 30: 108–112.

Rutowski, R. L. 1984. Sexual selection and the evolution of butterfly mating behavior. *J. Res. Lepid.* 23: 125–142.

Rutowski, R. L., G. W. Gilchrist, and B. Terkanian. 1987. Female butterflies mated with recently mated males show reduced reproductive output. *Behav. Ecol. Sociobiol.* 20: 319–322.

Ryan, M. J. 1980a. Female mate choice in a neotropical frog. *Science* 209: 523–525.

Ryan, M. J. 1980b. The reproductive behavior of the bullfrog (*Rana catesbeiana*). *Copeia* 1980: 108–114.

Ryan, M. J. 1983. Sexual selection and communication in a neotropical frog, *Physalaemus pustulosus*. *Evolution* 37: 261–272.

Ryan, M. J. 1985. *The Tungara Frog: A Study in Sexual Selection and Communication*. University of Chicago Press, Chicago.

Ryan, M. J. 1986. Neuroanatomy influences speciation rates among anurans. *Proc. Natl. Acad. Sci. USA* 83: 1379–1382.

Ryan, M. J. 1988a. Energy, calling, and selection. *Amer. Zool.* 28: 885–898.

Ryan, M. J. 1988b. Constraints and patterns in the evolution of anuran acoustic communication. In B. Fritzsch, M. Ryan, W. Wilczynski, T. E. Hetherington, and W. Walkowiak, eds., *The Evolution of the Amphibian Auditory System*, 637–677. Wiley, New York.

Ryan, M. J. 1990a. Signals, species, and sexual selection. *Am. Sci.* 78: 46–52.

Ryan, M. J. 1990b. Sexual selection, sensory systems and sensory exploitation. *Oxford Surv. Evol. Biol.* 7: 157–195.

Ryan, M. J., and A. Keddy-Hector. 1992. Directional patterns of female mate choice and the role of sensory biases. *Am. Nat.* 139: s4-s35.

Ryan, M. J., and A. S. Rand. 1990. The sensory basis of sexual selection for complex calls in the túngara frog, *Physalaemus pustulosus* (sexual selection for sensory exploitation). *Evolution* 44: 305–314.

Ryan, M. J., and W. E. Wagner, Jr. 1987. Asymmetries in mating preferences between species: Female swordtails prefer heterospecific males. *Science* 236: 595–597.

Ryan, M. J., and W. Wilczynski. 1988. Coevolution of sender and receiver: Effect on local mate preference in cricket frogs. *Science* 240: 1786–1788.

Ryan, M. J., M. D. Tuttle, and L. K. Taft. 1981. The costs and benefits of frog chorusing behavior. *Behav. Ecol. Sociobiol.* 8: 273–278.

Ryan, M. J., M. D. Tuttle, and A. S. Rand. 1982. Bat predation and sexual advertisement in a neotropical anuran. *Am. Nat.* 119: 136–139.

Ryan, M. J., J. H. Fox, W. Wilczynski, and A. S. Rand. 1990. Sexual selection for sensory exploitation in the frog *Physalaemus pustulosus*. *Nature* 343: 66–67.

Saether, B.-E., J. A. Kålås, L. Löfaldli, and R. Andersen. 1986. Sexual size dimorphism and reproductive ecology in relation to mating system in waders. *Biol. J. Linn. Soc.* 28: 273–284.

Safina, C. 1984. Selection for reduced male size in raptorial birds: The possible roles of female choice and mate guarding. *Oikos* 43: 159–164.

Sakaluk, S. K. 1984. Male crickets feed females to ensure complete sperm transfer. *Science* 223: 609–610.

Sakaluk, S. K. 1986a. Sperm competition and the evolution of nuptial feeding behavior in the cricket, *Gryllodes supplicans* (Walker). *Evolution* 40: 584–593.

Sakaluk, S. K. 1986b. Is courtship feeding by male insects parental investment? *Ethology* 73: 161–166.

Sakaluk, S. K. 1988. Inheritance of male parental investment in an insect. *Am. Nat.* 132: 594–601.

Sakaluk, S. K., and J. J. Belwood. 1984. Gecko phonotaxis to cricket calling song: A case of satellite predation. *Anim. Behav.* 32: 659–662.

Sakaluk, S. K., and W. H. Cade. 1980. Female mating frequency and progeny production in singly and doubly mated house and field crickets. *Can. J. Zool.* 58: 404–411.

Sandegren, F. E. 1976. Agonistic behaviour in the male northern elephant seal. *Behaviour* 57: 136–158.

Sanderson, N. 1989. Can gene flow prevent reinforcement? *Evolution* 43: 1223–1235.

Sandhall, Å. 1987. *Trollsländor i Europa*. Interpublishing, Stockholm.

Sappington, T. W., and O. R. Taylor. 1990. Disruptive sexual selection in *Colias eurytheme* butterflies. *Proc. Natl. Acad. Sci. USA* 87: 6132–6135.

Sargent, R. C. 1982. Territory quality, male quality, courtship intrusions, and female nest-choice in the threespine stickleback, *Gasterosteus aculeatus*. *Anim. Behav.* 30: 364–374.

Sargent, R. C. 1988. Paternal care and egg survival both increase with clutch size in the fathead minnow, *Pimephales promelas*. *Behav. Ecol. Sociobiol.* 23: 33–37.

Sargent, R. C., and J. B. Gebler. 1980. Effects of nest site concealment on hatching success, reproductive success, and paternal behavior of the threespine stickleback, *Gasterosteus aculeatus*. *Behav. Ecol. Sociobiol.* 7: 137–142.

Sargent, R. C., M. R. Gross, and E. P. van den Berghe. 1986. Male mate choice in fishes. *Anim. Behav.* 34: 545–550.

Savage, J. M. 1966. An extraordinary new toad (*Bufo*) from Costa Rica. *Rev. Biol. Trop.* 14: 153–167.

Sawyer, S., and D. L. Hartl. 1981. On the evolution of behavioral reproductive isolation: The Wallace effect. *Theor. Pop. Biol.* 19: 261–273.

Schaeffer, S. W., C. J. Brown, and W. W. Andersson. 1984. Does mate choice affect fitness? *Genetics* 107: 94.

Schaffer, W. M., and M. V. Schaffer. 1977. The adaptive significance of variation in reproductive habit in Agavaceae. In B. Stonehouse and C. Perrins, eds., *Evolutionary Ecology*. Macmillan, London.

Schal, C., and W. J. Bell. 1982. Ecological correlates of paternal investment of urates in a tropical cockroach. *Science* 218: 170–172.

Schall, J. J. 1986. Prevalence and virulence of a haemogregarine parasite of the Aruban whiptail lizard, *Cnemidophorus arubensis*. *J. Herpetol.* 20: 318–324.

Schaller, G. 1972. *The Serengeti Lion*. University of Chicago Press, Chicago.

Schatral, A. 1990. Body size, song frequency and mating success of male bush crickets *Requena verticalis* (Orthoptera, Tettigoniidae, Listrocelidinae) in the field. *Anim. Behav.* 40: 982–984.

Schemske, D. W. 1980a. Evolution of floral display in the orchid *Brassavola nodosa*. *Evolution* 34: 489–493.

Schemske, D. W. 1980b. Floral ecology and hummingbird pollination of *Combretum farinosum* in Costa Rica. *Biotropica* 12: 169–181.

Schemske, D. W., and C. Fenster. 1983. Pollen-grain interactions in a neotropical *Costus*: Effects of clump size and competitors. In D. L. Mulcahy and E. Ottaviano, eds., *Pollen: Biology and Implications for Plant Breeding*, 405–410. Elsevier, Amsterdam.

Schilcher, F. von., and M. Dow. 1977. Courtship behaviour in *Drosophila*: Sexual isolation or sexual selection. *Z. Tierpsychol.* 43: 304–310.

Schlessman, M. A. 1988. Gender diphasy ("sex choice"). In J. Lovett Doust and L. Lovett Doust, eds., *Plant Reproductive Ecology*, 139–153. Oxford University Press, New York.

Schlichting, C. D., and B. Devlin. 1989. Male and female reproductive success in the hermaphroditic plant *Phlox drummondii*. *Am. Nat.* 133: 212–227.

Schluter, D. 1988. Estimating the form of natural selection on a quantitative trait. *Evolution* 42: 849–861.

Schmale, M. C. 1981. Sexual selection and reproductive success in males of the bicolor damselfish, *Eupomacentrus partitus* (Pisces: Pomacentridae). *Anim. Behav.* 29: 1172–1184.

Schmid-Hempel, P., and B. Speiser. 1988. Effects of inflorescence size on pollination in *Epilobium angustifolium*. *Oikos* 53: 98–104.

Schmidt-Bey, W. 1913. Neckereien der Raubvögel nebst Gedanken über die Entstehung ihrer sekundären Geschlechtsunterschiede. *Ornit. Monatsschr.* 38: 400–414.

Schneider, V. H. 1982. Phonotaxis bei Weibchen des kanarischen Laubfrosches, *Hyla meridionalis*. *Zool. Anz.* 208: 161–174.

Schodde, R., and I. Mason. 1980. *Nocturnal Birds of Australia*. Lansdowne, Melbourne.

Schoen, D. J., and M. T. Clegg. 1985. The influence of flower color on outcrossing rate and male reproductive success in *Ipomoea purpurea*. *Evolution* 39: 1242–1249.

Schoen, D. J., and S. C. Stewart. 1986. Variation in male reproductive investment and male reproductive success in white spruce. *Evolution* 40: 1109–1120.

Schoener, T. W. 1967. The ecological significance of sexual size dimorphism in the lizard *Anolis conspersus*. *Science* 155: 474–477.

Schoener, T. W. 1977. Competition and the niche. In C. Gans and D. W. Tinkle, eds., *Biology of the Reptilia*, 35–136. Academic Press, London.

Schwabl, H. 1992. Winter and breeding territorial behaviour and levels of reproductive hormones of migratory European robins. *Ornis Scand.* 23: 271–276.

Schwagmeyer, P. L. 1979. The Bruce effect: An evaluation of male/female advantages. *Am. Nat.* 114: 932–938.

Schwagmeyer, P. L. 1988. Scramble-competition polygyny in an asocial mammal: Male mobility and mating success. *Am. Nat.* 131: 885–892.

Schwagmeyer, P. L., and C. H. Brown. 1983. Factors affecting male-male competition in thirteen-lined ground squirrels. *Behav. Ecol. Sociobiol.* 13: 1–6.

Schwagmeyer, P. L., and G. A. Parker. 1990. Male mate choice as predicted by sperm competition in thirteen-lined ground squirrels. *Nature* 348: 62–64.

Schwagmeyer, P. L., and S. J. Woontner. 1985. Mating competition in an asocial ground squirrel, *Spermophilus tridecemlineatus*. *Behav. Ecol. Sociobiol.* 17: 291–296.

Schwagmeyer, P. L., and S. J. Woontner. 1986. Scramble competition polygyny in thirteen-lined ground squirrels: The relative contributions of overt conflict and competitive mate searching. *Behav. Ecol. Sociobiol.* 19: 359–364.

Schwartz, J. J. 1986. Male calling behavior and female choice in the neotropical treefrog *Hyla microcephala*. *Ethology* 73: 116–127.

Schwartz, J. J. 1987a. The function of call alternation in anuran amphibians: A test of three hypotheses. *Evolution* 41: 461–471.

Schwartz, J. J. 1987b. The importance of spectral and temporal properties in species and call recognition in a neotropical treefrog with a complex vocal repertoire. *Anim. Behav.* 35: 340–347.

Scott, D. K., and T. H. Clutton-Brock. 1990. Mating systems, parasites and plumage dimorphism in waterfowl. *Behav. Ecol. Sociobiol.* 26: 261–273.

Searcy, W. A. 1979a. Female choice of mates: A general model for birds and its application to red-winged blackbirds (*Agelaius phoeniceus*). *Am. Nat.* 114: 77–100.

Searcy, W. A. 1979b. Morphological correlates of dominance in captive male red-winged blackbirds. *Condor* 81: 417–420.

Searcy, W. A. 1979c. Male characteristics and pairing success in red-winged blackbirds. *Auk* 96: 353–363.

Searcy, W. A. 1979d. Sexual selection and body size in male red-winged blackbirds. *Evolution* 33: 649–661.

Searcy, W. A. 1982. The evolutionary effects of mate selection. *Ann. Rev. Ecol. Syst.* 13: 57–85.

Searcy, W. A. 1984. Song repertoire size and female preferences in song sparrows. *Behav. Ecol. Sociobiol.* 14: 281–286.

Searcy, W. A. 1988. Dual intersexual and intrasexual functions of song in red-winged blackbirds. *Acta XIX Congr. Int. Orn.* 2: 1373–1381.

Searcy, W. A. 1989. Pseudoreplication, external validity and the design of playback experiments. *Anim. Behav.* 38: 715–717.

Searcy, W. A. 1990. Species recognition of song by female red-winged blackbirds. *Anim. Behav.* 40: 1119–1127.

Searcy, W. A. 1992. Song repertoire and mate choice in birds. *Amer. Zool.* 32: 71–80.

Searcy, W. A., and M. Andersson. 1986. Sexual selection and the evolution of song. *Ann. Rev. Ecol. Syst.* 17: 507–533.

Searcy, W. A., and E. A. Brenowitz. 1988. Sexual differences in species recognition of avian song. *Nature* 332: 152–154.

Searcy, W. A., and P. Marler. 1981. A test for responsiveness to song structure and programming in female sparrows. *Science* 213: 926–928.

Searcy, W. A., and P. Marler. 1984. Interspecific differences in the response of female birds to song repertoires. *Z. Tierpsychol.* 66: 128–142.

Searcy, W. A., and K. Yasukawa. 1981. Does the "sexy son" hypothesis apply to mate choice in red-winged blackbird? *Am. Nat.* 117: 343–348.

Searcy, W. A., and K. Yasukawa. 1983. Sexual selection and red-winged blackbirds. *Am. Sci.* 71: 166–174.

Searcy, W. A., and K. Yasukawa. 1989. Alternative models of territorial polygyny in birds. *Am. Nat.* 134: 323–343.

Searcy, W. A., and K. Yasukawa. 1990. Use of song repertoire in intersexual and intrasexual contexts by male red-winged blackbirds. *Behav. Ecol. Sociobiol.* 27: 123–128.

Searcy, W. A., and K. Yasukawa. 1994. *Polygyny and Sexual Selection in Red-winged Blackbirds*. Princeton University Press, Princeton, N.J., in press.

Searcy, W. A., M. H. Searcy, and P. Marler. 1982. The response of swamp sparrows to acoustically distinct song types. *Behaviour* 80: 70–83.

Searcy, W. A., D. Eriksson, and A. Lundberg. 1991. Deceptive behavior in pied flycatchers. *Behav. Ecol. Sociobiol.* 29: 167–175.

Seger, J. 1985. Unifying genetic models for the evolution of female choice. *Evolution* 39: 1185–1193.

Seger, J., and R. Trivers. 1986. Asymmetry in the evolution of female mating preferences. *Nature* 319: 771–773.

Sekulic, R. 1982. The function of howling in red howler monkeys (*Alouatta seniculus*). *Behaviour* 81: 38–54.

Selander, R. K. 1965. On mating systems and sexual selection. *Am. Nat.* 99: 129–141.

Selander, R. K. 1966. Sexual dimorphism and differential niche utilization in birds. *Condor* 68: 113–151.

Selander, R. K. 1972. Sexual selection and dimorphism in birds. In B. Campbell, ed., *Sexual Selection and the Descent of Man*, 180–230. Aldine, Chicago.

Semler, D. E. 1971. Some aspects of adaptation in a polymorphism for breeding colours in the threespine stickleback (*Gasterosteus aculeatus*). *J. Zool., London* 165: 291–302.

Severinghaus, L., B. H. Kurtak, and G. C. Eickwort. 1981. The reproductive behavior of *Anthidium manicatum* (Hymenoptera: Megachilidae) and the significance of size for territorial males. *Behav. Ecol. Sociobiol.* 9: 51–58.

Shaklee, W. E., C. W. Knox, and S. J. Marsden. 1952. Inheritance of the sex difference of body weight in turkeys. *Poultry Sci.* 31: 822–825.

Shapiro, D. Y. 1987. Sexual differentiation, social behavior and the evolution of sex change in coral reef fishes. *Bioscience* 37: 490–497.

Shapiro, D. Y. 1989. Inapplicability of the size-advantage model to coral reef fishes. *TREE* 4: 272.

Shapiro, L. E., and D. A. Dewsbury. 1986. Male dominance, female choice and male copulatory behavior in two species of voles (*Microtus ochrogaster* and *Microtus montanus*). *Behav. Ecol. Sociobiol.* 18: 267–274.

Sherman, P. W. 1981. Reproductive competition and infanticide in Belding's ground squirrels and other animals. In R. D. Alexander and D. W. Tinkle, eds., *Natural Selection and Social Behavior: Recent Research and New Theory*, 311–331. Chiron Press, New York.

Sherman, P. W. 1989. Mate guarding as paternity insurance in Idaho ground squirrels. *Nature* 338: 418–420.

Sherman, P. W., and M. L. Morton. 1984. Demography of Belding's ground squirrels. *Ecology* 65: 1617–1628.

Sherman, P. W., and D. F. Westneat. 1988. Multiple mating and quantitative genetics. *Anim. Behav.* 36: 1545–1547.

Shields, W. M. 1982. *Philopatry, Inbreeding and the Evolution of Sex*. State University of New York Press, Albany.

Shine, R. 1978. Sexual size dimorphism and male combat in snakes. *Oecologia* 33: 269–277.

Shine, R. 1979. Sexual selection and sexual dimorphism in the Amphibia. *Copeia* 1979: 297–306.

Shine, R. 1987. Sexual selection in amphibians: A reply to Halliday and Verrell. *Herpet. J.* 1: 202–204.

Shine, R. 1988. The evolution of large body size in females: A critique of Darwin's "fecundity advantage" model. *Am. Nat.* 131: 124–131.

Shine, R. 1989. Ecological causes for the evolution of sexual dimorphism: A review of the evidence. *Q. Rev. Biol.* 64: 419–461.

Shine, R. 1990. Proximate determinants of sexual differences in adult body size. *Am. Nat.* 135: 278–283.

Shine, R., and L. Schwarzkopf. 1992. The evolution of reproductive effort in lizards and snakes. *Evolution* 46: 62–75.

Short, L. L. 1972. Systematics and behavior of South American flickers (Aves, *Colaptes*). *Bull. Am. Mus. Nat. Hist.* 149: 1–109.

Short, L. L. 1982. *Woodpeckers of the World*. Delaware Museum of Natural History, Greenville.

Short, R. V. 1979. Sexual selection and its component parts, somatic and genital selection, as illustrated by man and the great apes. *Adv. Study Behav.* 9: 131–158.

Shuster, S. M. 1981. Sexual selection in the Socorro isopod, *Thermosphaeroma thermophilum* (Cole) (Crustacea: Peracarida). *Anim. Behav.* 29: 698–707.

Shuster, S. M. 1989. Male alternative reproductive strategies in a marine isopod crustacean (*Paracerceis sculpta*): The use of genetic markers to measure differences in fertilization success among α-, β-, and γ-males. *Evolution* 43: 1683–1698.

Shuster, S. M. 1990. Courtship and female mate selection in a marine isopod crustacean, *Paracerceis sculpta*. *Anim. Behav.* 40: 390–399.

Shuster, S. M. 1992. The reproductive behaviour of α-, β-, and γ-male morphs in *Paracerceis sculpta*, a marine isopod crustacean. *Behaviour* 121: 231–258.

Shuster, S. M., and M. J. Wade. 1991a. Equal mating success among male reproductive strategies in a marine isopod. *Nature* 350: 608–610.

Shuster, S. M., and M. J. Wade. 1991b. Female copying and sexual selection in a marine isopod crustacean, *Paracerceis sculpta*. *Anim. Behav.* 41: 1071–1078.

Shutler, D., and P. J. Weatherhead. 1990. Targets of sexual selection: Song and plumage of wood warblers. *Evolution* 44: 1967–1977.

Sibley, C. G., and J. E. Ahlquist. 1990. *Phylogeny and Classification of Birds: A Study in Molecular Evolution*. Yale University Press, New Haven.

Sigmund, W. R. 1983. Female preference for *Anolis carolinensis* males as a function of dewlap color and background coloration. *J. Herpet.* 17: 137–143.

Sigurjónsdóttir, H. 1981. The evolution of sexual size dimorphism in gamebirds, waterfowl and raptors. *Ornis Scand.* 12: 249–260.

Sigurjónsdóttir, H., and G. A. Parker. 1981. Dung fly struggles: Evidence for assessment strategy. *Behav. Ecol. Sociobiol.* 8: 219–230.

Sikkel, P. C. 1989. Egg presence and developmental stage influence spawning-site choice by female garibaldi. *Anim. Behav.* 38: 447–456.

Silberglied, R. E. 1984. Visual communication and sexual selection among butterflies. In R. I. Vane-Wright and P. R. Ackery, eds., *The Biology of Butterflies*, 207–223. Academic Press, London.

Silberglied, R. E., and O. R. Taylor. 1978. Ultraviolet reflection and its behavioral role in courtship of sulfur butterflies *Colias eurytheme* and *Colias philodice* (Lepidoptera, Pieridae). *Behav. Ecol. Sociobiol.* 3: 203–243.

Sillén-Tullberg, B. 1981. Prolonged copulation: A male "post-copulatory" strategy in a promiscuous species, *Lygaeus equestris* (Heteroptera: Lygaeidae). *Behav. Ecol. Sociobiol.* 9: 283–289.

Silverin, B. 1980. Effects of long-acting testosterone treatment on free-living pied flycatchers *Ficedula hypoleuca*. *Anim. Behav.* 28: 906–912.

Silverman, H. B., and M. J. Dunbar. 1980. Aggressive tusk use by the narwhal (*Monodon monoceros L.*). *Nature* 284: 57–58.

Simmons, L. W. 1986a. Inter-male competition and mating success in the field cricket, *Gryllus bimaculatus (De Geer)*. *Anim. Behav.* 34: 567–579.

Simmons, L. W. 1986b. Female choice in the field cricket *Gryllus bimaculatus* (De Geer). *Anim. Behav.* 34: 1463–1470.

Simmons, L. W. 1987a. Heritability of a male character chosen by females of the field cricket *Gryllus bimaculatus*. *Behav. Ecol. Sociobiol.* 21: 129–133.

Simmons, L. W. 1987b. Female choice contributes to offspring fitness in the field cricket, *Gryllus bimaculatus* (De Geer). *Behav. Ecol. Sociobiol.* 21: 313–322.

Simmons, L. W. 1988a. The calling song of the field cricket, *Gryllus bimaculatus* (De Geer): Constraints on transmission and its role in intermale competition and female choice. *Anim. Behav.* 36: 380–394.

Simmons, L. W. 1988b. Male size, mating potential and lifetime reproductive success in the field cricket, *Gryllus bimaculatus* (De Geer). *Anim. Behav.* 36: 372–379.

Simmons, L. W. 1990. Post-copulatory guarding, female choice and the levels of gregarine infections in the field cricket, *Gryllus bimaculatus*. *Behav. Ecol. Sociobiol.* 26: 403–409.

Simmons, L. W. 1992. Quantification of role reversal in relative parental investment in a bush cricket. *Nature* 358: 61–63.

Simmons, L. W., and W. J. Bailey. 1990. Resource influenced sex roles of zaprochiline tettigonids (Orthoptera: Tettigonidae). *Evolution* 44: 1853–1868.

Simmons, L. W., and G. A. Parker. 1989. Nuptial feeding in insects: Mating effort versus paternal investment. *Ethology* 81: 332–343.

Simmons, R. 1988a. Honest advertising, sexual selection, courtship displays, and body condition of polygynous male harriers. *Auk* 105: 303–307.

Simmons, R. 1988b. Food and the deceptive acquisition of mates by polygynous male harriers. *Behav. Ecol. Sociobiol.* 23: 83–92.

Simpson, G. G. 1953. *The Major Features of Evolution*. Columbia University Press, New York.

Singer, M. C. 1982. Sexual selection for small size in male butterflies. *Am. Nat.* 119: 440–443.

Sivinski, J. 1980. Sexual selection and insect sperm. *Florida Entomol.* 63: 99–111.

Sivinski, J. 1984. Effect of sexual experience on male mating success in a lek forming tephritid (*Anastrepha suspensa*) (Loew). *Florida Entomol.* 67: 126–130.

Sivinski, J., T. Burk, and J. C. Webb. 1984. Acoustic courtship signals in the Caribbean fruit fly, *Anastrepha suspensa* (Loew). *Anim. Behav.* 32: 1011–1016.

Slagsvold, T. 1986. Nest settlement by the pied flycatcher: Does the female choose her mate for the quality of his house or himself? *Ornis Scand.* 17: 210–220.

Slagsvold, T., and J. T. Lifjeld. 1988. Plumage colour and sexual selection in the pied flycatcher (*Ficedula hypoleuca*). *Anim. Behav.* 36: 395–407.

Slagsvold, T., and J. T. Lifjeld. 1992. Plumage color is a condition-dependent sexual trait in male pied flycatchers. *Evolution* 46: 825–828.

Slagsvold, T., and G.-P. Saetre. 1991. Evolution of plumage color in male pied flycatchers (*Ficedula hypoleuca*): Evidence for female mimicry. *Evolution* 45: 910–917.

Slagsvold, T., J. T. Lifjeld, G. Stenmark, and T. Breiehagen. 1988. On the cost of searching for a mate in female pied flycatchers *Ficedula hypoleuca*. *Anim. Behav.* 36: 433–442.

Slagsvold, T., T. Amundsen, S. Dale, and H. Lampe. 1992. Female-female aggression explains polyterritoriality in male pied flycatchers. *Anim. Behav.* 43: 397–407.

Slater, P.J.B. 1981. Chaffinch song repertoires: Observations, experiments and a discussion of their significance. *Z. Tierpsychol.* 56: 1–24.

Slater, P.J.B., and F. A. Clements. 1981. Incestuous mating in zebra finches. *Z. Tierpsychol.* 57: 201–208.

Slatkin, M. 1978. Spatial patterns in the distribution of polygenic characters. *J. theor. Biol.* 70: 213–228.

Slatkin, M. 1984. Ecological causes of sexual dimorphism. *Evolution* 38: 622–630.

Smith, B. H., and M. Ayasse. 1987. Kin-based male mating preferences in two species of halictine bee. *Behav. Ecol. Sociobiol.* 20: 313–318.

Smith, D.A.S. 1984. Mate selection in butterflies: Competition, coyness, choice and chauvinism. In R. I. Vane-Wright and P. R. Ackery, eds., *The Biology of Butterflies*, 225–244. Academic Press, London.

Smith, D. G. 1972. The role of epaulets in the red-winged blackbird, *Agelius phoeniceus* social system. *Behaviour* 41: 251–267.

Smith, D. G. 1976. An experimental analysis of the function of red-winged backbird song. *Behaviour* 56: 136–156.

Smith, D. G. 1979. Male singing ability and territory integrity in red-winged blackbirds (*Agelaius phoeniceus*). *Behaviour* 68: 193–206.

Smith, H. G., and R. Montgomerie. 1991. Sexual selection and the tail ornaments of North American barn swallows. *Behav. Ecol. Sociobiol.* 28: 195–201.

Smith, H. G., R. Montgomerie, T. Poldmaa, B. N. White, and P. T. Boag. 1991. DNA fingerprinting reveals relation between tail ornaments and cuckoldry in barn swallows, *Hirundo rustica*. *Behav. Ecol.* 2: 90–97.

Smith, J.N.M. 1988. Determinants of lifetime reproductive success in the song sparrow. In T. H. Clutton-Brock, ed., *Reproductive Success*, 154–172. University of Chicago Press, Chicago.

Smith, L. H. 1965. Changes in the tail feathers of the adolescent lyrebird. *Science* 147: 510–513.

Smith, L. H. 1982. Moulting sequences in the development of the tail plumage of the Superb lyrebird, *Menura novaehollandiae*. *Aust. Wildl. Res.* 9: 311–330.

Smith, R. L. 1979a. Paternity assurance and altered roles in the mating behaviour of a giant water bug, *Abedus herberti* (Heteroptera: Belastomatidae). *Anim. Behav.* 27: 716–725.

Smith, R. L. 1979b. Repeated copulation and sperm precedence: Paternity assurance for a male brooding water bug. *Science* 205: 1029–1031.

Smith, R. L. 1980. Evolution of exclusive postcopulatory paternal care in the insects. *Forida Entomol.* 63: 65–78.

Smith, R. L., ed. 1984. *Sperm Competition and the Evolution of Animal Mating Systems*. Academic Press, Orlando, Florida.

Smith, R. L., and W. M. Langley. 1978. Cicada stress sound: An assay of its effectiveness as a predator defense mechanism. *Southw. Nat.* 23: 187–196.

Smith, S. M. 1980. Demand behavior: A new interpretation of courtship feeding. *Condor* 82: 291–295.

Smith, S. M. 1982. Raptor "reverse" dimorphism revisited: A new hypothesis. *Oikos* 39: 118–121.

Smith, T. B. 1990a. Natural selection on bill characters in the two bill morphs of the African finch *Pyrenestes ostrinus*. *Evolution* 44: 832–842.

Smith, T. B. 1990b. Resource use by bill morphs of an African finch: Evidence for intraspecific competition. *Ecology* 71: 1246–1257.

Smith, T. B. 1993. Disruptive selection and the genetic basis of bill size polymorphism in the African finch *Pyrenestes. Nature* 363: 618–620.

Smith, W. J. 1977. *The Behavior of Communicating: An Ethological Approach.* Harvard University Press, Cambridge, Mass.

Snow, A. A. 1986. Pollination dynamics in *Epilobium canum* (Onagracae): Consequences for gametophytic selection. *Am. J. Bot.* 73: 139–151.

Snow, A. A., and S. J. Mazer. 1988. Gametophytic selection in *Raphanus raphanistrum*: A test for heritable variation in pollen competitive ability. *Evolution* 42: 1065–1075.

Snow, A. A., and T. P. Spira. 1991a. Pollen vigour and the potential for sexual selection in plants. *Nature* 352: 796–797.

Snow, A. A., and T. P. Spira. 1991b. Differential pollen-tube growth rates and nonrandom fertilization in *Hibiscus moscheutos* (Malvaceae). *Amer. J. Bot.* 78: 1419–1426.

Snow, A. A., and D. F. Whigham. 1989. Costs of flower and fruit production in *Tipularia discolor* (Orchidaceae). *Ecology* 70: 1286–1293.

Snow, D. W. 1971. Social organization of the blue-backed manakin. *Wilson Bull.* 83: 35–38.

Snow, D. W. 1982. *The Cotingas.* Oxford University Press, Oxford.

Snyder, N.F.R., and J. W. Wiley. 1976. Sexual size dimorphism in hawks and owls of North America. *Orn. Monogr.* 20: 1–96.

Sober, E. 1984. *The Nature of Selection.* MIT Press, Cambridge, Mass.

Soper, R. S., G. E. Shewell, and D. Tyrrell. 1976. *Colcondamyia auditrix* nov. sp. (Diptera: Sarcophagidae). A parasite which is attracted by the mating song of its host, *Okanagana rimosa* (Homoptera: Cicadidae). *Can. Ent.* 108: 61–68.

Southwick, E. E. 1984. Photosynthate allocation to floral nectar: A neglected energy investment. *Ecology* 65: 1775–1779.

Spencer, H. G., B. H. McArdle, and D. M. Lambert. 1986. A theoretical investigation of speciation by reinforcement. *Am. Nat.* 128: 241–262.

Spieth, H. T. 1974. Mating behavior and evolution of the Hawaiian *Drosophila*. In M.J.D. White, ed., *Genetic Mechanisms of Speciation in Insects*, 94–101. Reidel, Dordrecht, Holland.

Spieth, H. T. 1981. *Drosophila heteroneura* and *Drosophila silvestris*: Head shapes, behavior and evolution. *Evolution* 35: 921–930.

Spurrier, M. F., M. S. Boyce, and F. J. Manly. 1991. Effects of parasites on mate choice by captive sage grouse. In J. E. Loye and M. Zuk, eds., *Bird-Parasite Interactions*, 389–398. Oxford University Press, Oxford.

Stacey, N. E., A. L. Kyle, and N. R. Liley. 1986. Fish reproductive pheromones. In D. Duvall, D. Müller-Schwarze, and R. M. Silverstein, eds., *Chemical Signals in Vertebrates 4*, 117–134. Plenum, New York.

Staddon, J.E.R. 1975. A note on the evolutionary significance of "supernormal" stimuli. *Am. Nat.* 109: 541–545.

Stamps, J. A. 1977. Social behavior and spacing patterns in lizards. In C. Gans and D. W. Tinkle, eds., *Biology of the Reptilia*, vol. 7, 169–204. Academic Press, New York.

Stamps, J. A. 1983. Sexual selection, sexual dimorphism, and territoriality. In R. B. Huey, E. R. Pianka, and T. W. Schoener, eds., *Lizard Ecology: Studies of a Model Organism*, 169–204. Harvard University Press, Cambridge, Mass.

Stamps, J. A., and S. M. Gon III. 1983. Sex-biased pattern variation in the prey of birds. *Ann. Rev. Ecol. Syst.* 14: 231–253.

Stanton, M. L., and R. E. Preston. 1988. Ecological correlates of petal size variation in wild radish, *Raphanus sativus* (Brassicaceae). *Am. J. Bot.* 75: 528–539.

Stanton, M. L., A. A. Snow, and S. N. Handel. 1986. Floral evolution: Attractiveness to pollinators increases male fitness. *Science* 232: 1625–1627.

Stanton, M. L., A. A. Snow, S. N. Handel, and J. Bereczky. 1989. The impact of a flower-color polymorphism on mating patterns in experimental populations of wild radish (*Raphanus raphanistrum* L.). *Evolution* 43: 335–346.

Stanton, M., H. J. Young, N. C. Ellstrand, and J. M. Clegg. 1991. Consequences of floral variation for male and female reproduction in experimental populations of wild radish, *Raphanus sativus* L. *Evolution* 45: 268–280.

Stearns, S. C., ed. 1987. *The Evolution of Sex and Its Consequences*. Birkhäuser, Basel.

Stearns, S. C. 1992. *The Evolution of Life Histories*. Oxford University Press, Oxford.

Stebbins, G. L. 1981. Why are there so many species of flowering plants? *Bioscience* 31: 573–577.

Steele, R. H. 1986a. Courtship feeding in *Drosophila subobscura*. I. The nutritional significance of courtship feeding. *Anim. Behav.* 34: 1087–1098.

Steele, R. H. 1986b. Courtship feeding in *Drosophila subobscura*. II. Courtship feeding by males influences female mate choice. *Anim. Behav.* 34: 1099–1108.

Steele, R. H., and L. Partridge. 1988. A courtship advantage for small males in *Drosophila subobscura*. *Anim. Behav.* 36: 1190–1197.

Stenmark, G., T. Slagvold, and J. T. Lifjeld. 1988. Polygyny in the pied flycatcher, *Ficedula hypoleuca*: A test of the deception hypothesis. *Anim. Behav.* 36: 1646–1657.

Stephens, D. W., and J. R. Krebs. 1986. *Foraging Theory*. Princeton University Press, Princeton, N.J.

Stephenson, A. G. 1979. An evolutionary examination of the floral display of *Catalpa speciosa* (Bignoniaceae). *Evolution* 33: 1200–1209.

Stephenson, A. G. 1981. Flower and fruit abortion: Proximate causes and ultimate functions. *Ann. Rev. Ecol. Syst.* 12: 253–279.

Stephenson, A. G., and R. I. Bertin. 1983. Male competition, female choice and sexual selection in plants. In L. Real, ed., *Pollination Biology*, 110–149. Academic Press, Orlando, Florida.

Stephenson, A. G., and J. A. Winsor. 1986. *Lotus corniculatus* regulates offspring quality throught selective fruit abortion. *Evolution* 40: 453–458.

Stephenson, A. G., J. A. Winsor, C. D. Schlichting, and L. E. Davis. 1988. Pollen competition, nonrandom fertilization, and progeny fitness: A reply to Charlesworth. *Am. Nat.* 132: 303–308.

Stirling, I. 1975. Factors affecting the evolution of social behaviour in the pinnipedia. *Rapp. Proc.-verb. Réun. Cons. Int. Expl. Mer* 169: 205–212.

Stoddart, D. M. 1974. The role of odor in the social biology of small mammals. In M. C. Birch, ed., *Pheromones*, 297–315. North-Holland, Amsterdam.

Stonehouse, B. 1968. Thermoregulatory function of growing antlers. *Nature* 218: 870–872.

Stoner, G., and F. Breden. 1988. Phenotypic differentiation in female preference

related to geographic variation in male predation risk in the Trinidad guppy (*Poecilia reticulata*). *Behav. Ecol. Sociobiol.* 22: 285–291.
Storer, R. W. 1966. Sexual dimorphism and food habits in three North American accipiters. *Auk* 83: 423–436.
Strain, J. G., and R. L. Mumme. 1988. Effects of food supplementation, song playback, and temperature on vocal territorial behavior of Carolina wrens. *Auk* 105: 11–16.
Strathmann, R. R. 1990. Why life histories evolve differently in the sea. *Am. Zool.* 30: 197–207.
Straughan, I. R. 1975. An analysis of the mechanisms of mating call discrimination in the frogs *Hyla regilla* and *H. cadaverina*. *Copeia* 1975: 415–424.
Studd, M. V., and R. J. Robertson. 1985a. Sexual selection and variation in reproductive strategy in male yellow warblers (*Dendroica petechia*). *Behav. Ecol. Sociobiol.* 17: 101–109.
Studd, M. V., and R. J. Robertson. 1985b. Evidence for reliable badges of status in territorial yellow warblers (*Dendroica petechia*). *Anim. Behav.* 33: 1102–1113.
Sturtevant, A. H. 1915. Experiments on sex recognition and the problem of sexual selection. *J. Anim. Behav.* 5: 351–366.
Stutchbury. B. J. 1991. The adaptive significance of male subadult plumage in purple martins: Plumage dyeing experiements. *Behav. Ecol. Sociobiol.* 29: 297–306.
Sullivan, B. K. 1982a. Sexual selection in Woodhouse's Toad (*Bufo woodhousei*) I. Chorus organization. *Anim. Behav.* 30: 680–686.
Sullivan, B. K. 1982b. Significance of size, temperature and call attributes to sexual selection in *Bufo woodhousei australis*. *J. Herpetol.* 16: 103–106.
Sullivan, B. K. 1983a. Sexual selection in the great plains toad (*Bufo cognatus*). *Behaviour* 84: 258–264.
Sullivan, B. K. 1983b. Sexual selection in Woodhouse's Toad (*Bufo woodhousei*). II. Female choice. *Anim. Behav.* 31: 1011–1017.
Sullivan, B. K. 1984. Sex dimorphism in anurans: A comment. *Am. Nat.* 123: 721–724.
Sullivan, B. K. 1987. Sexual selection in Woodhouse's toad (*Bufo woodhousei*). III. Seasonal variation in male mating success. *Anim. Behav.* 35: 912–919.
Sullivan, B. K. 1989. Passive and active female choice: A comment. *Anim. Behav.* 37: 692–694.
Sullivan, B. K., and G. E. Walsberg. 1985. Call rate and aerobic capacity in Woodhouse's toad (*Bufo woodhousei*). *Herpetologica* 41: 404–407.
Sullivan. M. S., and N. Hillgarth. 1993. Mating system correlates of tarsal spurs in the Phasianidae. *J. Zool., London* 231: 203–214.
Sullivan, M. S., P. A. Robertson, and N. Aebischer. 1993. Fluctuating asymmetry measurement. *Nature* 361: 409–410.
Summers, K. 1989. Sexual selection and intrafemale competition in the green poison dart frog, *Dendrobates auratus*. *Anim. Behav.* 37: 797–805.
Summers, K. 1992a. Mating strategies in two species of dart-poison frogs: A comparative study. *Anim. Behav.* 43: 907–919.
Summers, K. 1992b. Dart-poison frogs and the control of sexual selection. *Ethology* 91: 89–107.

Sutherland, S. 1986. Patterns of fruit-set: What controls fruit-flower ratios in plants? *Evolution* 40: 117–128.

Sutherland, S. 1987. Why hermaphroditic plants produce many more flowers than fruits: Experimental tests with *Agave mckelveyana*. *Evolution* 41: 750–759.

Sutherland, S., and L. F. Delph. 1984. On the importance of male fitness in plants: Patterns of fruit-set. *Ecology* 65: 1093–1104.

Sutherland, W. J. 1985a. Measures of sexual selection. *Oxford Surv. Evol. Biol.* 1: 90–101.

Sutherland, W. J. 1985b. Chance can produce a sex difference in variance in mating success and explain Bateman's data. *Anim. Behav.* 33: 1349–1352.

Sutherland, W. J. 1987. Random and deterministic components of variance in mating success. In J. W. Bradbury and M. B. Andersson, eds., *Sexual Selection: Testing the Alternatives*, 209–219. Wiley, Chichester, U.K.

Sutherland, W. J., and M.C.M. De Jong. 1991. The evolutionarily stable strategy for secondary sexual characters. *Behav. Ecol.* 2: 16–20.

Suttie, J. M. 1980. Influence of nutrition on growth and sexual maturation of captive red deer stags. In E. Reimers, E. Gaare, and S. Skjenneberg, eds., *Proceedings of the Second International Reindeer/Caribou Symposium*, 341–349. Direktoratet for Vilt og Ferskvannsfisk, Trondheim, Norway.

Suttie, J. M., and R.N.B. Kay. 1983. The influence of nutrition and photoperiod on the growth of antlers of young red deer. In R. D. Brown, ed., *Antler Development in Cervidae*, 61–71. Caesar Kleiberg Wildlife Research Institute, Kingsville, Texas.

Svärd, L. 1985. Paternal investment in a monandrous butterfly, *Pararge aegeria*. *Oikos* 45: 66–70.

Svärd, L., and C. Wiklund. 1988. Fecundity, egg weight and longevity in relation to multiple matings in females of the monarch butterfly. *Behav. Ecol. Sociobiol.* 23: 39–43.

Svensson, B. G., and E. Pettersson. 1987. Sex-role reversed courtship behaviour, sexual dimorphism and nuptial gifts in the dance fly, *Empis borealis* (L.). *Ann. Zool. Fennici* 24: 323–334.

Svensson, B. G., and E. Pettersson. 1988. Non-random mating in the dance fly *Empis borealis*: The importance of male choice. *Ethology* 79: 307–316.

Svensson, I. 1988. Reproductive costs in two sex-role reversed pipefish species (*Syngnathidae*). *J. Anim. Ecol.* 57: 929–942.

Sweeney, B. W., and R. L. Vannote. 1982. Population synchrony in mayflies: A predator satiation hypothesis. *Evolution* 36: 810–821.

Taigen, T. L., and K. D. Wells. 1985. Energetics of vocalizations by an anuran amphibian. *J. Comp. Physiol.* 155: 163–170.

Tasker, C. R., and J. A. Mills. 1981. A functional analysis of courtship feeding in the red-billed gull, *Larus novaehollandiae scopulinus*. *Behaviour* 77: 222–241.

Taylor, C. E., A. D. Pereda, and J. A. Ferrari. 1987. On the correlation between mating success and offspring quality in *Drosophila melanogaster*. *Am. Nat.* 129: 721–729.

Taylor, P. D., and G. C. Williams. 1982. The lek paradox is not resolved. *Theor. Pop. Biol.* 22: 392–409.

Telford, S. R. 1985. Mechanisms and evolution of inter-male spacing in the painted reedfrog (*Hyperolius marmoratus*). *Anim. Behav.* 33: 1353–1361.

Telford, S. R., and J. Van Sickle. 1989. Sexual selection in an African toad (*Bufo gutteralis*): The roles of morphology, amplexus displacement and chorus participation. *Behaviour* 110: 62–75.

Temeles, E. J. 1985. Sexual size dimorphism of bird-eating hawks: The effect of prey vulnerability. *Am. Nat.* 125: 485–499.

Templeton, A. R. 1979. Once again, why 300 species of Hawaiian *Drosophila*? *Evolution* 33: 513–517.

Templeton, A. R. 1981. Mechanisms of speciation—a population genetic approach. *Ann. Rev. Ecol. Syst.* 12: 23–48.

Templeton, A. R. 1989. The meaning of species and speciation: A genetic perspective. In D. Otte and J. A. Endler, eds., *Speciation and Its Consequences*, 3–27. Sinauer, Sunderland, Mass.

Temrin, H. 1986. Singing behaviour in relation to polyterritorial polygyny in the wood warbler (*Phylloscopus sibilatrix*). *Anim. Behav.* 34: 146–152.

Temrin, H. 1989. Female pairing options in polyterritorial wood warblers *Phylloscopus sibilatrix*: Are females deceived? *Anim. Behav.* 37: 579–586.

Temrin, H. 1991. Deceit of mating status in passerine birds: An evaluation of the deception hypothesis. *Curr. Orn.* 8: 247–271.

ten Cate, C., and P. Bateson. 1988. Sexual selection: The evolution of conspicuous characteristics in birds by means of imprinting. *Evolution* 42: 1355–1358.

ten Cate, C., and P. Bateson. 1989. Sexual imprinting and a preference for "supernormal" partners in Japanese quail. *Anim. Behav.* 38: 356–358.

Testa, J. W., and D. B. Siniff. 1987. Population dynamics of Weddell seals (*Leptonychotes weddelli*) in McMurdo Sound, Antarctica. *Ecol. Monogr.* 57: 149–165.

Thiessen, D. D., and M. Rice. 1976. Mammalian scent gland marking and social behavior. *Psych. Bull.* 83: 505–539.

Thomas, A.L.R. 1993. On the aerodynamics of bird's tails. *Phil. Trans. R. Soc. Lond.* B 340: 361–380.

Thompson, C. W., and M. C. Moore. 1991. Throat colour reliably signals status in male tree lizards, *Urosaurus ornatus*. *Anim. Behav.* 42: 745–753.

Thompson, D. J., and J. T. Manning. 1981. Male selection by *Asellus* (Crustacea: Isopoda). *Behaviour* 78: 178–187.

Thompson, S. 1986. Male spawning success and female choice in the mottled triplefin, *Forsterygion varium* (Pisces: Tripterygiidae). *Anim. Behav.* 34: 580–589.

Thomson, J. D., and J. Brunet. 1990. Hypotheses for the evolution of dioecy in seed plants. *TREE* 5: 11–16.

Thomson, J. D., and B. A. Thomson. 1989. Dispersal of *Erythronium grandiflorum* pollen by bumblebees: Implications for gene flow and reproductive success. *Evolution* 43: 657–661.

Thornhill, R. 1976a. Sexual selection and paternal investment in insects. *Am. Nat.* 110: 153–163.

Thornhill, R. 1976b. Sexual selection and nuptial feeding behavior in *Bittacus apicalis* (Insecta: Mecoptera). *Am. Nat.* 110: 529–548.

Thornhill, R. 1978. Sexually selected predatory and mating behavior of the hangingfly, *Bittacus stigmaterus* (Mecoptera: Bittacidae). *Ann. Entomol. Soc. Am.* 71: 597–601.

Thornhill, R. 1979. Male and female sexual selection and the evolution of mating systems in insects. In M. S. Blum and N. A. Blum, eds., *Sexual Selection and Reproductive Competition in Insects*, 81–121. Academic Press, New York.

Thornhill, R. 1980a. Mate choice in *Hylobittacus apicalis* (Insecta: Mecoptera) and its relation to some models of female choice. *Evolution* 34: 519–538.

Thornhill, R. 1980b. Rape in *Panorpa* scorpionflies and a general rape hypothesis. *Anim. Behav.* 28: 52–59.

Thornhill, R. 1980c. Sexual selection in the black-tipped hangingfly. *Sci. Am.* 242: 138–145.

Thornhill, R. 1980d. Competitive, charming males and choosy females: Was Darwin correct? *Florida Entomol.* 63: 5–30.

Thornhill, R. 1980e. Sexual selection within mating swarms of the lovebug *Plecia nearctica* (Diptera: Bibionidae). *Anim. Behav.* 28: 405–412.

Thornhill, R. 1981. *Panorpa* (Mecoptera: Panorpidae) Scorpionflies: Systems for understanding resource-defense polygyny and alternative male reproductive efforts. *Ann. Rev. Ecol. Syst.* 12: 355–386.

Thornhill, R. 1983. Cryptic female choice and its implications in the scorpionfly *Harpobittacus nigriceps*. *Am. Nat.* 122: 765–788.

Thornhill, R. 1984. Alternative female choice tactics in the scorpionfly *Hylobittacus apicalis* (Mecoptera) and their implications. *Amer. Zool.* 24: 367–383.

Thornhill, R. 1986a. Relative parental contribution of the sexes to their offspring and the operation of sexual selection. In M. H. Nitecki and J. A. Kitchell, eds., *Evolution of Animal Behavior. Paleontological and Field Approaches*, 113–136. Oxford University Press, Oxford.

Thornhill, R. 1986b. Early history of sexual selection theory. *Evolution* 40: 446–447.

Thornhill, R. 1992a. Female preference for the pheromone of males with low fluctuating asymmetry in the Japanese scorpionfly (*Panorpa japonica*: Mecoptera). *Behav. Ecol.* 3: 277–283.

Thornhill, R. 1992b. Fluctuating asymmetry, interspecific aggression and male mating tactics in two species of Japanese scorpionflies. *Behav. Ecol. Sociobiol.* 30: 357–363.

Thornhill, R. 1992c. Fluctuating asymmetry and the mating system of the Japanese scorpionfly, *Panorpa japonica*. *Anim. Behav.* 44: 867–879.

Thornhill, R., and J. Alcock. 1983. *The Evolution of Insect Mating Systems*. Harvard University Press, Cambridge, Mass.

Thornhill, R., and D. T. Gwynne. 1986. The evolution of sexual differences in insects. *Am. Sci.* 74: 382–389.

Thornhill, R., and K. P. Sauer. 1992. Genetic sire effects on the fighting ability of sons and daughters and mating success of sons in the scorpionfly (*Panorpa vulgaris*). *Anim. Beh.* 43: 255–264.

Thorpe, W. H. 1961. *Bird-Song*. Cambridge University Press, Cambridge, U.K.

Thorson, G. 1950. Reproductive and larval ecology of marine bottom invertebrates. *Biol. Rev.* 25: 1–45.

Thresher, R. E., and J. T. Moyer. 1983. Male success, courtship complexity, and patterns of sexual selection in three congeneric species of sexually monochromatic and dichromatic damselfishes (Pisces: Pomacentridae). *Anim. Behav.* 31: 113–127.

Tilley, S. G. 1968. Size-fecundity relationships and their evolutionary implications in five desmognathine salamanders. *Evolution* 22: 806–816.

Tinbergen, N. 1939. The behaviour of the snow bunting in spring. *Trans. Linn. Soc. N.Y.* 5: 1–94.

Tinbergen, N. 1951. *The Study of Instinct*. Clarendon Press, Oxford.

Tinbergen, N., B.J.D. Meeuse, L. K. Boerema, and W. W. Varossieau. 1942. Die Balz des Samtfalters, *Eumenis* (=*Satyrus*) *semele* (L.). *Z. Tierpsychol.* 5: 182–226.

Tinsley, R. C. 1990. The influence of parasite infection on mating success in spadefoot toads, *Scaphiopus couchii*. *Am. Zool.* 30: 313–324.

Titman, R. D., and J. K. Lowther. 1975. The breeding behaviour of a crowded population of mallards. *Can. J. Zool.* 53: 1270–1283.

Todd, J. W., and W. J. Lewis. 1976. Incidence and oviposition patterns of *Trichopoda pennipes* (F.), a parasite of the southern green stink bug, *Nezara viridula* (L.). *J. Georgia Entomol. Soc.* 11: 50–54.

Tomlinson, I.P.M. 1988. Diploid models of the handicap principle. *Heredity* 60: 283–293.

Trail, P. W. 1985a. Territoriality and dominance in the lek-breeding Guianan cock-of-the-rock. *Nat. Geogr. Res.* 1: 112–124.

Trail, P. W. 1985b. Courtship disruption modifies mate choice in a lek-breeding bird. *Science* 227: 778–780.

Trail, P. W. 1985c. The intensity of selection: Intersexual and interspecific comparisons require consistent measures. *Am. Nat.* 126: 434–439.

Trail, P. W. 1990. Why should lek-breeders be monomorphic? *Evolution* 44: 1837–1852.

Trail, P. W., and E. Adams. 1989. Active mate choice at cock-of-the-rock leks: Tactics of sampling and comparison. *Behav. Ecol. Sociobiol.* 25: 283–292.

Trail, P. W., and D. L. Koutnik. 1986. Courtship disruption at the lek in the Guianan cock-of-the-rock. *Ethology* 73: 197–218.

Travis, J., S. Emerson, and M. Blouin. 1987. A quantitative genetic analysis of larval life history traits in *Hyla crucifer*. *Evolution* 41: 145–156.

Trivers, R. L. 1972. Parental investment and sexual selection. In B. Campbell, ed., *Sexual Selection and the Descent of Man, 1871–1971*, 136–179. Heinemann, London.

Trivers, R. L. 1976. Sexual selection and resource-accruing abilities in *Anolis garmani*. *Evolution* 30: 253–269.

Trivers, R. L. 1985. *Social Evolution*. Benjamin/Cummings, Menlo Park, Calif.

Trivers, R. L., and D. E. Willard. 1973. Natural selection of parental ability to vary the sex ratio of offspring. *Science* 179: 90–92.

Tsubaki, Y., and T. Ono. 1987. Effects of age and body size on the male territorial system of the dragonfly, *Nannophya pygmaea* Rambur (Odonata: Libellulidae). *Anim. Behav.* 35: 518–525.

Turelli, M. 1984. Heritable genetic variation via mutation-selection balance: Lerch's zeta meets the abdominal bristle. *Theor. Pop. Biol.* 25: 138–193.

Turner, J.R.G. 1978. Why male butterflies are non-mimetic: Natural selection, sexual selection, group selection, modification and sieving. *Biol. J. Linn. Soc.* 10: 385–432.

Tuttle, M. D., and M. J. Ryan. 1981. Bat predation and the evolution of frog vocalizations in the neotropics. *Science* 214: 677–678.

Ulagaraj, S. M., and T. J. Walker. 1975. Responses of flying mole crickets to three parameters of synthetic songs broadcast outdoors. *Nature* 253: 530–532.

Ullrey, D. E. 1983. Nutrition and antler development in white-tailed deer. In R. D. Brown, ed., *Antler Development in Cervidae*, 49–59. Caesar Kleberg Wildlife Research Institute, Kingsville, Texas.

Unger, L. M., and R. C. Sargent. 1988. Allopaternal care in the fathead minnow, *Pimephales promelas*: Females prefer males with eggs. *Behav. Ecol. Sociobiol.* 23: 27–32.

Val, F. C. 1977. Genetic analysis of the morphological differences between two interfertile species of Hawaiian *Drosophila*. *Evolution* 31: 611–629.

van den Assem, J., J. J. A. van Iersel, and R. L. Los-Den Hartogh. 1989. Is being large more important for female than for male parasitic wasps? *Behaviour* 108: 160–195.

van den Berghe, E. P. 1988. Piracy as an alternative reproductive tactic for males. *Nature* 334: 697–698.

van den Berghe, E. P., and M. R. Gross. 1989. Natural selection resulting from female breeding competition in a Pacific salmon (Coho: *Oncorhynchus kisutch*). *Evolution* 43: 125–140.

Vane-Wright, R. I. 1975. An integrated classification for polymorphism and sexual dimorphism in butterflies. *J. Zool., London* 177: 329–337.

van Noordwijk, A. 1987. Quantitative ecological genetics of great tits. In F. Cooke and P. A. Buckley, eds., *Avian Genetics: A Population and Ecological Approach*, 363–380. Academic Press, London.

van Noordwijk, A. J., J. H. van Balen, and W. Sharloo. 1980. Heritability of ecologically important traits in the great tit. *Ardea* 68: 193–203.

van Oordt, G. J., and G.C.A. Junge. 1936. Die hormonale Wirkung der Gonaden auf Sommer- und Prachtkleid. III. Der Einfluss der Kastration auf männliche Kampfläufer (*Philomachus pugnax*). *W. Roux' Archiv f. Entwicklungsmechanik* 134: 7–121.

Van Ramshorst, J. D., and A. Van den Nieuwenhuisen. 1978. *Aquarium Encyclopedia of Tropical Freshwater Fish*. Elsevier, Lausanne.

van Rhijn, J. 1973. Behavioural dimorphism in male ruffs (*Philomachus pugnax*). *Behaviour* 47: 153–229.

van Rhijn, J. 1991. *The Ruff*. Poyser, London.

Van Someren, V. D. 1946. The dancing display and courtship of Jackson's whydah (*Coliuspasser jacksoni* Sharpe). *J. E. Afr. Nat. Hist. Soc.* 18: 131–141.

Vehrencamp, S. L., and J. W. Bradbury. 1984. Mating systems and ecology. In J. R. Krebs and N. B. Davies, eds., *Behavioural Ecology: An Evolutionary Approach*, 2d ed., 251–278. Blackwell, Oxford.

Vehrencamp, S. L., J. W. Bradbury, and R. M. Gibson. 1989. The energetic cost of display in male sage grouse. *Anim. Behav.* 38: 885–896.

Veiga, J. P. 1990. Infanticide by male and female house sparrows. *Anim. Behav.* 39: 496–502.

Vepsäläinen, K., and M. Nummelin. 1985. Male territoriality in the waterstrider *Limnoporus rufoscutellatus*. *Ann. Zool. Fennici* 22: 441–448.

Vermeij, G. J. 1988. The evolutionary success of passerines: A question of semantics? *Syst. Zool.* 37: 69–71.

Verner, J. 1964. The evolution of polygamy in the long-billed marsh wren. *Evolution* 18: 252–261.

Verner, J., and G. H. Engelsen. 1970. Territories, multiple nest building, and polygyny in the long-billed marsh wren. *Auk* 87: 557–567.

Verner, J., and M. F. Willson. 1966. The influence of habitats on mating systems of North American passerine birds. *Ecology* 47: 143–147.

Verner, J., and M. F. Willson. 1969. Mating systems, sexual dimorphism and the role of male North American passerine birds in the nesting cycle. *Orn. Monogr.* 9: 1–76.

Vernon, C. J. 1971. Notes on the biology of the black coucal. *Ostrich* 42: 242–258.

Verrell, P. A. 1982a. Male newts prefer large females as mates. *Anim. Behav.* 30: 1254–1255.

Verrell, P. A. 1982b. The sexual behaviour of the red-spotted newt, *Notophthalmus viridescens*. *Anim. Behav.* 30: 1224–1236.

Verrell, P. A. 1985. Male mate choice for large fecund females in the red-spotted newt, *Notophthalmus viridescens*: How is size assessed? *Herpetologica* 41: 382–386.

Verrell, P. A. 1986. Male discrimination of larger, more fecund females in the smooth newt, *Triturus vulgaris*. *J. Herpetol.* 20: 416–422.

Verrell, P. A. 1987. Limited male mating capacity in the smooth newt, *Triturus vulgaris vulgaris*. *J. Comp. Psych.* 100: 291–295.

Verrell, P. A. 1989. Male mate choice for fecund females in a plethodontid salamander. *Anim. Behav.* 38: 1086–1088.

Verrell, P. A., and T. R. Halliday. 1985. Reproductive dynamics of a population of smooth newts, *Triturus vulgaris*, in southern England. *Herpetologica* 41: 386–395.

Vitt, L. J., and W. E. Cooper, Jr. 1985. The evolution of sexual dimorphism in the skink *Eumeces laticeps*: An example of sexual selection. *Can. J. Zool.* 63: 995–1002.

Vollrath, F. 1980a. Why are some spider males small? A discussion including observations on *Nephila clavipes*. *8 Int. Arachn.-Kongr., Verhandl.*, 165–169.

Vollrath, F. 1980b. Male body size and fitness in the web-building spider *Nephila clavipes*. *Z. Tierpsychol.* 53: 61–78.

Vollrath, F., and G. A. Parker. 1992. Sexual dimorphism and distorted sex ratios in spiders. *Nature* 360: 156–159.

von Haartman, L. 1956. Territory in the pied flycatcher *Muscicapa hypoleuca*. *Ibis* 98: 460–475.

von Schantz, T., and I. N. Nilsson. 1981. The reversed size dimorphism in birds of prey: A new hypothesis. *Oikos* 36: 129–132.

von Schantz, T., G. Göransson, G. Andersson, I. Fröberg, M. Grahn, A. Helgée, and H. Wittzell. 1989. Female choice selects for a viability-based male trait in pheasants. *Nature* 337: 166–169.

von Schilcher, F., and M. Dow. 1977. Courtship behaviour in *Drosophila*: Sexual isolation or sexual selection? *Z. Tierpsychol.* 43: 304–310.

Waage, J. K. 1975. Reproductive isolation and the potential for character displacement in the damselflies, *Calopteryx maculata* and *C. aequabilis* (Odonata: Calopterygidae). *Syst. Zool.* 24: 24–36.

Waage, J. K. 1979. Reproductive character displacement in *Calopteryx*. *Evolution* 33: 104–116.

Waage, J. K. 1986. Evidence for widespread sperm displacement ability among Zygoptera (Odonata) and the means for predicting its presence. *Biol. J. Linn. Soc.* 28: 285–300.

Wade, M. J. 1979. Sexual selection and variance in reproductive success. *Am. Nat.* 114: 742–747.

Wade, M. J. 1987. Measuring sexual selection. In J. W. Bradbury and M. B. Andersson, eds., *Sexual Selection: Testing the Alternatives*, 197–207. Wiley, Chichester, U.K.

Wade, M. J., and S. J. Arnold. 1980. The intensity of sexual selection in relation to male sexual behavior, female choice and sperm precedence. *Anim. Behav.* 28: 446–461.

Wade, M. J., and S. G. Pruett-Jones. 1990. Female copying increases the variance in male mating success. *Proc. Natl. Acad. Sci. USA* 87: 5749–5753.

Wagner, W. E. 1989. Fighting, assessment, and frequency alteration in Blanchard's cricket frog. *Behav. Ecol. Sociobiol.* 25: 429–436.

Wakelin, D. M., and J. M. Blackwell. 1988. *Genetics of Resistance to Bacterial and Parasite Infection*. Taylor and Francis, London.

Walker, E. M., and P. S. Corbet. 1975. *The Odonata of Canada and Alaska*. Vol. 3, *Anisoptera*. University of Toronto Press, Toronto.

Walker, T. J. 1974. Character displacement and acoustic insects. *Amer. Zool.* 14: 1137–1150.

Walker, T. J. 1983. Diel patterns of calling in nocturnal Orthoptera. In D. T. Gwynne and G. K. Morris, eds., *Orthopteran Mating Systems*, 240–267. Westview, Boulder, Colo.

Wall, S. C., A. Mathis, R. G. Jaeger and W. G. Gergits 1989. Male salamanders with high-quality diets have faeces attractive to females. *Anim. Behav.* 38: 546–548.

Wallace, A. R. 1889. *Darwinism*. 2d ed. Macmillan, London.

Wallace, B. 1987. Ritualistic combat and allometry. *Am. Nat.* 129: 775–776.

Wallace, C. S., and P. W. Rundel. 1979. Sexual dimorphism and resource allocation in male and female shrubs of *Simmondsia chinensis*. *Oecologia* 44: 34–39.

Walsh, N. E., and D. Charlesworth. 1992. Evolutionary interpretations of differences in pollen tube growth rates. *Q. Rev. Biol.* 67: 19–37.

Walter, H. 1979. *Eleonora's Falcon: Adaptations to Prey and Habitat in a Social Raptor*. University of Chicago Press, Chicago.

Walther, R. 1958. Zum Kampf- und Paarungsverhalten einiger Antilopen. *Z. Tierpsychol.* 15: 340–380.

Waltz, E. C. 1982. Alternative mating tactics and the law of diminishing returns: The satellite threshold model. *Behav. Ecol. Sociobiol.* 10: 75–83.

Waltz, E. C., and L. L. Wolf. 1984. By jove!! Why do alternative mating tactics assume so many different forms? *Am. Zool.* 24: 333–343.

Wang, G., M. D. Greenfield, and T. E. Shelly. 1990. Inter-male competition for high-quality host-plants: The evolution of protandry in a territorial grasshopper. *Behav. Ecol. Sociobiol.* 27: 191–198.

Ward, P. I. 1983. Advantages and a disadvantage of large size for male *Gammarus pulex* (Crustacea: Amphipoda). *Behav. Ecol. Sociobiol.* 14: 69–76.

Ward, P. I. 1984. The effects of size on the mating decisions of *Gammarus pulex* (Crustacea, Amphipoda). *Z. Tierpsychol.* 64: 174–184.

Ward, P. I. 1986. A comparative field study of the breeding behaviour of a stream and a pond population of *Gammarus pulex* (Amphipoda). *Oikos* 46: 29–36.

Ward, P. I. 1988a. Sexual selection, natural selection, and body size in *Gammarus pulex* (Amphipoda). *Am. Nat.* 131: 348–359.

Ward, P. I. 1988b. Sexual dichromatism and parasitism in British and Irish freshwater fish. *Anim. Behav.* 36: 1210–1215.

Ward, P. I. 1989. Sexual showiness and parasitism in freshwater fish: Combined data from several isolated water systems. *Oikos* 55: 428–429.

Warner, R. R. 1975. The adaptive significance of sequential hermaphroditism in animals. *Am. Nat.* 109: 61–86.

Warner, R. R. 1978. The evolution of hermaphroditism and unisexuality in aquatic and terrestrial vertebrates. In E. S. Reese and F. J. Lighter, eds., *Contrasts in Behavior*, 77–101. Wiley, New York.

Warner, R. R. 1984a. Deferred reproduction as a response to sexual selection in a coral reef fish: A test of the life historical consequences. *Evolution* 38: 148–162.

Warner, R. R. 1984b. Mating behavior and hermaphroditism in coral reef fishes. *Am. Sci.* 72: 128–136.

Warner, R. R. 1987. Female choice of sites versus mates in a coral reef fish, *Thalassoma bifasciatum*. *Anim. Behav.* 35: 1470–1478.

Warner, R. R. 1988a. Traditionality of mating-site preferences in a coral reef fish. *Nature* 335: 719–721.

Warner, R. R. 1988b. Sex change in fishes: Hypotheses, evidence, and objections. *Env. Biol. Fish.* 22: 81–90.

Warner, R. R. 1988c. Sex change and the size-advantage model. *TREE* 3: 133–136.

Warner, R. R. 1990. Male versus female influences on mating-site determination in a coral reef fish. *Anim. Behav.* 39: 540–548.

Warner, R. R., and R. K. Harlan. 1982. Sperm competition and sperm storage as determinants of sexual dimorphism in the dwarf surfperch, *Micrometrus minimus*. *Evolution* 36: 44–55.

Warner, R. R., and S. G. Hoffman. 1980. Local population size as a determinant of mating system and sexual composition in two tropical marine fishes (*Thalassoma* spp.). *Evolution* 34: 508–518.

Warner, R. R., and D. R. Robertson. 1978. Sexual patterns in the labroid fishes of the Western Caribbean. I. The wrasses (Labridae). *Smiths. Contrib. Zool.* 254: 1–27.

Warner, R. R., D. R. Robertson, and E. G. Leigh, Jr. 1975. Sex change and sexual selection. *Science* 190: 633–639.

Waser, N. M. 1983a. Competition for pollination and floral character differences among sympatric plant species: A review of evidence. In C. E. Jones and R. J. Little, eds., *Handbook of Experimental Pollination Biology*, 277–293. Van Nostrand Reinhold, New York.

Waser, N. M. 1983b. The adaptive nature of floral traits: Ideas and evidence. In L. Real, ed., *Pollination Biology*, 241–285. Academic Press, Orlando, Florida.

Waser, N. M. 1993. Population structure, optimal outbreeding, and assortative mating in angiosperms. In N. W. Thornhill, ed., *The Natural History of Inbreeding and Outbreeding: Theoretical and Empirical Perspectives*. University of Chicago Press, Chicago, in press.

Waser, N. M., and M. V. Price. 1981. Pollinator choice and stabilizing selection for flower color in *Delphinium nelsonii*. *Evolution* 35: 376–390.

Waser, N. M., and M. V. Price. 1983a. Pollinator behaviour and natural selection for flower colour in *Delphinium nelsonii*. *Nature* 302: 422–424.

Waser, N. M., and M. V. Price. 1983b. Optimal and actual outcrossing in plants, and the nature of plant-pollinator interaction. In C. E. Jones and R. J. Little, eds., *Handbook of Experimental Pollination Biology*, 341–359. Van Nostrand Reinhold, New York.

Waser, N. M., and M. V. Price. 1989. Optimal outcrossing in *Ipomopsis aggregata*: Seed set and offspring fitness. *Evolution* 43: 1097–1109.

Waser, N. M., and M. V. Price. 1991. Outcrossing distance effects in *Delphinium nelsonii*: Pollen loads, pollen tubes, and seed set. *Ecology* 72: 171–179.

Waser, N. M., and M. V. Price. 1993. Crossing distance effects on prezygotic performance in plants: An argument for female choice. *Oikos* 68: in press.

Waser, N. M., M. V. Price, A. M. Montalvo, and R. N. Gray. 1987. Female mate choice in a perennial herbaceous wildflower, *Delphinium nelsonii*. *Evol. Trends in Plants* 1: 29–33.

Wasserman, F. E. 1977. Mate attraction function of song in the white-throated sparrow. *Condor* 79: 125–127.

Wasserman, M., and H. R. Koepfer. 1977. Character displacement for sexual isolation between *Drosophila mojavensis* and *Drosophila arizonensis*. *Evolution* 31: 812–823.

Watson, A. 1967. Territory and population regulation in the red grouse. *Nature* 215: 1274–1275.

Watson, A. 1970. Territorial and reproductive behaviour of red grouse. *J. Reprod. Fert. Suppl.* 11: 3–14.

Watson, P. J. 1986. Transmission of a female sex pheromone thwarted by males in the spider *Linyphia litigiosa* (Linyphiidae). *Science* 233: 219–221.

Watson, P. J. 1990. Female-enhanced male competition determines the first mate and principal sire in the spider *Linyphia litigiosa*. *Behav. Ecol. Sociobiol.* 26: 77–90.

Watson, P. J., and R. Thornhill. 1994. Fluctuating asymmetry and sexual selection. *TREE* 9: 21–25.

Watt, W. B. 1983. Adaptation at specific loci. II. Demographic and biochemical elements in the maintenance of the *Colias* PGI polymorphism. *Genetics* 103: 691–724.

Watt, W. B., P. A. Carter, and S. M. Blower. 1985. Adaptation at specific loci. IV. Differential mating success among glycolytic allozyme genotypes of *Colias* butterflies. *Genetics* 109: 157–175.

Watt, W. B., P. A. Carter, and K. Donohue. 1986. Females' choice of "good genotypes" as mates is promoted by an insect mating system. *Science* 233: 1187–1190.

Weary, D. M., and R. E. Lemon. 1988. Evidence against the continuity-versatility relationship in bird song. *Anim. Behav.* 36: 1379–1383.

Weatherhead, P. J. 1990. Secondary sexual traits, parasites, and polygyny in red-winged blackbirds, *Agelaius phoeniceus*. *Behav. Ecol.* 1: 125–130.

Weatherhead, P. J., and D. J. Hoysak. 1984. Dominance structuring of a red-winged blackbird roost. *Auk* 101: 551–555.

Weatherhead, P. J., and R. J. Robertson. 1979. Offspring quality and the polygyny threshold: "The sexy son hypothesis." *Am. Nat.* 113: 201–208.

Weatherhead, P. J., and K. L. Teather. 1991. Are skewed fledgling sex ratios in sexually dimorphic birds adaptive? *Am. Nat.* 138: 1159–1172.

Weatherhead, P. J., G. F. Bennett, and D. Shutler. 1991a. Sexual selection and parasites in wood-warblers. *Auk* 108: 147–152.

Weatherhead, P. J., D. J. Hoysak, K. J. Metz, and C. G. Eckert. 1991b. A retrospective analysis of red-band effects on red-winged blackbirds. *Condor* 93: 1013–1016.

Webb, J. C., T. Burk, and J. Sivinski. 1983. Attraction of female Caribbean fruit flies, *Anastrepha suspensa* (Diptera: Tephritidae), to the presence of males and male-produced stimuli in field cages. *Ann. Ent. Soc. Am.* 76: 996–998.

Webster, M. S. 1992. Sexual dimorphism, mating system and body size in New World blackbirds (Icterinae). *Evolution* 46: 1621–1641.

Webster, T. P., and J. M. Burns. 1973. Dewlap color variation and electrophoretically detected sibling species in a Haitian lizard, *Anolis brevirostris*. *Evolution* 27: 368–377.

Wedell, N. 1993. Evolution of Nuptial Gifts in Bushcrickets. Ph.D. diss., University of Stockholm, Sweden.

Wedell, N., and A. Arak. 1989. The wartbiter spermatophore and its effect on female reproductive output (Orthoptera: Tettigoniidae, *Decticus verrucivorus*). *Behav. Ecol. Sociobiol.* 24: 117–125.

Weidmann, U. 1990. Plumage quality and mate choice in mallards (*Anas platyrhynchos*). *Behaviour* 115: 127–141.

Weldon, P. J., and G. M. Burghardt. 1984. Deception divergence and sexual selection. *Z. Tierpsychol.* 65: 89–102.

Wells, K. D. 1977a. Territoriality and male mating success in the green frog (*Rana clamitans*). *Ecology* 58: 750–762.

Wells, K. 1977b. The social behaviour of anuran amphibians. *Anim. Behav.* 25: 666–693.

Wells, K. D. 1978. Territoriality in the green frog (*Rana clamitans*): Vocalizations and agonistic behaviour. *Anim. Behav.* 26: 1051–1063.

Wells, K. 1980. Social behavior and communication of a dendrobatid frog (*Colostethus trinitatis*). *Herpetologica* 36: 189–199.

Wells, K. D. 1988. The effect of social interactions on anuran vocal behavior. In B. Fritzsch, M. Ryan, W. Wilczynski, T. E. Hetherington, and W. Walkowiak, eds., *The Evolution of the Amphibian Auditory System*, 433–454. Wiley, New York.

Wells, K. D., and K. M. Bard. 1987. Vocal communication in a neotropical treefrog, *Hyla ebraccata*: Responses of females to advertisement and aggressive calls. *Behaviour* 101: 200–210.

Wells, K. D., and J. J. Schwartz. 1984. Vocal communication in a neotropical treefrog, *Hyla ebraccata*: Advertisement calls. *Anim. Behav.* 32: 405–420.

Wells, K. D., and T. L. Taigen. 1989. Calling energetics of a neotropical treefrog, *Hyla microcephala*. *Behav. Ecol. Sociobiol.* 25: 13–22.

Werner, D. I. 1978. On the biology of *Tropidurus delanonis* Baur (Iguanidae). *Z. Tierpsychol.* 47: 337–395.

West, M. J., A. P. King, and D. H. Eastzer. 1981. Validating the female bioassay of cowbird song: Relating differences in song potency to mating success. *Anim. Behav.* 29: 490–501.

West-Eberhard, M. J. 1979. Sexual selection, social competition, and evolution. *Proc. Am. Phil. Soc.* 123: 222–234.

West-Eberhard, M. J. 1983. Sexual selection, social competition, and speciation. *Q. Rev. Biol.* 58: 155–183.

West-Eberhard, M. J. 1984. Sexual selection, competitive communication and species-specific signals in insects. In T. Lewis, ed., *Insect Communication*, 283–324. Academic Press, New York.

West-Eberhard, M. J., et al. 1987. Conflicts between and within the sexes in sexual selection. In J. W. Bradbury and M. B. Andersson, eds., *Sexual Selection: Testing the Alternatives*, 180–195. Wiley, Chichester, U.K.

Westermarck, E. 1891. *The History of Human Marriage*. Macmillan, London.

Westneat, D. F. 1990. Genetic parentage in the indigo bunting: A study using DNA fingerprinting. *Behav. Ecol. Sociobiol.* 27: 67–76.

Westneat, D. F., P. W. Sherman, and M. L. Morton. 1990. The ecology and evolution of extra-pair copulations. *Curr. Orn.* 7: 331–369.

Wettstein, R. 1935. *Handbuch der systematischen Botanik*. 4th ed. Franz Deuticke, Leipzig.

Wheeler, P., and P. J. Greenwood. 1983. The evolution of reversed sexual dimorphism in birds of prey. *Oikos* 40: 145–149.

White, G. 1789. *The Natural History and Antiquities of Selborne*. London.

Whitehead, H., and T. Arnbom. 1987. Social organization of sperm whales off the Galapagos Islands, February–April 1985. *Can. J. Zool.* 65: 913–919.

Whitney, C. L., and J. R. Krebs. 1975a. Mate selection in Pacific tree frogs. *Nature* 255: 325–326.

Whitney, C. L., and J. R. Krebs. 1975b. Spacing and calling in Pacific tree frogs, *Hyla regilla*. *Can. J. Zool.* 53: 1519–1527.

Whittingham, L. A., A. Kirkconnell, and L. M. Ratcliffe. 1992. Differences in song and sexual dimorphism between Cuban and North American red-winged blackbirds *Agelaius phoeniceus*. *Auk* 109: 928–933.

Wickler, W. 1967. Socio-sexual signals and their intra-specific imitation among primates. In D. Morris, ed., *Primate Ethology*, 69–147. Weidenfield and Nicolson, London.

Wickler, W. 1985. Stepfathers in insects and their pseudo-parental investment. *Z. Tierpsychol.* 69: 72–78.

Widén, P. 1984. Reversed sexual size dimorphism in birds of prey: Revival of an old hypothesis. *Oikos* 43: 259–263.

Wiens, D. 1984. Ovule survivorship, brood size, life history, breeding systems, and reproductive success in plants. *Oecologia* 64: 47–53.

Wiens, D., C. L. Calvin, C. A. Wilson, C. I. Davern, D. Frank, and S. R. Seavey. 1987. Reproductive success, spontaneous embryo abortion, and genetic load in flowering plants. *Oecologia* 71: 501–509.

Wiernasz, D. C. 1989. Female choice and sexual selection of male wing melanin pattern in *Pieris occidentalis* (Lepidoptera). *Evolution* 43: 1672–1682.

Wiernasz, D. C., and J. G. Kingsolver. 1992. Wing melanin pattern mediates species recognition in *Pieris occidentalis*. *Anim. Behav.* 43: 89–94.

Wiewandt, T. A. 1982. Evolution of nesting patterns in Iguanine lizards. In G. M. Burghardt and A. S. Rand, eds., *Iguanas of the World: Their Behavior, Ecology, and Conservation*, 119–141. Noyes, Park Ridge, N.J.

Wiggins, D. A., and R. D. Morris. 1986. Criteria for female choice of mates: Courtship feeding and parental care in the common tern. *Am. Nat.* 128: 126–129.

Wiklund, C., and T. Fagerström. 1977. Why do males emerge before females? A hypothesis to explain the incidence of protandry in butterflies. *Oecologia* 31: 153–158.

Wiklund, C. G., and J. Stigh. 1983. Nest defence and evolution of reversed sexual size dimorphism in snowy owls (*Nyctea scandiaca*). *Ornis Scand.* 14: 58–62.

Wiklund, C. G., and A. Village. 1992. Sexual and seasonal variation in territorial behaviour of kestrels, *Falco tinnunculus*. *Anim. Behav.* 43: 823–830.

Wiley, R. H. 1973. Territoriality and non-random mating in sage grouse, *Centrocercus urophasianus*. *Anim. Behav. Monogr.* 6: 85–169.

Wiley, R. H. 1974. Evolution of social organization and life-history patterns among grouse. *Q. Rev. Biol.* 49: 201–227.

Wiley, R. H. 1991. Lekking in birds and mammals: Behavioral and evolutionary issues. *Adv. Study Behav.* 20: 201–291.

Wiley, R. H., and K. N. Rabenold. 1984. The evolution of cooperative breeding by delayed reciprocity and queing for favorable social position. *Evolution* 38: 609–621.

Wiley, R. H., and D. G. Richards. 1982. Adaptations for acoustic communication in birds: Sound transmission and signal detection. In D. E. Kroodsma and E. H. Miller, eds., *Acoustic Communication in Birds*, 131–170. Academic Press, New York.

Wilkinson, G. S. 1987. Equilibrium analysis of sexual selection in *Drosophila melanogaster*. *Evolution* 41: 11–21.

Wilkinson, G. S., et al. 1987. The empirical study of sexual selection. In J. W. Bradbury and M. B. Andersson, eds., *Sexual Selection: Testing the Alternatives*, 235–245. Wiley, Chichester, U.K.

Wilkinson, G. S., K. Fowler, and L. Partridge. 1990. Resistance of genetic correlation structure to directional selection in *Drosophila melanogaster. Evolution* 44: 1990–2003.

Wilkinson, P. F., and C. C. Shank. 1977. Rutting-fight mortality among musk oxen on Banks Island, Northwest Territories, Canada. *Anim. Behav.* 24: 756–758.

Williams, D. M. 1983. Mate choice in the mallard. In P. Bateson, ed., *Mate Choice*, 297–310. Cambridge University Press, Cambridge, U.K.

Williams, E. E., and A. S. Rand. 1977. Species recognition, dewlap function and faunal size. *Am. Zool.* 17: 261–270.

Williams, G. C. 1966. *Adaptation and Natural Selection: A Critique of Some Current Evolutionary Thought*. Princeton University Press, Princeton, N.J.

Williams, G. C. 1975. *Sex and Evolution*. Princeton University Press, Princeton, N.J.

Williams, G. C. 1978. Mysteries of sex and recombination. *Q. Rev. Biol.* 53: 287–289.

Williams, G. C. 1992. *Natural Selection: Domains, Levels, and Challenges*. Oxford University Press, Oxford.

Williams, M. B. 1978. Sexual selection, adaptation, and ornamental traits: The advantage of seeming fitter. *J. theor. Biol.* 72: 377–383.

Williams, N. H. 1983. Floral fragrances as cues in animal behavior. In C. E. Jones and R. J. Little, eds., *Handbook of Experimental Pollination Biology*, 50–71. Van Nostrand Reinhold, New York.

Willoughby, E. J., and T. J. Cade. 1964. Breeding behavior of the American kestrel (sparrow hawk). *Living Bird* 3: 75–96.

Willson, M. F. 1979. Sexual selection in plants. *Am. Nat.* 113: 777–790.

Willson, M. F. 1983. *Plant Reproductive Ecology*. Wiley, New York.

Willson, M. F. 1990. Sexual selection in plants and animals. *TREE* 5: 210–214.

Willson, M. F. 1991. Sexual selection, sexual dimorphism and plant phylogeny. *Evol. Ecol.* 5: 69–87.

Willson, M. F., and N. Burley. 1983. *Mate Choice in Plants: Tactics, Mechanisms, and Consequences*. Princeton University Press, Princeton, N.J.

Willson, M. F., and P. W. Price. 1977. The evolution of inflorescence size in *Asclepias* (Asclepiadaceae). *Evolution* 31: 495–511.

Willson, M. F., and B. J. Rathcke. 1974. Adaptive design of the floral display in *Asclepias syriaca* L. *Amer. Midl. Nat.* 92: 47–57.

Wilson, D. S., and A. Hedrick. 1982. Speciation and the economics of mate choice. *Evol. Theor.* 6: 15–24.

Wilson, E. O. 1975. *Sociobiology: The New Synthesis*. Harvard University Press, Cambridge, Mass.

Wilson, E. O. 1987. Kin recognition: An introductory synopsis. In D.J.C. Fletcher and C. D. Michener, eds., *Kin Recognition in Animals*, 7–18. Wiley, Chichester, U.K.

Wimberger, P. H. 1988. Food supplement effects on breeding time and harem size in the red-winged blackbird (*Agelaius phoeniceus*). *Auk* 105: 799–802.

Wingfield, J. C. 1980. Fine temporal adjustment of reproductive functions. In S. E. Mikami, S. Ishii, and M. Wada, eds., *Avian Endocrinology*, 367–389. Academic Press, New York.

Wingfield, J. C. 1984. Androgens and mating systems: Testosterone-induced polygyny in normally monogamous birds. *Auk* 101: 665–671.

Wingfield, J. C., G. F. Ball, A. M. Dufty, R. E. Hegner, and M. Ramenofsky. 1987. Testosterone and aggressive behaviour in birds. *Am. Sci.* 75: 602–608.

Wingfield, J. C., R. E. Hegner, A. M. Dufty, Jr., and G. F. Ball. 1990. The "challenge hypothesis": Theoretical implications for patterns of testosterone secretion, mating systems, and breeding strategies. *Am. Nat.* 136: 829–846.

Winkel, W., D. Richter, and R. Berndt. 1970. Über Beziehungen zwischen Farbtyp und Lebensalter männlicher Trauerschnäpper (*Ficedula hypoleuca*). *Vogelwelt* 95: 60–70.

Winter, A. J., and T. Rollenhagen. 1990. The importance of male and female acoustic behaviour for reproductive isolation in *Ribautodelphax* planthoppers (Homoptera: Delphacidae). *Biol. J. Linn. Soc.* 40: 191–206.

Wirtz, P., M. Szabados, H. Pethig, and J. Plant. 1988. An extreme case of interspecific territoriality: Male *Anthidium manicatum* (Hymenoptera, Megachilidae) wound and kill intruders. *Ethology* 78: 159–167.

Wishart, R. A. 1983. Pairing chronology and mate selection in the American wigeon. *Can. J. Zool.* 61: 1733–1743.

Witschi, E. 1961. Sex and secondary sexual characters. In A. J. Marshall, ed., *Biology and Comparative Physiology of Birds*, Academic Press, New York. 115–168.

Witt, P. N., and J. S. Rovner, eds., 1982. *Spider Communication*. Princeton University Press, Princeton, N.J.

Wittenberger, J. F. 1979. A model for delayed reproduction in iteroparous animals. *Am. Nat.* 114: 439–446.

Wittenberger, J. F. 1981. *Animal Social Behavior*. Duxbury Press, Boston.

Wittenberger, J. F. 1983. Tactics of mate choice. In P. Bateson, ed., *Mate Choice*, 435–447. Cambridge University Press, Cambridge, U.K.

Wittzell, H. 1991a. Directional selection on morphology in the pheasant, *Phasianus colchicus*. *Oikos* 61: 394–400.

Wittzell, H. 1991b. Natural and Sexual Selection in the Pheasant *Phasianus cochichus*. Ph. D. diss., University of Lund, Sweden.

Wolff, J. O., and D. M. Cicirello. 1989. Field evidence for sexual selection and resource competition infanticide in white-footed mice. *Anim. Behav.* 38: 637–642.

Woodward, B. 1982a. Male persistence and mating success in Woodhouse's toad (*Bufo woodhousei*). *Ecology* 63: 583–585.

Woodward, B. D. 1982b. Sexual selection and nonrandom mating patterns in desert anurans (*Bufo woodhousei*, *Scaphiopus couchi*, *S. multiplicatus* and *S. bombifrons*). *Copeia* 1982: 351–355.

Woodward, B. D. 1986. Paternal effects on juvenile growth in *Scaphiopus multiplicatus* (the New Mexico spadefoot toad). *Am. Nat.* 128: 58–65.

Woodward, B. D. 1987. Paternal effects on offspring traits in *Scaphiopus couchi* (Anura: Pelobatidae). *Oecologia* 73: 626–629.

Woodward, B. D., J. Travis, and S. Mitchell. 1988. The effects of the mating system on progeny performance in *Hyla crucifer* (Anura: Hylidae). *Evolution* 42: 784–794.

Woolbright, L. L. 1983. Sexual selection and size dimorphism in anuran amphibia. *Am. Nat.* 121: 110–119.

Woolbright, L. L. 1985. Anuran size dimorphism: Reply to Sullivan. *Am. Nat.* 125: 741–743.

Wrangham, R. W. 1980. Female choice of least costly males: A possible factor in the evolution of leks. *Z. Tierpsychol.* 54: 357–367.

Wu, C.-I. 1985. A stochastic simulation study on speciation by sexual selection. *Evolution* 39: 66–82.

Wunderle, J. M. 1991. Age-specific foraging proficiency in birds. *Curr. Ornith.* 8: 273–324.

Wyatt, R. 1983. Pollinator-plant interactions and the evolution of breeding systems. In L. Real, eds., *Pollination Biology*, 51–95. Academic Press, New York.

Yamamoto, T. 1975. The medaka, *Oryzias latipes*, and the guppy, *Lebistes reticulatus*. In R. C. King, eds., *Handbook of Genetics*, 133–149. Plenum, New York.

Yamazaki, K., E. A. Boyse, V. Mike, H. T. Thaler, B. J. Mathieson, J. Abbott, J. Boyse, and Z. A. Zayas. 1976. Control of mating preferences in mice by genes in the major histocompatibility complex. *J. Exp. Med.* 144: 1324–1335.

Yamazaki, K., G. K. Beauchamp, D. Kupniewski, J. Bard, L. Thomas, and E. A. Boyse. 1988. Familial imprinting determines H-2 selective mating preferences. *Science* 240: 1331–1332.

Yampolsky, E., and H. Yampolsky. 1922. Distribution of sex forms in the phanerogamic flora. *Bibl. Genet.* 3: 1–62.

Yanagimachi, R. 1961. The life-cycle of *Peltogasterella* (Cirripedia, Rhizocephala). *Crustaceana* 2: 183–186.

Yasukawa, K. 1978. Aggressive tendencies and levels of a graded display: Factor analysis of response to song playback in the red-winged blackbird (*Agelaius phoeniceus*). *Behav. Biol.* 23: 446–459.

Yasukawa, K. 1981a. Song repertoires in the red-winged blackbird (*Agelaius phoeniceus*): A test of the Beau Geste hypothesis. *Anim. Behav.* 29: 114–125.

Yasukawa, K. 1981b. Song and territory defense in the red-winged blackbird. *Auk* 98: 185–187.

Yasukawa, K. 1981c. Male quality and female choice of mate in the red-winged blackbird (*Agelaius phoeniceus*). *Ecology* 62: 922–929.

Yasukawa, K., and W. A. Searcy. 1985. Song repertoires and density assessment in red-winged blackbirds: Further tests of the Beau Geste hypothesis. *Behav. Ecol. Sociobiol.* 16: 171–175.

Yasukawa, K., and W. A. Searcy. 1986. Simulation models of female choice in red-winged blackbirds. *Am. Nat.* 128: 307–318.

Yasukawa, K., J. L. Blank, and C. B. Patterson. 1980. Song repertoires and sexual selection in the red-winged blackbird. *Behav. Ecol. Sociobiol.* 7: 233–238.

Young, H. J., and M. L. Stanton. 1990a. Influence of environmental quality on pollen competitive ability in wild radish. *Science* 248: 1631–1633.

Young, H. J., and M. L. Stanton. 1990b. Influences of floral variation on pollen removal and seed production in wild radish. *Ecology* 7: 536–547.

Zahavi, A. 1975. Mate selection—a selection for a handicap. *J. theor. Biol.* 53: 205–214.

Zahavi, A. 1977a. Reliability in communication systems and the evolution of altruism. In B. Stonehouse and C. M. Perrins, eds., *Evolutionary Ecology*, 253–259. Macmillan, London.

Zahavi, A. 1977b. The cost of honesty (further remarks on the handicap principle). *J. theor. Biol.* 67: 603–605.

Zahavi, A. 1981. Natural selection, sexual selection and the selection of signals. In G.G.E. Scudder and J. H. Reveals, eds., *Evolution Today*, 133–138. Carnegie-Mellon University Press, Pittsburgh.

Zahavi, A. 1987. The theory of signal selection and some of its implications. In V. P. Delfino, ed., *Proceedings of International Symposium on Biology and Evolution*, 305–325. Adriatrica Editrica, Bari.

Zahavi, A. 1991. On the definition of sexual selection, Fisher's model, and the evolution of waste and of signals in general. *Anim. Behav.* 42: 501–503.

Zakon, H., and W. Wilczynski. 1988. The physiology of the VIIIth nerve. In B. Fritzsch, M. Ryan, W. Wilczynski, T. Hetherington, and W. Walkowiak, eds., *The Evolution of the Amphibian Auditory System*, 125–155. Wiley, New York.

Zeh, D. W. 1987a. Life history consequences of sexual dimorphism in a chernetid pseudoscorpion. *Ecology* 68: 1495–1501.

Zeh, D. W. 1987b. Aggression, density, and sexual dimorphism in chernetid pseudoscorpions (Arachnida: Pseudoscorpionida). *Evolution* 41: 1072–1087.

Zeh, D. W., and R. L. Smith. 1985. Paternal investment by terrestrial arthropods. *Am. Zool.* 25: 785–805.

Zeh, D. W., and J. A. Zeh. 1988. Condition-dependent sex ornaments and field tests of sexual-selection theory. *Am. Nat.* 132: 454–459.

Zimmerman, J. L. 1966. Polygyny in the dickcissel. *Auk* 83: 534–546.

Zimmerman, J. L. 1971. The territory and its density dependent effect in *Spiza americana*. *Auk* 88: 591–612.

Zimmerman, M. 1980. Reproduction in *Polemonium*: Competition for pollinators. *Ecology* 61: 497–501.

Zimmerman, M. 1984. Reproduction in *Polemonium*: A five year study of seed production and implications for competition for pollinator service. *Oikos* 42: 225–228.

Zimmerman, M. 1988. Nectar production, flowering phenology, and strategies for pollination. In J. Lovett Doust and L. Lovett Doust, eds., *Plant Reproductive Ecology*, 157–178. Oxford University Press, New York.

Zimmerman, M., and G. H. Pyke. 1988. Reproduction in *Polemonium*: Assessing the factors limiting seed set. *Am. Nat.* 131: 723–738.

Zuk, M. 1987a. The effects of gregarine parasites, body size and time of day on spermatophore production and sexual selection in field crickets. *Behav. Ecol. Sociobiol.* 21: 65–72.

Zuk, M. 1987b. Variability in attractiveness of male field crickets (Orthoptera: Gryllidae) to females. *Anim. Behav.* 35: 1240–1248.

Zuk, M. 1988. Parasite load, body size, and age of wild-caught male field crickets (Orthoptera: Gryllidae): Effects on sexual selection. *Evolution* 42: 969–976.

Zuk, M. 1989. Validity of sexual selection in birds. *Nature* 340: 104–105.

Zuk, M. 1991a. Sexual ornaments as animal signals. *TREE* 6: 228–231.

Zuk, M. 1991b. Parasites and bright birds: New data and a new prediction. In J. E.

Loye and M. Zuk, eds., *Bird-Parasite Interactions*, 317–327. Oxford University Press, Oxford.

Zuk, M. 1992. The role of parasites in sexual selection: Current evidence and future directions. *Adv. Study Behav.* 21: 39–68.

Zuk, M., K. Johnson, R. Thornhill, and J. D. Ligon. 1990a. Mechanisms of female choice in red jungle fowl. *Evolution* 44: 477–485.

Zuk, M., K. Johnson, R. Thornhill, and J. D. Ligon. 1990b. Parasites and male ornaments in free-ranging and captive red jungle fowl. *Behaviour* 114: 232–248.

Zuk, M., R. Thornhill, J. D. Ligon, and K. Johnson. 1990c. Parasites and mate choice in red jungle fowl. *Am. Zool.* 30: 235–244.

Zuk, M., R. Thornhill, J. D. Ligon, K. Johnson, S. Austad, S. H. Ligon, N. W. Thornhill, and C. Costin. 1990d. The role of male ornaments and courtship behavior in female mate choice of Red Jungle Fowl. *Am. Nat.* 136: 459–473.

Zuk, M., J. D. Ligon, and R. Thornhill. 1992. Effects of experimental manipulation of male secondary sex characters on female mate preference in red jungle fowl. *Anim. Behav.* 44: 999–1006.

Zumpe, D. 1965. Laboratory observations on the aggressive behaviour of some butterfly fishes (Chaetodontidae). *Z. Tierpsychol.* 22: 226–236.

Author Index

Subject Index

Taxonomic Index